T0136087

BIOSURFACTANTS
Production and Utilization—Processes, Technologies, and Economics

BIOSURFACTANTS
Production and Utilization—Processes, Technologies, and Economics

Edited by
Naim KOSARIC
University of Western Ontario

Fazilet VARDAR-SUKAN
Ege University

CRC Press
Taylor & Francis Group
Boca Raton London New York

CRC Press is an imprint of the
Taylor & Francis Group, an **informa** business

CRC Press
Taylor & Francis Group
6000 Broken Sound Parkway NW, Suite 300
Boca Raton, FL 33487-2742

First issued in paperback 2021

© 2015 by Taylor & Francis Group, LLC
CRC Press is an imprint of Taylor & Francis Group, an Informa business

No claim to original U.S. Government works

ISBN 13: 978-1-03-223658-2 (pbk)
ISBN 13: 978-1-4665-9669-6 (hbk)

Version Date: 20140916

This book contains information obtained from authentic and highly regarded sources. Reasonable efforts have been made to publish reliable data and information, but the author and publisher cannot assume responsibility for the validity of all materials or the consequences of their use. The authors and publishers have attempted to trace the copyright holders of all material reproduced in this publication and apologize to copyright holders if permission to publish in this form has not been obtained. If any copyright material has not been acknowledged please write and let us know so we may rectify in any future reprint.

Except as permitted under U.S. Copyright Law, no part of this book may be reprinted, reproduced, transmitted, or utilized in any form by any electronic, mechanical, or other means, now known or hereafter invented, including photocopying, microfilming, and recording, or in any information storage or retrieval system, without written permission from the publishers.

For permission to photocopy or use material electronically from this work, please access www.copyright.com (http://www.copyright.com/) or contact the Copyright Clearance Center, Inc. (CCC), 222 Rosewood Drive, Danvers, MA 01923, 978-750-8400. CCC is a not-for-profit organization that provides licenses and registration for a variety of users. For organizations that have been granted a photocopy license by the CCC, a separate system of payment has been arranged.

Trademark Notice: Product or corporate names may be trademarks or registered trademarks, and are used only for identification and explanation without intent to infringe.

Publisher's Note
The publisher has gone to great lengths to ensure the quality of this reprint but points out that some imperfections in the original copies may be apparent.

Library of Congress Cataloging-in-Publication Data

Biosurfactants (CRC Press)
   Biosurfactants : production and utilization--processes, technologies, and economics / edited by Naim Kosaric and Fazilet Vardar-Sukan.
       pages cm. -- (Surfactant science ; 159)
   Includes bibliographical references and index.
   ISBN 978-1-4665-9669-6 (hardback)
   1. Biosurfactants. I. Kosaric, Naim, 1928- editor. II. Vardar-Sukan, Fazilet, 1956- editor. III. Title.

TP248.B57B583 2015
668'.14--dc23   2014024231

Visit the Taylor & Francis Web site at
http://www.taylorandfrancis.com

and the CRC Press Web site at
http://www.crcpress.com

Contents

Contributors vii

SECTION I Production

Chapter 1   Types and Classification of Microbial Surfactants 3
Rudolf Hausmann and Christoph Syldatk

Chapter 2   Sophorolipids: Microbial Synthesis and Application 19
I.N.A. Van Bogaert, K. Ciesielska, B. Devreese, and W. Soetaert

Chapter 3   Biosurfactants versus Chemically Synthesized Surface-Active Agents 37
Steve Fleurackers

Chapter 4   Biosurfactants Produced by Genetically Manipulated Microorganisms:
Challenges and Opportunities 49
Kamaljeet K. Sekhon Randhawa

Chapter 5   Production of Biosurfactants from Nonpathogenic Bacteria 73
*Roger Marchant, S. Funston, C. Uzoigwe, P.K.S.M. Rahman,
and Ibrahim M. Banat*

Chapter 6   The Prospects for the Production of Rhamnolipids on Renewable Resources:
Evaluation of Novel Feedstocks and Perspectives of Strain Engineering 83
Marius Henkel, Christoph Syldatk, and Rudolf Hausmann

Chapter 7   Utilization of Palm Sludge for Biosurfactant Production 101
Parveen Jamal, Wan Mohd Fazli Wan Nawawi, and Zahangir Alam

Chapter 8   Bioreactors for the Production of Biosurfactants 117
Janina Beuker, Christoph Syldatk, and Rudolf Hausmann

Chapter 9   Purification of Biosurfactants 129
Andreas Weber and Tim Zeiner

Chapter 10 Cost Analysis of Biosurfactant Production from a Scientist's Perspective 153
Gunaseelan Dhanarajan and Ramkrishna Sen

SECTION II Applications

Chapter 11 Patents on Biosurfactants and Future Trends.. 165

      E. Esin Hames, Fazilet Vardar-Sukan, and Naim Kosaric

Chapter 12 Industrial Applications of Biosurfactants..245

      Letizia Fracchia, Chiara Ceresa, Andrea Franzetti, Massimo Cavallo,
      Isabella Gandolfi, Jonathan Van Hamme, Panagiotis Gkorezis,
      Roger Marchant, and Ibrahim M. Banat

Chapter 13 Biological Applications of Biosurfactants and Strategies to Potentiate
      Commercial Production 269

      Mohd Sajjad Ahmad Khan, Brijdeep Singh, and Swaranjit Singh Cameotra

Chapter 14 Perspectives on Using Biosurfactants in Food Industry 295

      Lívia Vieira de Araujo, Denise Maria Guimarães Freire, and Márcia Nitschke

Chapter 15 Biosurfactant Applications in Agriculture 313

      Rengathavasi Thavasi, Roger Marchant, and Ibrahim M. Banat

Chapter 16 Biosurfactants and Soil Bioremediation...327

      Edwan Kardena, Qomarudin Helmy, and Naoyuki Funamizu

Chapter 17 Biosurfactant Use in Heavy Metal Removal from Industrial Effluents
      and Contaminated Sites... 361

      Andrea Franzetti, Isabella Gandolfi, Letizia Fracchia, Jonathan Van Hamme,
      Panagiotis Gkorezis, Roger Marchant, and Ibrahim M. Banat

Index... 371

Contributors

Zahangir Alam
Department of Biotechnology Engineering
International Islamic University Malaysia
Kuala Lumpur, Malaysia

Ibrahim M. Banat
School of Biomedical Sciences
University of Ulster
Ulster, Northern Ireland, United Kingdom

Janina Beuker
Institute of Food Science and Biotechnology
University of Hohenheim
Stuttgart, Germany

Swaranjit Singh Cameotra
Environmental Biotechnology and Microbial
    Biochemistry Laboratory
Institute of Microbial Technology
Chandigarh, India

Massimo Cavallo
Department of Pharmaceutical Sciences
Università del Piemonte Orientale "Amedeo
    Avogadro"
Novara, Italy

Chiara Ceresa
Department of Pharmaceutical Sciences
Università del Piemonte Orientale "Amedeo
    Avogadro"
Novara, Italy

K. Ciesielska
Laboratory for Protein Biochemistry and
    Biomolecular Engineering
Ghent University
Gent, Belgium

Lívia Vieira de Araujo
Department of Biochemistry
Universidade Federal do Rio de Janeiro
Rio de Janeiro, Brazil

B. Devreese
Laboratory for Protein Biochemistry and
    Biomolecular Engineering
Ghent University
Ghent, Belgium

Gunaseelan Dhanarajan
Department of Biotechnology
Indian Institute of Technology Kharagpur
West Bengal, India

Steve Fleurackers
Research and Development Department
Ecover
Malle, Belgium

Letizia Fracchia
Department of Pharmaceutical Sciences
Università del Piemonte Orientale "Amedeo
    Avogadro"
Novara, Italy

Andrea Franzetti
Department of Earth and Environmental
    Sciences
University of Milano-Bicocca
Milano, Italy

Denise Maria Guimarães Freire
Department of Biochemistry
Universidade Federal do Rio de Janeiro
Rio de Janeiro, Brazil

Naoyuki Funamizu
Division of Built Environment
Hokkaido University
Kita-ku Sapporo, Japan

S. Funston
School of Biomedical Sciences
University of Ulster
Ulster, Northern Ireland, United Kingdom

Isabella Gandolfi
Department of Earth and Environmental
    Sciences
University of Milano-Bicocca
Milano, Italy

Panagiotis Gkorezis
Department of Environmental Biology
CMK-Hasselt University
Diepenbeek, Belgium

E. Esin Hames
Department of Bioengineering
Ege University
Izmir, Turkey

Rudolf Hausmann
Department of Bioprocess Engineering
and
Institute of Food Science and Biotechnology
University of Hohenheim
Stuttgart, Germany

Qomarudin Helmy
Water and Wastewater Engineering Research
    Group
Institut Teknologi Bandung
West Java, Indonesia

Marius Henkel
Institute of Process Engineering in Life
    Sciences
Karlsruhe Institute of Technology (KIT)
Karlsruhe, Germany

Parveen Jamal
Department of Biotechnology Engineering
International Islamic University Malaysia
Kuala Lumpur, Malaysia

Edwan Kardena
Water and Wastewater Engineering Research
    Group
Institut Teknologi Bandung
West Java, Indonesia

Mohd Sajjad Ahmad Khan
Environmental Biotechnology and Microbial
    Biochemistry Laboratory
Institute of Microbial Technology
Chandigarh, India

Naim Kosaric
Department of Chemical and Biochemical
    Engineering
University of Western Ontario
London, Ontario, Canada

Roger Marchant
School of Biomedical Sciences
University of Ulster
Ulster, Northern Ireland, United Kingdom

Márcia Nitschke
Physical Chemistry Department
Universidade de São Paulo
São Carlos, Brazil

P.K.S.M. Rahman
School of Science and Engineering
Teesside University
Middlesbrough, England, United Kingdom

Kamaljeet K. Sekhon Randhawa
Department of Plant and Environmental
    Sciences
University of Copenhagen
Frederiksberg, Denmark

Ramkrishna Sen
Department of Biotechnology
Indian Institute of Technology Kharagpur
Kharagpur, West Bengal, India

Brijdeep Singh
Dr. B.R. Ambedkar Medical College
Bangalore, Karnataka, India

W. Soetaert
Laboratory of Industrial Biotechnology and
    Biocatalysis
Ghent University
Ghent, Belgium

Christoph Syldatk
Institute of Process Engineering in Life
    Sciences
Karlsruhe Institute of Technology (KIT)
Karlsruhe, Germany

Rengathavasi Thavasi
Jeneil Biotech Inc.
Saukville, Wisconsin

C. Uzoigwe
School of Science and Engineering
Teesside University
Middlesbrough, England, United Kingdom

I.N.A. Van Bogaert
Laboratory of Industrial Biotechnology and
  Biocatalysis
Ghent University
Ghent, Belgium

Jonathan Van Hamme
Department of Biological Sciences
Thompson Rivers University
Kamloops, British Columbia, Canada

Fazilet Vardar-Sukan
Department of Bioengineering
Ege University
Izmir, Turkey

Wan Mohd Fazli Wan Nawawi
Department of Biotechnology Engineering
International Islamic University Malaysia
Kuala Lumpur, Malaysia

Andreas Weber
Department of Biochemical Engineering
Technical University Dortmund
Dortmund, Germany

Tim Zeiner
Department of Biochemical Engineering
Technical University Dortmund
Dortmund, Germany

Section I

Production

1 Types and Classification of Microbial Surfactants

Rudolf Hausmann and Christoph Syldatk

CONTENTS

1.1 Low-Molecular Weight Biosurfactants..5
1.2 Fatty Acids and Phospholipids 5
1.3 Glycolipids...5
1.4 Polyketideglycosids..8
1.5 Isoprenoide and Carotenoid Glycolipids..9
1.6 Lipopeptides 9
1.7 Spiculisporic Acid..11
1.8 High-Molecular Weight Biosurfactants...11
1.9 Proteins 12
1.10 Conclusions...12
References...13

Microorganisms produce manifold metabolites that do not seem to be necessary for their growth and survival. These metabolites are differentiated from primary metabolites and are usually designated as special metabolites, secondary metabolites, or natural products.

Many life processes require the presence of amphiphilic substances. Specifically, amphiphilics, for example, phospholipids, form the basis of all biological membranes. Although many cell components, such as fatty acids and phospholipids, generally, lower interfacial tension, additional specific compounds that lower interfacial tension are known from many microorganisms; they usually comprise unique structures. Such metabolites that lower interfacial tension are often secreted by microorganisms either into the culture medium or are integrated into the cell wall, thus permitting them to grow on or to take up hydrophobic substrates. They are often designated as "biosurfactants." These biosurfactants are among the few known microbial metabolites with bio-physically useful properties.

Owing to the lipid moiety, the extracellular compounds are assigned to the exolipids or "free" lipids. The majority of these exolipids is only formed under special, usually limiting growth conditions. A large number of type-specific, partially very unusual glycolipids, lipopolysaccharides, lipopeptides, and proteins are known. Despite the diversity in structures, all of these metabolites that lower surface tension are designated as biosurfactants.

The designation of biosurfactants for amphiphilic substances of microbial origin is to differentiate them from conventional synthetic surfactants. All in all, about 2000 different amphoteric structures of biological origin have been described. These substances were mainly interesting due to their antibiotics properties.

However, the term biosurfactant is sometimes synonymously used to refer to any natural surfactant or those obtained by chemical bonding of polar head groups and the hydrophobic tails, obtained from a natural source. Well-known examples of biosurfactants in a broader sense are soybean and egg yolk lecithins obtained from plant and animal sources, respectively, and alkyl polyglucosides (APGs) for chemically obtained surfactants from renewable sources.

All biosurfactants comprise at least one hydrophilic and one hydrophobic part due to their amphiphilic character. The molecular structure often also contains several hydrophobic and corresponding hydrophilic parts. The hydrophobic part usually comprises saturated or unsaturated fatty acids, hydroxyl fatty acids, or fat alcohols, with various other structures such as isoprenoids being possible as well. The chain length usually comprises between 8 and 18 carbon atoms. The hydrophilic part may be made up of either structurally relatively simple ester, hydroxyl, phosphate, or carboxyl groups, or of carbohydrates—such as mono, oligo, or polysaccharides—peptides or proteins. Many anionic and neutral biosurfactants are known. Cationic biosurfactants, in contrast, have been described extremely rarely, probably because they have a toxic effect, just like cationic surfactants in general.

Within the biosurfactants, the glycolipids form the greatest share, with the non-sugar component, the aglycone, being highly versatile. These structures are particularly interesting since many biosurfactants exhibit high efficiency at concurrently good biological degradability. They can also be produced from renewable resources.

Generally, biosurfactants are assigned the following properties beneficial for industrial use:

- Great structure diversity (about 2000 described biosurfactants)
- Beneficial surfactant properties
- Low eco-toxicity
- Antibiotic or bioactive effects
- Complete biological degradability
- Production from renewable resources

Although the biotechnological production of microbial surfactants has already been established so far, they have only been used in niche areas due to high production costs. A drastic reduction of production costs is, therefore, necessary to establish microbial surfactants as a general alternative to conventional surfactants also outside of the previous market niches.

There are numerous books, reviews, and original papers covering near-exhaustive aspects of natural surfactants and biosurfactants ranging from their application fields, microbial ecological, biotechnological, to chemical structure analysis.

A few of the selected books are those by Lang and Trowitsch-Kienast (2002), Sen (2010), Soberón-Chávez (2010), and general reviews are given by Satpute et al. (2010), Gutnick et al. (2011), and Merchant and Banat (2012). A thorough review of the chemical structures in the broadest sense covering natural surfactants is given in a series presented by Dembitsky (2004a,b, 2005a,b,c,d,e, 2006). This review focuses on low-molecular weight microbial surfactants with a well-defined structure prepared by fermentation covering the various types and classification of surfactants. The term biosurfactant is applied in its strictest sense referring exclusively to surfactants taken directly from microbial sources, without any organic synthesis.

In addition to the fact that they can be produced by renewable resources, biosurfactants are superior to their synthetic counterparts mainly by two essential characteristics: their structural diversity and the specific biological activity of many structures. It is evident that such additional properties exceeding pure reduction of surface tension make them particularly interesting for some applications.

Detailed consideration of the different biosurfactants regarding their actual application capacity is not possible for most biosurfactants, since the chemical structures and physical characterization of the surfactant properties alone are no indication of the performance properties in product formulations. For this, the corresponding biosurfactants must be available in quantities of about 0.1–1 kg. Thus, industrial product development of biosurfactants is limited to some few biosurfactants, including spiculisporic acid, sophorolipids, rhamnolipids, and mannosylerithritollipids.

Below, the best-known biosurfactants and, using some examples, the structural diversity and potential of microbial biosurfactants, in general, are illustrated based on selected structures.

1.1 LOW-MOLECULAR WEIGHT BIOSURFACTANTS

Usually, low-molecular weight biosurfactants are glycolipids or lipopeptides, but may also belong to the groups of simple fatty acids and free phospholipids. The best-examined glycolipids that reduce surface tension are acylated disaccharides with long-chain fatty acid or hydroxyl fatty acid residues. Lipopeptides comprise a peptide moiety that is synthesized by non-ribosomal peptidsynthases, linked to fatty acid or hydroxyl fatty acid residues.

1.2 FATTY ACIDS AND PHOSPHOLIPIDS

Some bacteria and fungi form free fatty acids or phospholipids when growing on *n*-alkanes (Desai and Banat 1997). Fatty acids can be produced by microbial oxidation of alkanes. A detailed overview of such fatty acids is provided by Rehm and Reiff (1981). The strongest reduction of surface and interface tensions is achieved by fatty acids with chain lengths of C12–C14. In addition to unbranched fatty acids, many more complex microbial fatty acids have been described that exhibit hydroxyl groups or other alkyl residues. Some of these complex fatty acids, such as corynomycol acids (Fujii et al. 1999), are strong surfactants.

Phospholipids are the main part of microbial membranes and are usually not present in an extracellular form. However, *Acinetobacter* sp. HO1-N secreted extracellular phospholipid vesicles were formed when growing on hexadecane. The strong surfactant effect of these vesicles was derived only indirectly via an optically clear micro-emulsion of hexadecane in water (Kappeli and Finnerty 1979, 1980). Kappeli and Finnerty (1979) also reported that some strains of *Aspergillus* produce phospholipids.

Rhodococcus erythropolis DSM 43215 also excreted phosphatidylethanolamines that lower surface tension and occur at growth on *n*-alkanes (Kretschmer et al. 1982). Phosphatidylethanolamine (Figure 1.1) is one of the most common phospholipids other than phosphatidylcholine and usually one of the main components of bacterial membranes.

1.3 GLYCOLIPIDS

Glycolipids comprising mono or oligosaccharides as well as lipid moieties form the most important group of low molecular weight biosurfactants. The saccharide part can comprise glucose, mannose, galactose, galactosesulfate, glucuronic acid, or rhamnose moieties. The lipid moiety comprises either saturated or unsaturated fatty acids, hydroxyl fatty acids or fat alcohols. The four biotechnologically important groups of microbial glycolipids are rhamnolipids, sophorolipids, trehaloselipids, and mannosylerytitollipids.

More than 250 glycolipids, including their chemical structures and biological activities, are described by Dembitsky (2004a,b).

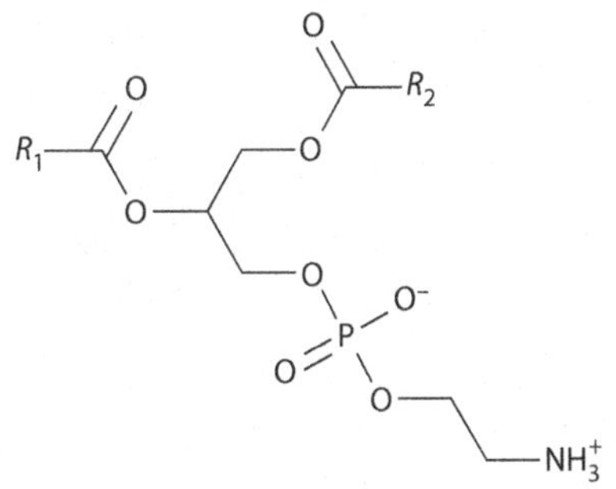

FIGURE 1.1 Phosphatidylethanolamine. R_1, R_2 = typically long, saturated or unsaturated, unbranched aliphatic chains.

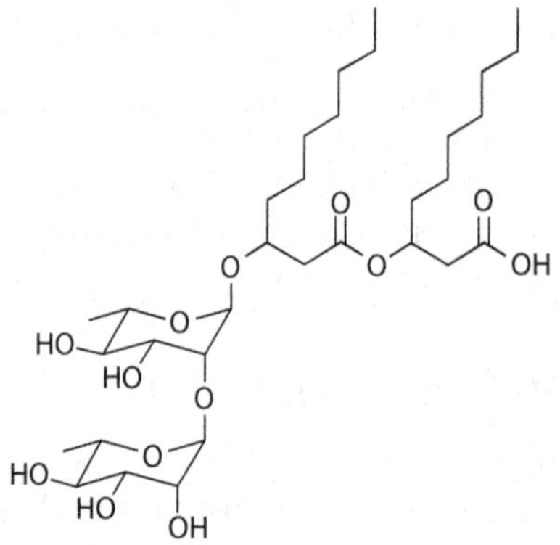

FIGURE 1.2 The characteristic di-rhamnolipid, α-L-rhamnopyranosyl-α-L-rhamnopyranosyl-3-hydroxy-decanoyl-3-hydroxydecanoate from *Pseudomonas aeruginosa*.

Rhamnolipids are mainly known from *Pseudomonas aeruginosa* and comprise one or two α-L-rhamnose units, linked via *O*-glycosidic linkage to one or two 3-hydroxyl fatty acid moieties. Natural rhamnolipids are present as mixtures of different congeners. The chain length of the 3-hydroxyl fatty acids varies between 8 and 16 carbon atoms, with 3-hydroxyl decanoic acid (Deziel et al. 2000) being predominant in *P. aeruginosa* and 3-hydroxyl tetradecanoic acid in *Burkholderia* species (Hörmann et al. 2010). The best-known rhamnolipid congener, being the α-L-rhamnopyranosyl-α-L-rhamnopyranosyl-3-hydroxydecanoyl-3-hydroxydecanoate, is displayed in Figure 1.2. A rare rhamnolipid with three hydroxyl fatty acid parts has been described only for *Burkholderia plantarii* (Andrä et al. 2006). Abdel-Mawgoud et al. (2010) provide an overview of the diversity of known rhamnolipids.

Sophorolipids contain the disaccharide sophorose and may be present in two forms, the lactonic form and the open acid form (Nunez et al. 2001). There are many sophorolipid structures (Asmer et al. 1988) that have been mainly described for *Candida bombicola* (teleomorph *Starmerella bombicola*) and *C. apicola*. Predominantly, the hydrophobic part comprises a glycosidically bound 17-hydroxyoleic acid that is usually connected lactonically with the 4″ position of the sophorose, as well as acetyl residue in the 6′ and 6″ positions (Figure 1.3).

Trehaloselipids are mainly known from *Mycobacterium* (Goren 1972), *Arthrobacter* (Suzuki et al. 1969), and *Rhodococcus* (Peng et al. 2007) species. They contain the disaccharide trehalose, which is acylated with long-chained, α-branched 3-hydroxyl fatty acids called mycol acids (Figure 1.4). These acyl groups are linked to the C6 and C6′ positions of trehalose in dimycolates, and at the C6 position in monomycolates. Additionally, other mycolic acid-containing glycolipids have been described (Lang and Philp 1998). The glycolipid 6,6′-dimycolyltrehalose is known as cord factor,

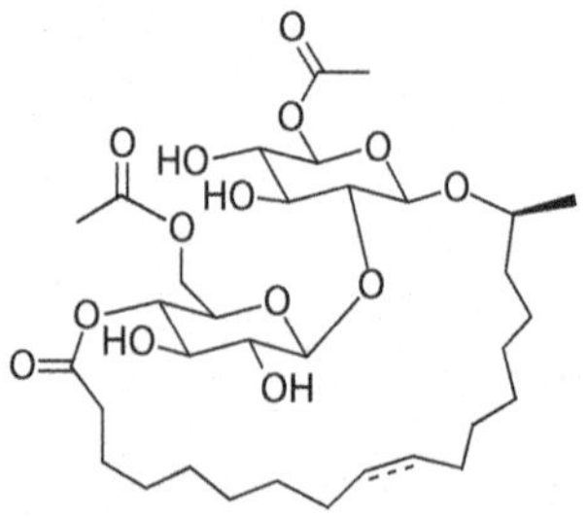

FIGURE 1.3 Characteristic lactonic sophorolipid from *Candida bombicola*, the 1,4″-lactone of 17-L-(2′-*O*-β-D-glucopyranosyl-β-D-glucopyranosyl)oxyoctadecanoic acid 6′,6″-diacetate.

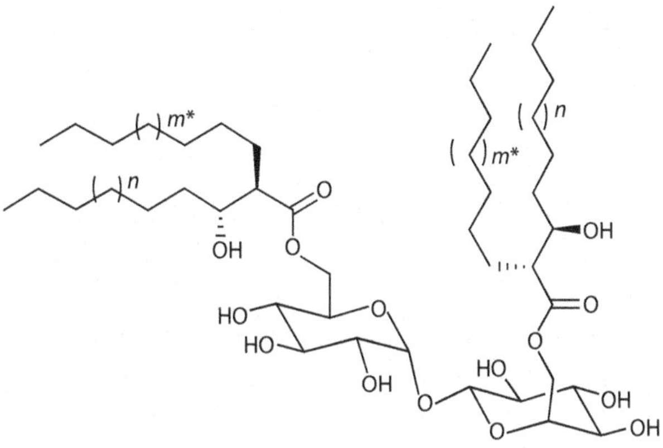

FIGURE 1.4 Trehalose-dicorynomycolate from *Rhodococcus erythropolis* $m + n = 18$–22. (*Partially unsaturated.)

and is an important virulence factor in mycobacteria infections (tuberculosis, leprosy) (Ryll et al. 2001; Kai et al. 2007). Asselineau and Asselineau (1978) provide an overview of mycobacterial trehaloselipids. The mycolic acids of the mycobacterial glycolipids are usually highly complex and comprise various functional groups, such as epoxy, ester, keto, methoxy, or cyclopropane groups (Barry et al. 1998).

Mannosylerythritollipids (MELs) comprise 4-*O*-β-D-mannopyranosyl-D-erythritol in their carbohydrate moiety, which may display diverse acylation patterns. The chain lengths of the acyl group vary considerably. MELs are mainly known from yeast species such as *Candida* (Kim et al. 1999) and *Pseudozyma* (formerly *Candida*) (Rau et al. 2005; Morita et al. 2008) and the closely related *Ustilago maydis* (Bolker et al. 2008). The typical MELs are illustrated in Figure 1.5, with the main components of the mixtures usually being MEL-A and MEL-B. Regarding the application potential, MELs are among the most promising glycolipids (Rau and Kitamoto 2008). One of the reasons for this is their suitability for pharmaceutical applications.

Glucoselipids are comparatively unusual glycolipids. Rubiwettin RG1 from *Serratia rubidaea* (Matsuyama et al. 1990) is a rhamnolipid-like exolipid comprising a glucose unit linked to two 3-hydroxyl fatty acids with chain lengths C14 and C10 as lipid main components. Another glucose lipid has been described by *Alcanivorax borkumensis* (Yakimov et al. 1998). The lipid moiety here comprises a 3-hydroxydecanacid tetramer (Schulz et al. 1991).

Cellobioselipids (ustilagin acids) are described as the second glycolipid group of *U. maydis* (Figure 1.6). The disaccharide cellobiose is glycosidically linked to the terminal hydroxyl group of a 15,16-dihydroxypalmitin acid (ustilagin acid A) or a 2,15,16-trihydroxypalmitin acid (ustilagin acid B). The cellobiose is substituted variably by acetyl or different acyl groups in the 6′ and 2″ positions (Bolker et al. 2008).

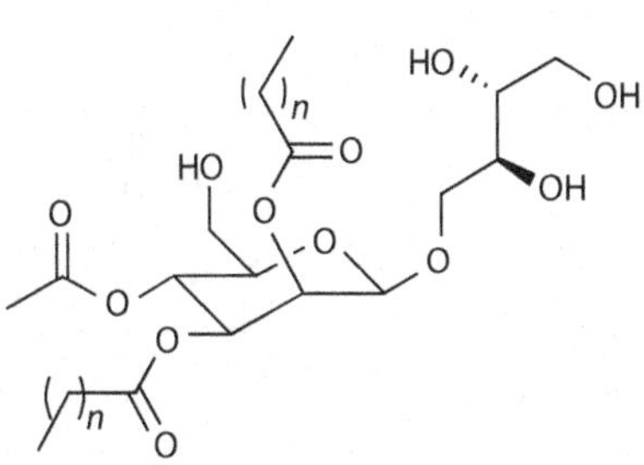

FIGURE 1.5 Typical mannosylerythritollipids (MELs) from *Pseudozyma* sp. $n = 6, 8, 10,$ or 12.

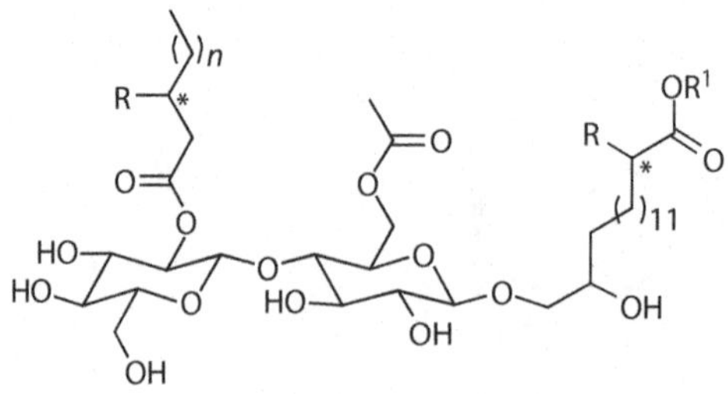

FIGURE 1.6 Cellobioselipide from *Ustilago maydis* R = OH or H; R^1 = H or CH$_3$; n = 1–14.

1.4 POLYKETIDEGLYCOSIDS

Although polyketides are typical secondary metabolites of microorganisms, comparatively few glycosylated polyketides have been described as biosurfactants. The polyketides with amphoteric properties include, among others, ionophoric and macrocyclical glycosides. The following lists two examples for microbially producible amphoteric polyketideglycosides.

The designation ionophore usually describes composite macrocyclical compounds that may form reversible chelates with ions and transport them as carriers through biological membranes otherwise impermeable to ions. The term is derived from "ion-carrying" (Greek "carrying" = phorós). Ionophorics are a functionally limited heterogeneous group of amphoteric molecules. In the narrower sense, they are not among the biosurfactants. Interesting glycosidic ionophores are the colopsinols A–E isolated from extracts of the marine dinoflagellate *Amphidinium* sp. (Y-5) (Kobayashi et al. 1999; Kubota et al. 1999, 2000; Kobayashi and Kubota 2007). The hydrophobic polyketide aglycone of colopsinol A (Figure 1.7) is formed by an aliphatic, linear C56 body that comprises two methyl, one methylide, two keto, five hydroxyl, two epoxy, and one sulfate-esterified tetrahydropyrane groups. The hydrophilic part is formed by the sugar component gentiobiose (6-*O*-β-D-glucopyranosyl-D-glucose) and the sulfate ester.

An interesting bioactive polyketide from the macrolide group is elaiophylin (see Figure 1.8), a dimeric makrodiolidglycoside with side chains folded into a cyclic hemiketal and glycolized with 2-desoxy-l-fucose (Arcamone et al. 1959; Kaiser and Kellerschierlein 1981). Elaiophyline and similar derivatives are formed by various *Streptomycetes* (specifically *Streptomyces melanosporus*) and can be produced by fermentation, partially in large quantities (Haydock et al. 2004).

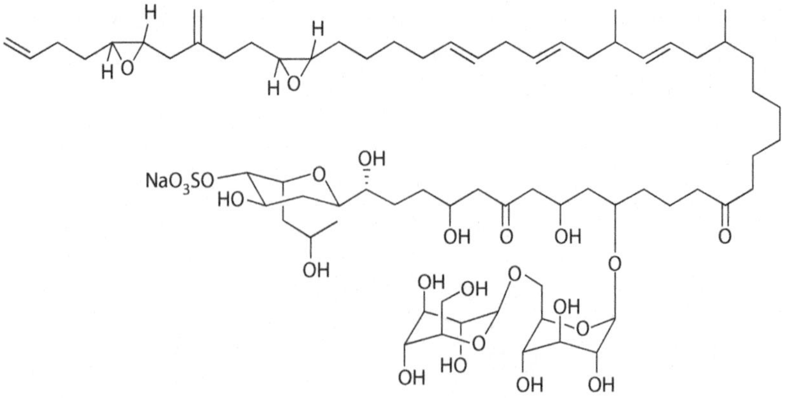

FIGURE 1.7 Colopsinol A from *Amphidinium* sp.

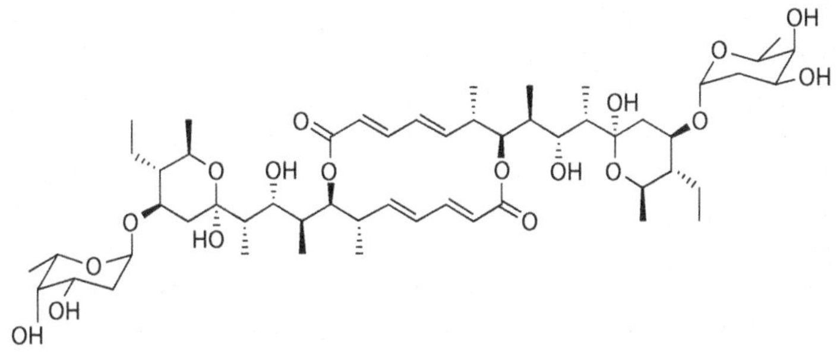

FIGURE 1.8 Elaiophylin from *Streptomyces melanosporus.*

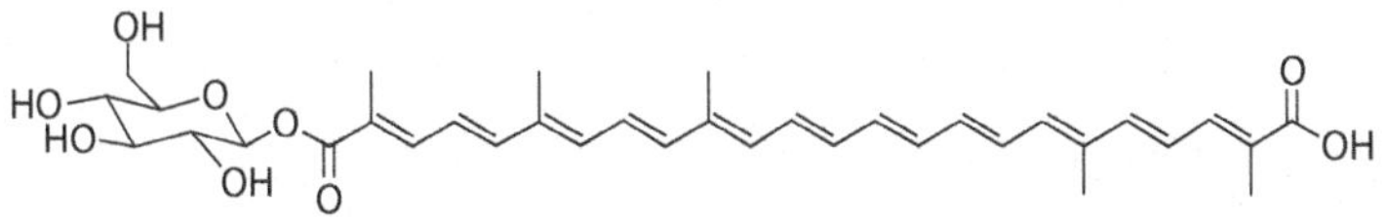

FIGURE 1.9 Carotenoid glycosiden from *Pseudomonas rhodos* and red-strain *Rhizobium lupine.*

1.5 ISOPRENOIDE AND CAROTENOID GLYCOLIPIDS

Microbial carotenoids are often formed by thermophilic bacteria. Therefore, it is assumed that the carotenoid formation is a protective mechanism to stabilize the cell membranes at high temperatures. Specifically in carotenoid glycosides (Figure 1.9), it is assumed that they are able to bridge the lipid double layer of microbial cell membranes to stabilize them at high temperatures (Yokoyama et al. 1995). As an example of this, the carotenoid glycosid b-D-glucosyl of the 4,4″-diapocarotene-6,6′-dioic acid (Figure 1.9) was isolated from the thermophilic microorganisms *Pseudomonas rhodos* and *Rhizobium lupini* (Liaaen-Jensen 1969; Kleinig et al. 1977; Kleinig and Broughton 1982a,b; Kleinig and Schmitt 1982).

1.6 LIPOPEPTIDES

Microbial lipopeptides are cyclic peptides that are acylated with a fatty acid. They are secreted into the growth medium by various microorganisms, including Gram-positive species, such as *Bacillus,* *Lactobacillus,* and *Streptomyces,* and Gram-negative species, such as *Pseudomonas* and *Serratia.* The natural lipopeptide that was first discovered was surfactin by *B. subtilis* (Arima et al. 1968). Many lipopeptides show not only a reduction of surface tension (Vater 1986) but also a strong antibiotic effect (Baltz et al. 2005). The best-known compounds of this class are certainly surfactin, polymyxin B, and the lipopeptidic antibiotic daptomycin by *Streptomyces roseosporus,* approved since 2003 (Eisenstein 2004; Baltz et al. 2005). Dexter et al. (2006) and Dexter and Middelberg (2007a,b) provide a current overview of the different groups of peptides and lipopeptides lowering surface tension. In addition to natural lipopeptides, synthetic lipopeptides have also been examined as antibiotics more frequently (Jerala 2007).

The nomenclature of the lipopeptides has been rather chaotic; hence, similar lipopeptides of one group are often called by different names, while other groups are summarized under one name. An example of this is the group of surfactins. In contrast to proteins, lipopeptides are not formed ribosomally by translation of an mRNA, but by special, non-ribosomal peptidsynthases, in which one module each leads to the addition of an amino acid, ring closure, and acylation (Peypoux et al. 1999).

Surfactin from *Bacillus subtilis* is a cyclic lipopeptide (Figure 1.10) comprising seven amino acids and different 3-hydroxyl fatty acids. The main component is the 3-hydroxyl-13-methyl-myristin

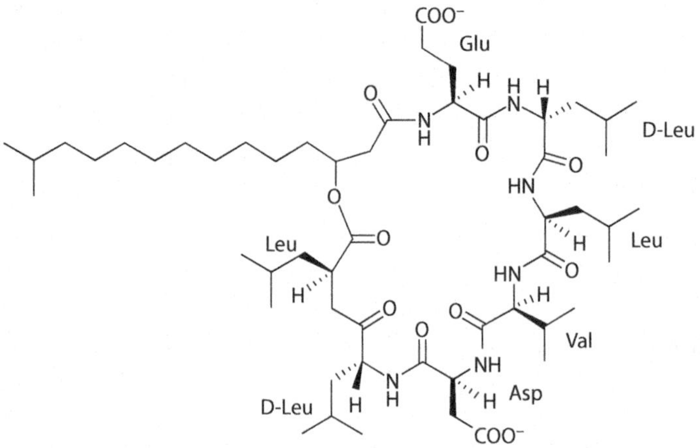

FIGURE 1.10 Representative surfactin from *B. subtilis.*

acid. Surfactin is a very good surfactant that also has antibacterial properties (Arima et al. 1968; Kakinuma et al. 1968; Peypoux et al. 1999; Ongena and Jacques 2008). It is synthesized by a linear, non-ribosomal peptide synthase, the surfactin synthase. When dissolved, it shows a characteristic saddle-like conformation that is essential for the wide bioactive range of surfactin (Hue et al. 2001). In addition to surfactin, *B. subtilis* produces two other lipopeptides as well, specifically Iturin and Fengycin.

Polymyxines are a group of cationic, branched, cyclic dekapeptides. The polymyxines A–E have been insulated from different strains of *Bacillus polymyxa* since 1947 (Stansly 1949). Polymyxine B is a decapeptide with eight amino acids forming a ring and linked to a branched fatty acid. The lipopeptid gained a certain importance as an antibiotic (Zavascki et al. 2007; Kwa et al. 2008; Landman et al. 2008).

Viscosin from *Pseudomonas fluorescens, P. libanensis*, and *P. viscosa* is a cyclic lipopeptide that reduces surface tension. The structure contains hydrophobic amino acids, linked to a fatty acid (Neu and Poralla 1990; Neu et al. 1990; Saini et al. 2008).

Serrawettin is a group of cyclodepsipeptides produced by *Serratia marcescens*. Serrawettin W1 or Serratamolide (Figure 1.11) is a rotationally symmetric cyclodepsipeptide, produced non-ribosomally by an aminolipid synthetase (Li et al. 2005). It comprises two serine and mainly two 3-hydroxydecan acid parts (Wasserman et al. 1962; Matsuyama et al. 1985). Serrawettin W2 is a cyclic lipopeptide composed of five different amino acids and 3-hydroxydecanoic acid and is required for swarming motility in *S. marcescens* (Matsuyama et al. 1992; Lindum et al. 1998). Serrawettin W2 shows antimicrobial properties and is sensed and avoided by nematodes (Pradel et al. 2007). Serratamolide and serrawettin W2 have been used in studies of therapeutic activity and proapoptotic effects in cancer research (Escobar-Diaz et al. 2005; Soto-Cerrato et al. 2005).

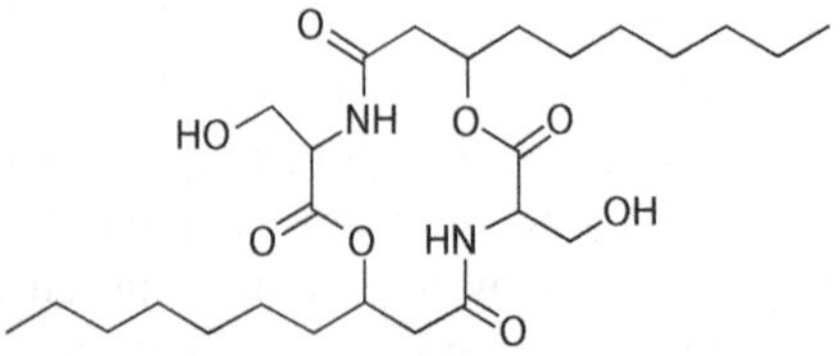

FIGURE 1.11 Serratamolide from *Serratia marcescens.*

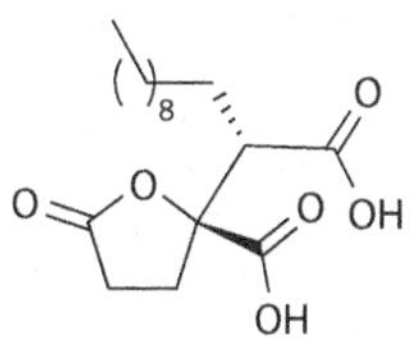

FIGURE 1.12 Spiculisporic acid from *Penicillium spiculisporum*.

1.7 SPICULISPORIC ACID

Spiculisporic acid (Figure 1.12) (4,5-dicarboxy-γ-oentadecanolacton) is formed as a secondary metabolite of *Penicillium spiculisporum* and accumulates in the culture fluid, from which needle-shaped crystals can be acquired by acid precipitation and subsequent recrystallization. The maximum titer reported by Tabuchi et al. (1977) is up to 110 g/L after a cultivation time of 10 days.

The commercial potential of spiculisporic acid has been examined comprehensively. It is not only interesting as a biosurfactant due to its availability, but also for its unique structure and environmental compatibility (Tabuchi et al. 1977; Ishigami et al. 2000).

1.8 HIGH-MOLECULAR WEIGHT BIOSURFACTANTS

High-molecular weight polymeric biosurfactants are produced by many bacteria of different species. They are polysaccharides, proteins, lipopolysaccharides, lipoproteins, or complex mixtures of these referred to as lipoheteropolysaccharides. The best-known high-molecular weight biosurfactants are emulsans that are formed by various prokaryotes, including *Archaea*, Gram-positive, and Gram-negative bacteria. The emulsan of *Acinetobacter* types is best known, however (Sar and Rosenberg 1983; Rosenberg and Ron 1997). Two of these *Acinetobacter* emulsans, RAG-1 (Figure 1.13) and BD4 emulsan, have been examined in more detail.

Emulsan is a highly effective emulsifier even in low concentrations of 0.01–0.001%. The emulsifying effect is relatively specific: pure aliphatics, aromatics, or cyclical hydrocarbons are not emulsified, while many mixtures of aliphatic and aromatic compounds are effectively emulsified (Rosenberg and Ron 1999).

The RAG-1 emulsan of *Acinetobacter* sp. ATCC 31012 (RAG-1) is not a defined polysaccharide, but a complex mix of high-molecular exopolysaccharides and lipopolysaccharides. The exopolysaccharide is probably a anionic polysaccharide with a molecular weight of 200–250 kDa (Dams-Kozlowska et al. 2008). The lipopolysaccharide of RAG-1 emulsan probably has a polysaccharide part of D-galactosamine, D-galactosaminuronic acid, and di-amino-6-deoxy-D-glucose at a ratio of 1:1:1.

The amino groups are either acetylated or amidically linked to a 3-hydroxyl butyric acid. The lipid moiety comprises singly unsaturated fatty acids with a chain length of 10–18 C-atoms, linked to the saccharid moiety either via *O*-acyl- or *N*-acyl links. The lipid share makes up the emulsan by up to 23% (w/w) (Zhang et al. 1999).

The *Acinetobacter calcoaceticus* BD4 emulsan, in contrast, is a complex protein–polysaccharide mixture. The polysaccharide moiety of the BD4 emulsan comprises repeating heptasaccharides that are built from L-rhamnose, D-glucose, D-glucuronic acid, and D-mannose at a ratio of 4:1:1:1. It is interesting that neither the extracellular protein nor the polysaccharide reduce surface tension in their pure forms (Kaplan et al. 1985, 1987). A similarly complex biosurfactant that also comprises polysaccharide and protein moieties is alasane from *Acinetobacter radioresistens* KA53 (Navon-Venezia et al. 1995, 1998). The polysaccharide moiety is rather unusual due to covalently bound alanine. The emulsifying effect of alasane is essentially due to one of the alasane proteins (45 kDa), which has a stronger emulsifying effect than the alasane as such (Toren et al. 2002).

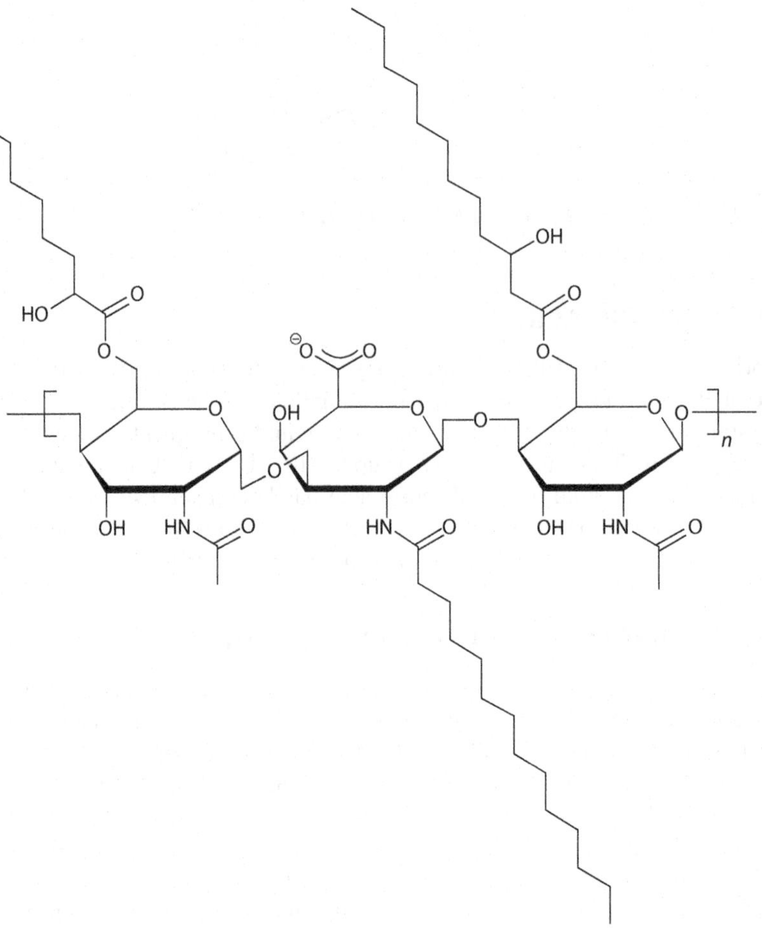

FIGURE 1.13 RAG-1 emulsan from *Acinetobacter* sp. ATCC 31012.

Gutierrez et al. (2008) describe another high-molecular glycoprotein with surfactant effect from *Pseudoalteromonas* sp. TG12, called PE12 by them. PE12 has a molecular weight in excess of 2000 kDa. PE12 is noticeable because it comprises xylose at an unusually high ratio (28%).

1.9 PROTEINS

Proteins are a very interesting class of little-explored and biotechnologically hardly used biosurfactants. Naturally occurring foams often contain protein degradation products or special foaming proteins. A particularly interesting example of this is the ranaspumines of tropical frogs that use foam nests to protect their eggs. These proteins comprise different structures with foam-stabilizing properties. Another strongly foaming protein is latherin, mainly known from horse sweat (Cooper and Kennedy 2010).

A better-known class is that of hydrophobines, small proteins secreted by fungi (Wessels 1997, 2000; Wosten and Wessels 1997; Linder 2009).

1.10 CONCLUSIONS

Microorganisms are able to form a wide range of metabolites that reduce surface tension. They are either secreted into the culture medium or integrated into the cell wall, usually, permitting growth on or in the reception of hydrophobic substrates. These "biosurfactants" usually show a very low critical micelle concentration as compared with chemically produced surfactants, are biodegradable, and often have interesting bio-active properties. Therefore, they are also very interesting for industrial applications in food, cosmetics, and pharmaceutics.

Many of the above biosurfactants appear to be naturally optimized for the corresponding microorganisms, but they are not perfect for industrial applications in the present form, and therefore only used in niche areas so far. Therefore, the search for microbial producers of new structures, for example, in unusual biotopes, is still very interesting. In the meantime, novel metagenom-based screening methods have come into use here as well.

In the still low number of established microbial production procedures, it turns out that modern molecular–biological methods may not only permit higher product concentrations, but also clearly improve molecule structures regarding specific applications. Both the above yeast-produced sophorolipids that have long been known in literature and the MELs have become important platform composites for many different industrial applications by modifying their basic structures in the meantime. This is expected for bacterial rhamnolipids in future as well.

Another still existing limitation in biosurfactants is the much higher production costs as compared with chemically produced surfactants, preventing widespread use of these interesting compounds. The economic efficiency of the production processes strongly depends on suitable and effective methods of isolation and cleaning of the products formed in all cases.

All in all, biosurfactants produced based on renewable resources are currently about to leave their niche position and become an industrial reality.

REFERENCES

Abdel-Mawgoud, A. M., F. Lépine et al. 2010. Rhamnolipids: Diversity of structures, microbial origins, and roles. *Applied Microbiology and Biotechnology*, **86**: 1323–1336.

Andrä, J., J. Rademann et al. 2006. Endotoxin-like properties of a rhamnolipid exotoxin from *Burkholderia* (*Pseudomonas*) *plantarii*: Immune cell stimulation and biophysical characterization. *Biological Chemistry* **387**(3): 301–310.

Arcamone, F. M., C. Bertazzoli et al. 1959. Melanosporin and elaiophylin, new antibiotics from *Streptomyces melanosporus* (sive *Melanosporofaciens*) n. sp. *Giorn Microbiol* **7**(3): 207–216.

Arima, K., A. Kakinuma et al. 1968. Surfactin a crystalline peptidelipid surfactant produced by *Bacillus subtilis*—Isolation characterization and its inhibition of fibrin clot formation. *Biochemical and Biophysical Research Communications* **31**(3): 488–494.

Asmer, H. J., S. Lang et al. 1988. Microbial-production, structure elucidation and bioconversion of sophorose lipids. *Journal of the American Oil Chemists Society* **65**(9): 1460–1466.

Asselineau, C. and J. Asselineau 1978. Specific lipids in mycobacteria. *Annales de Microbiologie* **A129**(1): 49–69.

Baltz, R. H., V. Miao et al. 2005. Natural products to drugs: Daptomycin and related lipopeptide antibiotics. *Natural Product Reports* **22**(6): 717–741.

Barry, C. E., R. E. Lee et al. 1998. Mycolic acids: Structure, biosynthesis and physiological functions. *Progress in Lipid Research* **37**(2–3): 143–179.

Bolker, M., C. W. Basse et al. 2008. *Ustilago maydis* secondary metabolism—From genomics to biochemistry. *Fungal Genetics and Biology* **45**: S88–S93.

Cooper, A. and M. W. Kennedy 2010. Biofoams and natural protein surfactants. *Biophysical Chemistry* **151**(3): 96–104.

Dams-Kozlowska, H., M. P. Mercaldi et al. 2008. Modifications and applications of the *Acinetobacter venetianus* RAG-1 exopolysaccharide, the emulsan complex and its components. *Applied Microbiology and Biotechnology* **81**(2): 201–210.

Dembitsky, V. M. 2004a. Astonishing diversity of natural surfactants: 1. Glycosides of fatty acids and alcohols. *Lipids* **39**(10): 933–953.

Dembitsky, V. M. 2004b. Chemistry and biodiversity of the biologically active natural glycosides. *Chemistry & Biodiversity* **1**(5): 673–781.

Dembitsky, V. M. 2005a. Astonishing diversity of natural surfactants: 2. Polyether glycosidic ionophores and macrocyclic glycosides. *Lipids* **40**(3): 219–248.

Dembitsky, V. M. 2005b. Astonishing diversity of natural surfactants: 3. Carotenoid glycosides and isoprenoid glycolipids. *Lipids* **40**(6): 535–557.

Dembitsky, V. M. 2005c. Astonishing diversity of natural surfactants: 4. Fatty acid amide glycosides, their analogs and derivatives. *Lipids* **40**(7): 641–660.

Dembitsky, V. M. 2005d. Astonishing diversity of natural surfactants: 5. Biologically active glycosides of aromatic metabolites. *Lipids* **40**(9): 869–900.

Dembitsky, V. M. 2005e. Astonishing diversity of natural surfactants: 6. Biologically active marine and terrestrial alkaloid glycosides. *Lipids* **40**(11): 1081–1105.

Dembitsky, V. M. 2006. Astonishing diversity of natural surfactants: 7. Biologically active hemi- and monoterpenoid glycosides. *Lipids* **41**(1): 1–27.

Desai, J. D. and I. M. Banat 1997. Microbial production of surfactants and their commercial potential. *Microbiology and Molecular Biology Reviews* **61**(1): 47–64.

Dexter, A. F., A. S. Malcolm et al. 2006. Reversible active switching of the mechanical properties of a peptide film at a fluid–fluid interface. *Nature Materials* **5**(6): 502–506.

Dexter, A. F. and A. P. J. Middelberg 2007a. Peptides as functional surfactants. *1st International Conference on Multiscale Structures and Dynamics of Complex Systems*, Beijing, People's Republic of China.

Dexter, A. F. and A. P. J. Middelberg 2007b. Switchable peptide surfactants with designed metal binding capacity. *Journal of Physical Chemistry C* **111**(28): 10484–10492.

Deziel, E., F. Lepine et al. 2000. Mass spectrometry monitoring of rhamnolipids from a growing culture of *Pseudomonas aeruginosa* strain 57RP. *Biochimica et Biophysica Acta-Molecular and Cell Biology of Lipids* **1485**(2–3): 145–152.

Eisenstein, B. I. 2004. Lipopeptides, focusing on daptomycin, for the treatment of Gram-positive infections. *Expert Opinion on Investigational Drugs* **13**(9): 1159–1169.

Escobar-Diaz, E., E. Lopez-Martin et al. 2005. AT514, a cyclic depsipeptide from *Serratia marcescens*, induces apoptosis of B-chronic lymphocytic leukemia cells: Interference with the Akt/NF-Î°B survival pathway. *Leukemia* **19**(4): 572–579.

Fujii, T., R. Yuasa et al. 1999. Biodetergent—IV. Monolayers of corynomycolic acids at the air–water interface. *Colloid and Polymer Science* **277**(4): 334–339.

Goren, M. B. 1972. Mycobacterial lipids—Selected topics. *Bacteriological Reviews* **36**(1): 33.

Gutierrez, T., T. Shimmield et al. 2008. Emulsifying and metal ion binding activity of a glycoprotein exopolymer produced by *Pseudoalteromonas* sp. strain TG12. *Applied and Environmental Microbiology* **74**(15): 4867–4876.

Gutnick, D. L., H. Bach et al. 2011. 3.59—*Biosurfactants. Comprehensive Biotechnology* (2nd edition). Burlington, Academic Press: pp. 699–715.

Haydock, S. F., T. Mironenko et al. 2004. The putative elaiophylin biosynthetic gene cluster in *Streptomyces* sp DSM4137 is adjacent to genes encoding adenosylcobalamin-dependent methylmalonyl CoA mutase and to genes for synthesis of cobalamin. *Journal of Biotechnology* **113**(1–3): 55–68.

Hörmann, B., M. M. Muller et al. 2010. Rhamnolipid production by *Burkholderia plantarii* DSM 9509(T). *European Journal of Lipid Science and Technology* **112**(6): 674–680.

Hue, N., L. Serani et al. 2001. Structural investigation of cyclic peptidolipids from *Bacillus subtilis* by high-energy tandem mass spectrometry. *Rapid Communications in Mass Spectrometry* **15**(3): 203–209.

Ishigami, Y., Y. J. Zhang et al. 2000. Spiculisporic acid. Functional development of biosurfactant. *Chimica Oggi-Chemistry Today* **18**(7–8): 32–34.

Jerala, R. 2007. Synthetic lipopeptides: A novel class of anti-infectives. *Expert Opinion on Investigational Drugs* **16**(8): 1159–1169.

Kai, M., Y. Fujita et al. 2007. Identification of trehalose dimycolate (cord factor) in *Mycobacterium leprae*. *Febs Letters* **581**(18): 3345–3350.

Kaiser, H. and W. Kellerschierlein 1981. Structure elucidation of elaiophylin—Spectroscopy and chemical degradation. *Helvetica Chimica Acta* **64**(2): 407–424.

Kakinuma, A., G. Tamura et al. 1968. Wetting of fibrin plate and apparent promotion of fibrinolysis by surfactin a new bacterial peptidlipid surfactant. *Experientia* **24**(11): 1120–1121.

Kaplan, N., E. Rosenberg et al. 1985. Structural studies of the capsular polysaccharide of *Acinetobacter calcoaceticus* Bd4. *European Journal of Biochemistry* **152**(2): 453–458.

Kaplan, N., Z. Zosim et al. 1987. Reconstitution of emulsifying activity of *Acinetobacter calcoaceticus* Bd4 emulsan by using pure polysaccharide and protein. *Applied and Environmental Microbiology* **53**(2): 440–446.

Kappeli, O. and W. R. Finnerty 1979. Partition of alkane by an extracellular vesicle derived from hexadecane-grown *Acinetobacter*. *Journal of Bacteriology* **140**(2): 707–712.

Kappeli, O. and W. R. Finnerty 1980. Characteristics of hexadecane partition by the growth-medium of *Acinetobacter* sp. *Biotechnology and Bioengineering* **22**(3): 495–503.

Kim, H. S., B. D. Yoon et al. 1999. Characterization of a biosurfactant, mannosylerythritol lipid produced from *Candida* sp SY16. *Applied Microbiology and Biotechnology* **52**(5): 713–721.

Kleinig, H. and W. J. Broughton 1982a. Carotenoids of rhizobia. 5. Carotenoid-pigments in a red strain of rhizobium from *Lotononis bainesii* Baker. *Archives of Microbiology* **133**(2): 164–164.

Kleinig, H. and W. J. Broughton 1982b. Carotenoid pigments in a red strain of Rhizobium, from *Lotononis bainesii* Baker. *Arch. Microbiol.* **133**: 164–174.

Kleinig, H., W. Heumann et al. 1977. Carotenoids of *Rhizobia*. 1. New carotenoids from *Rhizobium lupini*. *Helvetica Chimica Acta* **60**(1): 254–258.

Kleinig, H. and R. Schmitt 1982. On the biosynthesis of C30 carotenoic acid glucosyl esters in *Pseudomonas rhodos*—Analysis of Car-mutants. *Zeitschrift Fur Naturforschung C-A Journal of Biosciences* **37**(9): 758–760.

Kobayashi, J., T. Kubota et al. 1999. Colopsinol A, a novel polyhydroxyl metabolite from marine dinoflagellate *Amphidinium* sp. *Journal of Organic Chemistry* **64**(5): 1478–1482.

Kobayashi, J. I. and T. Kubota 2007. Bioactive macrolides and polyketides from marine dinoflagellates of the genus *Amphidinium*. *Journal of Natural Products* **70**(3): 451–460.

Kretschmer, A., H. Bock et al. 1982. Chemical and physical characterization of interfacial-active lipids from *Rhodococcus erythropolis* grown on normal-alkanes. *Applied and Environmental Microbiology* **44**(4): 864–870.

Kubota, T., M. Tsuda et al. 1999. Colopsinols B and C, new long chain polyhydroxy compounds from cultured marine dinoflagellate Amphidinium sp. *Journal of the Chemical Society-Perkin Transactions* **1**(23): 3483–3487.

Kubota, T., M. Tsuda et al. 2000. Colopsinols D and E, new polyhydroxyl linear carbon chain compounds from marine dinoflagellate *Amphidinium* sp. *Chemical & Pharmaceutical Bulletin* **48**(10): 1447–1451.

Kwa, A. L., V. H. Tam et al. 2008. Polymyxins: A review of the current status including recent developments. *Annals Academy of Medicine Singapore* **37**(10): 870–883.

Landman, D., C. Georgescu et al. 2008. Polymyxins revisited. *Clinical Microbiology Reviews* **21**(3): 449–465.

Lang, S. and J. C. Philp 1998. Surface-active lipids in rhodococci. *Antonie Van Leeuwenhoek International Journal of General and Molecular Microbiology* **74**(1–3): 59–70.

Lang, S. and W. Trowitsch-Kienast 2002. *Biotenside*. Stuttgar, Leipzig, Wiesbaden, B.G. Teubner.

Li, H., T. Tanikawa et al. 2005. *Serratia marcescens* gene required for surfactant serrawettin W1 production encodes putative aminolipid synthetase belonging to nonribosomal peptide synthetase family. *Microbiology and Immunology* **49**(4): 303.

Liaaen-Jensen, S. 1969. Selected examples of structure determination of natural carotenoids. *Pure and Applied Chemistry* **20**: 421–448.

Linder, M. B. 2009. Hydrophobins: Proteins that self assemble at interfaces. *Current Opinion in Colloid & Interface Science* **14**(5): 356–363.

Lindum, P. W., U. Anthoni et al. 1998. N-Acyl-L-homoserine lactone autoinducers control production of an extracellular lipopeptide biosurfactant required for swarming motility of *Serratia liquefaciens* MG1. *Journal of Bacteriology* **180**(23): 6384–6388.

Matsuyama, T., M. Fujita et al. 1985. Wetting agent produced by *Serratia marcescens*. *FEMS Microbiology Letters* **28**(1): 125–129.

Matsuyama, T., K. Kaneda et al. 1990. Surface-active novel glycolipid and linked 3-hydroxy fatty-acids produced by *Serratia rubidaea*. *Journal of Bacteriology* **172**(6): 3015–3022.

Matsuyama, T., K. Kaneda et al. 1992. A novel extracellular cyclic lipopeptide which promotes flagellum-dependent and -independent spreading growth of *Serratia marcescens*. *Journal of Bacteriology* **174**(6): 1769–1776.

Merchant, R. and I. M. Banat 2012. Microbial biosurfactants: Challenges and opportunities for future exploitation. *Trends in Biotechnology* **30**(11): 558–565.

Morita, T., M. Konishi et al. 2008. Production of glycolipid biosurfactants, mannosylerythritol lipids, by *Pseudozyma siamensis* CBS 9960 and their interfacial properties. *Journal of Bioscience and Bioengineering* **105**(5): 493–502.

Navon-Venezia, S., E. Banin et al. 1998. The bioemulsifier alasan: Role of protein in maintaining structure and activity. *Applied Microbiology and Biotechnology* **49**(4): 382–384.

Navon-Venezia, S., Z. Zosim et al. 1995. Alasan, a new bioemulsifier from *Acinetobacter radioresistens*. *Applied and Environmental Microbiology* **61**(9): 3240–3244.

Neu, T. R., T. Hartner et al. 1990. Surface-active properties of viscosin—A peptidolipid antibiotic. *Applied Microbiology and Biotechnology* **32**(5): 518–520.

Neu, T. R. and K. Poralla 1990. Emulsifying agents from bacteria isolated during screening for cells with hydrophobic surfaces. *Applied Microbiology and Biotechnology* **32**(5): 521–525.

Nunez, A., R. Ashby et al. 2001. Analysis and characterization of sophorolipids by liquid chromatography with atmospheric pressure chemical ionization. *Chromatographia* **53**(11–12): 673–677.

Ongena, M. and P. Jacques 2008. *Bacillus* lipopeptides: Versatile weapons for plant disease biocontrol. *Trends in Microbiology* **16**(3): 115–125.

Peng, F., Z. Liu et al. 2007. An oil-degrading bacterium: *Rhodococcus erythropolis* strain 3C-9 and its biosurfactants. *Journal of Applied Microbiology* **102**(6): 1603–1611.

Peypoux, F., J. M. Bonmatin et al. 1999. Recent trends in the biochemistry of surfactin. *Applied Microbiology and Biotechnology* **51**(5): 553–563.

Pradel, E., Y. Zhang et al. 2007. Detection and avoidance of a natural product from the pathogenic bacterium *Serratia marcescens* by *Caenorhabditis elegans*. *Proceedings of the National Academy of Sciences* **104**(7): 2295–2300.

Rau, U. and D. Kitamoto. 2009. Mannosylerythritol lipids: Production and downstream processing. In: D. Hayes et al. (eds.), *Biobased Surfactants and Detergents: Synthesis, Properties, and Applications*. Vol. 51–77. AOCS Press, Urbana, USA.

Rau, U., L. A. Nguyen et al. 2005. Fed-batch bioreactor production of mannosylerythritol lipids secreted by *Pseudozyma aphidis*. *Applied Microbiology and Biotechnology* **68**(5): 607–613.

Rehm, H. J. and I. Reiff 1981. *Mechanisms and Occurrence of Microbial Oxidation of Long-chain Alkanes*. Berlin/Heidelberg, Springer.

Rosenberg, E. and E. Z. Ron 1997. Bioemulsans: Microbial polymeric emulsifiers. *Current Opinion in Biotechnology* **8**(3): 313–316.

Rosenberg, E. and E. Z. Ron 1999. High- and low-molecular-mass microbial surfactants. *Applied Microbiology and Biotechnology* **52**(2): 154–162.

Ryll, R., Y. Kumazawa et al. 2001. Immunological properties of trehalose dimycolate (cord factor) and other mycolic acid-containing glycolipids—A review. *Microbiology and Immunology* **45**(12): 801–811.

Saini, H. S., B. E. Barragan-Huerta et al. 2008. Efficient purification of the biosurfactant viscosin from *Pseudomonas libanensis* strain M9–3 and its physicochemical and biological properties. *Journal of Natural Products* **71**(6): 1011–1015.

Sar, N. and E. Rosenberg 1983. Emulsifier production by *Acinetobacter calcoaceticus* strains. *Current Microbiology* **9**(6): 309–313.

Satpute, S. K., A. G. Banpurkar et al. 2010. Methods for investigating biosurfactants and bioemulsifiers: A review. *Critical Reviews in Biotechnology* **30**(2): 127–144.

Schulz, D., A. Passeri et al. 1991. Marine biosurfactants.1. Screening for biosurfactants among crude-oil degrading marine microorganisms from the North-Sea. *Zeitschrift für Naturforschung C-A Journal of Biosciences* **46**(3–4): 197–203.

Sen, R. 2010. *Biosurfactants*. New York, Springer.

Soberón-Chávez, G. 2010. *Biosurfactants: From Genes to Applications*. Springer.

Soto-Cerrato, V., B. Montaner et al. 2005. Cell cycle arrest and proapoptotic effects of the anticancer cyclodepsipeptide serratamolide (AT514) are independent of p53 status in breast cancer cells. *Biochemical Pharmacology* **71**(1–2): 32–41.

Stansly, P. G. 1949. The polymyxins—A review and assessment. *American Journal of Medicine* **7**(6): 807–818.

Suzuki, T., K. Tanaka et al. 1969. Trehalose lipid and alpha-branched-beta-hydroxy fatty acid formed by bacteria grown on N-alkanes. *Agricultural and Biological Chemistry* **33**(11): 1619.

Tabuchi, T., I. Nakamura et al. 1977. Factors affecting production of open-ring acid of spiculisporic acid by *Penicillium spiculisporum*. *Journal of Fermentation Technology* **55**(1): 43–49.

Toren, A., E. Orr et al. 2002. The active component of the bioemulsifier alasan from *Acinetobacter radioresistens* KA53 is an OmpA-like protein. *Journal of Bacteriology* **184**(1): 165–170.

Vater, J. 1986. Lipopetides, an attractive class of microbial surfactants. In: Kilian, H.-G. and Weiss, A. (eds.), *Progress in Colloid and Polymer Science*. Berlin/Heidelberg, Verlag, Springer: **72**: 12–18.

Wasserman, H. H., J. J. Keggi et al. 1962. Structure of serratamolide. *Journal of the American Chemical Society* **84**(15): 2978–2982.

Wessels, J. G. H. 1997. Hydrophobins: Proteins that change the nature of the fungal surface. *Advances in Microbial Physiology, Vol. 38*. R. K. Poole. **38:** 1–45.

Wessels, J. G. H. 2000. Hydrophobins, unique fungal proteins. *Mycologist* **14**(4): 153–159.

Wosten, H. A. B. and J. G. H. Wessels 1997. Hydrophobins, from molecular structure to multiple functions in fungal development. *Mycoscience* **38**(3): 363–374.

Yakimov, M. M., P. N. Golyshin et al. 1998. *Alcanivorax borkumensis* gen. nov., sp. nov., a new, hydrocarbon-degrading and surfactant-producing marine bacterium. *International Journal of Systematic Bacteriology* **48**: 339–348.

Yokoyama, A., G. Sandmann et al. 1995. Thermozeaxanthins, new carotenoid-glycoside-esters from thermophilic eubacterium *Thermus thermophilus*. *Tetrahedron Letters* **36**(27): 4901–4904.

Zavascki, A. P., L. Z. Goldani et al. 2007. Polymyxin B for the treatment of multidrug-resistant pathogens: A critical review. *Journal of Antimicrobial Chemotherapy* **60**(6): 1206–1215.

Zhang, J. W., S. H. Lee et al. 1999. Surface properties of emulsan-analogs. *Journal of Chemical Technology and Biotechnology* **74**(8): 759–765.

2 Sophorolipids
Microbial Synthesis and Application

I.N.A. Van Bogaert, K. Ciesielska, B. Devreese,
and W. Soetaert

CONTENTS

2.1 Introduction 19
2.2 Structure and Properties 19
2.3 Supramolecular Assembly 21
2.4 Sophorolipid-Producing Organisms 21
2.5 Sophorolipid Biosynthesis 23
2.6 Regulation of Sophorolipid Biosynthesis... 24
2.7 The Sophorolipid Production Process 25
    2.7.1 Fermentation Parameters.. 25
    2.7.2 Substrates.. 27
    2.7.3 Product Recovery.. 27
    2.7.4 Costs 28
2.8 Strain Modification.. 28
2.9 Applications... 29
    2.9.1 Cleaning.. 29
    2.9.2 Cosmetics and Dermatological Applications.. 29
    2.9.3 Plant Protection... 29
    2.9.4 Feed and Food 30
    2.9.5 Bioremediation... 30
    2.9.6 Medical Applications.. 30
    2.9.7 Enzyme Induction... 31
    2.9.8 Source of Chemical Compounds 31
    2.9.9 Nanotechnology 32
References.. 32

2.1 INTRODUCTION

Sophorolipids are glycolipidic biosurfactants extracellularly produced by several yeast species. Owing to the nonpathogenic character of the production host and the high yields, there is quite some commercial interest in these molecules. Indeed, to date sophorolipids are available in the market and find applications in real-life products. Further research is currently being conducted in order to broaden their application range, either by evaluating their behavior in new applications or by modifying the sophorolipid chemical structure.

2.2 STRUCTURE AND PROPERTIES

Sophorolipids are glycolipid biosurfactants consisting of a sophorose sugar head and a hydrophobic fatty acid tail. Sophorose is a glucose disaccharide with a β-1,2 bond and can be acetylated at the

6′ and/or 6″ positions. To the sophorose, a single terminal or subterminal hydroxylated fatty acid (C16 or C18) is β-glycosidically linked. The carboxylic end of this fatty acid can be free (acidic/open form) or internally esterified at the 4″ position of the sophorose head (lactone form) (Figure 2.1). Rarely, the esterification occurs at the 6′- or 6″ position. The hydroxyl fatty acid can contain one or more unsaturated bonds (Asmer et al. 1988). The main producers are yeast species belonging to the *Starmerella* clade. They synthesize sophorolipids as a mixture of molecules that differ in the fatty acid part (chain length, saturation, and position of hydroxylation), in the acetylation pattern as well in lactonization (Davila et al. 1993).

The exact structure of the sophorolipid isoform has immense influence on the physicochemical properties. Lactonic sophorolipids have enhanced biological properties (e.g., inhibiting effects), and have a better capacity to lower the surface tension, whereas the acidic sophorolipids are more soluble and are better foam formers (Van Bogaert et al. 2011). Despite the fact that di- or monoacetylated sophorolipids are less soluble, they have better antibacterial (Lang et al. 1989), antiviral, and cytokine stimulating effects (Shah et al. 2005). Concerning the surfactant properties however, one must bear in mind that there is a natural synergy among the different compounds occurring in the natural blend (Hirata et al. 2009a).

Sophorolipids lower the surface tension from 72.8 mN/m down to 40–30 mN/m, even in the presence of salts (Hirata et al. 2009b) and in a wide temperature range (Nguyen et al. 2010). The reported critical micelle concentration (CMC) values range from 11 to 250 mg/L, depending on the applied measuring methods and conditions (Develter and Lauryssen 2010). Even with this variation, these values are about two orders of magnitude lower compared with chemical-derived surfactants, adding to their environmental friendly profile and in certain applications circumventing the problems related to a higher price, as less product is needed.

The minimal dynamic surface tension is 32.1 mN/m, but is only achieved in semi-static conditions (at least 40 s) and high concentrations. Nevertheless, one can observe a fast drop in surface tension even at short bubble lifetimes of 30 ms, rendering sophorolipids potentially suitable for dynamic applications such as spray-on coating and cleaning, this is in contrast to other glycolipids (Develter and Lauryssen 2010). Sophorolipids are low-foaming surfactants even at high concentrations (Hirata et al. 2009b). Moreover, they are good wetting agents illustrated by their ability to decrease the contact angle of water on polyvinyl chloride (PVC) from 110° to 80° at a minimal concentration of 36 mg/L (Develter and Lauryssen 2010). Both features render sophorolipids ideal components for hard surface cleaning products and dishwashing rinse aids (EP1445302).

As can be expected from biosurfactants, sophorolipids were demonstrated to be readily biodegradable as determined by the standard manometric respirometry and stable metabolite studies OECD 301C and 301F (Hirata et al. 2009b; Renkin 2003). Aquatic toxicity is 10-fold less compared with conventional surfactants and *Daphnia* reproduction is not affected at all (Renkin 2003).

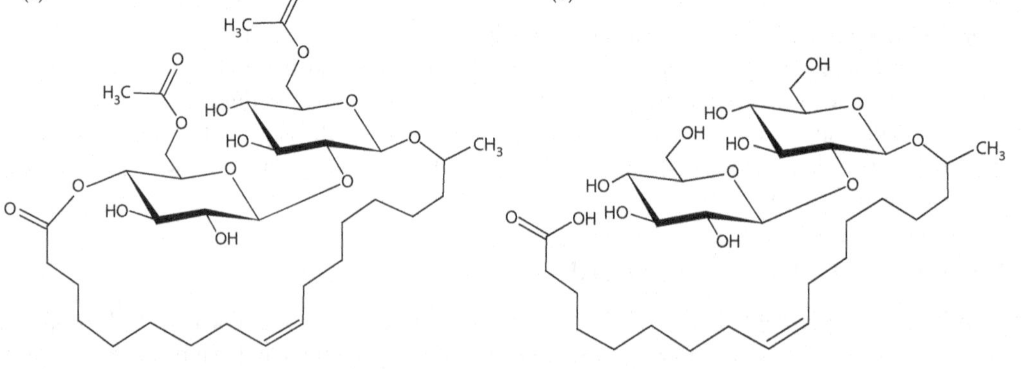

FIGURE 2.1   (a) Lactonic diacetylated sophorolipid. (b) Acidic nonacetylated sophorolipid.

Furthermore, tests with acidic sophorolipids and the natural blend pointed out that they are not irritating to the skin, do not trigger allergic reactions, and have an oral safety level that is greater than or equal to 5 mL/kg weight (US 5756471). Cytotoxicity was evaluated by the dimethylthiazol-diphenyltetrazoliumbromide method with human epidermal keratinocytes and was proven to be low (Hirata et al. 2009b).

While sophorolipids are demonstrated to be noncytoxic and do not influence aquatic systems, they inhibit growth of some fungi and Gram-positive bacteria, such as *Bacillus subtilis*, *Staphylococcus epidermidis*, *Staphylococcus aureus*, *Streptococcus faecium*, *Propionibacterium acnes*, and *Corynebacterium xerosis*. These latter ones are the causal agents of acne and dandruff, respectively, rendering them attractive ingredients for cosmetic, hygienic, and pharmaco-dermatological products as described below (Hommel et al. 1987; Kim et al. 2002; Lang et al. 1989; Mager et al. 1987). The mode of action most likely involves interactions with the cellular membrane, as demonstrated for *B. subtillis* where increased leakage of intracellular enzymes was detected upon treatment (Kim et al. 2002). Further biological activities are discussed in the application part.

2.3 SUPRAMOLECULAR ASSEMBLY

Self-assembly is a process in which single components organize themselves into a structure or pattern as a result of specific local interactions, without external forces. Self-assembly of sophorolipids is entropically driven and takes place above the CMC. Beyond this point, surfactants organize themselves in water such that their polar head groups become oriented toward the water, whereas the hydrophobic tails cluster together. This orientation leads to the formation of various superstructures such as micelles, vesicles, or multilayers.

Acidic sophorolipids, especially, received a lot of attention because of their unique structural features including an asymmetrical polar head size (disaccharide vs. COOH) and a kinked hydrophobic core (*cis*-9-octadecenoic chain; Zhou et al. 2004). Different techniques have been employed to investigate the various structures of the self-assembly including light microscopy, small- and wide-angle x-ray scattering, Fourier transform infrared spectroscopy, and dynamic laser light scattering. The degree of ionization of the -COOH group greatly influences the sophorolipid self-assembly (Baccile et al. 2012). When the degree of ionization increases, negative charges at the micellar surfaces are introduced, which initiate changes in shape, aggregation state, and surface properties. Micelles are formed at low (pH < 5) and medium (5 < pH < 8) degrees of ionization. At high ionization (pH > 8), large net-like aggregates are observed.

In the same study of Baccile et al., the morphology of the assemblies was evaluated as a function of sophorolipid concentration. For sophorolipid concentrations <1 wt%, spherical micelles having an average radius of 3.0 nm existed. At a sophorolipid concentrations ≥1 wt%, micelles were no longer spherical and started to elongate from $c > 0.5$ wt%, and at $c = 5$ wt%, a cylindrical micelle shape was observed.

There are some contradictory reports on the formation of giant twisted and helical ribbons. Zhou et al. (2004) report on the formation of giant ribbons depending on the pH and time for the C18:1-*cis* sophorolipids, while Baccile et al. (2012) did not observe these structures and Dhasaiyan et al. (2013) reported giant ribbons exclusively for C18:1-*trans* sophorolipids.

2.4 SOPHOROLIPID-PRODUCING ORGANISMS

Candida apicola was the first species described to have produced sophorolipids (Gorin et al. 1961; formerly called *Torulopsis magnolia*). However, for industrial purposes, the most applied producer is *Starmerella* (syn. *Candida*) *bombicola*, a nonpathogenic yeast isolated from the honey of *Bombus* sp. (the bumble-bee), by Spencer in 1970 (Spencer et al. 1970). The name *Starmerella bombicola* was proposed by taxonomists as they discovered the new clade *Starmerella* to which *Candida bombicola* was classified based on high 18S rDNA identity (Rosa and Lachance 1998). Strains from the

Starmerella clade are fermentative and utilize a few carbon sources, such as glucose, galactose, raffinose, and sucrose. They are osmotolerant, which indicates a specialization toward a microenvironment with a high osmotic pressure such as nectar. In 2010, Kurtzman discovered new members of the *Starmerella* genus producing sophorolipids: *C. stellata, C. riodocensis*, and *Candida* sp. *NRRL Y-27208* (Kurtzman et al. 2010). All these species produce predominantly acidic sophorolipids in contrast to *S. bombicola* and *C. apicola*, which produce mostly the lactone form. Interestingly, in *Candida* sp. *NRRL Y-27208* a novel form of dimeric and trimeric sophorose containing sophorolipids was identified by MALDI-TOFMS (Figure 2.2). Surprisingly, some of these compounds were also detected in very minor amounts in the *S. bombicola* sophorolipid mixture (Price et al. 2012). Furthermore, Konishi et al. (2008) discovered *C. batistae*, which produces mainly diacetylated acidic sophorolipids containing mostly terminally hydroxylated octadecanoic acid as the lipid tail. Imura et al. (2010) described *C. floricola* TM 1502 as a new sophorolipid producer, secreting mainly diacetylated acidic sophorolipids. Generally, in the last 3 years, the interest in the *Starmerella* clade of organisms increased. Other new strains were isolated from flowers and fruits, such as *Starmerella* sp. nov. (Sipiczki 2013), *Starmerella jinningensis* sp. nov. (Li et al. 2013), and *Candida kuoi* sp. (Kurtzman). Only the last one was shown to produce acidic sophorolipids similar to *C. batistae, C. riodocensis,* and *C. stellata.*

Also other microorganisms, not belonging to the *Starmerella* clade, were reported to produce sophorolipids. *Wickerhamiella domercqiae*, a strain isolated from oil waste, was shown to secrete

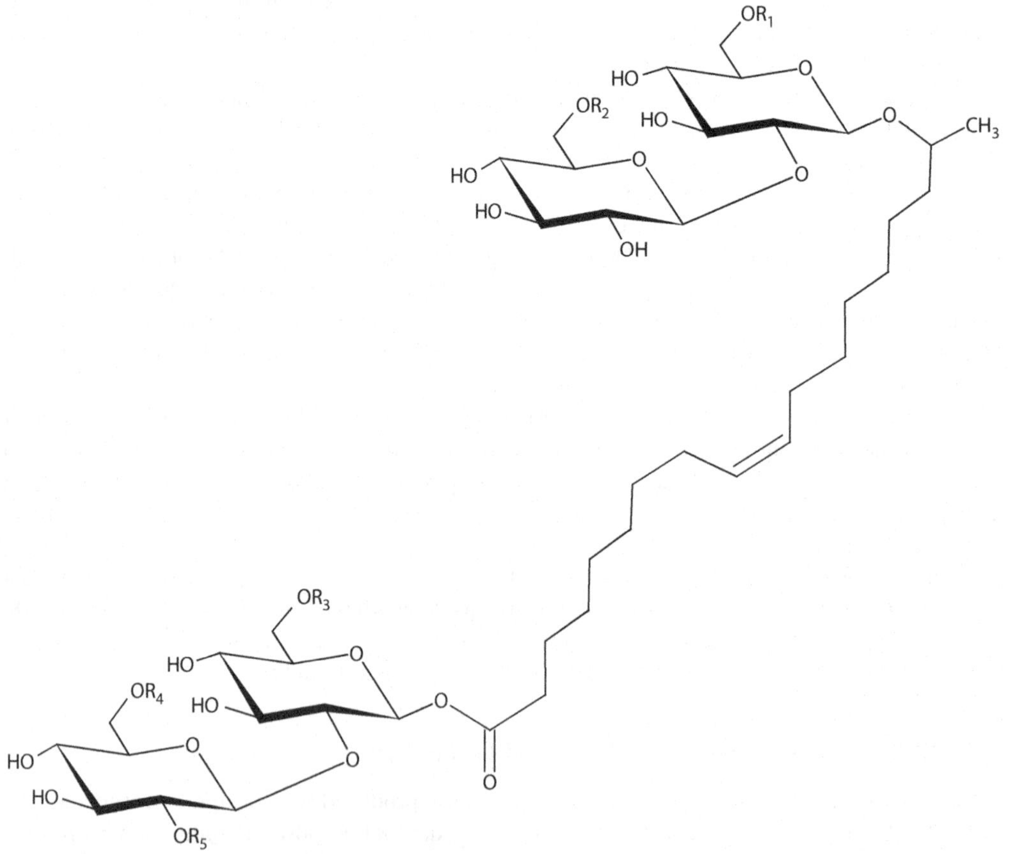

FIGURE 2.2 The structure of dimeric and trimeric sophorolipids. $R_1, R_2, R_3, R_4 =$ –H or –COCH$_3$ independently of each other. For *Starmerella bombicola* $R_5 =$ –H, for *Candida* sp. *NRRL Y-27208* $R_5 =$ –H, a fatty acid, a fatty acid with a sophorose or a fatty acid with a sophorose and another fatty acid attached to it.

molecules that are almost identical to the major components of sophorolipids produced by *S. bombicola* and *C. apicola* (Chen et al. 2006a). Moreover, *W. domercqiae* is also able to produce lactonic diacetylated sophorolipids, with 17-hydroxyoctadecanoic acid as a lipid, in a high yield. Sophorolipid production was also reported in the thermotolerant yeast *Pichia anomala*. However, the yield was very low and the structural properties of the products have not been described (Thaniyavarn et al. 2008). Finally, Tulloch et al. found a new form of sophorolipids produced by *Candida* (now *Rhodotorula*) *bogoriensis*. Its structure differs from the sophorolipids of *S. bombicola* in the hydroxy fatty acid moiety, which is 13-hydroxyl dodecosanoic acid (C22; Tulloch et al. 1968).

2.5 SOPHOROLIPID BIOSYNTHESIS

Generally, sophorolipid yields increase extremely when both hydrophilic and hydrophobic carbon sources are present in the medium (e.g., glucose and vegetable oil, respectively). This can be explained by the fact that when only a hydrophilic substrate is present, sophorolipid synthesis requires *de novo* fatty acid synthesis at the cost of additional energy, and the efficiency of the process drops. However, when a hydrophobic substrate is present, it is directly incorporated into the sophorolipids (Hommel et al. 1994b). Different hydrophobic carbon sources have already been tested: from alkanes, fatty acids, alcohols to esters. Interestingly, only those with a chain length similar to *de novo* produced sophorolipids (C16–C18) seem to be easily incorporated and result in high fermentation yields. Among the most favored are rapeseed oil/esters that contain mainly C18:1 and C18:2 fatty acids (Davila et al. 1994). In contrast, a rather poor integration of oils originating from coconut and meadow foam is observed, probably because they contain either medium or very long chain fatty acids (Van Bogaert et al. 2010).

This preference for C16 and C18 fatty acids can be explained by the specificity of the cytochrome P450 monooxygenase, which is proposed to be the first enzyme in the sophorolipid pathway and which hydroxylates mainly C16:0, C18:1, and C18:0 fatty acids. Therefore, in case of a hydrophobic source with a different carbon chain length, the fungus needs to either shorten or lengthen the fatty acid chains, or, most likely, completely degrade them. In this case, *de novo* synthesis of the sophorolipid hydrophobic moiety is required.

Biologically derived sophorolipids can be modified by chemo-enzymatic processes as described for some of the applications. Yet, all existing methods are expensive and time consuming. The availability of genetically modified strains able to produce new-to-nature molecules with different properties would bypass the need for these additional steps and purifications. However, in order to create such strains, the knowledge about sophorolipid biosynthesis needs to be improved.

Sophorolipid biosynthetic enzymes are organized in a gene cluster, a feature quite typical for fungal secondary metabolites (Van Bogaert et al. 2013; Figure 2.3b). The initial step involves the terminal or subterminal hydroxylation of a fatty acid by an endoplasmic reticulum (ER)-associated cytochrome P450 enzyme (Figure 2.3a). Its expression during sophorolipid production was confirmed at the transcription level by real-time reversed transcription PCR (Van Bogaert et al. 2010). Next, a glucose molecule is bound to the hydroxylated fatty acid-forming glucolipids. Then, a second glucose is attached creating an acidic sophorolipid. As described by Saerens et al. (2011a,c), those two reactions are carried out by two glucosyltransferases, UGTA1 and UGTB1, which use UDP-glucose as donor. The obtained acidic nonacetylated sophorolipids can be secreted as such or, alternatively, they first undergo an acetylation on the sophorose 6' and/or 6'' hydroxyl groups by an acetyl-CoA-dependent acetyltransferase before secretion (Saerens et al. 2011b). Secretion in the extracellular environment is mediated by active transport: an ATP-dependent multidrug resistance protein is responsible for majority of the translocation (Van Bogaert et al. 2013). Outside the cell, the molecules then can undergo lactonization catalyzed by a specific lactone esterase (Ciesielska 2013; Van Bogaert et al. 2011). The gene is not localized in the sophorolipid biosynthetic cluster and most likely evolved independently, which is supported by the fact that some organisms do not produce lactonic sophorolipids at all and that its expression profile in *S. bombicola* differs from the profile of the cluster genes.

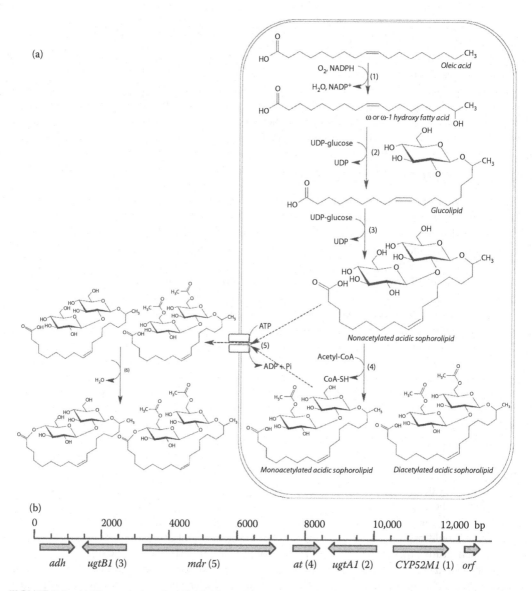

(a)

(b)

FIGURE 2.3 (a) Theoretical sophorolipids biosynthesis pathway. (b) Sophorolipid gene cluster. (1) Cytochrome P450 monooxygenase CYP52M1, (2) UDP-glucosyltransferase *ugtA1*, (3) UDP-glucosyltransferase *ugtB1*, (4) acetyltransferase *at*, (5) sophorolipid transporter *mdr*, *adh*: putative alcohol dehydrogenase, *orf*: open-reading frame with unknown function.

A proteomic comparison between sophorolipid inducing and noninducing conditions in *S. bombicola* demonstrated a coordinated synthesis of the enzymes of the sophorolipid biosynthetic cluster. In the stationary phase, their levels were two orders of magnitude higher than in exponentially grown cells. The same clear upregulation was detected in RNA-sequencing experiments (Ciesielska et al. 2013).

2.6 REGULATION OF SOPHOROLIPID BIOSYNTHESIS

Sophorolipids are mainly secreted during the stationary phase and their presence is not obligatory for cell viability. Hence, they can be considered as secondary metabolites. The molecules are produced at certain limiting conditions in excess of a carbon source and, therefore, it is suggested that they act as an external carbon sink.

In other fungal organisms producing glycolipids, for example, MELS and cellobiose lipids, the expression of the clustered genes is controlled by a transcription factor that is activated in response to a changing environment. For example, in *Ustilago maydis,* the production of biosurfactants occurs after activation upon nitrogen limitation of a transcription factor present in the cluster (Teichmann et al. 2010). In the *S. bombicola* sophorolipid cluster, no transcription factor was detected (Van Bogaert et al. 2013). However, for *C. apicola* it is suggested that the ammonium ion concentration influences for sophorolipid production (Hommel et al. 1994a). Also Davila et al. (1992) described that sophorolipid production in *S. bombicola* is connected with nitrogen limitation. Later, Albrecht et al. (1996) followed the nitrogen and phosphate concentration during *S. bombicola* growth and concluded that sophorolipid production occurred at total phosphate exhaustion and nitrogen limitation. It was also proposed that under these conditions NAD(+)- and NADP(+)-dependent isocitrate dehydrogenase (NAD/P-ICDH) has a declined specific activity. This, together with a normal isocitrate synthase activity, causes the release of excessive citrate into the cytosol where ATP-dependent citrate lyase converts it into oxaloacetate and acetyl-CoA, the building block of fatty acids. Albrecht et al. followed the activity of NAD and NADP-ICDH during growth in a bioreactor and also tested the cofactor's influence on the activity of those enzymes. It was suggested that the decline in specific activity of NAD/P-ICDH is not regulated at the enzyme activity but at the enzyme synthesis level.

2.7 THE SOPHOROLIPID PRODUCTION PROCESS

2.7.1 Fermentation Parameters

As mentioned above, sophorolipid synthesis in not associated with actively growing cells, but with stationary phase. Hence, production only starts after a nonproductive growth phase of 1 or 2 days, but can be maintained at good production rates for over 1 week when the carbon sources are kept available by, for example, a fed-batch fermentation. Indeed, as explained in Section 2.5, production is optimal when both glucose (hydrophilic carbon source) and vegetable oil (hydrophobic C-source) are supplied. A correct feeding rate of both turns out to be important for productivity and influences the sophorolipid composition (balance lactonic vs. acidic form) as well (Davila et al. 1997). Besides the carbon sources, the medium should contain an (organic) nitrogen source, such as yeast extract or corn steep liquor, and favorable, but not essential, buffering agents or small amounts of minerals.

In general, most fermentations are carried out at 25°C or 30°C. The amount of obtained sophorolipid is nearly identical for both temperatures, whereas for fermentations at 25°C, biomass growth is lower, and the glucose consumption rate is higher as compared with the fermentation at 30°C (Casas and Garcia-Ochoa 1999).

After initiation of the fermentation process, pH is allowed to drop spontaneously till 3.5 and should be maintained at this point for optimal sophorolipid production (Gobbert et al. 1984). Throughout the whole fermentation process, the culture broth should be supplied with sufficient oxygen; the yeast cells are very sensitive to oxygen limitation during their exponential growth and cannot grow anaerobically. Furthermore, good aeration conditions are important for sophorolipid production as the cytochrome P450 monooxygenase uses molecular oxygen.

When looking at general product yield, values of over 400 g/L can be achieved (Daniel et al. 1998; Pekin et al. 2005). In the referred cases, the volumetric productivities were 0.92 and 0.7 g/L/h, due to the prolonged incubation times. However, higher volumetric values of, for instance, 1.9 and 2.1 g/L/h can be obtained with a respective yield of 365 and 317 g/L sophorolipids (Davila et al. 1997; Kim et al. 2009). Recently, Gao et al. could further increase productivity till 3.7 g/L/h, thanks to fermentations run at very high cell density (Gao et al. 2012). An overview of the most remarkable productivities is given in Table 2.1.

TABLE 2.1

Overview of Remarkable Sophorolipid Production Methods with Respect to Yields

Characteristics	Production (g/L)	C-Yield (g/g)	Time (h)	Volumetric Productivity (g/L/h)	References
10% glucose 10% animal fat	120	0.58	68	2.4	Deshpande and Daniels (1995)
10% glucose 10.5% safflower oil	137		192	0.7	Zhou and Kosaric (1993)
10% glucose 10.5% rapeseed oil	160		216	0.7	Zhou and Kosaric (1995)
30% rapeseed oil 10% deproteinized whey concentrate lactose not consumed	280		280	1.0	Daniel et al. (1998)
Single cell-oil from *Cryptococcus curvatus* grown on deproteinized whey concentrate 40% rapeseed oil added after single cell oil consumption	422		145 +410	0.8–1.0	Daniel et al. (1998b)
Turkish corn oil Glucose Honey added when glucose depleted	>400	>0.6	436	>0.9	Pekin et al. (2005)
Fed batch of glucose and octadecane	175	0.33	165	1.1	Davila et al. (1994)
Fed batch of glucose and rapeseed FAEE	340	0.65	165	2.1	
Fed batch of glucose and rapeseed oil	255	0.53	165	1.5	
11% glucose 10% soybean oil fed batch	120	0.6	110	1.1	Lee and Kim (1993)
10% glucose 10% sunflower oil resting cells	120	0.6	200	0.6	Casas and Garcia-Ochoa (1999)
Glucose Oleic acid Fed batch Crystals if limited oil feeding	180		200	0.9	Rau et al. (1996)
Glucose Rapeseed oil Fed batch	300	0.68	125	2.4	Rau et al. (2001)
4% oleic acid and 10% glucose at start additional fed batch focus on aeration (50–80 mM O$_2$/L/h)	350		>week	1–1.5	Guilmanov et al. (2002)
Glucose: 30–40 g/L Rapeseed oil	365		192	1.9	Kim et al. (2009)
Glucose Tallow fatty acid residue Fed batch	120	0.41	240	0.5	Felse et al. (2007)
Glucose Rapeseed oil High dry cell weight	200		54	3.7	Gao et al. (2012)

FAEE = fatty acid ethyl ester.

2.7.2 Substrates

Glucose is the hydrophilic substrate of choice. Other mono- or disaccharides can also act as substrate: fructose, mannose, saccharose, maltose, raffinose, sucrose, and lactose, yet at the expense of productivity (Gobbert et al. 1984; Klekner et al. 1991; Zhou and Kosaric 1993). In this respect, cheaper substrates or waste streams were evaluated as well. Again, glycerol can be used, but has a negative effect on the overall yield (Ashby et al. 2006). The same is true for sugarcane or soy molasses, which contain a number of sugars that are nonfermentable for *S. bombicola* (e.g., stachyose). However, molasses is a source of protein and nitrogen and can (partially) replace the conventional nitrogen source (Solaiman et al. 2007; Takahashi et al. 2011).

Several types of molecules can act as hydrophobic carbon source: oils, fats, fatty acids, their corresponding esters, alkanes, or waste streams containing one of them. In general, fatty acid methyl- or ethyl esters derived from vegetable oils result in improved yields compared to their corresponding oils and both perform better than alkanes (Table 2.1; Davila et al. 1994). However, the presence of residual hydrophobic carbon source in the final sophorolipids is a bigger issue for fatty acid esters due to their corrosive and irritating properties.

Free fatty acids of a specific length can be used as well, although there are some constraints such as a melting temperature above room temperature for the saturated fatty acids and effects on the cell's electron balance. As can be expected from the enzyme preferences and the native sophorolipid fatty acid tail lengths, oleic acid performs bests (Asmer et al. 1988). Consequently, vegetable oils rich in oleic acid, such as rapeseed oil, promote sophorolipid production (US 5900366). In this respect, also the high oleic sunflower oil would be a good choice. As this oil came to be widely available only recently due to expiring of a patent, no reports on fermentations with this substrate are consultable yet, but the fatty acid profile with over 80% oleic acid (vs. 55–75 for rapeseed oil) looks promising.

Analogue to fatty acids, the effectiveness of alkanes depends on their chain length. Hexadecane, heptadecane, and octadecane achieve the best production yields and appear to be directly converted into hydroxy fatty acids and incorporated into the sophorolipid molecules, this way strongly influencing the fatty acid composition of the sophorolipid mixture (Cavalero and Cooper 2003; Davila et al. 1994).

Nonincorporated hydrophobic substrates are mainly oxidized to CO_2 by beta-oxidation and for the best-performing substrates a carbon conversion yield between 60% and 70% is obtained (see also Table 2.1).

2.7.3 Product Recovery

On laboratory scale, recuperation of sophorolipids from the culture broth is generally done by organic solvent extraction with, for example, ethyl acetate. Residual hydrophobic carbon source is co-extracted and can be removed by additional treatment with hexane. For some applications, sophorolipids are extracted with pentanol (Baccile et al. 2013) or *t*-butyl methyl ether (Rau et al. 2001).

For larger scale and industrial applications, physical separation methods are preferred. Sophorolipids are heavier than water, allowing to centrifuge them down or to just decant them from the fermentation medium after heating (salting out). Further elimination of water and impurities may be required. In general, the fed-batch fermentation needs to be stopped when too high concentrations are reached resulting in high viscosity with stirring and aeration problems. In this respect, separation strategies integrated with the production process could prolong the production phase. Ultrasound separation was evaluated and cells could be retained at high efficiencies and viability, but the sophorolipid purification yield can still be improved (Palme et al. 2010). Foam fractionation is another option often suggested for biosurfactants, but this process has not been established as yet for sophorolipids. Cyclic fermentation with intermediate sedimentation and harvesting steps can be applied as well (US5879913).

Under optimal fermentation and feeding conditions, most of the sophorolipids will be present in the lactonic diacetylated form. This isoform is not very water soluble and has a rather rigid structure resulting in the formation of crystals, which facilitates the purification process. Crystallization can be mimicked to a large extent when using sugarcane molasses: sophorolipids tend to aggregate with the sugarcane fibers (Yang et al. 2012).

If one requires a specific sophorolipid isoform, chromatographic methods with silica or preparative reversed-phase columns are required (Lin 1996). Yet, this is only feasible at small scale.

2.7.4 Costs

Obviously, production price depends on several factors: the type of substrates used and the hereto connected yields, product recovery strategy and required purity, scale, and so on. From these factors, scale has the main influence on the production price; large-scale production allows big investments in equipment and maintenance bringing down the cost per kg of the product. Ashby et al. (2013) made a recent price calculation based on a plant with an annual production capacity of 90.7 million kg per year; representing a 2% share of the surfactant market. They worked with a process yielding 100 g/L sophorolipids in 5 days (requiring two additional days for loading, unloading, cleaning, and sterilization) based on glucose, yeast extract, urea, and either oleic acid or high oleic acid sunflower oil. Investment in the equipment made up M$ 51.41, a large extent caused by equipment for product recovery and processing which involves a cell separation step by centrifugation, spray drying of the crude product, ethyl acetate washing and filtration, and another drying step. Taking into account costs for raw materials, utilities, labor, and depreciation, the price for sophorolipids produced from oleic acid or high oleic acid sunflower oil is respectively 2.54 or 2.95 $/kg. Again, these calculations were made based on a very large market share, hence costs linked to the current production volumes will be higher. In addition, no removal step for residual hydrophobic carbon sources was included. This is no problem for most detergent applications and if one has a well-controlled fermentation process reducing the relicts, but absence of these residues can be required for certain applications.

2.8 STRAIN MODIFICATION

Classic strain improvement methods using mutagenesis can be applied on sophorolipid-producing organisms as well. For instance, low energy ion beam implantation resulted in several *W. domercqiae* strains with increased sophorolipid production; the best strain produced 85% more sophorolipids compared with the wild type. The relative proportions of the different sophorolipid isoforms were influenced as well, yielding strains producing either more acidic or lactonic sophorolipids (Li et al. 2012). However, the genetic background of the mutants could not be determined.

In contrast to this black-box approach, targeted modifications can be implemented as well. Prerequisite for this is profound knowledge of the sophorolipid pathway and its regulation and the development of genetic engineering, something which was established during the last years for *S. bombicola* (Van Bogaert et al. 2008a,b, 2013). This allowed the creation of several defined mutants. Analogue to the above-mentioned *W. domercqiae* mutants, strains with a defined isoform profile could be created by knocking out the lactone esterase gene, a strain producing solely acidic sophorolipids, in amounts comparable to the wild type, was obtained. Contrary to the mutagenized mutants, no remaining lactonic forms are present and the genetic background is exactly known. If, however, an additional copy of the lactone esterase gene controlled by a constitutive promoter is introduced, more lactonic sophorolipids are present in the blend (WO2013092421).

Furthermore, a *S. bombicola* strain knocked out in beta-oxidation can be applied for the increased production of medium-chain sophorolipids starting from unconventional substrates (Van Bogaert et al. 2009) or the production of 20-hydroxyeicosatetraenoic acid, a compound of interest in medical studies (Van Bogaert et al. 2013). Other strains producing tailored glycolipids were developed as well: a *S. bombicola* strain producing glucolipids, cellobioselipids, and nonacetylated sophorolipids (Roelants et al. 2013; Saerens et al. 2011b,c). Although all strains were as vital as the wild type, the overall glycolipid production yields were lower, requiring further strain improvement.

2.9 APPLICATIONS

2.9.1 CLEANING

As such, sophorolipids are applied in different household products currently available on the global market: the Japanese company Saraya (http://worldwide.saraya.com/) includes them in their dishwasher products; they also find application as hard surface cleaners and rinse aids in several formulations of the Belgian company Ecover (http://www.ecover.com/). While about 10 years ago, the use of sophorolipids was mainly supported by smaller companies linked to specific niche markets, recently multinationals too show interest in these biosurfactants. Henkel, for instance, applies sophorolipids in some of its glass-cleaning products and both Unilever and Evonik-Degussa invested in research and patents on sophorolipids, but it is not clear whether and how these findings are commercialized. Weatholeo was founded in 2010, and is just as the cosmetic ingredient supplier Soleance, discussed below, a subsidiary of ARD in France. This company targets the large-scale manufacturing of several types of bio-surfactants with the goal of supplying to the detergent industry. Sophorolipids can be purchased under the brand name Sophoclean (http://www.wheatoleo.com).

2.9.2 COSMETICS AND DERMATOLOGICAL APPLICATIONS

The French company Soliance (http://www.soliance.com/) sells sophorolipids because of their excellent emulsifying and antibacterial properties. These glycolipids act against acne, dandruff, and body odors. Recently, Soliance introduced a new product in the market, Sophogreen, a plant-based solubilizer containing a high concentration of sophorolipids. Also in Asia, sophorolipids are commercialized; Intobio sells them in personal care products such as Sopholine mask sheets, acne soap, and liquid hypoallergenic soap.

Furthermore, sophorolipids are also claimed to have a protective effect on skin, hair, and nails, and therefore they are interesting substrates for cosmetic and dermatological products (Morya et al. 2013). Their use as a percutaneous absorption controller to facilitate administration of an active compound such as lactoferrin is claimed as well (JP2009062288). Moreover, they stimulate dermal fibroblast metabolism, collagen neosynthesis (Borzeix 1999) and participate in wound-healing processes (Maingault 1999). They can be applied in wound dressings as well in order to enhance gas permeation and might contribute to better wound healing due to some of the above-mentioned beneficial properties as well (WO2013112875). Moreover, sophorolipids act against dermatophytes causing ringworm infections (CN101199539).

2.9.3 PLANT PROTECTION

Many insecticides, fungicides, and herbicides are hydrophobic compounds, requiring the presence of emulsifiers or spreading agents in the formulation. Traditionally, chemical-derived surfactants are used, but with the growing share of organic farming, biosurfactants are checked out as well (WO2011039014). Sophorolipids are not only added as adjuvant, but can be applied as the active compound as well. As discussed previously, sophorolipids display inhibiting activities against specific organisms, among others against the plant pathogenic fungi *Phylophthora* sp., *Pythium* sp., and *Botrytis cineria* (Kim et al. 2002; Yoo et al. 2005). The activity against plant pathogenic bacteria can be1000-fold increased by using modified sophorolipid alkyl esters (methyl, ethyl, propyl, butyl, or hexyl) of the fatty acid carboxyl group (WO2013052615).

2.9.4 FEED AND FOOD

It has been demonstrated in the past that sophorolipids are useful emulsifiers; they can, for instance, improve the quality of wheat flour products (JP61205449). Yet, their commercialization in food products is hampered due to regulatory and administrative obstacles. However, there are few patent applications on a similar use in aquatic or animal feed (CN102696895, CN102696880, JP2010220516).

2.9.5 BIOREMEDIATION

Sophorolipids' emulsifying properties can be used in the petroleum industry, especially in secondary oil recovery and for removing the hydrocarbons from drill material (US5326407, US5900366). Furthermore, sophorolipids are claimed to be helpful in decontaminating soils and waters from hydrocarbons (US5654192) and in removing heavy metals from sediments (Mulligan et al. 2001). A formulation including sophorolipids can be applied for heavy metal absorption and flocculation in water as well (CN102764632). However, due to the good biodegradability of these biosurfactants, their application in *in situ* soil remediation is controversial as they tend to degrade or are being metabolized too fast.

2.9.6 MEDICAL APPLICATIONS

As mentioned earlier, sophorolipids display some biological activities, which differ depending on the exact structure of composition of the mixture. Although sophorolipids have some antibacterial activity, these actions are mild and not broad spectrum. Hence, sophorolipids cannot be considered as antibiotics *sensu stricuto*. Yet, it might be possible that the inhibitory effect on some bacteria plays a role in the action of sophorolipids as septic shock antagonist in animal models (Bluth et al. 2006; Sleiman et al. 2009), this in combination with a suggested interaction with the lipopolysaccharide (LPS)-mediated gene activation cascade due to interference with LPS-binding (Watson et al. 2012) and their claimed effect on the immune response as IgE suppressors (Bluth et al. 2008). This later property decreases pulmonary inflammation in a mouse asthma model and renders them promising anti-inflammatory or antiallergy agents (Hagler et al. 2007).

The first report on anticancer properties of sophorolipids was published in the 1990s: sophorolipids trigger cell differentiation instead of proliferation by specific interactions with the plasma membrane for the human promyelocytic leukemia cell line HL60 (Isoda et al. 1997). Also, the sophorolipids of *W. domercqiae* (the main product was identified as the diacetylated lactonic C18:1) were described to have some anticancer properties toward the human cancer cell lines H7402 (liver), A549 (lung), and HL60 and K562 (both leukemia lines) connected with their special interaction with the plasma membrane (Chen et al. 2006b). Induction of apoptosis was observed in the H7402 line (Chen et al. 2006a). In a later study, the effect of pure compounds on the human esophageal cancer cell lines KYSE109 and KYSE450 was investigated (Shao et al. 2012). It turned out that the diacetylated lactonic C18:1 had the strongest inhibitory effect compared to its monoacetylated counterpart, while no effect was observed for acidic sophorolipids, independent of their degree of acetylation. In contrast, Fu et al. (2008) found that the acidic sophorolipids were more effective against pancreatic cancer cells. Yet, the conflicting results could be attributed to a different mode of action for the specific cell lines. Also, the saturation profile influences the anticancer behavior: C18:1 should be preferred above C18:2 and C18:0 displayed only a weak activity (Shao et al. 2012). Illustrating the anticancer characteristics, sophorolipids also turned up in an experiment screening for histone deacetylase inhibitors, which are promising anticancer therapeutics (Wegener et al. 2008). Furthermore, native sophorolipids can be modified to enhance their anticancer effects. Methyl esters were demonstrated to be the most suitable for this purpose (Fu et al. 2008).

These methyl esters, but also the native sophorolipids and other esters, such as ethyl, butyl, and hexyl, can be applied as antibacterial, antiviral, and/or antispermidical agents (WO2004044216, WO2007130738). The US-based company Synthezyme is commercializing these compounds, and although the technology is based on the innovations of Gross who patented the medical applications, the company mainly focuses on their use as antimicrobials or biopesticides (e.g., WO2013052615).

Deuterated molecules or drugs have specific potential as they display a longer residence time in the body with lower metabolic rates, allowing lower dosage resulting in reduced toxicity risks for the patient. Deuterated sophorolipids can be produced by supplying the yeast cells with d-isostearic acid. In contrast to rhamnolipids, supplementation of deuterated D_2O has hardly any effect (Smyth et al. 2010).

2.9.7 Enzyme Induction

Sophorose is recognized as a good cellulase synthesis inducer in *Trichoderma* species. Yet, the costs associated with pure sophorose have limited its use in industrial cellulase production. Native sophorolipids were demonstrated to induce cellulase production in *Hypocrea jecorina*; the fungus degrades the sophorolipids resulting in the release of the inducer sophorose. Cellulase synthesis could be further increased when *S. bombicola* was cocultured with *Hypocrea jecorina* (Lo and Ju 2009). Sophorolipids also induce amylase production in *B. subtilis* and laccase and manganese peroxidase production in *Pleurotus ostreatus* (WO2007073371). The inducing effect of sophorose and sophorolipids can be exceeded by applying acid-treated sophorolipids; partial acidification gives rise, among others, to the active compounds with a varying degree of acetylation, conservation of the former lactone ester bond, but with breakage of the link between sophorose and the fatty acid hydroxyl group (Figure 2.4; WO2013003291).

2.9.8 Source of Chemical Compounds

As mentioned above, sophorolipids are a direct or indirect source of sophorose. Recently, a chemical method was described to obtain a protected sophorose bromide suitable for the creation of natural occurring plant sophorose glycosides such as zizybeoside, phenethyl glycoside, and ebracteatoside, displaying anti-inflammatory and anti-hepatitis properties (Hoffmann et al. 2012). Enzymatic release of sophorose is not possible as glycosidic enzymes tend to split the glucose moieties one by one. However, this method can be applied to create either glycolipids or hydroxy fatty acids (Imura et al. 2010; Rau et al. 1999; Saerens et al. 2009). These later compounds can be obtained by acidic hydrolysis as well and are from a chemical point of view interesting molecules for further modification. Straightforward reactions are lactonization to create macrocyclic esters for the perfume industry or polymerization (US4201844). More complex reactions are possible when converting the hydroxyl groups into azides or alkynes allowing click chemistry. In this way, TAG-based molecules

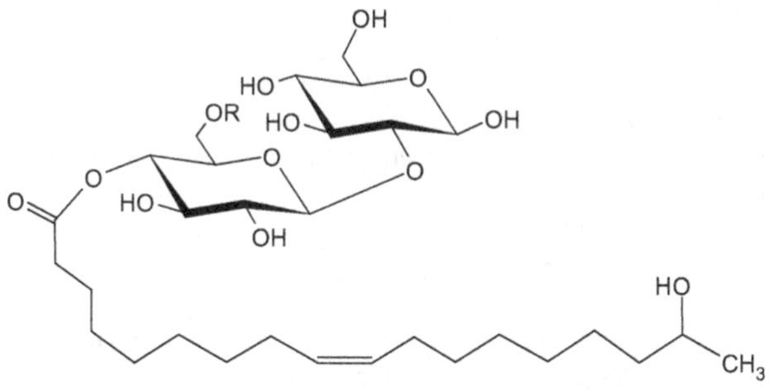

FIGURE 2.4 One of the molecules able to induce cellulase production in certain fungal species. The compound was obtained by partial acidification. R = –H or –COCH$_3$.

useful for the creation of active membrane-associated materials can be generated (Zerkowski et al. 2009). The hydroxyl function can also be converted into a ω-2 alkene, suitable for cross-metathesis and introduction of various functional groups (Zerkowski and Solaiman 2012).

2.9.9 NANOTECHNOLOGY

Sophorolipids recently found their way to nanotechnological applications as well. Just as traditional surfactants, sophorolipids can be applied as structure-directing agent. Indeed, acidic sophorolipids form nm-size micelles in water and depending on their concentration and pH, various geometries can be obtained (see Structure and Properties section). This characteristic allows usage as a structure-directing agent in the synthesis of nanostructured silica thin films. These nanomaterials aroused a growing interest because of their extremely high-specific surface area and tunable pore size distribution, rendering them ideal for various applications such as catalysis, filtration, sensing, and photovoltaic electrodes (Baccile et al. 2010).

Furthermore, sophorolipids can be applied as capping agents for the formation of nanoparticles.

Metal nanoparticles find application in various fields such as mechano- and electrical applications, catalysis, and biomedical use. In general, these particles are stabilized with a capping agent to prevent aggregation and allow dispersion. Acidic de-acetylated sophorolipids are extremely suitable for this purpose as their fatty acid tail will interact with the metal particle while the sugar moiety is exposed to the solvent, governing dispersion. In addition, these sophorolipids also act as a reducing agent, eliminating the necessity for an exogenous reducing agent (Kasture et al. 2007). Various metals can be capped this way. For instance, sophorolipid-coated silver and gold nanoparticles are useful in the medical field as carriers for various bioactive molecules; they turned out to be cyto-compatible up to 100 μM, with the gold particles displaying better cyto- and geno-compatible properties compared with the silver ones (Singh et al. 2010). Singh et al. (2009) demonstrated the antibacterial activity of sophorolipid-coated silver nanoparticles as such against both Gram-positive and Gram-negative bacteria. Iron oxide nanoparticles are interesting for medical applications as well; where their heat transferring and/or magnetic properties can be applied. The use of sophorolipids in either a one-step or two-step synthesis procedure shows a strong effect on the final material's structure; in the one-step synthesis poorly crystalline ferrihydrite nanoparticles are obtained, instead of magnetite (or maghemite) ones for the two-step system (Baccile et al. 2013).

REFERENCES

Albrecht, A, U Rau, and F Wagner. 1996. Initial steps of sophoroselipid biosynthesis by *Candida bombicola* ATCC 22214 grown on glucose. *Applied Microbiology and Biotechnology* 46 (1): 67–73.

Ashby, R, D K Y Solaiman, and T A Foglia. 2006. The use of fatty acids to enhance free acid sophorolipid synthesis. *Biotechnology Letters* 28: 253–260.

Ashby, R D, A J McAloon, D K Y Solaiman, W C Yee, and M Reed. 2013. A process model for approximating the production costs of the fermentative synthesis of sophorolipids. *Journal of Surfactants and Detergents* 16 (5): 683–691.

Asmer, H J, S Lang, F Wagner, and V Wray. 1988. Microbial-production, structure elucidation and bioconversion of sophorose lipids. *Journal of the American Oil Chemists Society* 65 (9): 1460–1466.

Baccile, N, F Babonneau, J Jestin, G Pehau-Arnaudet, and I V Bogaert. 2012. Unusual, pH-Induced, self-assembly of sophorolipid biosurfactants. *ACS Nano* 6 (6): 4763–4776.

Baccile, N, A-S Cuvier, C Valotteau, and I Van Bogaert. 2013. Practical methods to reduce impurities for gram-scale amounts of acidic sophorolipid biosurfactants. *European Journal of Lipid Science and Technology* 115: e–pages.

Baccile, N, N Nassif, L Malfatti, I N A Van Bogaert, W Soetaert, G Pehau-Arnaudet, and F Babonneau. 2010. Sophorolipids: A yeast-derived glycolipid as greener structure directing agents for self-assembled nanomaterials. *Green Chemistry* 12 (9): 1564–1567.

Baccile, N, R Noiville, L Stievano, and I V Bogaert. 2013. Sophorolipids-functionalized iron oxide nanoparticles. *Physical Chemistry Chemical Physics* 15 (5): 1606–1620.

Bluth, M H, S L Fu, A Fu, A Stanek, T A Smith-Norowitz, S R Wallner, R A Gross, M Nowakowski, and M E Zenilman. 2008. Sophorolipids decrease asthma severity and ova-specific IgE production in a mouse asthma model. *Journal of Allergy And Clinical Immunology* 121 (2): 6.

Bluth, M H, E Kandil, C M Mueller, V Shah, Y Y Lin, H Zhang, L Dresner et al. 2006. Sophorolipids block lethal effects of septic shock in rats in a cecal ligation and puncture model of experimental sepsis. *Critical Care Medicine* 34 (1): 188–195.

Borzeix, C F. 1999. Use of sophorolipids comprising diacetyl lactones as agent for stimulating skin fibroblast metabolism. World Patent WO 99/62479.

Casas, J A, and F Garcia-Ochoa. 1999. Sophorolipid production by *Candida bombicola*: Medium composition and culture methods. *Journal of Bioscience and Bioengineering* 88 (5): 488–494.

Cavalero, D A, and D G Cooper. 2003. The effect of medium composition on the structure and physical state of sophorolipids produced by *Candida bombicola* ATCC 22214. *Journal of Biotechnology* 103 (1): 31–41.

Chen, J, X Song, H Zhang, Y B Qu, and J Y Miao. 2006a. Sophorolipid produced from the new yeast strain *Wickerhamiella domercqiae* induces apoptosis in H7402 human liver cancer cells. *Applied Microbiology and Biotechnology* 72 (1): 52–59.

Chen, J, X Song, H Zhang, Y B Qu, and J Y Miao. 2006b. Production, structure elucidation and anticancer properties of sophorolipid from *Wickerhamiella domercqiae*. *Enzyme and Microbial Technology* 39 (3): 501–506.

Ciesielska, K. 2013. *Proteomic Study of the Sophorolipid Producer Starmerella bombicola*. Ghent University, Ghent, Belgium.

Ciesielska, K, B Li, S Groeneboer, INA Van Bogaert, Y-C. Lin, W. Soetaert, Y Van de Peer, and B Devreese. 2013. SILAC-based proteome analysis of *Starmerella bombicola* sophorolipid production. *Journal of Proteome Research* 12 (10): 4376–4392.

Daniel, H J, R T Otto, M Reuss, and C Syldatk. 1998. Sophorolipid production with high yields on whey concentrate and rapeseed oil without consumption of lactose. *Biotechnology Letters* 20 (8): 805–807.

Daniel, H J, M Reuss, and C Syldatk. 1998. Production of sophorolipids in high concentration from deproteinized whey and rapeseed oil in a two stage fed batch process using *Candida bombicola* ATCC 22214 and *Cryptococcus curvatus* ATCC 20509. *Biotechnology Letters* 20 (12): 1153–1156.

Davila, A M, R Marchal, N Monin, and J P Vandecasteele. 1993. Identification and determination of individual sophorolipids in fermentation products by gradient elution high-performance liquid-chromatography with evaporative light-scattering detection. *Journal of Chromatography* 648 (1): 139–149.

Davila, A M, R Marchal, and J P Vandecasteele. 1992. Kinetics and balance of a fermentation free from product inhibition—Sophorose lipid production by *Candida-bombicola*. *Applied Microbiology and Biotechnology* 38 (1): 6–11.

Davila, A M, R Marchal, and J P Vandecasteele. 1994. Sophorose lipid production from lipidic precursors—Predictive evaluation of industrial substrates. *Journal of Industrial Microbiology* 13 (4): 249–257.

Davila, A M, R Marchal, and J P Vandecasteele. 1997. Sophorose lipid fermentation with differentiated substrate supply for growth and production phases. *Applied Microbiology and Biotechnology* 47 (5): 496–501.

Deshpande, M, and L Daniels. 1995. Evaluation of sophorolipid biosurfactant production by *Candida bombicola* using animal fat. *Bioresource Technology* 54 (2): 143–150.

Develter, D, and L M L Lauryssen. 2010. Properties and industrial applications of sophorolipids. *European Journal of Lipid Science and Technology* 112: 628–638.

Dhasaiyan, P, A Banerjee, N Visaveliya, and B L V Prasad. 2013. Influence of the sophorolipid molecular geometry on their self-assembled structures. *Chemistry—An Asian Journal* 8 (2): 369–372.

Felse, P A, V Shah, J Chan, K J Rao, and R A Gross. 2007. Sophorolipid biosynthesis by *Candida bombicola* from industrial fatty acid residues. *Enzyme and Microbial Technology* 40 (2): 316–323.

Fu, S L, S R Wallner, W B Bowne, M D Hagler, M E Zenilman, R Gross, and M H Bluth. 2008. Sophorolipids and their derivatives are lethal against human pancreatic cancer cells. *Journal of Surgical Research* 148 (1): 77–82.

Gao, R, M Falkeborg, X Xu, and Z Guo. 2012. Production of sophorolipids with enhanced volumetric productivity by means of high cell density fermentation. *Applied Microbiology and Biotechnology* 97: 1103–1111.

Gobbert, U, S Lang, and F Wagner. 1984. Sophorose lipid formation by resting cells of *Torulopsis-bombicola*. *Biotechnology Letters* 6 (4): 225–230.

Gorin, P A, J F T Spencer, and A P Tulloch. 1961. Hydroxy fatty acid glycosides of sophorose from *Torulopsis magnoliae*. *Canadian Journal of Chemistry—Revue Canadienne De Chimie* 39 (4): 846–855.

Guilmanov, V, A Ballistreri, G Impallomeni, and R A Gross. 2002. Oxygen transfer rate and sophorose lipid production by *Candida bombicola*. *Biotechnology and Bioengineering* 77 (5): 489–494.

Hagler, M, T A Smith-Norowitz, S Chice, S R Wallner, D Viterbo, C M Mueller, R Gross et al. 2007. Sophorolipids decrease IgE production in U266 Cells by downregulation of BSAP (Pax5), TLR-2, STAT3 and IL-6. *Journal of Allergy and Clinical Immunology* 119 (1): S263–S263.

Hirata, Y, M Ryu, K Igarashi, A Nagatsuka, T Furuta, S Kanaya, and M Sugiura. 2009a. Natural synergism of acid and lactone type mixed sophorolipids in interfacial activities and cytotoxicities. *Journal of Oleo Science* 58 (11): 565–572.

Hirata, Y, M Ryu, Y Oda, K Igarashi, A Nagatsuka, T Furuta, and M Sugiura. 2009b. Novel characteristics of sophorolipids, yeast glycolipid biosurfactants, as biodegradable low-foaming surfactants. *Journal of Bioscience and Bioengineering* 108 (2): 142–146.

Hoffmann, N, J Pietruszka, and C Soeffing. 2012. From sophorose lipids to natural product synthesis. *Advanced Synthesis & Catalysis* 354 (6): 959–963.

Hommel, R, O Stuwer, W Stuber, D Haferburg, and H P Kleber. 1987. Production of water-soluble surface-active exolipids by *Torulopsis-apicola*. *Applied Microbiology and Biotechnology* 26 (3): 199–205.

Hommel, R K, S Stegner, L Weber, and H P Kleber. 1994a. Effect of ammonium-ions on glycolipid production by *Candida (Torulopsis) apicola*. *Applied Microbiology and Biotechnology* 42 (2–3): 192–197.

Hommel, R K, L Weber, A Weiss, U Himmelreich, O Rilke, and H P Kleber. 1994b. Production of sophorose lipid by *Candida (Torulopsis) apicola* grown on glucose. *Journal of Biotechnology* 33 (2): 147–155.

Imura, T, Y Masuda, H Minamikawa, T Fukuoka, M Konishi, T Morita, H Sakai, M Abe, and D Kitamoto. 2010. Enzymatic conversion of diacetylated sophoroselipid into acetylated glucoselipid: Surface-active properties of novel bolaform biosurfactants. *Journal of Oleo Science* 59 (9): 495–501.

Isoda, H, D Kitamoto, H Shinmoto, M Matsumura, and T Nakahara. 1997. Microbial extracellular glycolipid induction of differentiation and inhibition of the protein kinase C activity of human promyelocytic leukemia cell line HL60. *Bioscience Biotechnology and Biochemistry* 61 (4): 609–614.

Kasture, M, S Singh, P Patel, P A Joy, A A Prabhune, C V Ramana, and B L V Prasad. 2007. Multiutility sophorolipids as nanoparticle capping agents: Synthesis of stable and water dispersible co nanoparticles. *Langmuir* 23 (23): 11409–11412.

Kim, K, D Yoo, Y Kim, B Lee, D Shin, and E K Kim. 2002. Characteristics of sophorolipid as an antimicrobial agent. *Journal of Microbiology and Biotechnology* 12 (2): 235–241.

Kim, Y B, H S Yun, and E K Kim. 2009. Enhanced sophorolipid production by feeding-rate-controlled fed-batch culture. *Bioresource Technology* 100 (23): 6028–6032.

Klekner, V, N Kosaric, and Q H Zhou. 1991. Sophorose lipids produced from sucrose. *Biotechnology Letters* 13 (5): 345–348.

Konishi, M, T Fukuoka, T Morita, T Imura, and D Kitamoto. 2008. Production of new types of sophorolipids by *Candida batistae*. *Journal of Oleo Science* 57 (6): 359–369.

Kurtzman, C P. 2012. *Candida kuoi* sp nov., an anamorphic species of the *Starmerella* yeast clade that synthesizes sophorolipids. *International Journal of Systematic and Evolutionary Microbiology* 62: 2307–2311.

Kurtzman, C P, N P J Price, K J Ray, and T M Kuo. 2010. Production of sophorolipid biosurfactants by multiple species of the *Starmerella (Candida) bombicola* yeast clade. *FEMS Microbiology Letters* 311 (2): 140–146.

Lang, S, E Katsiwela, and F Wagner. 1989. Antimicrobial effects of biosurfactants. *Fett Wissenschaft Technologie—Fat Science Technology* 91 (9): 363–366.

Lee, K H, and J H Kim. 1993. Distribution of substrates carbon in sophorose lipid production by *Torulopsis-bombicola*. *Biotechnology Letters* 15 (3): 263–266.

Li, H, X Ma, L Shao, J Shen, and X Song. 2012. Enhancement of sophorolipid production of *Wickerhamiella domercqiae* var. sophorolipid CGMCC 1576 by low-energy ion beam implantation. *Applied Biochemistry and Biotechnology* 167 (3): 510–523.

Li, SL, ZY Li, LY Yang, XL Zhou, MH Dong, P Zhou, YH Lai, and CQ Duan. 2013. *Starmerella jinningensis* sp. nov., a yeast species isolated from flowers of *Erianthus rufipilus*. *International Journal of Systematic and Evolutionary Microbiology* 63: 388–392.

Lin, S C. 1996. Biosurfactants: Recent advances. *Journal of Chemical Technology and Biotechnology* 66 (2): 109–120.

Lo, C M, and L K Ju. 2009. Sophorolipids-induced cellulase production in cocultures of *Hypocrea jecorina* rut C30 and *Candida bombicola*. *Enzyme and Microbial Technology* 44 (2): 107–111.

Mager, H, R Röthlisberger, and F Wzgner. 1987. Use of sophorolose-lipid lactone for the treatment of dandruffs and body odour. European Patent 0209783.

Maingault, M. 1999. Utilization of sophorolipids as therapeutically active substances or cosmetic products, in particular for the treatment of the skin. US Patent US 5981497.

Morya, V K, C Ahn, S Jeon, and E K Kim. 2013. Medicinal and cosmetic potentials of sophorolipids. *Mini Reviews in Medicinal Chemistry* 13: 1761–1768.

Mulligan, C N, R N Yong, and B F Gibbs. 2001. Heavy metal removal from sediments by biosurfactants. *Journal of Hazardous Materials* 85 (1–2): 111–125.

Nguyen, T T L, A Edelen, B Neighbors, and D A Sabatini. 2010. Biocompatible lecithin-based microemulsions with rhamnolipid and sophorolipid biosurfactants: Formulation and potential applications. *Journal of Colloid and Interface Science* 348 (2): 498–504.

Palme, O, G Comanescu, I Stoineva, S Radel, E Benes, D Develter, V Wray, and S Lang. 2010. Sophorolipids from *Candida bombicola*: Cell separation by ultrasonic particle manipulation. *European Journal of Lipid Science and Technology* 112 (6): 663–673.

Pekin, G, F Vardar-Sukan, and N Kosaric. 2005. Production of sophorolipids from *Candida bombicola* ATCC 22214 using Turkish corn oil and honey. *Engineering In Life Sciences* 5 (4): 357–362.

Price, N P J, K J Ray, K E Vermillion, C A Dunlap, and C P Kurtzman. 2012. Structural characterization of novel sophorolipid biosurfactants from a newly identified species of *Candida yeast. Carbohydrate Research* 348: 33–41.

Rau, U, S Hammen, R Heckmann, V Wray, and S Lang. 2001. Sophorolipids: A source for novel compounds. *Industrial Crops and Products* 13 (2): 85–92.

Rau, U, R Heckmann, V Wray, and S Lang. 1999. Enzymatic conversion of a sophorolipid into a glucose lipid. *Biotechnology Letters* 21 (11): 973–977.

Rau, U, C Manzke, and F Wagner. 1996. Influence of substrate supply on the production of sophorose lipids by *Canidida bombicola* ATCC 22214. *Biotechnology Letters* 18 (2): 149–154.

Renkin, M. 2003. Environmental profile of sophorolipid and rhamnolipid biosurfactants. *La Rivista Italiana Delle Sostanze Grasse* 58: 249–252.

Roelants, S, K Saerens, T Derycke, I N A Van Bogaert, and W Soetaert. 2013. *Candida bombicola* as a platform organism for the production of tailor-made biomolecules. *Biotechnology and Bioengineering* 88: 501–509.

Rosa, C A, and M A Lachance. 1998. The yeast genus *Starmerella* gen. nov. and *Starmerella bombicola* sp. nov., the teleomorph of *Candida bombicola* (Spencer, Gorin & Tullock) Meyer & Yarrow. *International Journal of Systematic Bacteriology* 48 (4): 1413–1417.

Saerens, K, I N A Van Bogaert, W Soetaert, and E J Vandamme. 2009. Production of glucolipids and specialty fatty acids from sophorolipids by *Penicillium decumbens* naringinase: Optimization and kinetics. *Biotechnololgy Journal* 4: 517–524.

Saerens, K M J, S L K W Roelants, I N A Van Bogaert, and W Soetaert. 2011a. Identification of the UDP-glucosyltransferase gene UGTA1, responsible for the first glucosylation step in the sophorolipid biosynthetic pathway of *Candida bombicola* ATCC 22214. *FEMS Yeast Research* 11 (1): 123–132.

Saerens, K M J, L Saey, and W Soetaert. 2011b. One-step production of unacetylated sophorolipids by an acetyltransferase negative *Candida bombicola. Biotechnology and Bioengineering* 108 (12): 2923–2931.

Saerens, K M J, J Zhang, L Saey, I N A Van Bogaert, and W Soetaert. 2011c. Cloning and functional characterisation of the UDP-glucosyltransferase UgtB1 involved in sophorolipid production by *Candida bombicola* and creation of a glucolipid producing yeast strain. *Yeast* 28 (4): 279–292.

Shah, V, G F Doncel, T Seyoum, K M Eaton, I Zalenskaya, R Hagver, A Azim, and R Gross. 2005. Sophorolipids, microbial glycolipids with anti-human immunodeficiency virus and sperm-immobilizing activities. *Antimicrobial Agents and Chemotherapy* 49 (10): 4093–4100.

Shao, L, X Song, X Ma, H Li, and Y Qu. 2012. Bioactivities of sophorolipid with different structures against human esophageal cancer cells. *Journal of Surgical Research* 173 (2) (April): 286–291.

Singh, S, V D'Britto, A A Prabhune, C V Ramana, A Dhawan, and B L V Prasad. 2010. Cytotoxic and genotoxic assessment of glycolipid-reduced and -capped gold and silver nanoparticles. *New Journal of Chemistry* 34 (2): 294–301.

Singh, S, P Patel, S Jaiswal, A A Prabhune, C V Ramana, and B L V Prasad. 2009. A direct method for the preparation of glycolipid-metal nanoparticle conjugates: Sophorolipids as reducing and capping agents for the synthesis of water re-dispersible silver nanoparticles and their antibacterial activity. *New Journal of Chemistry* 33 (3): 646–652.

Sipiczki, M. 2013. *Starmerella caucasica* sp. nov., a novel anamorphic yeast species isolated from flowers in the *Caucasus. Journal of General and Applied Microbiology* 59 (1): 67–73.

Sleiman, J N, S A Kohlhoff, P M Roblin, S Wallner, R Gross, M R Hammerschlag, M E Zenilman, and M H Bluth. 2009. Sophorolipids as antibacterial agents. *Annals of Clinical and Laboratory Science* 39 (1): 60–63.

Smyth, T J, A Perfumo, R Marchant, I M Banat, M L Chen, R K Thomas, J Penfold, P S Stevenson, and N J Parry. 2010. Directed microbial biosynthesis of deuterated biosurfactants and potential future application to other bioactive molecules. *Applied Microbiology and Biotechnology* 87 (4): 1347–1354.

Solaiman, D K Y, R D Ashby, J A Zerkowski, and T A Foglia. 2007. Simplified soy molasses-based medium for reduced-cost production of sophorolipids by *Candida bombicola. Biotechnology Letters* 29 (9): 1341–1347.

Spencer, J F T, P A J Gorin, and A P Tulloch. 1970. *Torulopsis bombicola* sp. N. *Antonie Van Leeuwenhoek International Journal of General and Molecular Microbiology* 36 (1): 129–133.

Takahashi, M, T Morita, K Wada, N Hirose, T Fukuoka, T Imura, and D Kitamoto. 2011. Production of sopho-rolipid glycolipid biosurfactants from sugarcane molasses using *Starmerella bombicola* NBRC 10243. *Journal of Oleo Science* 60 (5): 267–273.

Teichmann, B, L D Liu, K O Schink, and M Bolker. 2010. Activation of the ustilagic acid biosynthesis gene cluster in *Ustilago maydis* by the C2H2 zinc finger transcription factor rua1. *Applied and Environmental Microbiology* 76 (8): 2633–2640.

Thaniyavarn, J, T Chianguthai, P Sangvanich, N Roongsawang, K Washio, M Morikawa, and S Thaniyavarn. 2008. Production of sophorolipid biosurfactant by *Pichia anomala*. *Bioscience Biotechnology and Biochemistry* 72 (8): 2061–2068.

Tulloch, A P, J F T Spencer, and M H Deinema. 1968. A new hydroxy fatty acid sophoroside from *Candida bogoriensis*. *Canadain Journal of Chemistry* 46 (3): 345–348.

Van Bogaert, I, G Zhang, J Yang, J-Y Liu, Y Ye, W Soetaert, and B D Hammock. 2014. Preparation of 20-HETE using multifunctional enzyme type 2 (MFE-2)-negative *Starmerella bombicola*. *Journal of Lipid Research* 54: 3215–3219.

Van Bogaert, I N A, K Ciesielska, S Roelants, S Groeneboer, B Devreese, and W Soetaert. 2011. A lactonase derived from *Candida bombicola* and uses thereof. *WO 2013/092421*.

Van Bogaert, I N A, S L De Maeseneire, D Develter, W Soetaert, and E J Vandamme. 2008a. Development of a transformation and selection system for the glycolipid-producing yeast *Candida bombicola*. *Yeast* 25(4): 273–278.

Van Bogaert, I N A, S L De Maeseneire, D Develter, W Soetaert, and E J Vandamme. 2008b. Cloning and char-acterisation of the glyceraldehyde 3-phosphate dehydrogenase gene of *Candida bombicola* and use of its promoter. *Journal of Industrial Microbiology & Biotechnology* 35 (10): 1085–1092.

Van Bogaert, I N A, M De Mey, D Develter, W Soetaert, and E J Vandamme. 2010. Importance of the cyto-chrome P450 monooxygenase CYP52 family for the sophorolipid-producing yeast *Candida bombicola* (vol 9, Pg 87, 2009). *FEMS Yeast Research* 10 (6): 791.

Van Bogaert, I N A, K Holvoet, B Roelants, SLKW, Li, Y-C Lin, Y Van de Peer, and W Soetaert. 2013. The biosynthetic gene cluster for sophorolipids: A biotechnological interesting biosurfactant produced by *Starmerella bombicola*. *Molecular Microbiology* 88: 501–509.

Van Bogaert, I N A, S Roelants, D Develter, and W Soetaert. 2010. Sophorolipid production by *Candida bom-bicola* on oils with a special fatty acid composition and their consequences on cell viability. *Biotechnology Letters* 32: 1509–1514.

Van Bogaert, I N A, J Sabirova, D Develter, W Soetaert, and E J Vandamme. 2009. Knocking out the MFE-2 gene of *Candida bombicola* leads to improved medium-chain sophorolipid production. *FEMS Yeast Research* 9 (4): 610–617.

Van Bogaert, I N A, J Zhang, and W Soetaert. 2011. Microbial synthesis of sophorolipids. *Process Biochemistry* 46 (4): 821–833.

Watson, Me, B Wu, M Bluth, and M Nowakowski. 2012. Mechanism of sophorolipid (SL) suppression of nitric oxide production in models of septic or endotoxic shock. *Journal of Immunology* 188: 169.9 (meeting abstract).

Wegener, D, C Hildmann, D Riester, A Schober, F-J Meyer-almes, H E Deubzer, I Oehme et al. 2008. Identification of novel small-molecule histone deacetylase inhibitors by medium-throughput screening using a fluorigenic assay. *Biochem J* 413 (1): 143–150.

Yang, X, L Zhu, C Xue, Yu Chen, Liang Qu, and Wenyu Lu. 2012. Recovery of purified lactonic sophorolipids by spontaneous crystallization during the fermentation of sugarcane molasses with *Candida albicans* O-13-1. *Enzyme and Microbial Technology* 51 (6–7): 348–353.

Yoo, D S, B S Lee, and E K Kim. 2005. Characteristics of microbial biosurfactant as an antifungal agent against plant pathogenic fungus. *Journal of Microbiology and Biotechnology* 15 (6): 1164–1169.

Zerkowski, J A, A Nunez, G D Strahan, and D K Y Solaiman. 2009. Clickable lipids: Azido and alkynyl fatty acids and triacylglycerols. *Journal of the American Oil Chemists Society* 86 (11): 1115–1121.

Zerkowski, J A, and D K Y Solaiman. 2012. Omega-functionalized fatty acids, alcohols, and ethers via olefin metathesis. *Journal of the American Oil Chemists Society* 89 (7): 1325–1332.

Zhou, Q H, and N Kosaric. 1993. Effect of lactose and olive oil on intracellular and extracellular lipids of *Torulopsis-bombicola*. *Biotechnology Letters* 15 (5): 477–482.

Zhou, Q H, and N Kosaric. 1995. Utilization of canola oil and lactose to produce biosurfactant with *Candida-bombicola*. *Journal of the American Oil Chemists Society* 72 (1): 67–71.

Zhou, S, C Xu, J Wang, W Gao, R Akhverdiyeva, V Shah, and R Gross. 2004. Supramolecular assemblies of a naturally derived sophorolipid. *Langmuir* 20 (19): 7926–7932.

Zhou, S Q, C Xu, J Wang, W Gao, R Akhverdiyeva, V Shah, and R Gross. 2004. Supramolecular assemblies of a naturally derived sophorolipid. *Langmuir* 20 (19): 7926–7932.

3 Biosurfactants versus Chemically Synthesized Surface-Active Agents

Steve Fleurackers

CONTENTS

3.1 Introduction 37
    3.1.1 Choosing Substances to Compare 37
    3.1.2 Other Chemically Derived Surface-Active Agents 38
        3.1.2.1 Esterquat: A Cationic Surfactant 39
        3.1.2.2 Cumene Sulfonate: A Hydrotrope...39
        3.1.2.3 Amphoteric Surfactants: Cocamidopropyl Betaine............................40
    3.1.3 Approaches Used for Comparison 40
3.2 Production: Sourcing and Synthesis...40
    3.2.1 Synthesis..41
        3.2.1.1 Polymerization of Carbohydrates to Alkyl Polyglycosides 41
        3.2.1.2 From Fatty Alcohol to Lauryl Ether Sulfate...............................42
        3.2.1.3 FAMEE: Ethoxylated Biodiesel...42
3.3 Usage: Formulation and Application 43
    3.3.1 A Practical Example: SL in Domestic Cleaning: Stability, pH, and Performance.....44
    3.3.2 Other Uses of Biosurfactants 44
3.4 Waste: Degradation and Environmental Impact...45
    3.4.1 Biodegradability 45
    3.4.2 Ecotoxicity 45
3.5 Final Conclusions...45
References..46

3.1 INTRODUCTION

A comparison between biosurfactants and chemically synthesized surfactants presents the unique challenge of comparing two types of compounds that ultimately have the similar function of being an amphiphilic molecule, but differ quite significantly with regard to their origins. These differing origins will have a profound impact on a variety of properties such as biodegradability, ecotoxicity, chemical stability, pricing, and so forth. It is the aim of this chapter to offer an insight into the various ways in which sourcing and synthesis will affect the final product. To this end, the three stages of the product's life cycle—production, usage, and waste—are compared to one another.

3.1.1 CHOOSING SUBSTANCES TO COMPARE

To date, a good number of different biosurfactants have been identified (for an overview, see Rosenberg and Ron, 1999), but relatively few have attracted a sustained commercial interest. An overview of biosurfactant patents by Shete et al. (2006) indicated that three of these biosurfactants

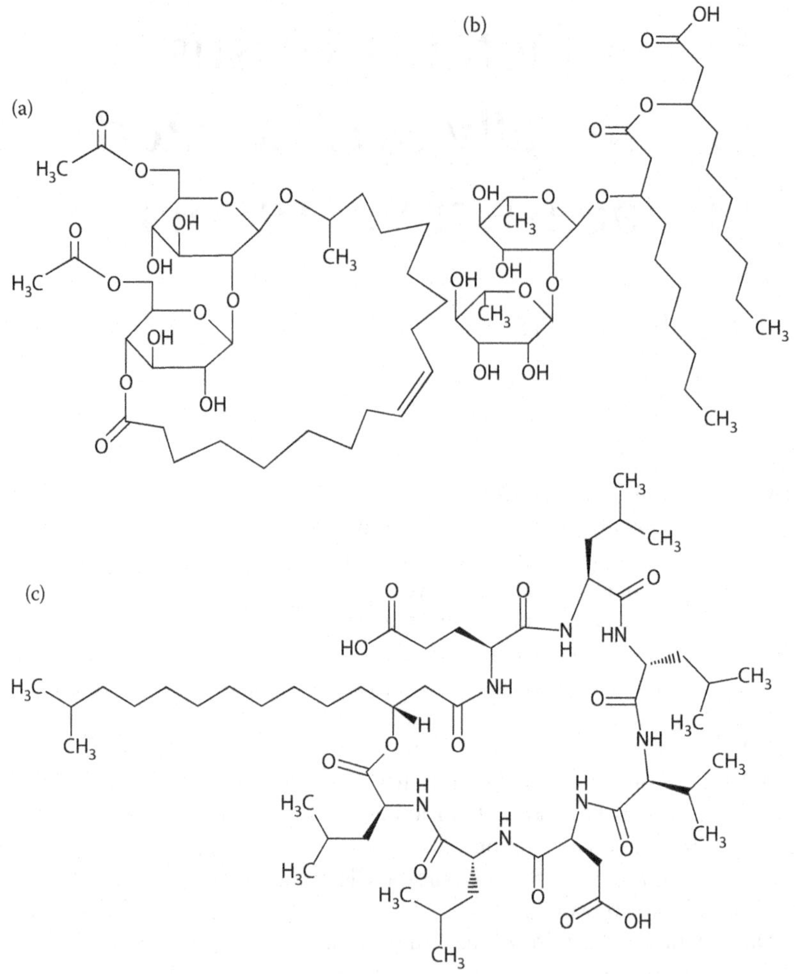

FIGURE 3.1 Chemical structures of biosurfactants which are used for comparison. (a) Lactone form of sophorolipids (SL), (b) di-rhamnose di-hydroxy decanoic acid rhamnolipid (RL), and (c) surfactin (SF).

account for about half of the total amount of patents filed. They are sophorolipids (SL), rhamnolipids, and surfactin, and will be used to represent biosurfactants from here on out. Examples of the chemical structures of the said biosurfactants are depicted in Figure 3.1.

The more traditional, chemically synthesized surfactants comprise an even more diverse group, especially when commercial availability is concerned. Anionic, cationic, nonionic, and amphoteric surfactants range from hydrotropes to emulsifiers, which have established markets and applications, whereas biosurfactants are just emerging. From this heterogeneous group, an admittedly somewhat arbitrary selection of three substances was made based on a combination of their market prevalence, structure, and application. These three are lauryl polyglycoside (APG), lauryl ether sulfate (LES), and fatty alcohol methyl ester ethoxylate (FAMEE) as shown in Figure 3.2.

3.1.2 Other Chemically Derived Surface-Active Agents

As previously stated, the range of chemically produced surfactants, which is currently available, far exceeds that of biosurfactants, both in volume and diversity. Certain applications have no direct biochemical counterpart and are therefore not included in the scope of this chapter. However, for

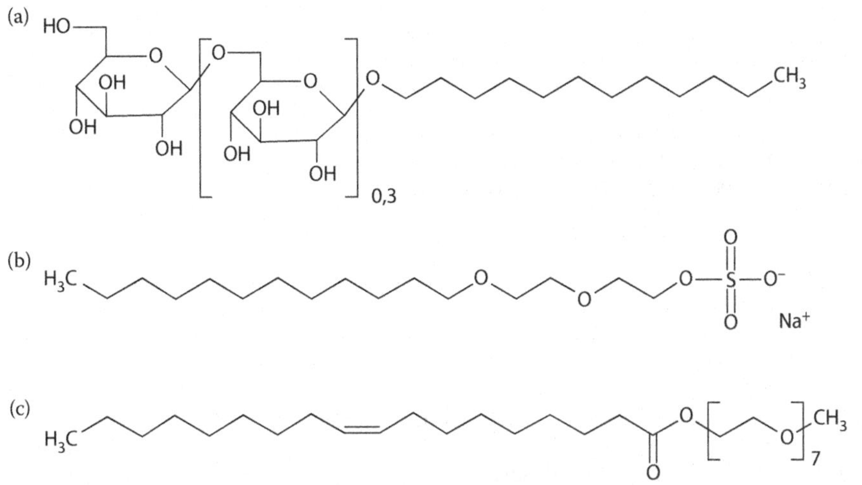

FIGURE 3.2 Chemical structures of chemically derived surfactants which are used for comparison. (a) Alkyl polyglycoside with an average of 1,3 glucose subunits (APG), (b) lauryl ether sulfate with an average of 2 ethylene oxide (EO) subunits (LES), and (c) fatty acid methyl ester ethoxylate based on oleic acid and having on average 7 EOs (FAMEE).

the sake of completeness, a selected number of well-known and commonly used chemically synthesized surfactants are presented here, together with their practical application.

3.1.2.1 Esterquat: A Cationic Surfactant

The structure of esterquat is depicted in Figure 3.3 (Adam and Hulshof, 1981); it shows a negative charge in the polar head, thus making it a cationic surfactant. It finds a common use as fabric softener, where the quaternary ammonium attaches itself to positive charges found on various textiles, those most commonly being a free carboxylic function. The lipophilic tail, in turn, provides a lubricating effect between the fibers or, in case of a conditioner, hair.

To date, no commercially viable cationic biosurfactants are known; including such a chemically derived surfactant in this comparison would therefore seem moot.

3.1.2.2 Cumene Sulfonate: A Hydrotrope

Depending on the length of the lipophilic moiety relative to the strength of the polar head, a surface-active agent may be classified, according to Griffin (1949), as a solubilizer or hydrotrope (short lipophilic chain), a detergent (medium length), or an emulsifier (long). Hydrotropes and solubilizers are terms that are used interchangeably, mostly depending on the function they perform. They are typically used in aquatic formulations to lower surface tension when foam formation is undesirable (hydrotrope), or in applications to assist in dissolving more hydrophobic compounds (solubilizer) and obtain a stable solution. One such example is cumene sulfonate, the structure of which is shown in Figure 3.4 (Griffin, 1949).

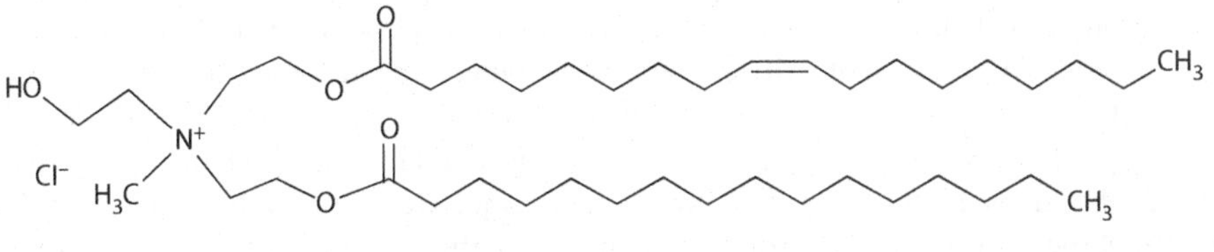

FIGURE 3.3 Chemical structure of esterquat—two fatty acid residues, oleic and palmitic, bound to methylated triethanol amine.

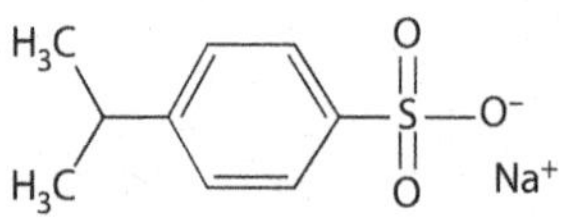

FIGURE 3.4 Chemical structure of the petrochemically derived solubilizer cumene sulfonate; note the shorter lipophilic moiety compared to the C_{12} tail of LES.

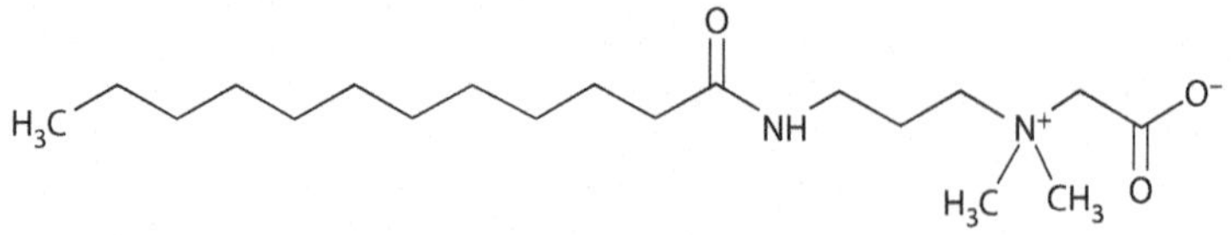

FIGURE 3.5   Cocamidopropyl betaine, an amphoteric surfactant.

Develter and Fleurackers (2007) describe a method by which the lipophilic moiety of SL may be shortened through ozonolysis to a length concurrent with that of a hydrotrope, but as of yet no commercial application exists. This, combined with the relatively small market share hydrotropes have on the whole of surfactants has excluded solubilizers.

3.1.2.3 Amphoteric Surfactants: Cocamidopropyl Betaine

Surfactants which have a zwitterionic polar head are classified as amphoteric. One such well-known surfactant is cocamidopropyl betaine and is depicted in Figure 3.5. It is used, among others, as a foam booster in personal care applications such as shampoo.

Again, no readily available biosurfactant amphoteric analogue exists and therefore this class of surfactants is excluded from this chapter.

3.1.3   Approaches Used for Comparison

Choosing a single method for comparing two such classes of molecules would not be suitable as there is no one approach that would evenly quantify aspects of production, usage, and disposal. A life cycle analysis may be suited for one specific application, for example, the use of a surfactant in the field of enhanced oil recovery, but in view of the fact that surfactants are used in a variety of fields, as well as the objective of comparing classes of surfactants rather than specific species, a different approach will be used to compare the different aspects. This is an attempt to yield a more balanced view. The production shall focus on the actual (bio)chemistry involved in making the substance, leaving the reader to evaluate the various processes with a calculation method that one thinks best. Usage will give an idea of the current market situation in which biosurfactants find themselves. Finally, wastage focuses on the environmental impact of both categories of surfactants.

3.2   PRODUCTION: SOURCING AND SYNTHESIS

The obvious difference between biosurfactants and their chemically synthesized counterparts is the manner in which they are produced. Biosurfactants are primarily synthesized by a microbial organism through biochemical action, whereas traditional surfactants are made by using more harsh chemical reactions. A second major difference can be found in the raw materials that are used in the production of the said compounds. Biosurfactant production invariably uses renewable precursors, although the production organisms involved are also able to assimilate aliphatic carbon chains from petrochemical sources. In the latter case, however, the objective is not product formation, but rather the removal of these carbon chains from the environment through assimilation, that is, bioremediation. Chemically synthesized surfactants, however, are predominantly made from petrochemical feedstock, either partially or wholly, as is more commonly the case. Growing concerns regarding the use of petrochemical resources—for example, their possible depletion in the near future and the risks to the environment associated with their extraction—and increased legal restrictions regarding degradability have marked an increase in the use of plant-based feedstocks.

3.2.1 Synthesis

The typical fermentation medium for biosurfactant production contains three main constituents: a carbon source, a nitrogen source, and various salts (e.g., buffers and trace elements). The production is operated around room temperature in a fermentation vessel typically under no more than one bar overpressure at a neutral to weak acidic pH. Such compositions, together with indicative fermentation conditions, are given in Table 3.1.

As the specific conditions for the production of various biosurfactants are discussed in more detail in the other chapters of this book, the following section focuses on various industrial techniques that are used in the chemical production of surfactants. Conversely, the chemical synthesis of surfactants typically uses more extreme conditions, an outline of which is given in Table 3.2.

3.2.1.1 Polymerization of Carbohydrates to Alkyl Polyglycosides

While alkyl polyglycosides are chemically made surfactants, they are fully based on naturally sourced raw materials, as are biosurfactants. These materials include corn starch and coconut/palm kernel oil, from which the C_{12}/C_{14} fatty acid fraction is converted into the concurrent fatty alcohol. Through an acid catalyzed transacetalization, the fatty alcohol is combined with one or more glucose subunits in a two-step reaction, as shown in Figure 3.6.

Depending on the carbohydrate source used, the associated costs for raw materials and plant equipment will vary. In general, the raw material cost will increase for the sequence: starch, glucose

---

TABLE 3.1

Examples of Selected Substrates and Conditions Used in Biosurfactant Fermentation Media

Biosurfactant	Sophorolipid (Zhou et al., 1992)	Rhamnolipid (Haba et al., 2000)	Surfactin (Al-Ajlani et al., 2007)
Production organism	*Starmerella bombicola*	*Pseudomonas aeruginosa*	*Bacillus subtilis*
Carbon source	Glucose, safflower oil	Glucose	Glucose
Nitrogen source	Urea, yeast extract	$NaNO_3$, yeast extract	Cotton seed flour
Temperature (°C)	30	30	35
pH	4.5	7.2	7.0

TABLE 3.2

Selected Feedstocks and Reaction Conditions for Chemically Synthesized Surfactants

Surfactant	APG (McCurry et al., 1993; Eskuchen and Nitsche, 1997)	LES (Raths et al., 1993, Maag, 1981)	FAMEE (Hreczuch, 2001)
Feedstock	Fatty alcohol, starch	Lauryl alcohol, ethylene oxide, SO_3	Rapeseed oil methyl ester, ethylene oxide
Catalyst	Dodecylbenzene-sulfonic acid	Calcium hydrotalcite	Proprietary calcium-based catalyst
Temperature (°C)	110–120	500[a]	135–185
Pressure (bar)	0.02–0.1	2	30–50

[a] Temperature required for activation of the ethoxylation catalyst.

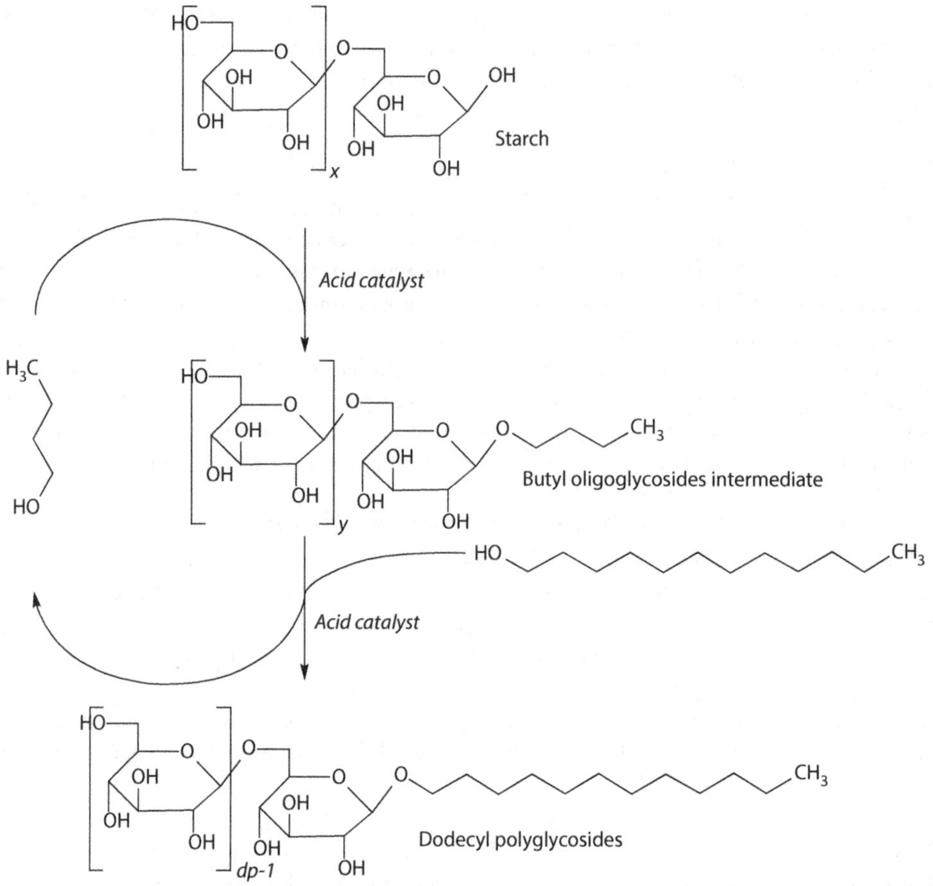

FIGURE 3.6 Formation reaction of APG; *dp* indicates the degree of polymerization, that is, the average amount of glucose subunits bound to the fatty alcohol. (Adapted from Eskuchen R., and Nitsche M., *Alkyl Polyglycosides*, 9–22. Weinheim: VCH Verlagsgesellschaft mbH, 1997.)

syrup, glucose monohydrate, and water-free glucose, while the requirements for plant equipment (and cost) decrease in the same order. Figure 3.7 shows a simplified flow diagram for the industrial synthesis of APG where these different carbohydrate sources are taken into account.

3.2.1.2 From Fatty Alcohol to Lauryl Ether Sulfate

LES, together with its non-ethoxylated cousin lauryl sulfate, is arguably one of the most commonly used anionic surfactants on the market today. A CESIO (European Committee for Surfactants and Their Organic Intermediates) estimate in the year 2000 places the European production volume around 300,000 tons per year (HERA, 2004). The surfactant is produced in two separate stages: a fatty alcohol in the C_{12} to C_{18} range is ethoxylated after which it is sulfated using SO_3 or chlorosulfonic acid, the reaction of which is depicted in Figure 3.8.

3.2.1.3 FAMEE: Ethoxylated Biodiesel

FAMEE is a relatively new surface-active agent on the market and as such, not as well known as either LES or APG (Renkin et al., 2005). It is, however, made from a recently established source material: biodiesel, that is, transesterified vegetable oil using methanol. Ethylene oxide is used to create a polar head at the esteric bond of approximately seven subunits (Hreczuch, 2001); this reaction is shown in Figure 3.9.

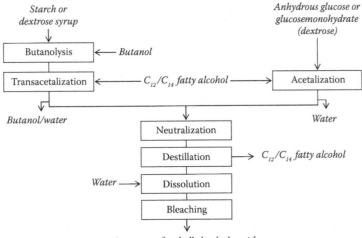

FIGURE 3.7 Schematic overview of the production of alkyl polyglycosides based on different carbohydrate sources—direct and indirect synthesis. (Adapted from Eskuchen R., and Nitsche M., *Alkyl Polyglycosides*, 9–22. Weinheim: VCH Verlagsgesellschaft mbH, 1997.)

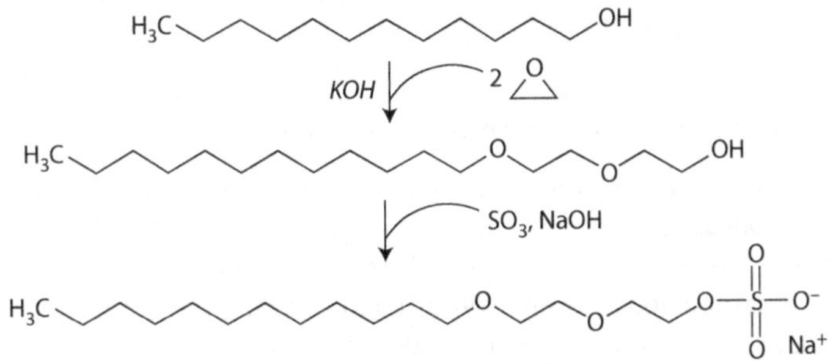

FIGURE 3.8 Two-stage synthesis of lauryl ether sulfate, first the fatty alcohol ethoxylate is formed, which in a second stage is sulfated. (Adapted from Maag H., *Tensid-Tashenbuch*, 85–167. München, Wien: Carl Hanser Verlag, 1981.)

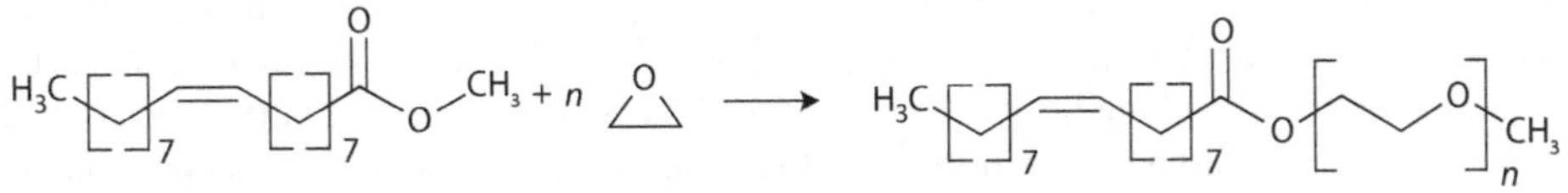

FIGURE 3.9 Ethoxylation reaction for the formation of FAMEE, a value for *n* of 7 is typical. (Adapted from Hreczuch W., *Tenside Surfactants Detergents* 38(2), 2001: 72–79.)

This reaction can be carried out in existing ethoxylation plants, as normally used for the ethoxylation of fatty alcohols as depicted in the first synthesis step for LES, but utilizes a different catalyst. A schematic setup of such an ethoxylation plant is shown in Figure 3.10.

3.3 USAGE: FORMULATION AND APPLICATION

As the market for biosurfactants is very much in its developmental stage, the currently available products do not differ much (or at all) from their native or wild-type form. When comparing the

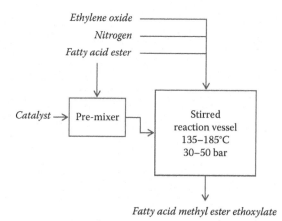

Fatty acid methyl ester ethoxylate

FIGURE 3.10 Schematic overview of the production of fatty acid methyl ester ethoxylates. (Adapted from Hreczuch W., *Tenside Surfactants Detergents* 38(2), 2001: 72–79.)

structures of Figure 3.1 with those of Figure 3.2, and more specifically the nature of the various bonds, it clearly shows that biosurfactants would hydrolyze more readily when subjected to, for example, acidic or basic conditions. Indeed, the softer conditions under which biosurfactants are made also result in a comparatively less stable molecule; esters and amides being weaker bonds when compared with ethers and the glycosidic version of acetals.

3.3.1 A Practical Example: SL in Domestic Cleaning: Stability, pH, and Performance

The aspect of decreased stability is not necessarily a drawback, as demonstrated in the case of a domestic cleaning application of SL. The esteric bonds contained within the said biosurfactant are susceptible to alkaline hydrolysis and as such, the product cannot be used for applications with a pH above 7 unless it has previously been hydrolyzed to its deacetylated acid form. As the acetylated lactone SL species are more performant, any water-based formula would therefore be limited to a pH between 3 and 7 if SLs are to be optimally utilized. An example of such a formula is the Held All Purpose Cleaner that contains, among others, a combination of SL and APG. The product was compared in a Swiss consumer test (Camenzind, 2011) to a conventional brand—Meister Proper, also known as Mr. Clean—and found to be equally performant on a variety of stains. In and of itself, this result would be relatively unremarkable, were it not for a marked difference in pH. Being a conventional all purpose cleaner, its pH is decidedly alkaline, standing at about 10. By contrast, the pH of the cleaner containing biosurfactants is almost neutral, being a mere 6. The effect of pH on cleaning efficiency has long been acknowledged, especially concerning fatty deposits, therefore it is quite remarkable to find that an ingredient that cannot tolerate alkalinity would compensate by preforming equally well under slightly acidic conditions. Comparable hard-surface cleaning formulations that include SL are currently also being marketed by Ecover as an all purpose cleaner and, in a more concentrated form, as a power cleaner.

3.3.2 Other Uses of Biosurfactants

When looking for other uses of biosurfactants, one may find a multitude of patents covering such diverse fields as medicine, soil remediation, enhanced oil recovery, and crop protection. However, actual products being used on the market that contain biosurfactants are few and far between. A couple of field tests in soil remediation have been carried out (see Ławniczak et al., 2013), but again no commercial process exists. Japanese producer Saraya Co., Ltd. was known to have marketed a number of domestic formulations containing SL, but has since withdrawn them, or at least stopped advertising the use of the said ingredient. However, two commercial sources of biosurfactants are known: Soliance (France) market SL and Jeneil Biotech Inc. (USA) have rhamnolipids in their product range.

3.4 WASTE: DEGRADATION AND ENVIRONMENTAL IMPACT

Depending on the application and/or the final phase of a surface-active substance existence is most commonly either discharge into a biological waste water treatment plant or release into the environment. Here, the two most relevant parameters regarding the waste phase—biodegradability and ecotoxicity—are discussed.

3.4.1 BIODEGRADABILITY

Although in the past the focus was on primary degradability, that is to test whether the functional form of a substance would disappear, current legislation regarding the use of surface-active agents primarily focuses on the complete or ultimate biodegradation to CO_2, water, and other small inorganics (see Article 4 of EC Regulation 648/2004) following a growing awareness regarding stable secondary metabolites. One such example is the ban on the use of nonylfenol ethoxylates (see Figure 3.11) where stable biodegradation fragments exhibit toxic and endocrine disruptive activity, as well as accumulate in the environment (Soaresa et al., 2008).

With this in mind, the three chemically synthesized surfactants that were chosen are all ultimately biodegradable (Renkin, 2003) as other synthetic surfactants not meeting the said criterion are being phased out and would not offer a relevant comparison with respect to this current trend. As the legislative awareness is growing, it would seem that in time complete biodegradability will be a requirement for all surface-active agents, and thus becoming a prerequisite rather than an advantage.

3.4.2 ECOTOXICITY

Another ecological concern regarding surfactants is their ecotoxicity, that is, the impact of the substance on the environment before it is degraded. When tested according to OECD Guideline 202, an acute immobilization study using the crustacean *Daphnia magna* (commonly called "water flea"), it was shown that biosurfactants are one order of magnitude less toxic to the aquatic environment than their chemically synthesized counterparts (Renkin, 2003). The fact that biosurfactants are produced by a biological system whereas synthetic surfactants are always xenobiotic, in some aspect or other, would offer a reasonable explanation for this observation even though both are fully biodegradable.

3.5 FINAL CONCLUSIONS

The first and foremost conclusion is that, for the moment at least, biosurfactants and their chemically synthesized counterparts are not as much in direct competition, as they complement each

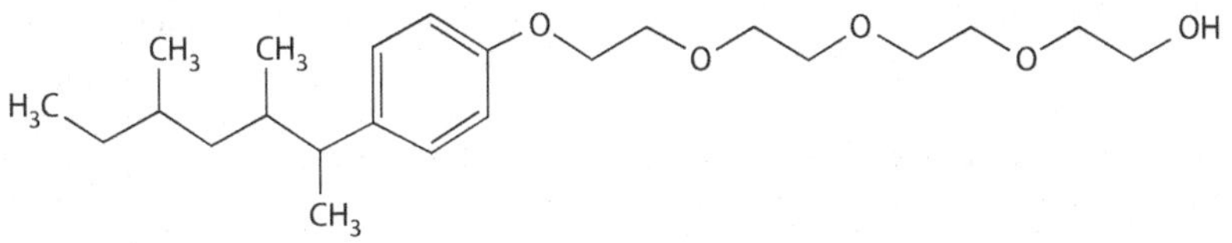

FIGURE 3.11 Example structure of the now banned nonylfenol ethoxylate that only partially degrades to stable, nonbiodegradable, and toxic nonylfenol. (Adapted from Soaresa A. et al., *Environment International* 34(7), 2008: 1033–1049.)

TABLE 3.3

Summarizing Overview of Selected Properties

Surfactant	SL	RL	SF	APG	LES	FAMEE
Ultimately biodegradable?	✓	✓	✓	✓	✓	✓
Wholly renewable?	✓	✓	✓	✓	✗[a]	✗[a]
Readily foaming?	✗	✓	✓	✓	✓	✗
Stable at an extreme pH?	✗[b]	✗[b]	✗	✓	✓[c]	✓[c]
Commercially available?	✓	✓	✗	✓✓[d]	✓✓[d]	✓

[a] Plant-based ethylene oxide is being marketed, but as of yet not used in the production of surfactants.

[b] Ester-free forms are resistant to alkaline and, up to a point, acid hydrolysis.

[c] Hydrolyzation may occur at extreme pH values which would be unsuitable for domestic use.

[d] Available in both different variants and from different suppliers—established product.

other. This is the result of the limited availability and application range of biosurfactants, as well as having a higher cost. They do, however, offer a more advantageous environmental profile over the more traditional surfactants currently on the market, both from an ecotoxicological and a sourcing point of view. As such, they are making inroads into areas where surfactants are either released directly into the environment, for example, bioremediation and crop protection, or in consumer applications for which ecological awareness has definite economical value. A comprehensive overview of these aspects is shown in Table 3.3.

Although the practical usage of biosurfactants shows a good deal of promise—as indicated by the available patent literature and one or two real-life instances—their actual market presence is, for the moment, negligible when compared with chemically synthesized surface-active agents.

REFERENCES

Adam W. E., and Hulshof W. T., Die Anwendung der kationische Tenside. In *Tensid-Taschenbuch*, 373–376. München, Wien: Carl Hanser Verlag, 1981.

Al-Ajlani M. M., Abid Sheikh M., Ahmad Z., and Hasnain S., Production of surfactin from *Bacillus subtilis* MZ-7 grown on pharmamedia commercial medium. *Microbial Cell Factories* 6(17), 2007. doi: 10.1186/1475-2859-6-17.

Camenzind B., Meister Proper kann Held nichts vormachen. *Saldo* 9, 2011: 18–21.

Develter D., and Fleurackers S., A method for the production of short chained glycolipids. *EP patent 1953237*, 2007.

EC Regulation no. 648/2004 of the European Parliament and of the Council on detergents, no. L108, amended by nos. L168, L354, L87 and L164: 26-6-2009.

Eskuchen R., and Nitsche M., Technology and production of alkyl polyglycosides. In *Alkyl Polyglycosides*, 9–22. Weinheim: VCH Verlagsgesellschaft mbH, 1997.

Griffin W. C., Classification of surface-active agents by 'HLB'. *Journal of the Society of Cosmetic Chemists* 1(5), 1949: 311–326.

Haba E., Espuny M.J., Busquets M., and Manresa A., Screening and production of rhamnolipids by *Pseudomonas aeruginosa* 47T2 NCIB 40044 from waste frying oils. *Journal of Applied Microbiology* 88, 2000: 379–387.

HERA—Human and Environmental Risk Assessment on ingredients of European household cleaning products Alcohol Ethoxysulphates (AES) Environmental Risk Assessment. *Environmental*, http://www.heraproject.com, retrieved on 1/10/2013, 2004.

Hreczuch W., Ethoxylated rapeseed oil acid methyl esters as new ingredients for detergent formulations. *Tenside Surfactants Detergents* 38(2), 2001: 72–79.

Ławniczak Ł., Marecik R., and Chrzanowski Ł., Contributions of biosurfactants to natural or induced bioremediation. *Applied Microbiology and Biotechnology* 97(6), 2013: 2327–2339.

Maag H., Herstellung und Eigenschaften der Tenside. In *Tensid-Tashenbuch*, 85–167. Stache, H. (ed.), München, Wien: Carl Hanser Verlag, 1981.

McCurry Jr, P. M., McDaniel R. S., Kozak W. G., Urfer A. D., and Howell G., Preparation of alkylpolyglycosides. *US patent 5266690*, 1993.

Raths H.-C., Endres H., Hensen H., and Tesmann H., Narrow-range fatty alcohol ethoxylates: Production and properties. *Henkel-Referate* 29, 1993: 84–90.

Renkin M., Environmental profile of sophorolipid and rhamnolipid biosurfactants. *La Revista Italiana delle Sostanze Grasse* LXXX, 2003: 249–252.

Renkin M., Fleurackers S., Szwach I., and Hreczuch W., Rapeseed methyl ester ethoxylates: A new class of surfactants of environmental and commercial interest. *Tenside Surfactants Detergents* 42(5), 2005: 280–287.

Rosenberg E., and Ron E. Z., High- and low-molecular-mass microbial surfactants. *Applied Microbiology and Biotechnology* 52, 1999: 154–162.

Shete A. M., Wadhawa G., Banat I. M., and Chopade B. A., Mapping of patents on bioemulsifier and biosurfactant: A review. *Journal of Scientific and Industrial Research* 65(2), 2006: 91–115.

Soaresa A., Guieysseb B., Jeffersona B., Cartmella E., and Lestera J. N., Nonylphenol in the environment: A critical review on occurrence, fate, toxicity and treatment in wastewaters. *Environment International* 34(7), 2008: 1033–1049.

Zhou Q. H., Klekner V., and Kosaric N., Production of sophorose lipids by *Torulopsis bombicola* from safflower oil and glucose. *Journal of the American Oil Chemists' Society* 69(1), 1992: 89–91.

4 Biosurfactants Produced by Genetically Manipulated Microorganisms
Challenges and Opportunities

Kamaljeet K. Sekhon Randhawa

CONTENTS

4.1 Biosurfactants 49
    4.1.1 Introduction 49
    4.1.2 Classification...50
4.2 Biosurfactant Production 54
    4.2.1 Factors Affecting Biosurfactant Production..54
        4.2.1.1 Nutritional Factors 55
        4.2.1.2 Environmental Factors...57
    4.2.2 Role of Recombinant DNA Technology and Genetically Modified Organisms
        in Enhanced Biosurfactant Production 57
    4.2.3 Extraction and Purification Methods...63
4.3 Applications of Biosurfactants 65
4.4 Challenges and Future Opportunities..65
References..67

4.1 BIOSURFACTANTS

4.1.1 INTRODUCTION

The term "bioremediation" was coined by scientists in the early 1980s, describing it as a process by which biological agents such as bacteria, fungi, or green plants are used to remove or neutralize the recalcitrant contaminants in soil or water. Bacteria and fungi work by breaking down pollutants such as polyaromatic hydrocarbons and polychlorinated biphenyls into harmless substitutes. Plants are used to aerate the polluted soil and to stimulate the microbial action. Plants have the unique property of absorbing pollutants such as salts and metals into their tissues, which are later harvested using extraction techniques and disposed. Phytoremediation is the term used to describe such remarkable property of plants for decontaminating polluted soil and water.

Bioremediation, in general, is a slow yet simple and less labor-intensive process. Biosurfactants, the surface active agents, became prominent as they played a significant role in the accelerated bioremediation of hydrocarbon or petroleum products by increasing their solubility and making them bioavailable. Biosurfactants are amphipathic molecules of microbial origin with both hydrophilic and hydrophobic moieties that partition preferentially at the interface between fluid phases with different degrees of polarity and H-bonding as oil–water or air–water interphase. They have some influence on the interfaces and bring down the interfacial tension between the two liquids. They play a role in emulsification, solubilization, desorption, and increasing the surface area of hydrocarbon contaminants for their microbial degradation. They have the unique property of reducing the surface and interfacial tensions and critical micelle concentration (CMC) in both aqueous solutions and hydrocarbon mixtures (Banat 1995, Rahman et al. 2002). Biosurfactants are extremely selective, less toxic, and biodegradable in comparison to their chemical counterparts.

According to the recent analysis published by Transparency Market Research, the global biosurfactants market is flourishing and is expected to reach USD 2210.5 million in 2018 with a volume of 476,512.2 tons. Out of this total volume, 21% consumption is assumed to be from developing provinces such as Asia, Africa, and Latin America. Europe is expected to maintain its prime position till 2018 in terms of volume and revenue followed by North America. Cost competitiveness has always been the bottleneck of biosurfactant production; however, the biosurfactant vendors vouch for their environment-friendly properties and specificity in action in order to maintain the market value.

Chemical surfactants, however, are cheap and easily available but they pose environmental threats. Apart from others, sodium triphosphate is one such surfactant used in many industrial and domestic products, for example, laundry detergents. Its discharge into natural waters has led to environmental problems like eutrophication and algal blooms, leading to hypoxic or oxygen-depleting conditions. The European Commission has banned the use of phosphates in consumer laundry detergent with effect from June 2013. Similar restraints are going to be applied to consumer dishwasher detergents from January 2017 (EC Press release, 2014). The negative environmental implications of chemical surfactants have led to the enormous faith in biosurfactants.

Time and again, scientific reports have emphasized and supported the environmental-friendly nature of biosurfactants. Recent data suggest that rhamnolipids are even superior to Triton X-100 since rhamnolipids are biodegradable under all conditions, whereas the surfactant Triton X-100 is partially biodegradable under aerobic and non-biodegradable under anaerobic, nitrate-reducing, and sulfate-reducing conditions (Mohan et al. 2006). Biosurfactants business is booming at the moment and the production companies have spread across the globe in the past two decades. Researchers are committed to the enhanced production of high-quality biosurfactants such as glycolipids, sophorolipids, cellobiose lipids, mannosylerythritol lipids (MELs), and rhamnolipids. They are trying to make biosurfactant production an economically feasible process by creating mutant or recombinant hyper-producing strains. This chapter sheds light on enhanced biosurfactant production by these strains and the major challenges in this endeavor.

4.1.2 CLASSIFICATION

The term "biosurfactant" is often used very loosely to refer to "emulsifying" and "dispersing agents" that do not considerably lower the surface tension of water but contains other properties of a classical surfactant (Desai and Desai 1993).

In the literature, the terms "surfactants" and "emulsifiers" are often used exchangeably. The term "bioemulsifier" is often used in an application-oriented manner to describe the combination of all the surface-active compounds that constitute the emulsion secreted by the cells to facilitate the uptake of an insoluble substrate. The surfactant molecules are able to form either micelles or inverted micelles or aggregate to form rod-shaped micelles, bilayers, and vesicles (Figure 4.1). They gather at interfaces and mediate between phases of different polarity, such as oil/water, and act as wetting agents on solid surfaces, namely, water/solid. This dynamic process is based on the ability of the surfactant to reduce the surface tension by governing the arrangement of liquid molecules, thus, influencing the formation of H bonds and other hydrophobic–hydrophilic interactions. The minimum surface tension value reached and the CMC needed are the parameters used to measure the efficiency of a biosurfactant (Rubina and Khanna 1995). For example, a lipopeptide biosurfactant from *Bacillus licheniformis* JF-2 at a CMC of 25 mg/L reduces the interfacial tension of saline against decane to about 10^{-3} dynes/cm, which is one of the lowest interfacial tensions reported for a microbial surfactant (Lin et al. 1994).

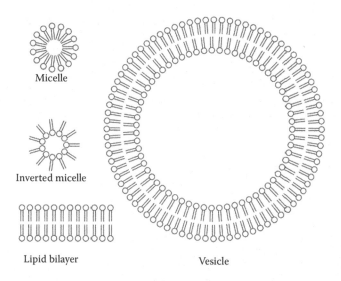

Micelle

Inverted micelle

Lipid bilayer

Vesicle

FIGURE 4.1 Different types of micelles.

The microorganisms that are capable of degrading hydrocarbons produce biosurfactants of diverse chemical nature and molecular size. Although the type of polar group present is the basis of classification of synthetic surfactants, biosurfactants are usually classified on the basis of their biochemical nature.

Biosurfactants can be broadly classified into five categories:

- *Glycolipids:* Glycolipid biosurfactants are usually carbohydrates in combination with long-chain aliphatic acids or hydroxy aliphatic acids. Glycolipids are generally low-molecular-weight biosurfactants (Ron and Rosenberg 2001) and the best known are rhamnolipids, trehalolipids, and sophorolipids.

- *Phospholipids and fatty acids:* Phospholipids are lipids containing one or more phosphate groups. Fatty acids are components of a long-chain aliphatic carboxylic acid found in natural fats and oils. Fatty acids also constitute membrane phospholipids and glycolipids. Certain hydrocarbon-degrading bacteria and yeasts produce appreciable amounts of phospholipids and fatty acids that act as biosurfactants.

- *Lipopeptides and lipoproteins:* Lipopeptides and lipoproteins are a heterogeneous class of biologically active peptides that serve to carry water-insoluble lipids in blood. The protein component alone is an apolipoprotein. Most of the lipopeptides produced by microorganisms are known for their antibiotic properties, but some have biosurfactant activity as well. Surfactin is one such distinctive lipopeptide that possess antibiotic and antiviral property and is a powerful biosurfactant. The polysaccharides, proteins, lipopolysaccharides, lipoproteins, or complex mixtures of these biopolymers act as high-molecular-weight biosurfactants (Ron and Rosenberg 2001).

- *Polymeric biosurfactants:* Polymeric biosurfactants have the ability to alter the rheological properties of aqueous solutions at low concentrations. Therefore, these biopolymers are used as thickeners and to stabilize emulsions, dispersions, and suspensions in aqueous systems. Polymeric biosurfactants are high-molecular-weight biopolymers with properties like high viscosity, high tensile strength, and resistance to shear. It is because of these properties that they have found a variety of industrial uses in pharmaceuticals, cosmetics, and food industries. Emulsan and biodispersan are the best examples.

- *Particulate biosurfactants:* Membrane or extracellular vesicles and whole microbial cells that play a role in hydrocarbon uptake act as "particulate" biosurfactants (Rosenberg 1986).

Apart from the five major categories, a new class of biosurfactants capable of lowering the surface tension of water to 26 dynes/cm with a CMC of 19 mg/L showing strong surface, and emulsifying activity was isolated from *Flavobacterium* sp. strain MTN11. These biosurfactants were named "flavolipids" (Bodour et al. 2004). The polar moiety of a flavolipid carries citric acid and two cadaverine molecules, which is different from the polar moieties of other classes of biosurfactants.

The chemical structures of some major biosurfactants are presented in Figure 4.2. A detailed classification of microbial surfactants is provided in Table 4.1.

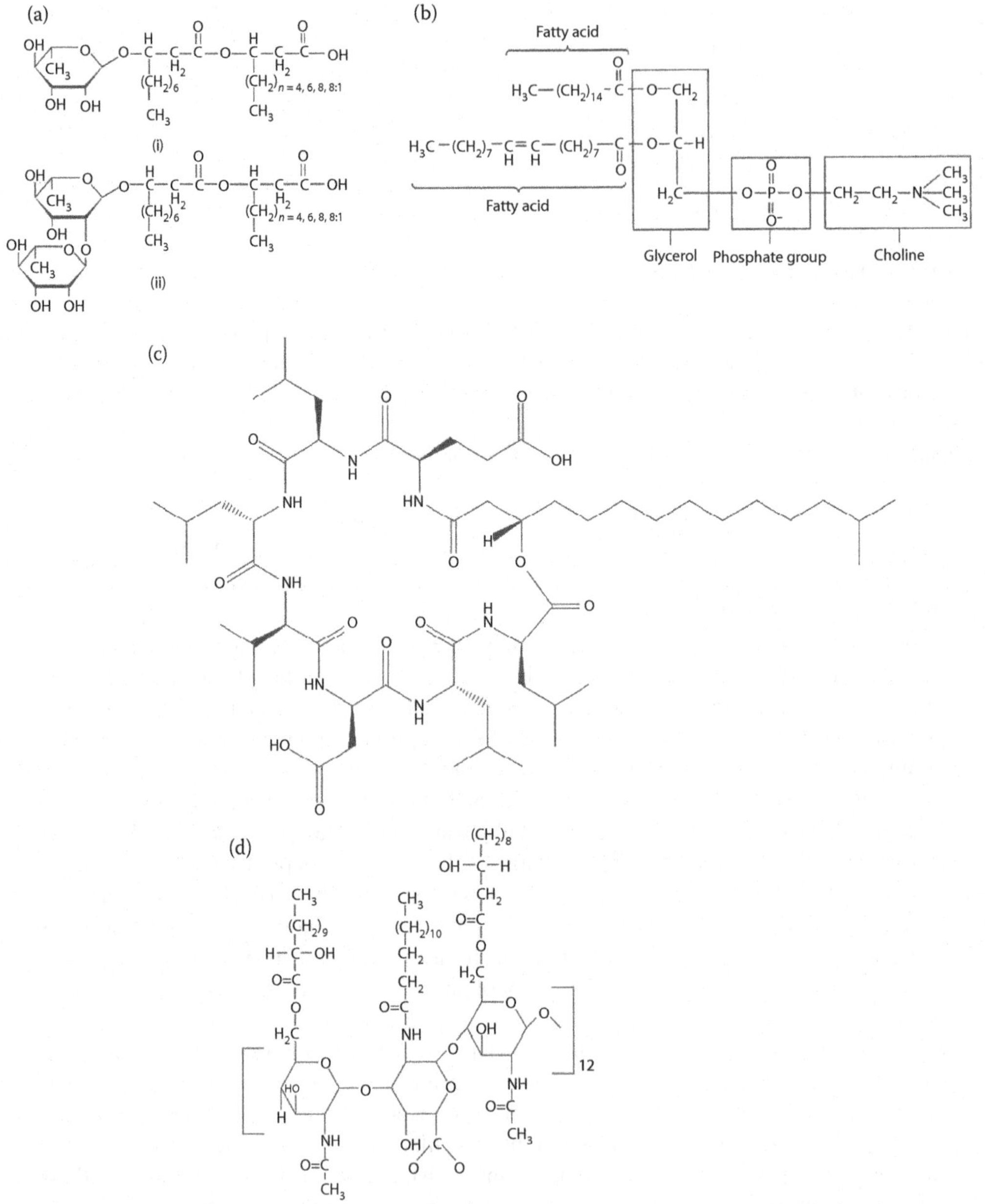

FIGURE 4.2 Chemical structures of some major biosurfactants: (a) (i) mono-rhamnolipid (L-rhamnosyl-β-hydroxydecanoyl-β-hydroxy(n + 4)oate, (ii) di-rhamnolipid (L-rhamnosyl-L-rhamnosyl-β-hydroxydecanoyl-β-hydroxy(n + 4)oate; (b) phospholipid; (c) surfactin; (d) emulsan.

TABLE 4.1

Classification of Biosurfactants

Biosurfactant Type	Producing Microbial Species
Glycolipids	
Trehalose mycolates	*Rhodococcus erythropolis, Arthrobacter paraffineu, Mycobacterium phlei, Nocardia erythropolis*
Trehalose esters	*Mycobacterium fortium, Micromonospora* sp., *M. smegmatis, M. paraffinicum, Rhodococcus erythropolis*
Trehalose mycolates of mono, di, trisaccharide	*Corynebacterium diptheriae, Mycobacterium smegmati, Arthrobacter* sp.
Rhamnolipids	*Pseudomonas* sp.
Sophorolipids	*Torulopsis bombicola/apicola, Torulopsis petrophilum, Candida* sp.
Rubiwettins R1 and RG1	*Serratia rubidaea*
Diglycosyl digyycerides	*Lactobacillus fermenti*
Schizonellins A and B	*Schizonella melanogramma*
Ustilipids	*Ustilago maydis* and *Geotrichum candidum*
Amino acid lipids	*Bacillus* sp.
Flocculosin	*Pseudomonas flocculosa*
Phospholipid and Fatty Acids	
Phospholipids, fatty acids	*Candida* sp., *Corynebacterium* sp., *Micrococcus* sp., *Acinetobacter* sp., *Thiobacillus thiooxidans, Asperigillus* sp., *Pseudomonas* sp., *Mycococcus* sp., *Penicillium* sp.
Lipopeptides and Lipoproteins	
Gramicidins	*Bacillus brevis*
Peptide lipids	*Bacillus licheniformis*
Polymyxin E1	*Bacillus polymyxa*
Ornithine-lipid	*Pseudomonas rubescens, Thiobacillus thiooxidans*
Viscosin	*Pseudomonas fluorescens*
Serrawettin W1, W2, W3	*Serratia marcescens*
Cerilipin	*Glucunobacter cerius*
Lysine-lipid	*Agrobacterium tumefaciens*
Surfactin, subtilysin, subsporin	*Bacillus subtilis*
Lichenysin G	*Bacillus licheniformisIM 1307*
Ornithine lipid	*Pseudomonas* sp., *Thiobacillus* sp., *Agrobacterium* sp., *Gluconobacter* sp.
Amphomycin	*Streptomyces canus*
Chlamydocin	*Diheterospora chlamydosporia*
Cyclosporin A	*Tolypocladium inflatum*
Enduracidin A	*Streptomyces fungicidicus*
Globomycin	*Streptomyces globocacience*
Bacillomycin L	*Bacillus subtilis*
Iturin A	*Bacillus subtilis*
Putisolvin I and II	*Pseudomonas putida*
Arthrofactin	*Arthrobacter*
Fengycin (Plipastatin)	*Bacillus thuringiensis CMB26*
Mycobacillin	*Bacillus subtilis*
Daptomycin	*Streptomyces roseosporus*
Syringomycin, Syringopeptin	*Pseudomonas syringae*
Polymeric Surfactants	
Lipoheteropolysaccharide (Emulsan)	*Acinetobacter calcoaceticus RAG-1, Arthrobacter calcoaceticus*
Heteropolysaccharide (Biodispersan)	*Acinetobacter calcoaceticus A$_2$*

continued

TABLE 4.1 (continued)
Classification of Biosurfactants

Biosurfactant Type	Producing Microbial Species
Polysaccharide protein	*Acinetobacter calcoaceticus strains*
Manno-protein	*Saccharomyces cerevisiae*
Carbohydrate-protein	*Candida petrophillum, Endomycopsis lipolytica*
Mannan-lipid complex	*Candida tropicalis*
Mannose/erythrose lipid	*Shizonella melanogramma, Ustiloga maydis*
Carbohydrate–protein–lipid complex	*Pseudomonas fluorescences, Debaryomyces polymorphus*
Liposan	*Candida lipolytica*
Alasan	*Acinetobacter calcoaceticus*
Protein PA	*Pseudomonas aeruginosa*
Particulate Biosurfactants	
Membrane vesicles	*Acinetobacter* sp. *H01-N*
Fimbriae, whole cells	*Acinetobacter calcoaceticus*

4.2 BIOSURFACTANT PRODUCTION

4.2.1 FACTORS AFFECTING BIOSURFACTANT PRODUCTION

The ability to produce a primary or secondary metabolite is regulated by the genetic makeup of the producer organism; however, nutritional and environmental factors have impact on the overall output of the metabolite. The sequence of events and the factors affecting biosurfactant production are orderly laid out in Figure 4.3.

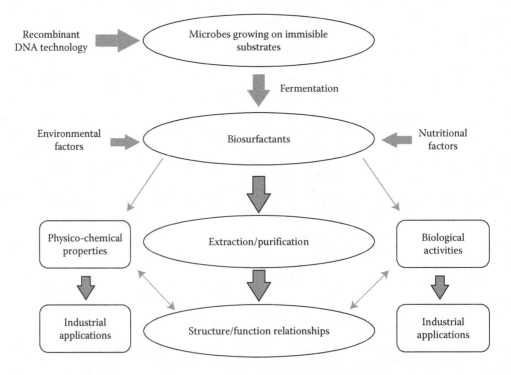

FIGURE 4.3 Diagram showing the chain of events in biosurfactant production.

4.2.1.1 Nutritional Factors

Carbon, nitrogen and phosphate sources, metal ions, and other additives used in the media formulation play a crucial role in the production and yield of biosurfactants.

4.2.1.1.1 Effect of Carbon Source

Research over the years has revealed the preference by microorganisms for hydrocarbon substrates over carbohydrates for biosurfactant production (Table 4.2). There is a sudden burst in the production of biosurfactants when microbes grow at the expense of water immiscible substrates. It is also observed that there is little biosurfactant production when cells grow on a readily available carbon source. The biosurfactant production was triggered when all the soluble carbon was consumed and when water-immiscible hydrocarbon was available in the medium (Banat 1995). Few examples are presented in this section. In *Pseudomonas* species, rhamnolipid production is regulated by the presence of *n*-alkanes in the medium while in *Pseudomonas aeruginosa* UW-1 higher production rates were achieved when grown on vegetable oil in comparison to liquid hydrocarbon (hexane) (Kosaric and Sukan 2000). It was also shown that rhamnolipid production in *P. aeruginosa* under denitrification conditions with carbon substrates palmitic acid, stearic acid, oleic acid, linoleic acid,

---

TABLE 4.2

Use of Low-Cost Carbon Sources for the Production of Biosurfactants by Various Microbial Species

Low Cost or Waste Raw Material	Biosurfactant Type	Producer Microbial Strain
Soybean oil refinery wastes	Rhamnolipids	*Pseudomonas aeruginosa* AT10
Curd whey and distillery wastes	Rhamnolipid	*Pseudomonas aeruginosa* BS2
Crude oil	Emulsan	*Acinetobacter calcoaceticus* RAG-1
Turkish corn oil	Sophorolipids	*Candida bombicola* ATCC 22214
Sunflower and soybean oil	Rhamnolipid	*Pseudomonas aeruginosa* DS10-129
Sunflower oil	Lipopeptide	*Serratia marcescens, Arthrobacter* sp. N3
Soybean oil	Mannosyl erythritol lipid	*Candida* sp. SY16
Soybean and olive oil WFO	Rhamnolipid	*Pseudomonas aeruginosa* 47T2 NCIB 40044
Soybean and sunflower oil soap stock waste	Rhamnolipid	*Pseudomonas aeruginosa* LBI
Vegetable oil	Glycolipid	*Pseudomonas fluorescens*
Oil refinery wastes	Glycolipids	*Candida antarctica and/or Candida apicola*
Rapeseed oil	Rhamnolipids	*Pseudomonas* sp. DSM 2874
Babassu oil	Sophorolipids	*Candida lipolytica* IA 1055
Potato process effluents	Lipopeptide	*Bacillus subtilis*
Cassava flour wastewater	Lipopeptide	*Bacillus subtilis* ATCC 21332, *Bacillus subtilis* LB5a
Olive oil	Lipopeptide, rhamnolipid	*Bacillus subtilis* SK320, *Pseudomonas fluorescens* Migula 1895-DSMZ
Sludge palm oil	Phospholipid	*Klebsiella pneumoniae* WMF02
Pre-treated molasses	Lipopeptide	*Brevibacterium aureum* MSA13
Diesel oil	Biosurfactant	*Saccharomyces lipolytica* CCT-0913, *B. subtilis, P. aeruginosa*
Cotton seed oil	Biosurfactant	*Candida glabrata* UCP 1002
Petrol	Surfactin	*Candida tropicalis* MTCC 230
Waste motor lubricant oil and peanut oil cake	Glycolipopeptide	*Corynebacterium kutscheri*
Soybean WFO	Rhamnolipid	*Pseudomonas putida* 33 and 300-B (wild and mutant strain)

Note: WFO—waste frying oil.

glycerol, vegetable oil, and glucose was successful and was free from problems such as foaming and respiration limitation (Chayabutra et al. 2000). The oil-degrading strain *P. aeruginosa* DS10-129 was used to optimize a substrate for maximum rhamnolipid production. It produced 4.31, 2.98, and 1.77 g/L rhamnolipid biosurfactant using soybean oil, safflower oil, and glycerol, respectively. Yield of biosurfactant increased even after the bacterial cultures reached the stationary phase of growth. Characterization of rhamnolipids using mass spectrometry revealed the presence of di-rhamno-lipids (Rha-Rha-C_{10}-C_{10}) (Rahman et al. 2002). Another *P. aeruginosa* strain showed maximum biosurfactant production in the presence of diesel among other carbon sources tested. *P. aeruginosa* supernatant even showed higher biosurfactant activity than *Bacillus subtilis* (Priya and Usharani 2009).

In another study, *Pseudomonas fluorescens* was able to grow efficiently on glucose and soybean oil as carbon source and the surface tension values obtained were 40.3 and 48.2 dynes/cm, respec-tively (Anyanwu and Okolo 2011). The surface tension values of *P. fluorescens* (isolated from waste-water) supernatants were in the range of 28.4 with phenanthrene to 49.6 dynes/cm with naphthalene and heptane as carbon sources, respectively (Stoimenova et al. 2009). Soybean waste frying oil as carbon source and glucose as growth initiator under fed-batch cultivation showed a rhamnolipid yield of 4.1 g/L by *Pseudomonas putida* 300-B mutant. *P. putida* 33 wild strain was subjected to gamma ray mutagenesis to produce this mutant (Ali et al. 2007). Another significant study by Thaniyavarn et al. (2006) further confirmed that the yield of rhamnolipid biosurfactant was affected by the carbon source. It came out to be 6.58, 2.91, and 2.93 g/L as rhamnose content when olive oil, palm oil, and coconut oil, respectively, were used as the substrates, whereas in case of *B. subtilis* the lipopeptide biosurfactant yield reached 4.92 g/L using glucose as carbon source (Ghribi and Ellouze-Chaabouni 2011). Olive oil has also been the carbon source of choice by Sekhon et al. (2009, 2011, 2012), where the study reflects the enhanced production of lipopeptides by the recombinants of *B. subtilis* SK320 in the presence of olive oil and the role of biosurfactant–esterase complex in emulsification.

4.2.1.1.2   Effect of Nitrogen Source

Reports emphasize that the nature and concentration of the N-source also affect biosurfactant pro-duction. Rhamnolipid production increased when nitrogen was exhausted in the medium. Nitrogen limitation is also important in the production of sophorose lipids.

According to Kosaric and Sukan (2000), the production of biosurfactants by the genera *Pseudomonas, Acinetobacter,* and *Torulopsis* can be regulated by the C/N ratio. Among the inor-ganic salts tested, ammonium salts and urea were preferred nitrogen sources for biosurfactant pro-duction by *Arthrobacter paraffineus* (Cooper and Paddock 1983), whereas nitrate ions supported maximum biosurfactant production in *P. aeruginosa* (MacElwee et al. 1990), *Rhodococcus* sp. (Abu-Ruwaida et al. 1991), and in *B. subtilis* (Makkar and Cameotra 1998). However, lichenysin-A production was reported to be enhanced two to fourfolds in *B. licheniformis* BAS50 by the addi-tion of L-glutamic acid and L-asparagine, respectively, to the medium (Yakimov et al. 1996). *P. aeruginosa* PA1 (isolated from oil production wastewater) showed higher production of rhamno-lipid, expressed by rhamnose (3.16 g/L) when glycerol and sodium nitrate were used as carbon and nitrogen sources, respectively, with a C/N ratio of 60/1 (Santa Anna et al. 2002). Likewise, a C/N (glucose/sodium nitrate) ratio of 22 was found to be appropriate for a rhamnolipid production of 5.46 g/L by *Pseudomonas nitroreducens* isolated from petroleum-contaminated soil (Onwosi and Odibo 2012). A C/N ratio of 3 (sucrose/ammonium nitrate) in case of *B. subtilis* (Fonseca et al. 2007), 10 (olive oil/ammonium nitrate) in *P. fluorescens* Migula 1895-DSMZ (Amrane et al. 2008), and 23 in *P. aeruginosa* LBI with a remarkable rhamnolipid yield of 16.9 g/L (Lovaglio et al. 2010), respectively, clearly indicated that C/N ratio is indeed an important parameter.

4.2.1.1.3   Effect of Other Factors/Sources

The influence of various nutrients or nutritional supplements on biosurfactant production cannot be ruled out. The amount of phosphate, iron, manganese, calcium, and trace elements in the medium greatly affect growth and biosurfactant production. Not only the metal supplements, but also the amino acids aspartic acid, aparagine, glutamic acid, valine, and lysine affect the final yield of bio-surfactants (Makkar and Cameotra 2002). To exemplify, supplementing commercial Pharmamedia (media made from embryo of cottonseed) with Fe^+ (4.0 mM) and sucrose (2 g/L) leads to maxi-mum production of surfactin (300 mg/L) by *B. subtilis* MZ-7 (Ajlani et al. 2007) and in case of *Rhodococcus erythropolis* biosurfactant production reached maximum concentration of 285 mg/L when phosphate buffer concentration was increased from 30 to 150 mmol/L (Pacheco et al. 2010). Besides, the yield of biosurfactant production is highly enhanced or inhibited by the addition of antibiotics such as penicillin or chloramphenicol (Kosaric and Sukan 2000).

4.2.1.2 Environmental Factors

There are many scientific reports where the individual effects of oxygen availability, salinity, pH, and temperature on biosurfactant production have been examined.

Study of the synergistic effects on *Lactobacillus pentosus* reflected the inter-dependence of these factors (Bello et al. 2012). In case of *B. subtilis* C9, a threefold higher yield of lipopeptides was observed under oxygen-limited conditions compared with oxygen-sufficient conditions (Kosaric and Sukan 2000) whereas, using a dissolved oxygen concentration of 30% in the medium showed a biosurfactant production of 4.92 g/L in *B. subtilis* SPB1 (Ghribi and Ellouze-Chaabouni 2011).

Analyzing other effects revealed that some biosurfactant producers were not affected by NaCl up to 10% (w/v) although reductions in the CMCs were detected (Abu-Ruwaida et al. 1991). For instance, the NaCl concentration of 8% and pH 6–8 worked best in case of halophilic bacteria *Halomonas* sp. BS4 (Donio et al. 2013), *B. subtilis* BBK-1 produces three types of lipopeptides—bacillomycin L, plipastatin, and surfactin in the presence of NaCl up to 8% (Roongsawang et al. 2002), a salt concentration between 2% and 10% was tolerable for a yield of 9 g/L glycolipid by *Candida sphaerica* (Luna et al. 2012), the rhamnolipid from *P. fluorescens* Migula 1895-DSMZ was stable at a temperature of 120°C, 10% NaCl concentration and a wide range of pH (Abouseoud et al. 2007) and a temperature between 4°C and 125°C and pH values in the range of 5–10 were suitable for *Arthrobacter* sp. N3 (Vilma et al. 2011).

4.2.2 ROLE OF RECOMBINANT DNA TECHNOLOGY AND GENETICALLY MODIFIED ORGANISMS IN ENHANCED BIOSURFACTANT PRODUCTION

Since the discovery of biosurfactants, the scientific community is eager to bring down their produc-tion cost. Even if the technology is commercially viable, to some extent research is still ongoing to create novel recombinant strains which can mass produce biosurfactants. There is a plethora of "natural" microbial strains inherently producing "natural" biosurfactants in the presence of immis-cible substrates.

Apart from having significant advantages like ecological safety (Vasileva-Tonkoval et al. 2001), less or no toxicity, biodegradability, compatibility with biogeochemical cycle, broad range of novel structural characteristics, high temperature stability, activity at extreme salinity and pH (Makkar and Cameotra 1997) over their chemical counterparts, biosurfactants enjoy a special privi-lege. Biosurfactant-producing microbes can be genetically modified and tailored to meet specific requirements (Figure 4.4). These recombinant or genetically modified organisms (GMOs) are then expected to provide an enhanced biosurfactant yield. The challenge of high cost of biosurfactant production, therefore, is dealt with the advancements made by the field of recombinant DNA tech-nology. To circumvent the production cost, synthetic media has also been replaced with cheap agro-industrial wastes (Haba et al. 2000, Makkar and Cameotra 2002, Sadouk et al. 2008, Kiran et al. 2010, Panesar et al. 2011).

There are several techniques of creating novel strains with desirable qualities. For instance, the mutant strains for enhanced biosurfactant production are produced using transposons, with chemi-cal mutagens such as *N*-methyl-*N*′nitro-*N*-nitrosoguanidine, using radiation or by selection on the

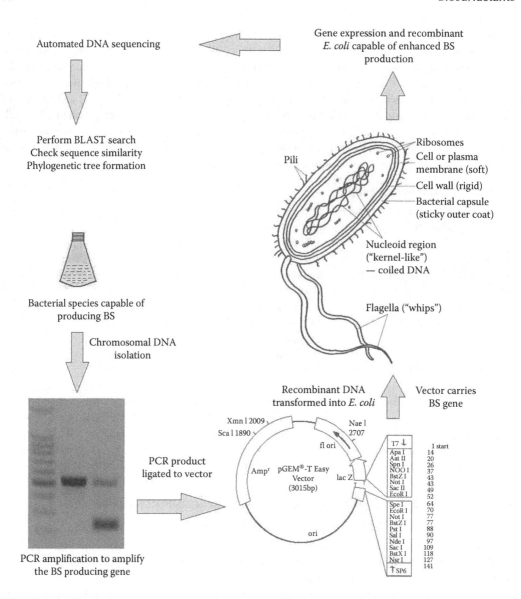

FIGURE 4.4 The basic molecular cloning strategy adopted for enhanced biosurfactant (BS) production.

basis of resistance to ionic detergents such as cetyl trimethyl ammonium bromide (Mukherjee et al. 2006). There are various non-pathogenic, high-yielding mutant and recombinant strains created for enhanced biosurfactant production over the years (Table 4.3).

Biosurfactant research received its major landmark with the isolation and characterization of the compound "surfactin" from *B. subtilis* by Arima et al. in 1968. Surfactin (mol. wt. 1036.34 Da), the complete structure given by Kakinuma et al. (1969) (Figure 4.2), is a cyclic lipopeptide characterized by a β-hydroxycarbonic acid moiety with strong surface activity and antibiotic properties. To date, surfactin is rated as the most biologically active and powerful biosurfactant with the capacity of lowering the surface tension of water from 72 to 27 dynes/cm with as low as 10 μM CMC. Over the years, the genes responsible for surfactin production have been rejigged by many research groups for the improved production of this lipopeptide (Mulligan and Gibbs 1989, Nakano and Zuber 1989, Mulligan and Chow 1991, Nakano et al. 1991, 1992, Tsuge et al. 1995, 1996, Yao et al. 2003, Yoneda et al. 2006, Lee et al. 2007, Sekhon et al. 2011). The initial yield of surfactin

TABLE 4.3

Genetically Manipulated Organisms and Mutants Developed for Enhanced Biosurfactant Production

Microorganism	Target Gene or Strategy Adopted	Type of Biosurfactant	Reference
Pseudomonas aeruginosa	*Escherichia coli lacZY* genes inserted	Rhamnolipid	Koch et al. (1988)
Bacillus subtilis	*srf*B	Surfactin	Zuber and Nakano (1989)
Bacillus subtilis Suf1	Ultraviolet mutation between *argC4* and *his*A1	Surfactin	Mulligan et al. (1989)
Bacillus subtilis	*srf*A	Surfactin	Nakano et al. (1991)
Pseudomonas putida PG201 mutant 59C7	Transposon Tn5-GM	Rhamnolipid	Koch et al. (1991)
Bacillus subtilis ATCC 21332	Mutation between *Arg4* and *His*A1 sites	Surfactin	Mulligan et al. (1991)
Bacillus subtilis	*sfp*	Surfactin	Nakano et al. (1992)
Bacillus subtilis ATCC 55033	Random mutagenesis with *N*-methyl-*N'*nitro-*N*-nitrosoguanidine	Surfactin	Carrera and Cosmina (1993)
Pseudomonas fluorescens ATCC 15453	*rhlR* and *rhlI* rhamnolipid regulatory elements	Rhamnolipid	Ochsner et al. (1995)
Pseudomonas putida strain KT2442	*rhlAB* genes	Rhamnolipid	Ochsner et al. (1995)
Pseudomonas aeruginosa EBN-8	Gamma ray-induced mutation	Rhamnolipid	Iqbal et al. (1995)
Bacillus subtilis MI 113	*lpa-14*	Surfactin	Ohno et al. (1995)
Bacillus subtilis YB8	*lpa, lpa-8*	Surfactin	Tsuge et al. (1992, 1996)
Bacillus subtilis MI113	*lpa-14*	Surfactin, iturin A	Nakayama et al. (1997)
Bacillus subtilis SB103	*Sfp*	Surfactin	Kim et al. (2000)
Recombinant *Bacillus subtilis* ATCC 21332	Contains recombinantly modified peptide synthetase	Surfactin	Symmank et al. (2002)
Bacillus subtilis 168	*srf*DB3, *asp*B3, *lpa*B3, *ycz*EB3	Surfactin, iturin A	Yao et al. (2003)
Pseudomonas aeruginosa PTCC1637	*N*-methyl-*N'*nitro-*N*-nitrosoguanidine	Rhamnolipid	Tahzibi et al. (2004)
Gordonia amarae	*Vitreoscilla* hemoglobin gene (*vgb*)	Trehalose lipids	Dogan et al. (2006)
Mutant *Bacillus subtilis* SD901	Chemical mutagenesis with *N*-methyl-*N'*nitro-*N*-nitrosoguanidine	Surfactin	Yoneda et al. (2006)
Bacillus subtilis C9-ET2 transformant	*srf*A operon	Surfactin	Lee et al. (2007)
Pseudomonas aeruginosa PAO1	RhlAB	Rhamnolipid	Wang et al. (2007)
Pseudomonas putida 300-B mutant	Gamma ray-induced mutation	Rhamnolipid	Ali Raza et al. (2007)
Mutant *Bacillus subtilis* E8	Ion beam implantation	Surfactin	Gong et al. (2009)
Rhodococcus erythropolis MTCC 2794	Artificial neural network (ANN) coupled with genetic algorithm (GA)	Glycolipid	Pal et al. (2009)
Bacillus subtilis fmbJ224	Sequential mutagenesis with lithium chloride, nitrosoguanidine, ethyl methane sulfonate, and streptomycin	Surfactin	Fengxia et al. (2011)
Bacillus subtilis SK320	*sfp, sfp0, srf*A	Lipopeptide	Sekhon et al. (2011)

as obtained by Arima et al. (1968) was reported to be 0.05–0.1 g/L of the crude product and 0.04–0.05 g/L of the purified product. In some preliminary studies, a surfactin yield of 0.7–0.8 g/L by *B. subtilis* ATCC 21332 (Cooper et al. 1981), 0.16 g/L by *B. subtilis* (Mulligan and Sheppard 1987, and 0.562 g/L by mutant of *B. subtilis* ATCC 21332 (Mulligan and Gibbs 1989) was obtained, which was not economically reasonable from a commercial point of view. A US Patent (No. 5227294) was obtained by Carrera and Cosmina (1993), for producing quantities of 1.2 to 2.0 g/L of purified surfactin (99%) by a mutant of *B. subtilis* ATCC 21332.

Subsequently, the production of this lipopeptide was carried out using a recombinant *B. subtilis* (Shoda et al. 1992). Surfactin yield by the recombinant strain was about 1½ times boosted when compared with the parent *B. subtilis* RB14, the strain in which the surfactin gene originated. Another recombinant *B. subtilis* MI 113 was created with the plasmid pC112 harboring *lpa-14*. This recombinant showed enhanced production of surfactin in solid-state fermentation on soybean curd residue (Ohno et al. 1995). *Lpa-14* was later reported to be a gene closely related to surfactin and iturin A production (Nakayama et al. 1997).

In subsequent years, modifications in the culture media also lead to the increased surfactin production ranging from 0.760 g/L (Sen and Swaminathan 1997) to 3.5 g/L (Wei and Chu 2002), 3.6 g/L (Yeh 2005) and 6.45 g/L (Yeh 2006), respectively. However, in a major breakthrough, chemical mutagenesis using *N*-methyl-*N'*nitro-*N*-nitrosoguanidine on recombinant strain *B. subtilis* MI 113 created a mutant *B. subtilis* SD901. This mutant when grown on soybean flour or its extract for 20–90 h gave the amount of surfactin production in the range of 8–50 g/L (Yoneda et al. 2006). This is the highest yield of surfactin ever reported. Another genetically modified strain (*B. subtilis* W1012) showed twofold increase in biosurfactant production besides producing green fluorescent protein when compared with the reference or parent strain (*B. subtilis* 1012) (Neves et al. 2007). A mutant, *B. subtilis* E8, obtained by ion beam implantation technique after 32 h cultivation produced 12.20 g/L concentration of crude surfactin with a CMC of 15 μM (Gong et al. 2009). A genetically stable strain *B. subtilis* fmbJ224 was also mutagenized sequentially with lithium chloride, nitrosoguanidine, ethyl methane sulphonate, and streptomycin to produce strain LN2-3 that showed 44.56-fold increase in surfactin yield (Fengxia 2011). Meanwhile, the recombinant strains *BioSa, BioSb,* and *BioSc,* produced by the molecular cloning of the biosurfactant genes from *B. subtilis* SK320 into *Escherichia coli,* showed twofold increase in the production with recovery of 2.13, 2.20, 2.45 g/L biosurfactant, respectively, when compared with the parent strain (1.2 g/L). The lipopeptides were potent emulsifiers exhibiting surface tension values of 38.4, 35, and 30.7 dynes/cm, respectively (Sekhon et al. 2011).

After 45 years of its discovery, research on surfactin has not decreased. The usage of surfactin as lipopeptide antibiotic, lysing agent for erythrocytes, and clotting inhibitor encourages the researchers to keep looking for that ideal or optimal approach to maximize its yield.

The equally intensively investigated genus in case of biosurfactant production enhancement and economics has been *Pseudomonas*. Notably, the discovery of rhamnolipids dates back to the mid-1940s. They were identified as oily glycolipids produced by *Pseudomonas pyocyanea* (now *P. aeruginosa*) grown on glucose by Bergström et al. (1946a,b). The glycolipid was named pyolipic acid. The chemical nature and structural units were identified as L-rhamnose sugar molecule and β-hydroxydecanoic acid (Bergström et al. 1946a,b, Jarvis and Johnson 1949, Edwards and Hayashi 1965) (Figure 4.2). Rhamnolipids display high surface activities and are easily produced with higher yields after relatively short incubation periods of the microbes (Abdel-Mawgoud et al. 2010). They are produced by various *Pseudomonas* species and are potentially applicable in enhanced oil recovery, biodegradation, and bioremediation. Regardless of the availability of good rhamnolipid overproducers, there has been a challenge for the sufficient yield of these biosurfactants.

In a study by Koch et al. (1991), a mutant 59C7 was generated and isolated by TnS-GM-induction of *P. aeruginosa* PG201. The mutant 59C7 grown in glucose-containing medium, was able to uptake and utilize hexadecane and produced double the amount of rhamnolipids than the wild-type strain. Likewise, a gamma-ray-induced mutant EBN-8 of *P. aeruginosa* strain S8 showed three to four times enhanced rhamnolipid production and emulsification when compared with the parent strain (Iqbal et al. 1995). The EBN-8 mutant gave a rhamnolipid yield of 4.1 and 6.3 g/L when grown on *n*-hexadecane and paraffin oil, respectively, in another study conducted by Raza et al. (2006). A mutant of *P. aeruginosa* designated as *Pseudomonas aeruginosa* PTCC1637 produced by random mutagenesis with *N*-methyl-*N'*nitro-*N*-nitrosoguanidine produced 10 times more rhamnolipid than the parent strain (Tahzibi et al. 2004).

In contrast, genetic engineering of the producer strains was also carried out to increase rhamnolipid production. The rhamnosyltransferase 1 complex (RhlAB) is the key enzyme responsible for rhamnolipid production. The genetic modification of rhamnolipid producing *P. aeruginosa* by inserting the *E. coli lacZY* genes into its chromosome was conducted so as to make it capable of growing and utilizing lactose from whey for enhanced biosurfactant production (Koch et al. 1988). However, as *P. aeruginosa* is an opportunistic pathogen in humans capable of producing wide variety of virulence factors, its industrial implications are limited. To overcome this problem, the *rhl*AB genes from *P. aeruginosa* were cloned and expressed in heterologous hosts (Ochsner et al. 1995). The chromosomal integration of RhlAB complex was also performed to obtain *P. aeruginosa* PEER02 and *E. coli* TnERAB, which produced rhamnolipid that was able to recover trapped oil from sand pack using as low as 250 mg/L concentration (Wang et al. 2007). Expression was under the control of *rhl*R and *rhl*I rhamnolipid regulatory elements or *tac* promoter. The recombinant strains *P. fluorescens* ATCC 15453 and *P. putida* KT2442 harboring the rhamnolipid gene cluster showed higher yields of 0.25 and 0.6 g/L, respectively, when compared with the parent strain. A similar strategy was adopted by Cha et al. (2008), wherein the opportunistic pathogen *P. aeruginosa* EMS1 (producing 5 g/L rhamnolipid) was replaced by *P. putida* by molecular cloning of the *rhlAB* rhamnosyltransferase gene into the later. Likewise, in a relatively recent study, *P. putida* KT24C1pVLT31_*rhlAB* strain was constructed as an alternative to the *P. aeruginosa* PAO1 in order to avoid the cumbrous product purification steps. Utilizing glucose as carbon source, the engineered strain was able to provide a high yield of rhamnolipids (Wittgens et al. 2011).

P. aeruginosa PAO1 might not be the suitable model organism due to its life-threatening opportunistic nature, but is displaying promising results when it comes to rhamnolipid yield. An astonishing yield of 39 g/L after 90 h of cultivation with sunflower oil as substrate and under nitrogen-limiting conditions has been reported by Muller et al. (2010, 2011). Conversely, a yield of 1 g/L of monorhamnolipids was obtained by growth of the non-pathogenic *Pseudomonas chlororaphis* on 2% glucose (Gunther et al. 2005).

To increase the rhamnolipid production *P. aeruginosa* (NRRL B-771) and its recombinant strain, PaJC were studied (Kahraman and Erenler 2012). *P. aeruginosa* expressed the *Vitreoscilla* hemoglobin gene (*vgb*) for rhamnolipid production and this gene was transferred to PaJC. Both the wild-type and recombinant strains started producing rhamnolipid at the stationary phase. The concentration in case of PaJC cells reached 8.373 g/L using 1% glucose-supplemented minimal medium. In another interesting study by Kryachko et al. (2013), the enhanced oil recovery effects of chemical surfactants, supernatants from native rhamnolipid producer *P. aeruginosa* PA14, *P. aeruginosa* PDO111 with repressed transcription of rhamnolipid production genes and recombinant *E. coli* carrying rhamnolipids production genes, were compared in sandpack columns. The enhancement of oil recovery was evident in the presence of rhamnolipids, cell surface parts, and/or bacterial cells.

Recently, an impressive study was conducted and a European Patent (EP2573172 A1) has been acquired by Blank et al. (2013), wherein a genetically modified *Pseudomonas* sp. is proficient in attaining a carbon yield of more than 0.18 Cmol rhamnolipid/Cmol substrate. The carbon yield coefficient is expressed as the C atoms comprised by a rhamnolipid in relation to the number of C atoms comprised by a substrate. The authors state that using glucose as the substrate, a yield of 0.728 mol rhamnolipid/mol glucose can be achieved. Other substrates used were glycerol, oleic acid, and soybean oil. The study claims to have achieved by far the maximum yield of mono- and di-rhamnolipids.

Rhamnolipid production by *Pseudomonas* genus has been thoroughly studied by many research groups. It is worth mentioning that the technology was successfully transferred and commercialized from laboratory to pilot scale by Jeneil Biosurfactant Company, subsidiary of Jeneil Biotech. Inc. (USA) by developing efficient bacterial strains. Rhamnolipid biosurfactants are certified as "greener" or "organic" throughout their life cycle.

There is no dearth of microbial species producing biosurfactants. For instance, the bacterial generas *Rhodococcus, Arthrobacter, Brevibacterium, Mycobacterim, Corynebacterium, Acinetobacter,* and the yeast belonging to genera *Torulopsis, Candida, Pseudozyma, Yarrowia, Kluveromyces,* and *Saccharomyces* have also been investigated. The biosurfactant produced by *R. erythropolis* has 94% (Ciapina et al. 2006), biosurfactant produced by *R. erythropolis* in another study has 97–99% (Pacheco et al. 2010), whereas the biosurfactant produced by *Rhodococcus* sp. strain TA6 has 70% (Shavandi et al. 2011) residual oil-recovering capacity, respectively.

To increase the yield, media optimization technique called artificial neural network (ANN) coupled with genetic algorithm (GA) was employed and was able to enhance the glycolipid production of *R. erythropolis* MTCC 2794 by 3.5-folds (Pal et al. 2009). ANN-GA-based models were also effective in sensitivity analysis. Similarly, for enhanced production of trehalose lipids, the *Vitreoscilla* hemoglobin gene (*vgb*) was electroporated, maintained, and expressed in *Gordonia amarae*. The recombinant strain was grown in rich medium with hexadecane and showed 1.4- and 2.4-fold productions with limited and normal aeration, respectively (Dogan et al. 2006). In another study, transposon mutants of *Acinetobacter calcoaceticus* RAG-1 produced structural variants of emulsan which showed enhanced emulsifying activity than the parents strain (Johri et al. 2002). A threefold increase in lipopeptide biosurfactant production using industrial and agro-industrial solid waste residues as substrate was reported in *Brevibacterim aureum* MSA13 (Seghal et al. 2010), a yield of 2.7 g/L was accomplished by *Corynebacterium alkanolyticum* ATCC 21511 in a self-cycling fermenter using hexadecane as substrate (Crosman et al. 2002), a biosurfactant with emulsification activity higher than Triton X-100 and yield of 6.4 g/L by *Corynebacterium kutscheri* (Thavasi et al. 2007) and a yield of 1.9 g/L biosurfactant from *Acinetobacter* species (Jagtap et al. 2010), respectively, indicates the potential of these generas.

The advantage of utilizing yeast for biosurfactant production is the GRAS (generally regarded as safe) status of yeasts. The examples include glycolipid biosurfactants produced by yeasts namely MELs isolated from *Pseudomonas antarctica* T-34 (renamed from *Candida antarctica* T-34) (Morita et al. 2013) and *Pseudozyma rugulosa* (Morita et al. 2006) show excellent surface active and vesicle-forming properties. After 144 h fermentation of *Candida bombicola* on glucose and sunflower oil the sophorolipids yield obtained was 120 g/L by Casas and Garcia-Ochoa (1999), whereas a similar yield of 120 g/L was reported by Deshpande and Daniels (1995) when *C. bombicola* was stimulated by the addition of animal fat in the culture media with 68 h fermentation. Similarly, yields of 21 and 79 g/L sophorolipids were obtained by *C. bombicola* after 168 h of fermentation using soy molasses as substrate and in the presence of glucose and oleic acid (Solaiman et al. 2004). These few examples quoting the yield of biosurfactants obtained express the promising future research possibilities linked with yeasts.

Studies also reflect that the biosurfactant production process involves a complex multicomponent enzyme system within the microbial cells and is equally dependent on the substrate availability. A new approach, therefore, was put forward by Kaar et al. (2009) in which "designed peptide surfactants" functional as surface active and foam-stabilizing agents can be produced recombinantly in microbial systems. The variations can be introduced by genetic manipulations. The data presented the production of the peptide biosurfactant GAM1 in recombinant *E. coli* with peptide purities exceeding 90%.

From the analysis of the significant investigations done so far and the methods that are available for improved biosurfactant production, it can be established that there is no shortage of genetic manipulation techniques. It would, however, be interesting to observe how the less-explored microbial strains are utilized for mass producing biosurfactants while keeping the overall production as economical as possible. The genetic manipulation of yeast and production of some recombinant varieties for mass production needs to be pursued further. The research until now accentuates that half the battle is won, however, the much-required breakthrough to increase the biosurfactant yield and making it commercially widespread is still awaited.

4.2.3 EXTRACTION AND PURIFICATION METHODS

The upstream supply chain and downstream recovery operations decide the overall yield of any product. The type and amount of carbon source used, nutritional and environmental factors, the genetics of the producer organism, and the extraction techniques adopted are all important parameters governing the production of biosurfactants.

The choice of method for the extraction of a particular biosurfactant depends on its ionic charge, solubility in water, whether the product is cell bound or extracellular and of course, the cost of recovery. The methods generally used for biosurfactant recovery include solvent extraction, adsorption followed by solvent extraction, precipitation, crystallization, centrifugation, and foam fractionation. Precipitation using ammonium sulfate, acetone, solvent extraction, ultrafiltration, ion exchange, dialysis, ultrafiltration, lyophilization, isoelectric focusing, and thin-layer chromatography are commonly used to recover biosurfactants (Satpute et al. 2010).

Most biosurfactants are secreted into the medium and thus are obtained after removal of cells from either culture filtrate or supernatant. The supernatant is extracted using a suitable solvent, lyophilized and total biosurfactant recovered as purified powder is reported as g/L (Figure 4.5). Three main factors that obstruct the commercialization of biosurfactants are (Mukherjee et al. 2006) (a) high cost of raw materials; (b) high recovery and purification costs; and (c) low yields obtained. It is also observed that downstream processes contribute 60% to the total production costs of biosurfactants. New methods and techniques of recovery and purification of biosurfactants are, therefore, regularly researched and introduced to check the cost of downstream processes.

Cooper et al. (1981) showed an enhanced production of the lipopeptide surfactin from *B. subtilis* by continuous product removal by foam fractionation. The surfactin could be easily recovered from the collapsed foam by acid precipitation. The yield was also improved by the addition of iron or manganese salts. Cultivation of bacteria in batch cultures for the production of biosurfactant is not commercially viable when the process is to be used at the industrial scale. Guerra-Santos et al. (1984) have shown that the yield of rhamnolipids was several folds higher when a continuous process was used compared to the batch culture for the mass production of biosurfactants. Mattei and Bertrand (1985) were able to obtain a large quantity (approx. 3 g/L) of biosurfactant continuously

FIGURE 4.5 The modern bench-top laboratory lyophilizer showing the freeze drying of biosurfactant samples.

by using a tangential flow filtration device. The production of biosurfactant AP-6 by *P. fluorescens* 378 (Persson et al. 1990) in a microcomputer-controlled multibatch fermentation system, which enabled simultaneous running of ten fermenters, greatly cut down the time and equipment costs, compared with traditional laboratory fermenters. Neu and Poralla (1990) isolated a biosurfactant from *Bacillus* sp. in which the foam produced was blown out of the fermenters, collected, and then centrifuged.

The biosurfactants produced by *Nocardia* sp. L-417 strain, when grown with *n*-hexadecane as sole carbon source, were purified by ammonium sulfate fractionation, chilled acetone, hexane treatments, silica gel column chromatography, and Sephadex LH-20 gel filtration (Kim et al. 2000). Foaming was successfully applied for the concentration of the lipopeptide biosurfactant surfactin from *B. subtilis* ATCC 21332 cell culture broths. Simultaneous high enrichments and recoveries of surfactin could not be obtained (around 70% of the total recovered) at a low concentration during the early stages of foaming. The use of low-stirrer speeds was essential in producing foam at a controlled rate. By collecting fractions of the foam produced between 10 and 30 h, from systems stirred at 166 and 146 rpm, a highly concentrated surfactin extract could be obtained. The surfactin concentration in the foam was 1.22 and 1.67 g/L, respectively, which represented enrichments and recovery of 60%. This study by Davis et al. (2001) points to the utility of foaming as a method for the recovery of surface-active fermentation products, particularly when used in an integrated production/recovery system.

Kuyukina et al. (2001) proposed methyl tertiary-butyl ether (MTBE) as a solvent for the extraction of biosurfactants from *Rhodococcus* bacterial cultures. After comparison with other well-known solvent systems used for biosurfactant extraction, it was found that MTBE was able to extract crude surfactant material with high product recovery (10 g/L), efficiency (CMC 130–170 mg/L), and good functional surfactant characteristics (surface tension values of 29 dynes/cm), respectively. The isolated surfactant complex contained 10% polar lipids, mostly glycolipids possessing maximal surface activity. Ultrasonic treatment of the extraction mixture increased the proportion of polar lipids in crude extract, resulting in increasing surfactant efficiency. Owing to certain characteristics of MTBE, such as relatively low toxicity, biodegradability, ease of downstream recovery, low flammability, and explosion safety, the use of this solvent as an extraction agent in industrial scale biosurfactant production is feasible.

The production and purification of the peptide biosurfactant GAM1 in recombinant *E. coli* was proposed by Kaar et al. (2009). The impurities were precipitated using solvent-, acid-, and heat-mediated precipitation methods; however, acid precipitation was the method of choice. Acid precipitation as proposed was a simple method with high recovery rate and high purification factor with peptide purities exceeding 90%. In another study, Sarachat et al. (2010) showed a rhamnolipid biosurfactant recovery of 97% and an enrichment ratio of 4 from *P. aeruginosa* SP4 using the technique of foam fractionation. The air flow rate of 30 mL/min, an initial foam height of 60 cm, 160–250 µm pore size of the sintered glass disk, an initial liquid volume of 25 mL, and an operation time of 4 h were the other parameters standardized for maximum recovery. A native *P. aeruginosa* RS29, isolated from crude oil, produced rhamnolipid as a primary metabolite and using chloroform:methanol (2:1) 6 g/L crude biosurfactant was extracted (Banat 2012). The chloroform:methanol (2:1) solvent extraction technique has been used by some other research groups as well (Anandraj and Thivakaran 2010, Aparna et al. 2011).

In another study, acid precipitation, ammonium sulfate precipitation, and organic solvent extraction techniques were used to recover rhamnolipid produced by *P. aeruginosa* USMAR-2. A recovery of 89.70% was obtained by organic solvent extraction method using methanol, 1-butanol, and chloroform as the solvent (Salleh et al. 2011). However, chilled acetone has been the solvent of choice by Sekhon et al. (2011) and Amrane et al. (2008) for precipitating biosurfactants from *B. subtilis* SK320 and its recombinants and *P. fluorescens* Migula 1895-DSMZ.

Studies point to the availability of various extraction or downstream processes for the optimum recovery of biosurfactants. However, at the industrial level these recovery techniques are worth spending a fortune on, if the biosurfactant producer organism (natural or genetically modified) is reliable and efficiently delivers higher yield.

4.3 APPLICATIONS OF BIOSURFACTANTS

During the past five decades or more, biosurfactant production by various microorganisms has been extensively studied by various research groups. Biosurfactants produced by the "native" and the "recombinant" strains are applicable in numerous sectors viz. environment, agriculture, medicine, detergents, cosmetics, food, and beverages. A detailed account of applications of biosurfactants is provided in Table 4.4.

4.4 CHALLENGES AND FUTURE OPPORTUNITIES

Biosurfactant production has gained impetus in the past two decades. The interest in biosurfactants research within the scientific community owes to their distinctive properties and various industrial applications. The commercial prosperity of any product, however, depends on its market demand, production cost, and ease of availability of raw material. The expensive downstream processing, low productivity, and the lack of proper understanding of the bioreactor systems used for their production are the major obstacles in case of biosurfactant production. Biosurfactants can be easily produced using cheap agro-industrial substrates and their market demand is high at the moment. Conversely, the cost of production is one such hurdle that prevents biosurfactants from capturing the chemical surfactants market.

Even though the natural biosurfactant-producing strains are widely distributed, the low-yield factor remains the matter of concern. Yield is an important parameter as it determines the expense of the substrate for any biotechnological product. In order to overcome this problem, many recombinants or GMOs are being created using transgenics targeting the very gene responsible for the production of biosurfactants. The genes responsible for biosurfactant production are placed under the regulation of strong promoters in non-pathogenic, heterologous hosts to augment production. These hyper-producing strains in combination with cheap raw material, optimized medium composition, and efficient downstream processing have to some extent made biosurfactants a commercially attractive and available product.

The current body of knowledge of biosurfactant production is more or less based on the genomic and proteomic studies of genera's *Bacillus* and *Pseudomonas*. The need of the hour is to explore the promising properties of other poorly researched microbes such as *Rhodococcus, Arthrobacter, Brevibacterium, Mycobacterim, Corynebacterium, Acinetobacter, Torulopsis, Candida, Pseudozyma, Yarrowia, Kluveromyces,* and *Saccharomyces* so as to achieve the required breakthrough for the biosurfactant mass production. Complete understanding of the molecular genetics of the hyper-producing strains in also necessary.

The conventional biosurfactants vis-à-vis chemical analogs are considered environmentally benign, human safe, and biodegradable. A recent study by Morita et al. (2013) shows the usage of glycolipid biosurfactants MELs produced by different basidiomycetous yeasts such as *Pseudozyma* in moisturization of dry skin, repair of damaged hair, and other skin care products and cosmetics. Similar cosmetic applications were confirmed by Williams (2009), where the study indicated the potential of sophorolipids in anti-aging, anti-wrinkle, skin smoothening, and acne treatment products. Singh and Cameotra (2004) applaud the application of microbial surfactants as antifungal, antiviral, antibacterial agents, on medical insertional materials, and probiotic preparations for urogenital infections. With the uncertainty and controversy surrounding the usage of GMOs in food and crops, the fate of biosurfactants in biomedical and as therapeutic or probiotic agents seems to be unpredictable. It is worth mentioning that the biosurfactants produced by natural biosurfactant producers are considered reliable; however, the ones produced by GMOs must go through the safety regulations to be used in products related to human contact, consumption, and healthcare.

TABLE 4.4
Some Major Applications of Biosurfactants

Applications	Role of Biosurfactants
Environmental	
Soil remediation and flushing	Emulsification through adherence to hydrocarbons, dispersion, foaming agent, detergent, soil flushing, crude oil-contaminated soil washing
Bioremediation	Emulsification of hydrocarbons, lowering of interfacial tension, metal sequestration, bioavailability and biodegradation of polyaromatic hydrocarbons and polychlorinated biphenyls, structural biofilm development
Agriculture	
Biocontrol	Facilitation of biocontrol mechanisms of microbes such as parasitism,
Phosphate fertilizers	antibiosis, competition, induced systemic resistance, and hypo-virulence
Spray application	Prevent caking during storage, wetting, dispersing, suspending of powered pesticides and emulsification of pesticide solutions, promote wetting, spreading and penetration of toxicant, useful as fungicides
Petroleum Production/Products	
Enhanced oil recovery	Improving oil drainage into well bore, stimulating release of oil entrapped by capillaries, wetting of solid surfaces, lowering of interfacial tension, dissolving of oil
De-emulsification	De-emulsification of oil emulsions, oil solubilization viscosity reduction, wetting agent
Drilling fluids	Emulsify oil, disperse solids, and modify rheological properties of drilling fluids for oil and gas wells
Biological Technology	
Microbiological	Physiological behavior such as cell mobility, cell communication, nutrient accession, cell–cell competition, plant, and animal pathogenesis
Biomedical and therapeutics	Antibacterial, antifungal, antiviral agents, antiadhesive agents to pathogens, probiotic agents, immuno modulatory molecules, vaccines, gene therapy, lyse erythrocytes, inhibit clot formation, lyse bacterial spheroplasts and protoplasts
Cosmetic	
Health and beauty products	Emulsifiers, foaming agents, solubilizers, wetting agents, cleansers, antimicrobial agents, mediators of enzyme action, moisturizer
Food and Beverages	
Emulsification and de-emulsification	Emulsifier, solubilizer, demulsifier, suspension, wetting, foaming, defoaming, thickener, lubricating agent
Functional ingredient	Interaction with lipids, proteins and carbohydrates, protecting agent For cleaning sanitizing
Food-processing plants	Improve removal of pesticides and in wax coating
Fruits and vegetables	Solubilize flavor oils, control consistency, retard staling
Bakery and ice cream	Improve washing, reduce processing time
Crystallization of sugar	Prevent spattering due to super heat and water
Elastomers/Plastics	
Emulsion polymerization	Solubilization, emulsification of monomers
Foamed polymers	Introduction of air, control of cell size
Latex adhesive	Promote wetting, improve bond strength
Plastic articles	Antistatic agents
Plastic coating/laminating	Wetting agents

TABLE 4.4 (continued)
Some Major Applications of Biosurfactants

Applications	Role of Biosurfactants
Industrial Cleaning	
Janitorial supplies	Detergents and sanitizers
De-scaling	Wetting agents and corrosion inhibitors in acid cleaning of boiler tubes and heat exchangers
Soft goods	Detergents for laundry and dry cleaning
Leather	
Skins	Detergent and emulsifier in degreasing
Tanning	Promote wetting and penetration
Hides	Emulsifiers in fat liquoring
Dyeing	Promote wetting and penetration

For example, if sodium lauryl sulfate used in toothpaste, soaps, hand lotions, and shampoos is to be replaced with biosurfactants from GMOs, safety standards must be taken into consideration. The knowledge of biosurfactants behavior/toxicity across different systems is vital before their use in various industries.

Apropos, the biosurfactant production process and the related innovations have come a long way, and the time is not far when the robust technology nailing down the production cost of these eco-friendly biomolecules will no longer be a dream.

REFERENCES

Abdel-Mawgoud, A.M., Lépine, F., and Déziel, E. 2010. Rhamnolipids: Diversity of structures, microbial origins and roles. *Appl. Microbiol. Biotechnol.* 86:1323–1336.

Abouseoud, M., Maachi, R., and Amrane, A. 2007. Biosurfactant production from olive oil by *Pseudomonas fluorescens. Comm. Cur. Res. Edu. Top. Trends App. Microbiol.*, A. Méndez-Vilas (Ed.) Formatex Research Center. Badajoz, Spain. 1:340–347.

Abu-Ruwaida, A.S., Banat, I.M., and Khamis, A. 1991. Nutritional requirements and growth characteristics of a biosurfactant producing *Rhodococcus* bacterium. *World J. Microbiol. Biotechnol.* 7:53–61.

Ajlani, M.A.M., Sheikh, M.A., Ahmad, Z., and Hasnain, S. 2007. Production of surfactin from *Bacillus subtilis* MZ-7 grown on pharmamedia commercial medium. *Microb. Cell. Fact.* 6:17.

Ali, R.Z., Saleem K.M., and Khalid, Z.M. 2007. Evaluation of distant carbon sources in biosurfactant production by a gamma-ray induced *Pseudomonas putida* mutant. *Process Biochem.* 42:686–692.

Amrane, A., Nabi, A., Abouseoud, M., Maachi, R., and Boudergua, S. 2008. Evaluation of different carbon and nitrogen sources in production of biosurfactant by *Pseudomonas fluorescens. Desalination.* 223:143–251.

Anandaraj, B., and Thivakaran, P. 2010. Isolation and production of biosurfactant producing organism from oil spilled soil. *Biosci. Tech.* 1(3):120–126.

Anyanwu, C.U., and Okolo, B.N. 2011. Effects of different carbon sources on biosurfactant production by a *Pseudomonas fluorescens* isolate. *Plant Prod. Res. J.* 15:7–11.

Aparna, A., Srinikethan, G., and Hegde, S. 2011. Effect of addition of biosurfactant produced by *Pseudomonas* sps. on biodegradation of crude oil. *2nd International Conference on Environmental Science and Technology IPCBEE.* Vol. 6, Singapore, IACSIT Press.

Arima, K., Kakinuma, A., and Tamura, G. 1968. Surfactin, a crystalline peptide lipid surfactant produced by *Bacillus subtilis*: Isolation, characterization and its inhibition of fibrin clot formation. *Biochem. Biophys. Res. Commun.* 31:488–494.

Banat, I.M. 1995. Biosurfactants production and possible uses in microbial enhanced oil recovery and oil pollution remediation: A review. *Biores Technol.* 51:1–12.

Banat, I.M. 2012. Isolation of biosurfactant-producing *Pseudomonas aeruginosa* RS29 from oil-contaminated soil and evaluation of different nitrogen sources in biosurfactant production. *Ann. Microbiol.* 62:753–763.

Bello, X.V., Devesa-Rey, R., Cruz, J.M., and Moldes, A.B. 2012. Study of the synergistic effects of salinity, pH, and temperature on the surface-active properties of biosurfactants produced by *Lactobacillus pentosus*. *J. Agric. Food Chem.* 60(5):1258–65.

Bergström, S., Theorell, H., and Davide, H. 1946a. On a metabolic product of *Ps. pyocyanea*. Pyolipic acid, active against *M. tuberculosis*. *Arkiv Chem. Mineral Geol.* 23A(13):1–12.

Bergström, S., Theorell, H., and Davide, H. 1946b. Pyolipic acid- A metabolic product of *Pseudomonas pyocyanea* active against *Mycobacterium tuberculosis*. *Arch. Biochem. Biophys.* 10:165–166.

Blank, L., Rosenau, F., Wilhelm, S., Wittgens, A., and Tiso, T. 2013. Means and methods for rhamnolipid production. European Patent No. EP2573172 A1.

Bodour, A.A., Guerrero-Barajas, C., Jiorle, B.V. et al. 2004. Structure and characterization of flavolipids, a novel class of biosurfactants produced by *Flavobacterium* sp. strain MTN11. *Appl. Environ. Microbiol.* 70(1):114–120.

Carrera, P., and Cosmina, P. 1993. Method of producing surfactin with the use of mutant of *Bacillus subtilis*. US Patent No. 5227294.

Casas, J.A., and Garcia-Ochoa, F. 1999. Sophorolipid production by *Candida bombicola*: Medium composition and culture methods. *J. Biosci. Bioeng.* 88(5):488–494.

Cha, M., Lee, N., Kim, M., Kim, M., and Lee, S. 2008. Heterologous production of *Pseudomonas aeruginosa* EMS1 biosurfactant in *Pseudomonas putida*. *Biores. Technol.* 99(7):2192–2199.

Chayabutra, C., Wu, J., and Ju, L.W. 2000. Rhamnolipid production by *Pseudomonas aeruginosa* under denitrification: Effect of limiting nutrients and carbon sources. *Biotechnol. Bioengg.* 72:25–33.

Ciapina, E.M.P., Melo, W.C., Santa Anna, L.M.M., Santos, A.S., Freire, D.M.G., and Pereira Jr, N. 2006. Biosurfactant production by *Rhodococcus erythropolis* grown on glycerol as sole carbon source. *Appl Biochem Biotechnol*. 880:129–132.

Cooper, D.G., MacDonald, C.R., Duff, S.J.B., and Kosaric, N. 1981. Enhanced production of surfactin from *Bacillus subtilis* by continuous product removal and metal cation additions. *Appl. Environ. Microbiol.* 42(3):408–412.

Cooper, D.G., and Paddock, D.A. 1983. *Torulopsis petrophilum* and surface activity. *Appl. Environ. Microbiol.* 46:1426–1429.

Crosman, J.T., Pinchuk, R.J., and Cooper, D.G. 2002. Enhanced biosurfactant production by Corynebacterium alkanolyticum ATCC 21511 using self-cycling fermentation. *J. Am. Oil Chem. Soc.* 79(5):467–472.

Davis, D.A., Lynch, H.C., and Varley, J. 2001. The application of foaming for the recovery of surfactin from *B. subtilis* ATCC 21332 cultures. *Enzyme Microbiol. Technol.* 28:346–354.

Desai, J.D., and Desai, A.J. 1993. Production of biosurfactants. In *Surfactant Science Series*, N. Kosaric (Ed.), Marcell Dekker Inc., New York. 48:62–97.

Deshpande, M., and Daniels, L. 1995. Evaluation of sophorolipid biosurfactant production by *Candida bombicola* using animal fat. *Biores. Technol.* 54(2):143–150.

Dogan, I., Pagilla, K.R., Webster, D.A., and Stark, B.C. 2006. Expression of *Vitreoscilla haemoglobin* in *Gordonia amarae* enhances biosurfactant production. *J. Ind. Microbiol. Biotechnol.* 33:693–700.

Donio, M.B.S., Ronica, F.A., Viji, V.T. et al. 2013. *Halomonas* sp. BS4, A biosurfactant producing halophilic bacterium isolated from solar salt works in India and their biomedical importance. *Springerplus*. 2(1):149.

Edwards, J.R., and Hayashi, J.A. 1965. Structure of a rhamnolipid from *Pseudomonas aeruginosa*. *Arch. Biochem. Biophys.* 111:415–421.

European Commission Press Release. 2014. http://europa.eu/rapid/press-release_IP-11-1542_en.htm

Fengxia, L. 2011. Breeding of high yield surfactin producing strain by a combined mutagenesis approach. *Food Sci.* 32(23):270–276.

Fonseca, R.R., Silva, A.J., De Franca, F.P. et al. 2007. Optimizing carbon/nitrogen ratio for biosurfactant production by a *Bacillus subtilis* strain. *Appl. Biochem. Biotechnol.* 137–140(1–12):471–86.

Ghribi, D., and Ellouze-Chaabouni, S. 2011. Enhancement of *Bacillus subtilis* lipopeptide biosurfactants production through optimization of medium composition and adequate control of aeration. *Biotechnol Res Intnl.* Article ID 653654.

Gong, G., Zheng, Z., Chen, H. et al. 2009. Enhanced production of surfactin by *Bacillus subtilis* E8 mutant obtained by ion beam implantation. *Food Technol. Biotechnol.* 47(1):27.

Guerra-Santos, L.H., Kappeli, O., and Fiechter, A. 1984. *Pseudomonas aeruginosa* biosurfactant production in continuous culture with glucose as carbon source. *Appl. Environ. Microbiol.* 48:301–305.

Gunther IV, N.W., Nunez, A., Fett, W., and Solaiman, D.K.Y. 2005. Production of rhamnolipids by *Pseudomonas chlororaphis*, a non-pathogenic bacterium. *Appl. Environ. Microbiol.* 71(5):2288–2293.

Haba, E., Bresco, O., Ferrer, C., Marques, A., Busquets, M., and Manresa, A. 2000. Isolation of lipase-screening bacteria by developing used frying oil as selective substrate. *Enzym. Microb. Technol.* 26:40–44.

Iqbal, S., Khalid, Z.M., and Malik, K.A. 1995. Enhanced biodegradation and emulsification of crude oil and hyperproduction of biosurfactants by a gamma ray-induced mutant of *Pseudomonas aeruginosa. Lett. Appl. Microbiol.* 21(3):176–179.

Jagtap, S., Yavankar, S., Pardesi, K., and Chopade, B. 2010. Production of bioemulsifier by *Acinetobacter* species isolated from healthy human skin. *Indian J. Exp. Biol.* 48(1):70–76.

Jarvis, F.G., and Johnson, M.J. 1949. A glyco-lipid produced by *Pseudomonas aeruginosa. J. Am. Chem. Soc.* 71:4124–4126.

Johri, A., Blank, W., and Kaplan, D. 2002. Bioengineered emulsans from *Acinetobacter calcoaceticus* RAG-1 transposon mutants. *Appl. Microbiol. Biotechnol.* 59(2–3):217–223.

Kaar, W., Hartmann, B.M., Fan, Y. et al. 2009. Microbial bio-production of a recombinant stimuli-responsive biosurfactant. *Biotechnol. Bioeng.* 102(1):176–187.

Kahraman, H., and Erenler, S.O. 2012. Rhamnolipid production by *Pseudomonas aeruginosa* engineered with the *Vitreoscilla hemoglobin* gene. *Prikl. Biokhim. Mikrobiol.* 48(2):212–217.

Kakinuma, A., Siguno, H., Isono, M., Tamura, G., and Arima, K. 1969. Determination of fatty acid in surfactin and elucidation of the total structure of surfactin. *Agric. Biol. Chem.*, 33:973–976.

Kim, S.H., Lim, E.J., Lee, S.O., Lee, J.D., and Lee, T.H. 2000. Purification and characterization of biosurfactants from *Nocardia* sp. L-417. *Biotechnol. Appl. Biochem.* 31:249–253.

Kiran, G., Anto Thomas, T., Selvin, J., Sabarathnam, B., and Lipton, A.P. 2010. Optimization and characterization of a new lipopeptide biosurfactant produced by marine *Brevibacterium aureum* MSA13 in solid state culture. *Bioresour. Technol.* 101:2389–2396.

Koch, A.K., Kappeli, O., Fiechter, A., and Reiser, J. 1991. Hydrocarbon assimilation and biosurfactant production in *Pseudomonas aeruginosa* mutants. *J. Bacteriol.* 17(13):4212–4219.

Koch, A.K., Reiser, J., Kappeli, O., and Fiechter, A. 1988. Genetic construction of lactose- utilizing strains of *Pseudomonas aeruginosa* and their application in biosurfactant production. *Nat. BioTechnol.* 6:1335–1339.

Kosaric, N., and Sukan, F.V. 2000. Biosurfactants. *Encyclopedia Microbiol.* 2nd ed. 1:618–635.

Kryachko, Y., Nathoo, S., Lai, P., Voordouw, J., Prenner, E.J., and Voordouw, G. 2013. Prospects for using native and recombinant rhamnolipid producers for microbially enhanced oil recovery. *Int. Biodet. Biodegrad.* 81:133–140.

Kuyukina, M.S., Ivshina, I.B., Philp, J.C., Christofi, N., Dunbar, S.A., and Ritchkova, M.I. 2001. Recovery of *Rhodococcus* biosurfactants using methyl tertiary-butyl ether extraction. *J. Microbiol. Methods.* 46:149–156.

Lee, S.C., Kim, S.H., Par, K,I.H., Chun, G.S.Y., and Choi, Y.l. 2007. Isolation and structural analysis of bamylocin A, novel lipopeptide from *Bacillus amyloliquefaciens* LP03 having antagonistic and crude oil-emulsifying activity. *Arch. Microbiol.* 188:307–312.

Lin, S.C., Carswell, K.S., Sharma, M.M., and Georgiou, G. 1994. Continous production of the lipopeptide biosurfactant of *Bacillus licheniformis* JF-2. *Appl. Microbiol. Biotechnol.* 41:281–285.

Lovaglio, R.B., Costa, S.G.V.A.O., Lima, C.J.B., Cortezi, M., and Contiero, J. 2010. Effect of C/N ratio and physicochemical conditions on the production of rhamnolipids by *Pseudomonas aeruginosa* LBI. *Res. J. Biotech.* 5(3):19–24.

Luna, J.M., Rufino, R.D., Campos-Takaki, G.M., and Sarubbo, L.A. 2012. Properties of the biosurfactant produced by *Candida sphaerica* cultivated in low-cost substrates. *Chem. Engg. Transac.* 27:67–72.

MacElwee, C.G., Lee, H., and Trevors, J.T. 1990. Production of extracellular emulsifying agent by *Pseudomonas aeruginosa* UG-1. *J. Ind. Microbiol.* 5:25–52.

Makkar, R.S., and Cameotra, S.S. 1997. Biosurfactant production by a thermophilic *Bacillus subtilis* strain. *J. Indust. Microbiol. Biotechnol.* 18:37–42.

Makkar, R.S., and Cameotra, S.S. 1998. Production of biosurfactant at mesophilic and thermophilic conditions by a strain of *Bacillus subtilis. J. Indust. Microbiol. Biotechnol.* 20:48–52.

Makkar, R.S., and Cameotra, S.S. 2002. An update on the use of unconventional substrates for biosurfactant production and their new applications. Mini review. *Appl. Microbiol. Biotechnol.* 58:428–434.

Mattei, G., and Bertrand, J.C. 1985. Production of biosurfactants by a mixed bacteria population grown in continuous culture on crude oil. *Biotechnol. Lett.* 7:217–222.

Mohan, P.K., Nakhla, G., and Yanful E.K. 2006. Biodegradability of surfactants under aerobic, anoxic, and anaerobic conditions. *J Environ. Engg.* 132(2):279–283.

Morita, T., Fukuoka, T., Imura, T., and Kitamoto, D. 2013. Production of mannosylerythritol lipids and their application in cosmetics. *Appl. Microbial. Biotechnol.* 97(11):4691–4700.

Morita, T., Koike, H., Koyama, Y. et al. 2013. Genome sequence of the basidiomycetous yeast *Pseudozyma antarctica* T-34, a producer of the glycolipid biosurfactants mannosylerythritol lipids. *Genome Announc.* 1(2):e00064–13.

Morita, T., Konishi, M., Fukuoka, T., Imura, T., and Kitamoto, D. 2006. Discovery of *Pseudozyma rugulosa* NBRC 10877 as a novel producer of the glycolipid biosurfactants, mannosylerythritol lipids, based on rDNA sequence. *Appl. Microbiol. Biotechnol.* 73:305–313.

Mukherjee, S., Das, P., and Sen, R. 2006. Towards commercial production of microbial surfactants. *Trend Biotech.* 24:509–515.

Muller, M.M., Hormann, B., Kugel, M., Syldatk, C., and Hausmann, R. 2011. Evaluation of rhamnolipid production capacity of *Pseudomonas aeruginosa* PAO1 in comparison to the rhamnolipid over-producer strains DSM 7108 and DSM 2874. *Appl. Microbiol. Biotechnol.* 89(3):585–592.

Muller, M.M., Hormann, B., Syldatk, C., and Hausmann, R. 2010. *Pseudomonas aeruginosa* PAO1 as a model for rhamnolipid production in bioreactor systems. *Appl. Microbiol. Biotechnol.* 87(1):167–174.

Mulligan, C.N., and Chow, T.Y. 1991. Enhanced production of biosurfactant through the use of a mutated *Bacillus subtilis* strain. US Patent No. 5037758.

Mulligan, C.N., and Gibbs, B.F. 1989. Correlation of nitrogen metabolism with biosurfactant production by *Pseudomonas aeruginosa*. *Appl. Environ. Microbiol.* 55:3016–3019.

Mulligan, C.N., and Sheppard, J.D. 1987. The production of surfactin by *Bacillus subtilis* grown on peat hydrolysate. *App. Microbiol. Biotechnol.* 27:110–116.

Nakano, M., Zuber, P., Corbell, N., and Besson, J. 1992. Isolation and characterization of *sfp:* A gene that functions in the production of the lipopeptide biosurfactant, surfactin, in *Bacillus subtilis*. *Mol. Gen. Genet.* 232:313–321.

Nakano, M.M., Magnuson, R., Myers, A., Curry, J., Grossman, and A.D., Zuber, P. 1991. *srfA* is an operon required for surfactin production, competence development, and efficient sporulation in *Bacillus subtilis*. *J. Bacteriol.* 173(5):1770–1778.

Nakano, M.M., and Zuber, P. 1989. Cloning and characterization of *srfB*, a regulatory gene involved in surfactin production and competence in *Bacillus subtilis*. *J. Bacteriol.* 171(10):5347–5353.

Nakayama, S., Takahashi, S., Hirai, M., and Shoda, M. 1997. Isolation of new variants of surfactin by a recombinant *Bacillus subtilis*. *Appl. Microbiol. Biotechnol.* 48:80–82.

Neu, T.R., and Poralla, K. 1990. Emulsifying agents from bacteria isolated during screening for cells with hydrophobic surfaces. *Appl. Microbiol. Biotechnol.* 32:521–525.

Neves, L.C.M.D., Miyamura, T.T.M.O., Kobayashi, M.J., Penna, T.C.V., and Converti, A. 2007. Production of biosurfactant by a genetically-modified strain of *Bacillus subtilis* expressing green florescent protein. *Ann. Microbiol.* 57(3):377–381.

Ochsner, U.A., Reiser, J., Fiechter, A., and Witholt, B. 1995. Production of *Pseudomonas aeruginosa* rhamnolipid biosurfactants in heterologous hosts. *Appl. Environ. Microbiol.* 61(9):3503–3506.

Ohno, A., Ano, T., and Shoda, M. 1995. Production of a lipopeptide antibiotic, surfactin, by recombinant *Bacillus subtilis* in solid state fermentation. *Biotechnol. Bioeng.* 47:209–214.

Onwosi, C.O., and Odibo, F.J. 2012. Effects of carbon and nitrogen sources on rhamnolipid biosurfactant production by *Pseudomonas nitroreducens* isolated from soil. *World J. Microbiol. Biotechnol.* 28(3):937–942.

Pacheco, G.J., Ciapina, E.M.P., Gomes, E.B., and Junior, N.P. 2010. Biosurfactant production by *Rhodococcus erythropolis* and its application to oil removal. *Brazilian J. Microbiol.* 41:685–693.

Pal, M.P., Vaidya, B.K., Desai, K.M., Joshi, R.M., Nene, S.N., and Kulkarni, B.D. 2009. Media optimization for biosurfactant production by *Rhodococcus erythropolis* MTCC 2794: Artificial intelligence versus a statistical approach. *J. Ind. Microbiol. Biotechnol.* 36:747–756.

Panesar, R., Panesar, P.S., and Bera, M.B. 2011. Development of low cost medium for the production of biosurfactants. *Asian J. Biotech.* 3(4):388–396.

Persson, A., Molin, G., Andersson, N., and Sjoholm, J. 1990. Biosurfactant yields and nutrient consumption of *Pseudomonas fluorescens* 378 studied in a microcomputer controlled multifermentation system. *Biotechnol. Bioengg.* 36:252–255.

Priya, T., and Usharani, G. 2009. Comparative study for biosurfactant production by using *Bacillus subtilis* and *Pseudomonas aeruginosa*. *Botany Res. Int.* 2(4):284–287.

Rahman, K.S.M., Rahman, T.J., Mc Clean, S., Marchant, R., and Banat, I.M. 2002. Rhamnolipid biosurfactant production by strains of *Pseudomonas aeruginosa* using low-cost raw materials. *Biotechnol. Prog.* 18(6):1277–1281.

Raza, Z.A., Khan, M.S., Khalid, Z.M., and Rehman, A. 2006. Production of biosurfactant using different hydrocarbons by *Pseudomonas aeruginosa* EBN-8 mutant. *Z. Naturforsch C.* 61(1–2):87–94.

Ron, E.Z., and Rosenberg, E. 2001. Natural roles of biosurfactants—Mini review. *Environ. Microbiol.* 3:229–236.

Roongsawang, N., Thaniyavarn, J., and Thaniyavarn, S. 2002. Isolation and characterization of a halotolerant *Bacillus subtilis* BBK-1 which produces three kinds of lipopeptides: Bacillomycin L, plipastatin, and surfactin. *Extremophiles* 6:499–506.

Rosenberg, E. 1986. Microbial surfactants. *CRC Crit. Rev. Biotechnol.* 3:109–132.

Rubina S., and Khanna S. 1995. Biosurfactants. *Ind. J. Microbiol.* 35:165–184.

Sadouk, Z., Hacene, H., and Tazerouti, A. 2008. Biosurfactants production from low cost substrate and degradation of diesel oil by a Rhodococcus strain. *Oil Gas Sci. Technol.* 63:747–753.

Salleh, S.M., Noh, N.A.M., and Yahya, A.R.M. 2011. Comparative study: Different recovery techniques of rhamnolipid produced by *Pseudomonas aeruginosa* USMAR-2. *International Conference on Biotechnology and Environment Management IPCBEE.* Vol. 18.

Santa Anna, L.M., Sebastian, G.V., Menezes, E.P., Alves, T.L.M., Santos, A.S., Pereira, N., and Freire, D.M.G. 2002. Production of biosurfactants from *Pseudomonas aeruginosa* PA1 isolated in oil environments. *Brazilian J. Chemic. Eng.* 19(2):159–166.

Sarachat, T.T., Pornsunthorntawee, O.O., Chavadej, S.S., and Rujiravanit, R.R. 2010. Purification and concentration of a rhamnolipid biosurfactant produced by *Pseudomonas aeruginosa* SP4 using foam fractionation. *Biores. Technol.* 101(1):324–330.

Satpute, S.K., Banpurkar, A.G., Dhakephalkar, P.K., Banat, I.M., and Chopade, B.A. 2010. Methods for investigating biosurfactants and bioemulsifiers: A review. *Crit. Rev. Biotechnol.* 30(2):127–144.

Seghal, K.G., Anto, T.T., Selvin, J., Sabarathnam, B., and Lipton, A.P. 2010. Optimization and characterization of a new lipopeptide biosurfactant produced by marine *Brevibacterium aureum* MSA13 in solid state culture. *Biores. Technol.* 101(7):2389–2396.

Sekhon, K.K., Khanna, S., and Cameotra, S.S. 2011. Enhanced biosurfactant production through cloning of three genes and role of esterase in biosurfactant release. *Micro. Cell Fact.* 10:49.

Sekhon, K.K., Khanna, S., and Cameotra, S.S. 2012. Biosurfactant production and potential correlation with esterase activity. *J. Pet Environ. Biotechnol.* 3:133.

Sekhon, K.K., Prakash, N.T., and Khanna, S. 2009. Cloning, expression and genetic regulation of a biosurfactant gene for bioremediation of hydrophobic chemical compounds. *J. Pure Appl. Microbiol.* 3(1):49–58.

Sen, R., and Swaminathan, T. 1997. Application of response surface methodology to evaluate the optimum environmental conditions for the enhanced production of surfactin. *Appl. Microbiol. Biotechnol.* 47:358–363.

Shavandi, M., Mohebalj, G., Haddadi, A., Shakarami, H., and Nuhi, A. 2011. Emulsification potential of a newly isolated biosurfactant-producing bacterium, *Rhodococcus* sp. strain TA6. *Colloids Surf. B. Biointerfaces* 82(2):477–482.

Shoda, M., Ohno, A., and Ano, T. 1992. Production of a lipopeptide antibiotic surfactin with recombinant *Bacillus subtilis*. *Biotech. Lett.* 14:1165–1168.

Singh, P., and Cameotra, S.S. 2004. Potential applications of microbial surfactants in biomedical sciences. *Trends Biotech.* 22(3):142–146.

Solaiman, D.K., Ashby, R.D., Nunez, A., and Foglia, T.A. 2004. Production of sophorolipids by *Candida bombicola* grown on soy molasses as substrate. *Biotechnol. Lett.* 26(15):1241–1245.

Stoimenova, E., Vasileva-Tonkova, E., Sotirova, A., Galabova, D., and Lalchev, Z. 2009. Evaluation of different carbon sources for growth and biosurfactant production by *Pseudomonas fluorescens* isolated from wastewaters. *Z. Naturforsch C.* 64(1–2):96–102.

Symmank, H., Franke, P., Saenger, W., and Bernhard, F. 2002. Modification of biologically active peptides: Production of a novel lipohexapeptide after engineering of *Bacillus subtilis* surfactin synthetase. *Protein Eng.* 15(11):913–921.

Tahzibi, A., Kamal, F., and Assadi, M.M. 2004. Improved production of rhamnolipids by a *Pseudomonas aeruginosa* mutant. *Iran. Biomed. J.* 8(1):25–31.

Thaniyavarn, J., Chongchin, A., Wanitsuksombut, N. et al. 2006. Biosurfactant production by *Pseudomonas aeruginosa* A41 using palm oil as carbon source. *J. Gen. Appl. Microbiol.* 52(2):215–22.

Thavasi, R., Jayalakshmi, S., Balasubramanian, T., and Banat, I.M. 2007. Biosurfactant production by *Corynebacterium kutscheri* from waste motor lubricant oil and peanut oil cake. *Lett. Appl. Microbiol.* 45(6):686–91.

Tsuge, K., Ano, T., and Shoda, M. 1995. Characterization of *Bacillus subtilis* YB8, coproducer of lipopeptides surfactin and plipastatin. *J Gen Appl Microbiol.* 41:541–545.

Tsuge, K., Ano, T., and Shoda, M. 1996. Isolation of a gene essential for biosynthesis of the lipopeptide antibiotics plipastatin B1 and surfactin in *Bacillus subtilis* YB8. *Arch. Microbiol.* 165:243–25.

Vasileva-Tonkoval, E., Galaboval, D., Karpenko, E., and Shulga, A. 2001. Biosurfactant—Rhamnolipid effects on yeast cells. *Lett. Appl. Microbiol.* 33:280–284.

Vilma, C., Saulius, G., Dovilė, S., and Egidijus, B. 2011. Production of biosurfactants by *Arthrobacter* sp. N3, a hydrocarbon degrading bacterium. *Proceedings of the 8th International Scientific and Practical Conference.* Rezekne, Latvia, 1:68–75.

Wang, Q., Fang, X., Bai, B. et al. 2007. Engineering bacteria for production of rhamnolipid as an agent for enhanced oil recovery. *Biotechnol. Bioeng.* 98(4):842–853.

Wei, Y.H., and Chu, I.M. 2002. Mn^{2+} improves surfactin production by *Bacillus subtilis. Biotechnol. Lett.* 24:479–482.

Williams, K. 2009. Biosurfactants for cosmetic application: Overcoming production challenges. *MMG 445 Basic Biotechnol.* 5:78–83.

Wittgens, A., Tiso, T., Arndt, T.T. et al. 2011. Growth independent rhamnolipid production from glucose using the non-pathogenic *Pseudomonas putida* KT2440. *Microb. Cell Fact.* 10:80.

Yakimov, M.M., Fredrickson, H.L., and Timmis, K.N. 1996. Effect of heterogeneity of hydrophobic moieties on surface activity of lichenysin A, a lipopeptide biosurfactant from *Bacillus licheniformis* BAS50. *Biotechnol. Appl. Biochem.* 23(1):13–18.

Yao, S., Gao, X., Fuchsbauer, N., Hillen, W., Vater, J., and Wang, J. 2003. Cloning, sequencing, and characterization of the genetic region relevant to biosynthesis of the lipopeptides Iturin A and Surfactin in *Bacillus subtilis. Curr. Microbiol.* 47:272–277.

Yeh, M.S., Wei, Y.H., and Chang, J.S. 2005. Enhanced production of surfactin from *Bacillus subtilis* by addition of solid carriers. *Biotechnol. Prog.* 21:1329–1334.

Yeh, M.S., Wei, Y.H., and Chang, J.S. 2006. Bioreactor design for enhanced carrier-assisted surfactin production with *Bacillus subtilis. Process Biochem.* 41:1799–1805.

Yoneda, T., Miyota, Y., Furuya, K., and Tsuzuki, T. 2006. Production process of surfactin. US patent 7011969.

5 Production of Biosurfactants from Nonpathogenic Bacteria

Roger Marchant, S. Funston, C. Uzoigwe, P.K.S.M. Rahman, and Ibrahim M. Banat

CONTENTS

5.1 Introduction 73
5.2 False Claims and the Need to Standardize Analytical Methods...74
5.3 Rhamnolipids...75
5.4 Other Glycolipids...77
5.5 Lipopeptide Biosurfactants...78
5.6 High Molecular Mass Bioemulsifiers 78
5.7 Concluding Remarks 79
Acknowledgment 79
References...79

5.1 INTRODUCTION

The annual worldwide consumption of chemical surfactants is estimated to be of the order of 13 million tons (Marchant and Banat 2012a); if biosurfactants can be used to replace even a small proportion of this market, the resulting production capacity would need to be very large (Marchant and Banat 2012b). As a result of this need for large-scale production, certain characteristics of the microbial biosurfactant-producing systems have become very important. These include substrate costs, product yield, and the potential costs of downstream processing, particularly where a mixture of congeners with different functionalities are produced. A wide range of biosurfactant-producing microorganisms have already been identified and investigated; these include bacteria, yeasts, and fungi. The majority of these organisms have little future as industrial scale producers, mainly due to the low yields of product that can be obtained in fermentations.

The outstanding exceptions to this are yeasts that produce glycolipids such as sophorolipids (see Chapter 2 by van Bogaert et al.) and mannsylerythritol lipids (MELs) (see Konishi et al. 2011). These are nonpathogenic organisms that give yields of 100–400 g/L of biosurfactant, which can be relatively easily harvested and can be produced from readily available feedstock sources such as rapeseed oil or soya oil. Because of the positive features of the glycolipid-producing systems, sophorolipids containing products are already constituents of cleaning, laundry, and personal care products that use these organisms in a number of countries worldwide. Despite this established use, the search continues to develop production systems for other biosurfactants with different functionalities in commercial products. One prime candidate already identified is rhamnolipid and it is in this area that much research activity is currently focused. Unfortunately, the literature in this area is beset by a large number of publications that do not meet minimal standards that would allow them to make a solid contribution to knowledge. The main problems focus around the often insecure methods used to identify the bacteria being used, that is, a lack of molecular identification; the inappropriate and incomplete methods used to isolate and characterize the biosurfactants produced; and no critical evaluation of yields. This chapter will not attempt to review every citation for a nonpathogenic bacterial production of biosurfactant, but will concentrate on surfactants and organisms that have been rigorously characterized and for which there is a real prospect of further exploitation in the future.

5.2   FALSE CLAIMS AND THE NEED TO STANDARDIZE ANALYTICAL METHODS

One of the main challenges currently faced in the area of rhamnolipid production is the need for a nonpathogenic alternative to *Pseudomonas aeruginosa* that can be used on an industrial scale. In recent years, there has been an increasing amount of research carried out into the screening, identification, and analysis of potential nonpathogenic rhamnolipid producers. One of the major issues that has recently become evident in this area is the publication of false claims of rhamnolipid production in nonpathogenic bacteria. This is primarily due to poor selection of analytical techniques, screening methods, and bacterial strain identification.

When screening potential nonpathogenic rhamnolipid-producing bacteria, it seems that a number of research groups tend to choose analytical methods, or only have available methods, that are not optimal to produce both accurate qualitative and quantitative analysis of the specific rhamnolipid congeners produced. Another problem that has been identified in this area of research is the lack of molecular biology techniques both in the identification of specific bacterial strains and in the identification of genes involved in rhamnolipid biosynthesis (i.e., *rhlA*, *rhlB*, and *rhlC* orthologs). On a number of occasions, our laboratory has discovered that when specific strains have been requested and received from other research groups, the 16S rRNA identification of these organisms has shown that they are not the species originally claimed by the authors. Another problem that we have encountered is that when the strain has been correctly identified, the results for rhamnolipid production from the original publication cannot be replicated, indicating that the claims made in the original paper may have been incorrect. On one occasion, some simple bioinformatics alignments showed that the strain in question completely lacked orthologs of the *rhlB* and *rhlC* genes, suggesting that claims of rhamnolipid production in the original publication were highly unlikely.

One way in which these incorrect claims can be eliminated would be to standardize the techniques used to identify and analyze rhamnolipid production in nonpathogenic bacteria. This would start with correct strain identification using polymerase chain reaction to amplify and sequence the 16S rRNA gene using the universal 27F/1492R primer set. Sequencing of other well-known housekeeping genes would also be acceptable as long as a significant-sized fragment was amplified (>1400 bp), sequenced, and a definitive BLAST alignment was obtained against a complete or draft genome sequence of the strain in question (Tayeb et al. 2005, 2008; Perfumo et al. 2013).

Once the specific strain has been correctly identified, specific techniques should be used to extract rhamnolipids from the culture media as well as provide a qualitative and quantitative assessment of the rhamnolipid congeners produced. Current techniques used to generate a crude extract containing rhamnolipids can vary; however, the acid precipitation and solvent extraction method described by Smyth et al. (2010a) has been shown to be an effective one. Once rhamnolipids have been extracted, accurate qualitative and quantitative analysis should be carried out using high-performance liquid chromatography or ultra-performance liquid chromatography coupled with mass spectrometry (MS) in a process commonly referred to as high-performance liquid chromatography/MS or ultra-performance liquid chromatography/MS. This allows for the specific identification of each rhamnolipid congener present in the crude extract as well as effective separation and quantification to provide an accurate individual and overall yield of rhamnolipid production. It is with these standardized methods of identification and analysis in place that false claims of rhamnolipid production and the overestimation of product yields can be eradicated in this emerging area of biosurfactant research.

5.3 RHAMNOLIPIDS

There have been several nonpathogenic bacterial strains reported to produce rhamnolipids, however, to date there have been no reports of strains producing this biosurfactant at a rate similar to *P. aeruginosa*. In recent times, however, the most-promising candidates have been identified as members of the closely related genus *Burkholderia*. Dubeau et al. (2009) reported that *Burkholderia thailandensis* E264, a nonpathogenic soil bacterium first isolated from the soils and stagnant waters of central and north eastern Thailand, carries two identical gene operons containing orthologs of the *rhlA*, *rhlB*, and *rhlC* genes. This bacterium is closely related to the pathogenic bacterium *B. pseudomallei*; however, *B. thailandensis* was shown to be nonpathogenic (Brett et al. 1998; Koh et al. 2012). Further experiments showed that this bacterium is capable of producing rhamnolipids, with both operons contributing to rhamnolipid biosynthesis. When glycerol and canola oil were used as a carbon source (at concentrations of 4%) rhamnolipid yields of 419.10 and 1473 mg/L were obtained, respectively. However, yields of more than 5 g/L have already been obtained in our laboratory with this organism, using glycerol as substrate (unpublished results). In addition to this, it was shown that the rhamnolipid congeners produced by this strain contained longer 3-hydroxy fatty acid side chain moieties, with the di-rhamnolipid Rha-Rha-C_{14}-C_{14} being produced most abundantly (Dubeau et al. 2009).

Current work with this strain has shown that unlike the complex mixture of congeners known to be produced by *P. aeruginosa*, there are relatively few congeners produced by *B. thailandensis*. This provides a significant advantage when it comes to the downstream purification and separation processes. In addition, it seems that *B. thailandensis* produces a product consisting of mostly di-rhamnolipid, with levels of each respective pre-cursor mono-rhamnolipid only observed in very low and trace quantities. One explanation for this could be that the rhamnosyltransferase II enzyme (encoded by *rhlC*) in *B. thailandensis* may have a significantly higher affinity for its substrate than the same enzyme in *P. aeruginosa*, resulting in all precursor mono-rhamnolipids being very quickly converted into di-rhamnolipids.

Another area that shows promise for *B. thailandensis* E264 is in the regulation of its rhamnolipid biosynthesis genes. In *P. aeruginosa,* rhamnolipid production is very tightly controlled by quorum-sensing systems involving transcriptional and posttranscriptional regulation factors (Reis et al. 2011). This makes the manipulation and overproduction of rhamnolipids very difficult as there are many contributing factors involved. In *B. thailandensis*, however, it seems that rhamnolipid production may not be controlled by such a complex system. Current work suggests that each rhamnolipid-producing operon may be controlled by a separate promoter region and that the expression of each operon could potentially be optimized and controlled individually (unpublished results). It has been shown that there are many genes controlled by quorum sensing in *B. thailandensis*, but whether rhamnolipid production is controlled in this way has yet to be determined (Ulrich et al. 2004).

In addition to *B. thailandensis*, rhamnolipid production has been reported in several other species of *Burkholderia* (Table 5.1). Tavares et al. showed that the nonpathogenic *B. kururiensis* was capable of rhamnolipid production at a rate of 0.78 g/L using glycerol as a carbon source (at a concentration of 3%). *B. kururiensis* was found to have the ability to degrade trichloroethylene and was originally isolated from an aquifer contaminated with trichloroethylene in Japan (Zhang et al. 2000). Unlike *B. thailandensis*, however, *B. kururiensis* was shown to produce a wide range of rhamnolipid congeners (mostly di-rhamnolipids) with varying-sized lipid moieties ranging from C_8-C_8 to C_{10}-C_{18}. This group then went on to produce a genetically modified strain of *B. kururiensis* in which they had inserted copies of the *rhlA* and *rhlB* genes isolated from *P. aeruginosa*. The resulting mutant strain, *B. kururiensis* LMM21, showed a significant increase in rhamnolipid production with the yield increasing from 0.78 to 5.67 g/L under the same growth conditions. This mutant strain was found to produce different rhamnolipid congeners to the wild-type strain with mostly mono-rhamnolipids being synthesized containing slightly shorter lipid moieties ranging from C_8-C_8 to C_{14}-C_{14}. Although there was a significant increase in the yield, the fact that such a wide range of congeners

TABLE 5.1

Burkholderia **Species Reported to Produce Rhamnolipids Including Carbon Sources, Yields Reported, and Most Frequent Congeners**

Burkholderia spp.	Carbon Source	Yield of Rhamnolipid Production	Most Abundant Congeners Produced	CMC Values	Source
Burkholderia thailandensis E264	Glycerol Canola oil	419.10 mg/L 1473 mg/L	Rha-Rha-C_{14}-C_{14} Rha-Rha-C_{14}-C_{14}	225 mg/L	Dubeau et al. (2009)
B. kururiensis KP23	Glycerol	780 mg/L	High variation (Mostly di-rhamnolipids)	180–220 mg/L	Tavares et al. (2013)
B. kururiensis LMM21 (engineered strain)	Glycerol	5670 mg/L	High variation (Mostly mono-rhamnolipids)	80–100 mg/L	Tavares et al. (2013)
B. plantarii DSM9509	Glucose	45.74 mg/L	Rha-Rha-C_{14}-C_{14}	15–20 mg/L (95% pure Rha-Rha-C_{14}-C_{14})	Hoermann et al. (2010)
B. glumae AU6208	Canola oil	1000.7 mg/L	High variation (mono- and di-rhamnolipids)	25–27 mg/L	Costa et al. (2011)

are produced may mean that downstream purification of individual congeners may be problematic if this bacterium were to be used on an industrial scale.

Another *Burkholderia* species that was shown to produce rhamnolipids is *B. plantarii*. This Gram-negative bacterium, first isolated in 1987, is found to cause seedling blight in rice plants; however, there have been no reports of it being pathogenic to humans or animals (Azegami et al. 1987). *B. plantarii* was grown in a parallel bioreactor system in nutrient broth supplemented with 10 g/L glucose, and rhamnolipid production was monitored over a period of 30 h. Results from this showed that, in a similar way to that of *B. thailandensis*, *B. plantarii* produces the di-rhamnolipid Rha-Rha-C_{14}-C_{14} in most abundance with other rhamnolipid congeners being produced in trace quantities. The yield of rhamnolipid produced by *B. plantarii*, however, was quite low at 45.74 mg/L with a CMC value of 15–20 mg/L (Hoermann et al. 2010). Although the rhamnolipid production yield of *B. plantarii* is quite low compared with other rhamnolipid-producing bacteria, the fact that a product of approximately 95% pure Rha-Rha-C_{14}-C_{14} was obtained from the culture medium is extremely encouraging. If growth conditions can be optimized to increase the yield, this bacterium may be a promising candidate for industrial application.

In a similar case to *B. plantarii*, the closely related *B. glumae* is also a pathogen of plants but is nonpathogenic to humans and animals. Work by Costa et al. (2011) showed that *B. glumae* can produce a wide variety of both mono- and di-rhamnolipid congeners with long chain lipid moieties ranging from C_{12}-C_{12} to C_{16}-C_{16}. When grown in a minimal medium supplemented with urea at 0.1 mol/L (as a nitrogen source) and canola oil 2% (as a carbon source), *B. glumae* produced a yield of 1000.7 mg/L after 144 h, which had a CMC value of 25–27 mg/L. The specific rhamnolipid congeners produced by *B. glumae*, however, were quite varied with the di-rhamnolipids Rha-Rha-C_{14}-C_{14} (32.68%), Rha-Rha-C_{12}-C_{14} (24.31%), and Rha-Rha-C_{12}-C_{12} (22.3%) found in most abundance among an array of other mono- and di-rhamnolipid congeners (Costa et al. 2011).

As we have now seen, some of the most promising nonpathogenic rhamnolipid-producing bacteria come from the genus *Burkholderia*. One of the main differences observed between these species of *Burkholderia* and the widely studied *P. aeruginosa* is in the structure of the rhamnolipids produced, primarily in the length of the lipid moieties. Although the exact functional differences between the long chain rhamnolipids (produced by *Burkholderia*) and the shorter chain rhamnolipids (produced by *P. aeruginosa*) have not yet been defined, there is some evidence to suggest that due to the increased hydrophobicity of the longer chain rhamnolipids, micelle formation occurs at a lower concentration and therefore results in a lower CMC value (Dubeau et al. 2009). Recently,

Wittgens et al. (2011) have cloned the *rhl A* and *rhlB* genes from *P. aeruginosa* strain PAO1 into the nonpathogenic *Pseudomonas putida* to give a strain able to produce the C_{10}-C_{10} monorhamnolipid typical of *P. aeruginosa* using glucose as the carbon source. These workers were able to show that *P. putida* is able to grow uninhibited at concentrations of rhamnolipid >90 g/L, but the actual production of rhamnolipid by the engineered strain was only 0.22 g/L, which increased to 1.5 g/L when the PHA pathway was knocked out. Although current work in this area is promising, further work is required to find and optimize rhamnolipid production in nonpathogenic bacteria to a level where industrial application would be feasible and cost effective.

5.4 OTHER GLYCOLIPIDS

Apart from the glycolipids produced by yeast and fungal species together with the rhamnolipids produced by bacteria, there are several other nonpathogenic bacterial producers. Perhaps, the best studied of these are the trehalolipids produced by species of *Rhodococcus*. In common with other biosurfactant producers, rhodococci produce a range of different congeners (Lang and Philp 1998) with different properties and, in the case of these bacteria, they are produced in response to different environmental and growth conditions. *Rhodococcus* species are among the most active hydrocarbon-degrading bacteria and when grown in a liquid culture the bacterial cells accumulate at the interface of the oil and water phases, which is a clear indication of the extreme hydrophobicity of the cell surface. The trehalose lipids that are produced by these organisms comprise a trehalose sugar moiety with a fatty acid portion that varies greatly in chain length from C_{20} to C_{90} and they remain predominantly anchored to the cell wall producing the extreme hydrophobicity of the cells. The structure of these trehalose lipids means that self-assembly in the liquid phase would not be favored; however, it has been estimated that about 10% is released into the growth medium to aid in the preliminary emulsification of the oil substrate (Lang and Philp 1998).

In addition to the nonionic biosurfactants produced by rhodococci (e.g., trehalose mycolates and derivatives), trehalose lipids can be synthesized in an ionic form as tetraesters (Rapp and Gabriel-Jürgens 2003), which are preferentially secreted into the medium to contribute to the emulsification of the oil substrate. Biosurfactants similar to those produced by rhodococci are also produced by the related organisms, *Corynebacterium*, *Mycobacterium*, *Nocardia*, and *Gordonia*, and it has been suggested that in *Gordonia* the cells are able to switch the access to oil substrates in response to growth conditions by modifying the type of glycolipids they produce, thus adjusting the hydrophobicity of the cell surface or the degree of emulsification of the substrate (Franzetti et al. 2008). It is clear from the studies carried out with all these active hydrocarbon-degrading organisms that they produce a wide range of surfactant molecules that function either anchored to the cell surface, to allow access to large oil drops, or transported outside the cell, to emulsify the substrate (Perfumo et al. 2010). However, as much of the glycolipid synthesized remains firmly attached to the cell wall, downstream processing to yield pure trehalolipid is likely to be difficult and costly, which suggests that these molecules are unlikely to become a major focus of industrial interest in the near future.

Although the majority of biosurfactant-producing microorganisms have been isolated from nonmarine sources, there are active hydrocarbon-degrading organisms that can produce glycolipid biosurfactants in the oceans. When spills of oil and hydrocarbons occur in the ocean, they stimulate the indigenous population of obligate hydrocarbonoclastic bacteria to become the dominant portion of the microbial community (Yakimov et al. 2007). There are very few genera of bacteria contributing to the hydrocarbon-degrader population with *Alcanivorax borkumensis* being the predominant organism. *A. borkumensis* can utilize alkanes as its sole carbon source while producing glycolipid biosurfactants. The major glycolipid molecule produced by this organism comprises a glucose residue linked to a tetrameric chain of fatty acids of C_6–C_{10} length (Yakimov et al. 1998), which is both cell-associated and secreted. Whether the glycolipid from *A. borkumensis* has potential for exploitation remains to be investigated, but the fact that at least a proportion is cell-associated suggests that there may be difficulties.

5.5   LIPOPEPTIDE BIOSURFACTANTS

Lipopeptide biosurfactants produced by various bacteria have been investigated for many years; they have a major ability to interact with cell membranes, giving them antimicrobial properties, but in the presence of hydrophobic compounds they can be effective surfactants. Members of the genus *Bacillus* are prominent producers of lipopeptide biosurfactants and their genome contains a large operon (*srfA*) responsible for lipopeptide synthesis interacting with non-ribosomal peptide synthetases. This situation accounts for the heterogeneity of molecular structures and activity that are produced. The basic structure of the lipopeptides comprises a cyclic peptide of 7–10 amino acids linked to a fatty acid tail. The individual amino acids, sequence and peptide cyclization, and the type, length (C_{13}—C_{18}), and branching of the fatty acid chain are all variables in a wide range of isoforms (Ongena and Jacques 2007). *Bacillus subtilis* is an ubiquitous Gram-positive rod-shaped bacterium found commonly in water, soil, and air, which contributes to nutrient cycling in the environment. This organism is industrially useful as it is one of the most widely used bacteria in the production of enzymes (amylases, proteases, inosine, ribosides, and amino acids) and speciality chemicals including biosurfactants (Erikson 1976). This bacterium has also been shown to produce a variety of antibacterial (Katz and Dermain 1977) and antifungal (Korzybski et al. 1978) compounds, including difficidin and oxydifficidin, with wide spectrum antibiotic activities against aerobic and anaerobic bacteria (Zimmerman et al. 1987).

Although *B. subtilis* has been associated with outbreaks of food poisoning (Gilbert et al. 1981; Kramer et al. 1982) and human infections, especially in hospitalized patients with surgical wounds, breast cancer, and leukemia (Logan 1988), the exact nature of its involvement in infections has not been established. This organism has, therefore, been classified as neither a human (Edberg 1991) nor plant (Claus and Berkeley 1986) pathogen, and is considered a class 1 containment agent, and industrial use in fermentation processes presents low risk of adverse effects to human health and environment. *B. subtilis* produces an effective and active cyclic lipopeptide biosurfactant known as surfactin (Cooper and Goldernberg 1987; Peypoux et al. 1999). Das and Mukherjee (2007) have also reported the production of lipopeptide surfactants by two strains of *B. subtilis* DM-03 and *B. subtilis* DM-04 on potato peels using both submerged and solid state fermentation techniques; however, the quantities produced were relatively small.

As with all biosurfactants, yield and cost of production will be critical in determining whether commercial exploitation will take place. Yeh et al. (2005) have reported an enhanced yield (36-fold) of surfactin using *B. subtilis* ATCC 21332 through the incorporation of activated carbon in the growth medium. Yields of 3.6 g/L of surfactin were achieved and the product reduced the surface tension of water to 27 mN/m with a CMC of 10 mg/L. Surfactin has been demonstrated as one of the most effective biosurfactants with greater efficiency in bioremediation and *in situ* microbial enhanced oil recovery than rhamnolipid and therefore may have potential for large-scale bio-industrial development.

5.6   HIGH MOLECULAR MASS BIOEMULSIFIERS

The biosurfactants we have considered in the early part of this chapter can be considered as relatively low-molecular mass molecules; however, many bacteria produce structurally complex biopolymers, polysaccharides, proteins, lipopolysaccharides, lipoproteins, and mixtures of these components that are amphiphilic (Smyth et al. 2010b). As these molecules may have many reactive groups exposed, they can bind tightly to hydrophobic molecules, making them very effective emulsifiers (Ron and Rosenberg 2002). The most extensively researched organisms are species of *Acinetobacter*, which produce biosurfactants such as emulsan and alasan. Unfortunately, members of this genus are often opportunistic human pathogens, placing them in the same category as *P. aeruginosa* and making them unwelcome additions to the manufacturing arsenal of microorganisms.

Less is known about the other polymeric emulsifiers produced by bacteria; however, there are a number of producers of exopolysaccharides (EPS) such as the sphingans produced by species of

Sphingomonas; but once again these are causative agents of nosocomial infections. A more promising source of EPS is the moderately halophilic bacterium *Halomonas eurihalina* (a safety class 1 organism) that produces an EPS of variable composition depending on the substrate on which it is growing. The EPS produced in the presence of hydrocarbons has less carbohydrate and protein and more uronic acids, acetyls, and sulfates (Calvo et al. 2002). The uronic acids in particular may be involved in the detoxification of hydrocarbons.

5.7 CONCLUDING REMARKS

Biosurfactant production by bacteria is distributed through many genera and the types of molecules produced are extremely varied. The role of these biosurfactants in the growth, physiology, and pathogenicity of bacteria is not fully understood, but it is clear that they play a part in accessing hydrophobic substrates, biofilm formation and maintenance, motility, and probably have many more functions. From an industrial exploitation viewpoint, the key initial consideration is whether a particular biosurfactant molecule has appropriate functionality in the product environment identified. The diversity of structure of biosurfactants probably ensures that the correct functionality exists somewhere in the spectrum; however, functionality is only the first step along the path to exploitation. Yield, cost, and ease of downstream processing then become the driving parameters. Unfortunately, many biosurfactants are produced in very low amounts (mg/L) and in complex mixtures that are difficult to separate. The biosurfactants that have already reached the market place in product formulations, for example, sophorolipids, are distinguished by the fact that fermentation yields can be measured in tens or hundreds of g/L, and this must be the goal for any aspiring new biosurfactant to reach the market place. The key to achieving success for an effective production process probably lies in a full understanding of the regulatory processes controlling the production of biosurfactant molecules in each organism.

ACKNOWLEDGMENT

The authors from Teesside University wish to thank the Commonwealth Commission for financial support for this project.

REFERENCES

Azegami, K., Nishiyama, K., Watanabe, Y., Kadota, I., Ohuchi, A., and Fukazawa, C. 1987. *Pseudomonas-plantarii* sp. nov., the causal agent of rice seedling blight. *Int. J. Syst. Bact.* 37(2): 144–152.

Brett, P., DeShazer, D., and Woods, D. 1998. *Burkholderia thailandensis* sp. nov., a *Burkholderia pseudomallei*-like species. *Int. J. Syst, Bact.* 48: 317–320.

Calvo, C., Martinez-Checa, F., Toledo, F.L., Porcel, J., and Quesada, E. 2002. Characteristics of bioemulsifiers synthesized in crude oil media by *Halomonas eurihalina* and their effectiveness in the isolation of bacteria able to grow in the presence of hydrocarbons. *Appl. Microbiol. Biotechnol.* 60(3): 347–351.

Claus, D. and Berkeley, R.C.W. 1986. Genus Bacillus Cohn 1872. In: P.H.A. Sneath et al. (eds.), *Bergey's Manual of Systematic Bacteriology*, Vol. 2. Williams & Wilkins Co., Baltimore, MD, pp. 1105–1139.

Cooper, D.G. and Goldernberg, B.G. 1987. Surface active agents from two *Bacillus* species. *Appl. Environ. Microbiol.* 53(2): 224–229.

Costa, S.G.V.A.O., Deziel, E., and Lepine, F. 2011. Characterization of rhamnolipid production by *Burkholderia glumae*. *Lett. Appl. Microbiol.* 53(6): 620–627.

Das, K. and Murkherjee, A.K. 2007. Differential utilization of pyrene as the sole source of carbon by *Bacillus subtilis* and *Pseudomonas aeruginosa* strains. Role of biosurfactants in enhancing bioavailability. *J. Appl. Microbiol.* 102: 195–203.

Dubeau, D., Deziel, E., Woods, D.E., and Lepine, F. 2009. *Burkholderia thailandensis* harbors two identical rhl gene clusters responsible for the biosynthesis of rhamnolipids. *BMC Microbiol.* 9: 263.

Edberg, S.C. 1991. *US EPA Human Health Assessment: Bacillus subtilis*. U.S. Environmental Protection Agency, Washington, DC.

Erikson, R.J. 1976. Industrial applications of the bacilli: A review and prospectus. In: D. Schlesinger (ed.), *Microbiology.* American Society for Microbiology, Washington, DC, pp. 406–419.

Franzetti, A., Bestetti, G., Caredda, P., La Colla, P., and Tamburini, E. 2008. Surface-active compounds and their role in the access to hydrocarbons in *Gordonia* strains. *FEMS Microbiol. Ecol.* 63: 238–248.

Gilbert, R.J., Turnbull, P.C.B., Parry, J.M., and Kramer, J.M. 1981. *Bacillus cereus* and other *Bacillus* species: Their part in food poisoning and other clinical infections. In: Berkeley R.C.W., Goodfellow M. (eds.), *The Aerobic Endospore-Forming Bacteria; Classification and Identification.* Academic Press, London, pp. 297–314.

Hoermann, B., Mueller, M.M., Syldatk, C., and Hausmann, R. 2010. Rhamnolipid production by *Burkholderia plantarii* DSM 9509(T). *Europ. J. Lipid Sci. Technol.* 112(6): 674–680.

Katz, E. and Demain, A.C. 1977. The peptide antibiotics of *Bacillus*: Chemistry, biogenesis, and possible functions. *Bacteriol. Rev.* 41: 449–474.

Koh, S.F., Tay, S.T., Sermswan, R., Wongratanacheewin, S., Chua, K.H., and Puthucheary, S.D. 2012. Development of a multiplex PCR assay for rapid identification of *Burkholderia pseudomallei, Burkholderia thailandensis, Burkholderia mallei* and *Burkholderia cepacia* complex. *J. Microbiol. Methods* 90: 305–308.

Konishi, M., Nagahama, T., Fukuoka, T., Morita, T., Imura, T., Kitamoto, D., and Hatada, Y. 2011. Yeast extract stimulates production of glycolipid biosurfactants, mannosylerythritol lipids, by Pseudozyma hubeiensis SY62. *J Bioscience & Bioengineering* 111: 702–705.

Korzybski, T., Kowszyk-Gindifer, Z., and Kurylowicz, W. 1978. Antibiotics isolated from the genus *Bacillus* (Bacillaceae). In: *Antibiotics—Origin, Nature and Properties*, Vol. III. American Society of Microbiology, Washington, DC, pp. 1529–1661.

Kramer, J.M., Turnbull, P.C.B., Munshi, G., and Gilbert, R.J. 1982. Identification and characterization of *Bacillus cereus* and other *Bacillus* species associated with foods and food poisoning. In: Corry J. E. L., Robert D., and Skinner F.A. (eds.), *Isolation and Identification Methods for Food Poisoning Organisms.* Academic Press, London, pp. 261–286.

Lang, S. and Philp, J.C. 1998. Surface-active lipids in rhodococci. *Antonie van Leeuwenhoek* 74: 59–70.

Logan, N.A. 1988. Bacillus species of medical and veterinary importance. *J. Med. Microbiol.* 25: 157–165.

Marchant, R. and Banat, I.M. 2012a. Microbial biosurfactants: Challenges and opportunities for future exploitation. *Trends Biotechnol.* 30: 558–565.

Marchant, R. and Banat, I.M. 2012b. Biosurfactants: A sustainable replacement for chemical surfactants? *Biotechnol. Lett.* 34: 1597–1605.

Ongena, M. and Jacques, P. 2007. *Bacillus* lipopeptides: Versatile weapons for plant disease biocontrol. *Trends Microbiol.* 16: 115–125.

Perfumo, A., Rudden, M., Smyth, T.J.P., Marchant, R., Stevenson, P.S., Parry, N.J., and Banat, I.M. 2013. Rhamnolipids are conserved biosurfactants molecules: Implications for their biotechnological potential. *Appl. Microbiol. Biotechnol.* 97: 7297–7306.

Perfumo, A., Smyth, T.J.P., Marchant, R., and Banat, I.M. 2010. Production and roles of biosurfactants and bioemulsifiers in accessing hydrophobic substrates. In: K.N. Timmis (ed.), *Handbook of Hydrocarbon and Lipid Microbiology*, Springer-Verlag, Berlin, Germany, pp. 1501–1512.

Peypoux, F., Bonmatin, J.M., and Wallach, J. 1999. Recent trends in the biochemistry of surfactin. *Appl. Microbiol. Biotechnol.* 51: 553–563.

Rapp, P. and Gabriel-Jürgens, L.H.E. 2003. Degradation of alkanes and highly chlorinated benzenes, and production of biosurfactants, by a psychrophilic *Rhodococcus* sp. and genetic characterization of its chlorobenzene dioxygenase. *Microbiology* 149: 2879–2890.

Reis, R.S., Pereira, A.G., Neves, B.C., and Freire, D.M.G. 2011. Gene regulation of rhamnolipid production in *Pseudomonas aeruginosa*—A review. *Biores. Technol.* 102: 6377–6384.

Ron, E.Z. and Rosenberg, E. 2002. Biosurfactants and oil bioremediation. *Curr. Opin. Biotechnol.* 8: 313–316.

Smyth, T.J.P., Perfumo, A., Marchant, R., and Banat, I.M. 2010a. Isolation and analysis of low molecular weight microbial glycolipids. In: K.N. Timmis (ed.), *Handbook of Hydrocarbon and Lipid Microbiology.* Springer-Verlag, pp. 3705–3723.

Smyth, T.J.P., Perfumo, A., McClean, S., Marchant, R., and Banat, I.M. 2010b. Isolation and analysis of lipopeptides and high molecular weight biosurfactants. In: K.N. Timmis (ed.), *Handbook of Hydrocarbon and Lipid Microbiology.* Springer-Verlag, Berlin, Germany, pp. 3689–3704.

Tavares, L.F.D., Silva, P.M., Junqueira, M., Mariano, D.C.O., Nogueira, F.C.S., Domont, G.B., Freire, D.M.G., and Neves, B.C. 2013. Characterization of rhamnolipids produced by wild-type and engineered *Burkholderia kururiensis. Appl. Microbiol. Biotechnol.* 97: 1909–1921.

Tayeb, L., Ageron, E., Grimont, F., and Grimont, P. 2005. Molecular phylogeny of the genus *Pseudomonas* based on rpoB sequences and application for the identification of isolates. *Res. Microbiol.* 156: 763–773.

Tayeb, L.A., Lefevre, M., Passet, V., Diancourt, L., Brisse, S., and Grimont, P.A.D. 2008. Comparative phylogenies of *Burkholderia, Ralstonia, Comamonas, Brevundimonas* and related organisms derived from rpoB, gyrB and rrs gene sequences. *Res. Microbiol.* 159: 169–177.

Ulrich, R., Hines, H., Parthasarathy, N., and Jeddeloh, J. 2004. Mutational analysis and biochemical characterization of the *Burkholderia thailandensis* DW503 quorum-sensing network. *J. Bact.* 186: 4350–4360.

Wittgens, A., Tiso, T., Arndt, T.T., Wenk, P., Hemmerich, J., Müller, C., Wichmann, R. et al. 2011. Growth independent rhamnolipid production from glucose using the non-pathogenic *Pseudomonas putida* KT2440. *Microbial. Cell Factories* 10: 80.

Yakimov, M.M., Golyshin, P.N., Lang, S., Moore, E.R.B., Abraham, W-R., Lünsdorf, H., and Timmis, K.N. 1998. *Alcanivorax borkumensis* gen. nov., sp. nov., a new hydrocarbon-degrading and surfactant-producing marine bacterium. *Int. J. Syst. Bacteriol.* 48: 339–348.

Yakimov, M.M., Timmis, K.N., and Golyshin, P.N. 2007. Obligate oil-degrading marine bacteria. *Curr. Opin. Biotechnol.* 18: 257–266.

Yeh, M.S., Wei, Y.H., and Chang, J.S. 2005. Enhanced production of surfactin from *Bacillus subtilis* by addition of solid carriers. *Biotechnol. Prog.* 21: 1329–1334.

Zhang, H., Hanada, S., Shigematsu, T., Shibuya, K., Kamagata, Y., Kanagawa, T., and Kurane, R. 2000. *Burkholderia kururiensis* sp nov., a trichloroethylene (TCE)-degrading bacterium isolated from an aquifer polluted with TCE. *Int. J. Syst. Evol. Microbiol.* 50: 743–749.

Zimmerman, S.B., Schwartz, C.D., Monaghan, R.L., Pleak, B.A., Weissberger, B., Gilfillan, E.C., Mochales, S. et al. 1987. Difficidin and oxydifficidin: Novel broad spectrum antibacterial antibiotics produced by *Bacillus subtilis*. *J. Antibiotics* 40: 1677–1681.

6 The Prospects for the Production of Rhamnolipids on Renewable Resources

Evaluation of Novel Feedstocks and Perspectives of Strain Engineering

Marius Henkel, Christoph Syldatk, and Rudolf Hausmann

CONTENTS

6.1 Introduction 83
6.2 Renewable Feedstocks for the Production of Rhamnolipids.......................................85
    6.2.1 Traditional and Alternative Substrates 85
    6.2.2 Carbon Sources...86
        6.2.2.1 Sugars, Polysaccharides, and Industrial Waste Streams...................86
        6.2.2.2 Glycerol...86
        6.2.2.3 Fats, Oils, and Free Fatty Acids...88
        6.2.2.4 Lignocellulosic Biomass 88
    6.2.3 Nitrogen Source 89
    6.2.4 Economic Assessment 89
6.3 Perspectives of Strain Engineering...90
    6.3.1 Alternative Nonpathogenic Strains...90
    6.3.2 Potential Targets for Strain Engineering 91
        6.3.2.1 Broadened Substrate Spectrum...92
        6.3.2.2 Metabolic Spectrum and Pathways...93
        6.3.2.3 By-Products and Optimized Carbon Yield.......................................93
        6.3.2.4 Influencing Regulatory Mechanisms...94
        6.3.2.5 Availability of Precursor Molecules 95
6.4 Conclusion and Outlook 95
References..95

6.1 INTRODUCTION

Environmental concerns have led to increased interest and demand for bio-based chemicals in the last years. Many industrially relevant surfactants are prominent examples of chemicals based on fossil resources, which may also be produced biotechnologically using renewable resources. The worldwide total annual production of surfactants is estimated to exceed 15 million tons (Van Bogaert et al. 2007). The applications of surfactants range from detergents for cleaning or washing, emulsifiers for the food industry, and enhanced oil recovery to specialized applications in the pharmaceutical and medical sectors. In the pre-petrochemical time, detergents, including soaps, were produced using renewable resources including vegetable oil or animal fat. Today, however, a substantial part of surfactants available are produced based on petrochemical sources (Van Bogaert et al. 2007). In addition, some of the surfactants today are only slowly or partially biodegradable and contribute to environmental impact.

These issues may be overcome by microbial surfactants, which can be produced biotechnologically using renewable resources. These microbial surfactants are generally believed to have less impact on the environment than conventional surfactants, due to the fact that they usually display much better biodegradability and lower toxicity than their chemically synthesized counterparts (Develter et al. 2007; Mohan et al. 2006). Even though most biosurfactants display very promising physicochemical properties, such as critical micellar concentration and reduction of surface tension, the real performance potential in commercial products such as detergents and other applications has not yet been fully assessed and biosurfactants, therefore, still offer a perspective for future exploitation. A few microbial biosurfactants also offer additional benefits as compared with traditional surfactants, such as antibiotic or antimycotic properties (Kim et al. 2000; Nielsen et al. 2006; Varnier et al. 2009). The interest in sustainable production processes has considerably increased in the last years. Biosurfactants are one potential target, as they show the potential to replace synthetic surfactants. So far, the application of biotechnologically produced surfactants is mainly restricted to certain specialized areas, as production processes, in general, cannot yet compete with synthetic surfactants from an economical point of view. This is mainly due to relatively low product yields, high-priced raw materials, and expensive downstream processing. Prominent members among these promising biosurfactants are rhamnolipids. Rhamnolipids are surface-active glycolipids mainly known from *Pseudomonas aeruginosa* (Figure 6.1).

In this chapter, current and alternative feedstocks, biotechnological processes and yields, as well as potential targets for strain engineering are presented and discussed using the example of rhamnolipids from *Pseudomonads*, both from a scientific and an economical point of view. The advantages and constraints of alternative feedstocks over established sources of carbon are also discussed. Potential targets for strain engineering, which may be used to improve current processes for rhamnolipid production, are presented to provide insight on rhamnolipid biosynthesis and the metabolism of different *Pseudomonads*.

Rhamnolipids were first described in 1946 (Bergström et al. 1946), and the chemical structure was unveiled in the following years (Edwards and Hayashi 1965; Jarvis and Johnson 1949). Rhamnolipids consist of one or two L-rhamnose units linked to one or two β-hydroxy fatty acids (Figure 6.1) and are known to be produced mainly by *P. aeruginosa*. Over the last decades, research on many different fields regarding rhamnolipid production has been conducted by several research groups: the rhamnolipid biosynthesis at the genetic level (Ochsner et al. 1994a; Rahim et al. 2000;

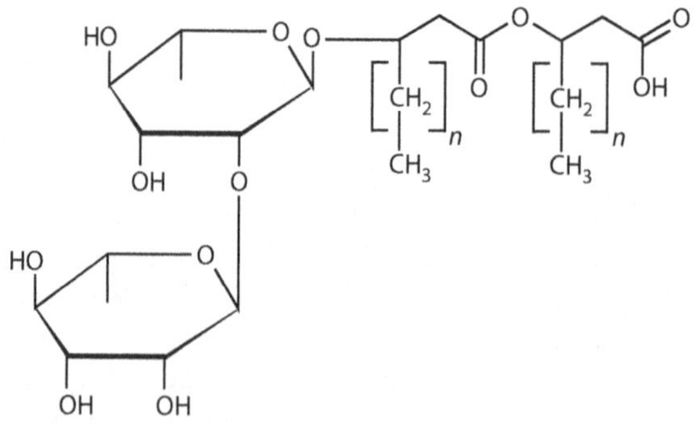

FIGURE 6.1 Chemical structure of di-rhamnolipids (Rha-Rha-C_n-C_n). $n = 4$–8.

Rehm et al. 2001; Zhu and Rock 2008), biosynthetic pathways related to rhamnolipid production (Burger et al. 1963; Déziel et al. 2003; Hauser and Karnovsky 1957, 1958), regulation of rhamnolipid production (Ochsner et al. 1994b; Pearson et al. 1997; Pesci and Iglewski 1997), quantification and analytics related to the detection of rhamnolipids (Déziel et al. 2000; Gartshore et al. 2000; Heyd et al. 2008; Mata-Sandoval et al. 1999; Schenk et al. 1995; Siegmund and Wagner 1991), as well as downstream processing (Walter et al. 2010). Besides rhamnolipid production by *P. aeruginosa*, *Burkholderia plantarii* is known to produce rhamnolipids containing three β-hydroxy-fatty acids (Andrä et al. 2006), ranging from 8 to 16 carbons in length. Although β-hydroxydecanoic acid is predominantly found in rhamnolipids from *P. aeruginosa* (Déziel et al. 2000), β-hydroxytetradecanoic acid is the major constituent found in rhamnolipids from *Burkholderia* sp. (Hörmann et al. 2010). Rhamnolipids are typically produced by *P. aeruginosa* as different congeners with predominant molecules Rha-Rha-C_{10}-C_{10} and Rha-C_{10}-C_{10}. Rhamnolipids potentially fulfill several functions in nature: solubilization and uptake of hydrophobic molecules, contact to hydrophobic surfaces, and intrusion into tissue due to hemolytic activity combined with a wide range of antimicrobial activity.

Compared with other biosurfactants, rhamnolipids can be produced in relatively high yields in comparably short times. They display relatively high surface activities and, as such, reduce the surface tension of water from 72 to 31 mN/m (Syldatk et al. 1985). The critical micellar concentration was determined to be between 20 and 225 mg/L in water (Dubeau et al. 2009; Syldatk et al. 1985). When comparing the biodegradability of rhamnolipids and sophorolipids to that of the synthetic surfactants Triton-X-100 and linear alkylbenzene sulfonates, it was shown that the synthetic surfactants were only partially degraded under the applied conditions, whereas rhamnolipids and sophorolipids were almost completely degraded (Develter et al. 2007). Furthermore, the aquatoxicity of rhamnolipids and sophorolipids was about 12 times lower than those of synthetic surfactants, as demonstrated by their EC_{50} values of 20–77 mg/L.

6.2 RENEWABLE FEEDSTOCKS FOR THE PRODUCTION OF RHAMNOLIPIDS

A wide variety of substrates served the biotechnological production of rhamnolipids in laboratories in the past, ranging from petrochemically derived substances to renewable substrates like plant or vegetable oils, sugars, and glycerol.

In addition to these feedstocks, it was demonstrated that several industrial waste streams may also potentially serve as appropriate substrates for the production of rhamnolipids. These industrial wastes include fatty acids (Abalos et al. 2001), waste frying oil (Haba et al. 2000), olive oil production effluents (Mercade et al. 1993), whey wastes (Dubey and Juwarkar 2001), and soap stock (Benincasa et al. 2002). Waste streams may become increasingly interesting for application in biotechnological processes, as they are, in general, less expensive and do not directly compete with food. In addition, the benefits of consuming waste substrates during biotechnological processes are not restricted to economic reasons; the reduced amount of waste is furthermore beneficial for the environment.

6.2.1 TRADITIONAL AND ALTERNATIVE SUBSTRATES

Current approaches for the production of rhamnolipids rely on the use of plant and vegetable oils as well as sugars, and, as such, draw resources from pools of human and animal feedstocks. The use of many agro-industrial by-products as substrates for biotechnological processes still indirectly competes with human food, since these by-products may also be used for other applications, for example, in animal feed. Using agro-industrial wastes for rhamnolipid production is furthermore accompanied by several disadvantages. Since agro-industrial wastes are usually locally not available in sufficiently large quantities, to sustain large-scale production processes, it is necessary to concentrate the wastes (e.g., the production of dry whey powder as a by-product of the dairy industry) to avoid high transport costs. This is furthermore complicated by the generally relatively low content of substrate in the waste streams (e.g., less than 5% dry matter in whey wastes; Dubey and

Juwarkar 2001). In addition, agro-industrial wastes usually vary in composition, which results in high variability of different batches of these potential feedstocks. Therefore, closer process monitoring may be required, due to different batches of substrate.

6.2.2 CARBON SOURCES

6.2.2.1 Sugars, Polysaccharides, and Industrial Waste Streams

Glucose has been used as a substrate for the production of rhamnolipids in the past (Guerra-Santos et al. 1984; Syldatk et al. 1985). However, the food-processing industry, most importantly sugar-processing plants and the dairy industry, produces a wide variety of by-products and waste streams that also contain different sugars. Whey, a liquid rich in nutrients (up to 75% of lactose and 15% of protein in dry matter, vitamins, and minerals), is the most important by-product of the dairy industry. The production of rhamnolipids by *P. aeruginosa* using whey as a source of carbon has been investigated in the past, and final concentrations of rhamnolipids of 1 g/L could be achieved (Table 6.1) (Dubey and Juwarkar 2001). However, even though whey is a potentially suitable substrate for biotechnological processes, still only half of the whey from the dairy industry is converted into valuable products, such as ingredients for food (Makkar et al. 2011). The major challenge regarding the application of whey is that only a few bacteria are able to use the galactose resulting from the cleavage of lactose. Therefore, the full potential is not achieved unless either microorganisms that are used for the production are able to metabolize galactose, or current strains are targeted by genetic engineering approaches to allow for the consumption of galactose. Wild-type strains of *P. aeruginosa* are unable to utilize galactose for growth. In particular, to establish galactose metabolism in *P. aeruginosa*, genes for galactokinase (*galK*) and UDP-glucose-hexose-1-phosphate uridyltransferase (*galT*) are required as key enzymes for the metabolization of galactose (Henkel et al. 2012). The sugar- and starch-processing industry produces large quantities of wastewater, which is rich in carbohydrates. This may potentially be used as a substrate for biotechnological production processes, for example, the production of enzymes on an industrial scale (Maneerat 2005). Processing of crops and other food, for example, washing, cracking, or peeling, usually results in significant losses of useable carbohydrates. Molasses, typical waste streams of the sugar-processing industry, are generated during crystallization of the sugar from liquid extracts of sugarcane or sugarbeet. They are discarded once further measures of extraction become uneconomical (Makkar and Cameotra 1997). However, liquid molasses still contain about 50% of sucrose in addition to other valuable compounds such as vitamins (Patel and Desai 1997), making molasses a suitable substrate to be used in biotechnological processes, which can also be obtained at a much lower price than pure sugar. Molasses have been successfully employed for the production of rhamnolipids in the past (Table 6.1) (Makkar and Cameotra 1997; Patel and Desai 1997). Wastewater from distilleries, also known as "spent wash," is an example of a waste stream that is difficult and costly to discard, due to its high content of organic and inorganic matter, high acidity, strong color, and usually unpleasant odor. Using this distillery waste as a sole source of carbon has been shown to allow for rhamnolipid production with *P. aeruginosa* in the past, however, with comparably low concentrations of 0.93 g/L (Table 6.1) (Babu et al. 1996). Final rhamnolipid concentrations in batch processes using up to 5.31 g/L of glucose have been reported in the past (Chen et al. 2007), however, with a low substrate-to-product yield of 0.13 g/g (Table 6.1). The maximum theoretical yield for rhamnolipid production with glucose was determined to be 0.52 g RL/g glucose (Henkel et al. 2012), which reveals that there is still room for significant optimization of the production process.

6.2.2.2 Glycerol

Glycerol is a widely used substrate in approaches for rhamnolipid production (Syldatk et al. 1985) and other biotechnological processes, since many different organisms are able to utilize glycerol as a source of carbon. Glycerol can typically be obtained from renewable resources, for example, by basic hydrolysis of triglycerides of animal fat or vegetable oil. Glycerol is furthermore a relevant

TABLE 6.1

Substrate-to-Product Conversion Yields ($Y_{P/S}$) and Maximum Product Concentrations (RL_{max}) for Different Renewable Substrates and Industrial Wastestreams Used for Rhamnolipid Production with *Pseudomonas aeruginosa*

Substrate	Average Content (If Applicable)	Strain	$Y_{P/S}$ [g/g]	RL_{max} [g/L]	Reference
Plant/Vegetable Oils and Fats					
Olive oil		44T1	0.38	7.65	Robert et al. (1989)
Sunflower oil		PAO1	0.23	39	Müller et al. (2010)
Corn oil		DSM 2659	0.224	8.94	Hembach (1994)
Soybean oil		DSM 7107	0.62	78	Giani et al. (1997)
Fish oil		BYK-2 KCTC	0.75	22.70	Lee et al. (2004)
Carbohydrates					
Glycerol		DSM2659	0.10	1.0	Santa Anna et al. (2002)
Glucose		S2	0.13	5.31	Chen et al. (2007)
Industrial Wastestreams					
Olive oil mill effluent (olive oil extraction)	Approx. 100 g/L sugars, nitrogen compounds, organic acids, oil	JAMM	0.058	1.4	Mercade et al. (1993)
Waste frying oil (food industry)	Oil, free fatty acids, up to 30% polar compounds	47T2	0.34	2.7	Haba et al. (2000)
Soap stock (oil refining)	Up to 50% free fatty acids	LBI	0.33	15.9	Benincasa et al. (2002)
Waste fatty acids (oil refining)		AT10	ND	9.5	Abalos et al. (2001)
Crude glycerol (biodiesel production)	Approx. 80% glycerol	DSM 2874	0.21	8.5	Syldatk, Lang, Matulovic et al. (1985)
Spent wash (distillery waste)	Approx. 50 g/L organic acids and sugars	BS2	2.73 (g/L)	0.91	Dubey and Juwarkar (2001)
Molasses (sugarcane refining)	Approx. 60% sucrose	EBN-8	0.392	1.45	Raza, Khan, and Khalid (2007)
Whey (dairy industry)	Approx. 25% sugars in dry mass	BS2	0.92 (g/L)	0.92	Dubey and Juwarkar (2001)

Note: ND—not determined.

by-product in biodiesel production, which is generated by transesterification of different vegetable oils with methanol. Furthermore, glycerol can be produced from propene obtained from petro-chemical processes and fossil sources (e.g., petroleum processing). As an abundant resource of many different origins, glycerol can be obtained in different purities. Crude glycerol, obtained as a by-product of biodiesel manufacturing, which contains approximately 90% glycerol and different organic acids and alcohols (Novaol), was used for the production of rhamnolipids in the past (de Sousa et al. 2011). However, as a market for trading crude glycerol has not been established so far, its prices are relatively high when compared with other wastes or by-products. With increased demand for renewable fuels, the production of biodiesel is expected to rise (EBB, 2010). This, in turn, will also account for higher amounts of waste glycerol being produced at the same time and may have an influence on availability and prices. When comparing the production of rhamnolipids by *P. aeruginosa* on pure glycerol with glycerol from biodiesel production, no loss in production rates could be observed due to the different compositions of the culture media when using crude glycerol (Walter 2009). This study indicates that crude glycerol may be employed directly for the production of rhamnolipids with little or no pretreatment. Processes for the production of rhamnolipids with glycerol typically output final product concentrations between 1.0 and 8.5 g/L, however, with low substrate-to-product conversion yields ($Y_{P/S} = 0.1–0.15$ g/g). This demonstrates that these processes have the potential to be significantly optimized, if the observed yield is compared with the maximum theoretical yield for rhamnolipid production of 0.59 g RL/g glycerol (Henkel et al. 2012).

6.2.2.3   Fats, Oils, and Free Fatty Acids

Plant and vegetable oils have been extensively used in processes for rhamnolipid production in the past. More than 3 million tons of oils and fats are produced every year (Nitschke et al. 2005), and three-fourths of the total production originates from plants and seeds (e.g., sunflower, rapeseed, palm, fish, coconut, soybean, and olive oil). Unrefined oil typically undergoes a series of different refinement steps before it is used as an ingredient in food or for food processing. During these processes, unwanted by-products of the extraction are removed (e.g., free fatty acids, pigments, sterols, hydrocarbons, and protein fragments) to counteract unwanted odor and taste, as well as for conservation purposes. The waste stream generated during this process of oil refinement is known as "soap stock." Its main uses today are, besides the production of soap, in animal feed formulations. Large amounts of vegetable oil are processed annually, and this accounts, in turn, for a large amount of soap stock being produced. As soap stock has a low degradability due to the high lipid content (Haba et al. 2000), its disposal is highly relevant both for the environment and for the financial interests of the oil-processing industries. Owing to its complex composition and high content of carbon as fatty acids, soap stock may serve as a cheap substrate for biotechnological processes (Zhu et al. 2007). The average price for soap stock is as low as about one-tenth of the price for refined oil (Nitschke et al. 2005). The production of rhamnolipids with *P. aeruginosa* using soap stock has been successfully performed in the past, with a final rhamnolipid concentration of 15.9 g/L (Benincasa et al. 2002), and a substrate-to-product yield of 0.33 g RL/g soap stock. Nitschke et al. (2005) furthermore compared soap stocks from different oil industries (e.g., soybean, cottonseed, corn) regarding their potential to serve as low-cost substrates for the production of rhamnolipids (Nitschke et al. 2010), with a final product concentration of 11.7 g/L. Food-processing industry wastes account for a major part of waste vegetable oils (e.g., deep frying and vegetable processing). These wastes are mainly used as supplements in animal feed formulations or for the production of biodiesel. It was shown that waste frying oil is a suitable substrate for rhamnolipid production, and the different compositions in terms of glycerides and fatty acids in used oil does not significantly impair growth or rhamnolipid yields (De Lima et al. 2007; Zhu et al. 2007). Today, plant and vegetable oils are the most efficient substrates used for rhamnolipid production, with generally substantially higher final concentrations and substrate-to-product conversion yields than non-hydrophobic substrates. However, it should be noted that most of the oils used are still in direct competition with human food, and, compared with other substrates, these oils are rather expensive substrates. Fatty acids in lower purities, referred to as "rubber grade," may be obtained at much lower prices (~one-tenth of the price), and may be worth considering as an alternative.

6.2.2.4   Lignocellulosic Biomass

As part of the world's plant life, lignocellulosic biomass belongs to the most abundant sources of organic carbon, and is a truly renewable substrate. As such, waste streams and by-products from many different industries and various other sources as well as agricultural wastes contain lignocellulosic material. Lignocellulose consists of cellulose, hemicellulose, and lignin, in varying proportions depending on the source. The polysaccharides (cellulose and hemicellulose) are linked to lignin via ester and ether linkages. Cellulose consists of linear β-(1-4)-linked D-glucose chains. Hemicellulose consists of several different heteropolysaccharides, which include the pentoses xylose and arabinose as well as hexoses (mainly mannose, galactose, and rhamnose) linked together in a branched structure. Lignin is a random polymer, which consists of phenylpropane units, with varying amounts of hydrogen substituted by methoxy groups. Lignocellulosic material can be classified into three categories (Dunlap and Chiang 1980): primary cellulosics, agricultural waste cellulosics, and municipal waste cellulosics. Primary cellulosics are obtained from plants that are cultivated specifically for the purpose of cellulose content. By-products and wastes from agricultural/food-processing units fall into the category of agricultural waste cellulosics. Municipal waste cellulosics are the paper or paper product-derived parts of municipal solid wastes. The use of lignocellulosic waste in biotechnological processes is scarce, and currently restricted to small-scale productions of cellulose–ethanol and other experimental processes. Therefore, its high potential currently remains almost unexploited. Investigations on the use of lignocellulosic sugars for the production of ethanol and organic acids demonstrated the feasibility to operate in an economical way (Taherzadeh and Karimi 2007). The main challenge regarding the application of lignocellulosic biomass is the usually intensive pretreatment, which is necessary to make the sugars available for the microorganisms. For this purpose, a series of mechanical and chemical pretreatments to release the sugars from the polymers are required. During this harsh treatment, however, several unwanted by-products are produced, for example, furfurals, which are known for their negative effect on microbial growth (Miller et al. 2009). In general, these by-products are removed prior to utilization in biotechnological processes, which is time and resource consuming. The production of rhamnolipids with *Pseudomonas* sp. may be possible using untreated lignocellulosic hydrolysates, because *Pseudomonas* sp. is able to degrade furfurals (DSMZ; Henkel et al. 2012). Most microorganisms, including *Pseudomonas* sp., used in biotechnological processes are unable to metabolize C5 sugars (xylose and arabinose, the major constituents of hemicellulose). Therefore, the full potential of lignocellulose-based biotechnological processes has yet to be reached. With the highest availability of all substrates (more than 500 million tons per year), a high maximum theoretical substrate-to-product conversion yield of 0.52 g RL/g lignocelluloses and a low price of approximately one-eighth of glucose (Henkel et al. 2012), lignocellulose shows a high potential for rhamnolipid production and biotechnological processes in general. However, either an alternative production host with the natural ability to consume arabinose and xylose along with C6 sugars or an existing strain engineered for the consumption of lignocellulose sugars is desirable when relying on this novel feedstock for rhamnolipid production.

6.2.3 Nitrogen Source

Many different sources of nitrogen have been investigated for rhamnolipid production, but the most frequently used substrates are nitrate salts and ammonia. In general, nitrate is believed to be the best substrate for rhamnolipid production that allows for higher yields as compared with other sources of nitrogen (Arino et al. 1996; Manresa et al. 1991). In contrast to the carbon source used in biotechnological processes, complex or less well-defined sources of nitrogen (e.g., yeast extract or protein hydrolysates) are more expensive than pure sources (e.g., nitrate, ammonia, or urea). However, one exception to this is a liquid by-product of corn wet-milling and processing, which is commonly referred to as "corn steep liquor." Corn steep liquor is a widely used source of nitrogen in industrial biotechnology, and it has also successfully been used in the past for the production of rhamnolipids (Patel and Desai 1997). This demonstrates that the search for cheaper or alternative substrates may not only be restricted to the carbon source, but also other constituents of the culture media may be worth considering.

6.2.4 Economic Assessment

The raw material costs for biosurfactants production are estimated to account for up to half of the total production costs (Mulligan and Gibbs 1993). Depending on the strain and the process, much more substrate is consumed during production than rhamnolipids are synthesized. This becomes evident when comparing the substrate-to-product conversion (Table 6.1). With values as low as 0.1 g of rhamnolipid per gram of substrate, this consequently means that the required amount of substrate consumed may be up to 10 times higher than the amount of rhamnolipid produced. This explains why using cheaper substrates and raw materials for the production process may have a significant impact on the overall production costs. During the development of a potential large-scale process for rhamnolipid production, ecological and economical as well as political aspects will need to be considered. Besides the pricing of the raw material, several other important factors may also need to be considered. The local availability of the substrate may play a major role in the calculations. This includes the fact that some resources are only generated at very few production sites, for example, waste glycerol from biodiesel production plants. Even if the local confinement of the resource would not be an issue, some resources, for example, dairy industry whey, may not be present in large enough quantities to sustain a large-scale production. Another major factor when using industrial wastes or other nonpure substances as a feedstock is that the average composition of the substrate needs to be investigated. For industrial wastes, it is usually possible to define a typical composition; however, this varies between different batches. It may, therefore, be required to supplement the feedstock on an as-required basis, which, depending on the process, may also require an advanced process control or frequent analytics or certification. For diluted resources, transport costs in terms of required volumes may be an issue, or the fact that it may also be required to concentrate them in order to obtain a useable substrate. Some industrial wastes may also require pretreatment in order to use them as substrates for microorganisms. This includes neutralization of acidic wastes (e.g., soap stock) or extensive mechanical and chemical pretreatment (e.g., lignocellulosic biomass). However, lignocellulosic biomass is the most abundant renewable source of potentially utilizable carbon, and the issue of pretreatment may be dramatically simplified by developing tailored strains, which may be able to produce sugar polymer-degrading enzymes (e.g., cellulases and xylosidases). As compared with these waste substrates, pure feedstocks typically perform well in industrial processes, as the composition is much better defined. However, many of these feedstocks either directly compete with food (e.g., plant and vegetable oils) or still indirectly compete with food (e.g., a substrate that may be suitable for application in animal feed).

6.3 PERSPECTIVES OF STRAIN ENGINEERING

Many microorganisms are known for the production of rhamnolipids, but most of them are different strains and isolates belonging to the species *P. aeruginosa*. However, mainly due to large differences in product yields and final total product concentrations, only a few strains are potentially relevant for industrial processes. The highest concentration of rhamnolipids to ever be reported in fed-batch process was 112 g/L in an industrial process for the microbial production of rhamnose from rhamnolipids by former Hoechst AG (Frankfurt-Hoechst, Germany). In this process, patented *P. aeruginosa* strains were used (Giani et al. 1997). However, it should be noted that these claims could never be verified on a laboratory scale, since critical details, especially regarding the quantification of rhamnolipids, are missing in the patent. Müller et al. (2011) reported a concentration of 35.7 g/L with these strains.

P. aeruginosa is relatively easy to cultivate, and the synthesis of rhamnolipids in particular has been the focus of research for decades, also because they are major components of virulence mechanisms. Furthermore, *P. aeruginosa* PAO1 has served as a model organism to study rhamnolipid production on several occasions, and the strain is furthermore fully sequenced and annotated (Davey et al. 2003; Müller et al. 2010; Stover et al. 2000). *P. aeruginosa* is an opportunistic pathogen for humans and classified as biosafety level L2. Therefore, this bacterium may not be the first choice for large-scale production due to the increased security measures that may be required, if nonpathogenic alternatives with acceptable yields existed.

6.3.1 Alternative Nonpathogenic Strains

The search for effective nonpathogenic rhamnolipid-producing strains is an ongoing field of research. Basically, there are two main concepts, which are investigated to obtain an alternative strain: the establishment of rhamnolipid biosynthesis in nonpathogenic bacteria and the production of rhamnolipids in novel nonpathogenic wild-type producers. Ochsner et al. introduced the *rhlAB* operon required for rhamnolipid biosynthesis in different host-strains, including *P. fluorescens*, *P. putida*, *Escherichia coli*, and *P. oleovorans*. The highest rhamnolipid levels could be observed in *P. putida*, with values up to 60 mg/L. In more recent work, a rhamnolipid concentration of 7.2 g/L was reported in a heterologous approach in *P. putida* (Cha et al. 2008; Ochsner et al. 1995). Nonpathogenic, alternative rhamnolipid-producing strains that could be identified so far mainly belong to other species of *Pseudomonas* or *Burkholderia*.

Even though many new systems and strains have been established and investigated in the past, none of them has shown the potential to replace *P. aeruginosa* strains for industrial processes, due to significantly lower yields and final concentrations by now. It may, therefore, be a favorable approach to further examine wild-type *P. aeruginosa* regarding yields and potential targets for genetic optimization. The comparably high yields of rhamnolipids achievable with these wild-type strains may provide a firm basis for further optimization by strain engineering.

6.3.2 POTENTIAL TARGETS FOR STRAIN ENGINEERING

There are several potential ways in which a rhamnolipid-producing strain may be optimized by genetic means. The following aspects may be considered as targets for strain engineering: besides the search for alternative nonpathogenic production strains, these targets include the substrate spectrum the strain is able to metabolize as well as the metabolic spectrum and yields, by-product formation and carbon yield, product formation and genetic regulation, and the availability of precursor molecules. Potential targets are summarized in Figure 6.2. Strain engineering approaches targeted on improved product yields usually require a complex interplay of genetics (e.g., deletions, insertions, alterations in promoter strength), analytics (e.g., metabolome and transcriptome analysis), and

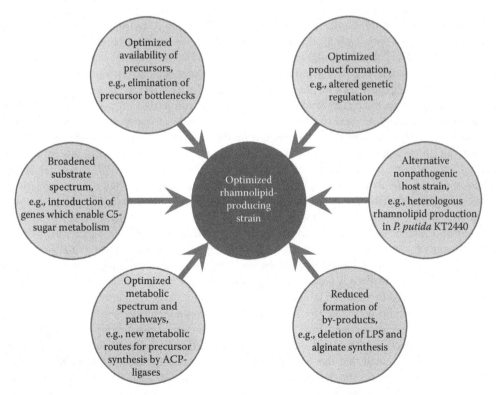

FIGURE 6.2 Potential targets for strain engineering of a rhamnolipid-producing strain.

systems biology (e.g., metabolic flux distributions and modeling), to quantify the effects of modification and identify potential bottlenecks (Müller and Hausmann 2011). The genetic and metabolic information in this section was obtained from the Kyoto Encyclopedia of Genes and Genomes (KEGG, available online at http://www.genome.jp/kegg/ (Kanehisa and Goto 2000)) and the National Center for Biotechnology Information (NCBI, available online at http://www.ncbi.nlm.nih.gov/).

6.3.2.1 Broadened Substrate Spectrum

Several alternative feedstocks consist of or contain components that are not accessible for *P. aeruginosa*. Strain engineering may, therefore, be applied to allow for enhanced yields or to allow for consumption of a specific substate, for example, by introducing genes that are required for the metabolization of a specific substrate. Koch et al. (1988) introduced the genes *lacY*, encoding for a β-galactoside permease and *lacZ*, encoding for a β-galactosidase, from *E. coli* to *P. aeruginosa* strains PAO1 and PG-201. The resulting strains were consequently able to grow on lactose and whey as a sole source of carbon. However, to reach the full potential of lactose consumption, the galactose resulting from the cleavage of lactose needs to be consumed as well, which leaves room for further optimization of a lactose-consuming strain (Koch et al. 1988). This is also the case when lignocellulosic hydrolysates are used as a source of carbon. As *Pseudomonads* are unable to metabolize C5 sugars from the hemicellulosic material and only consume glucose from cellulose, the full potential of lignocelluloses for rhamnolipid production is not reached. It has been demonstrated in the past that it is possible to alter *P. putida* S12 so that it utilizes xylose as a substrate for growth by expressing genes encoding for a xylose isomerase (*xylA*) and a xylulose kinase (*xylB*) (*Meijnen, de Winde, and Ruijssenaars 2008; Meijnen, de Winde, and Ruijssenaars 2009*). The resulting strain was consequently able to grow on xylose as a sole source of carbon. Interestingly, however, the resulting strain was able to metabolize L-arabinose as well. Meijnen et al. (2008) attributed this observation to a non-specific activity of the introduced genes *xylA* and *xylB* for L-arabinose. Another drawback of the application of lignocelluloses is the conversion of the raw material into useable sugar monomers. Typical methods include mechanical/chemical pretreatment followed by separation (e.g., filtration, precipitation) and acid/basic hydrolysis, which is usually carried out under high pressure and temperatures. Technical and environmental aspects of different processes for the extraction and purification of arabinoxylans have been evaluated in the past (Jacquemin et al. 2012). Some microorganisms such as yeasts and fungi provide a set of specific enzymes, which allow for conversion of sugar polymers into sugar monomers (e.g., xylosidases, arabinofuranohydrolases, and xyloglucanases). These enzymes may potentially be used so that most of the processing of lignocellulosic biomass is performed either directly by adding the recombinantly produced enzymes or by a recombinant approach in the host strain. As opposed to chemical treatment, this may reduce the overall costs for pretreatment and influence the economics of the overall process. A xylosidase (*pslG*) that catalyzes the breakdown of 1,4-β-D-xylane polymers into xylose monomers as well as a xylulose kinase (*xylB*) that catalyzes the phosphorylation of D-xylulose to D-xylulose-5-P are produced by *P. aeruginosa* PAO1. In addition, there is evidence for an endoxylanase in the PAO1 genome (locus tag PA2783, information obtained from the National Center for Biotechnology Information, NCBI), which shows distinct patterns for carbohydrate-binding and xylan-binding domains. This demonstrates that *P. aeruginosa* PAO1 may be a potentially suitable target for engineering toward a functioning xylose metabolism, as many crucial genes are already present in the genome. From a genetic point of view, only a xylose isomerase (*xylA*) would be required to allow for metabolizing xylose, which would be converted to D-xylulose (*xylA*) and phosphorylated to D-xylulose-5P, which is an intermediate of the pentose phosphate pathway. Since *P. aeruginosa* PAO1 possesses a complete and functioning pentose phosphate pathway, this introduction of xylulose isomerase may be sufficient to establish growth on xylose. However, it should be noted that this has never been investigated to date. *P. putida* KT2440 may be a promising candidate for the degradation of cellulosic material, since it contains genes related to the degradation of cellulosic material. A gene encoding for a protein with endoglucanase activity (locus tag PP_2637, information obtained from the Kyoto Encyclopedia of Genes and Genomes,

KEGG, Kanehisa and Goto 2000), which cleaves cellulose into smaller 1,4-β-D-glucane polymers to create more free ends for exo-cellobiohydrolases, is present in *P. putida* KT2440. *P. putida* KT2440 also expresses a periplasmic β-glucosidase (*bglX*), which cleaves cellobiose to glucose monomers. Furthermore, this strain is classified as biosafety level L1, which makes it a potentially interesting candidate for industrial production of rhamnolipids.

6.3.2.2 Metabolic Spectrum and Pathways

Precursors consumed during rhamnolipid biosynthesis are produced by *de novo* synthesis. This includes the sugar moiety, dTDP-L-rhamnose, as well as the fatty acids required for the synthesis of the hydrophobic counterpart, 3-(3-hydroxyalkanoyloxy)alkanoic acid (HAA) (Rehm et al. 2001). During cultivation on plant oils, free fatty acids can be found in the culture broth due to hydrolysis of triglycerides by lipases. By providing means of direct incorporation of these fatty acids or parts thereof directly in the synthesis of HAA, the process of precursor biosynthesis may be significantly faster and less energy will be lost due to a significantly reduced amount of required reactions, both in the catabolism and the anabolism of fatty acids. A process where precursors are used as a feed is established for the production of sophorolipids by *Candida bombicola* (Van Bogaert et al. 2011). During this process, glucose and a hydrophobic carbon source (e.g., plant oil) are simultaneously fed to provide precursors for product synthesis. The fatty acid needs to be linked to an acyl-carrier protein (ACP) to be used for the synthesis of rhamnolipid precursors, a reaction that is carried out by enzymes known as ACP ligases. ACP ligases exist with different specificities for the chain length of the fatty acid, and the introduction of a medium-chain length ACP-ligase may be suitable for this purpose (Figure 6.3).

P. aeruginosa is not able to degrade glucose via glycolysis. Instead, glucose is introduced into the Entner–Doudoroff (ED) pathway, and oxidized to glyceraldehyde-3-phosphate (gap) and pyruvate, via 2-keto-3-desoxy-6-phosphogluconate. In addition to *P. aeruginosa*, the ED pathway is also found in several other Gram-negative bacteria, including *Agrobacterium* sp. and *Rhizobium* sp. Some microorganisms, however, such as *E. coli*, use both pathways. Besides the regular glycolysis, this allows for the metabolization of different organic acids (e.g., gluconate), which are not degraded via glycolysis. In the catabolism of *P. aeruginosa*, only a phosphofructokinase, which catalyzes the phosphorylation of fructose-6-phosphate (fru-6P) to fructose-1,6-bisphosphate (fru-1,6-2P), would be required to establish a functioning glycolysis. *P. putida* furthermore lacks a second gene encoding for a hexokinase, which phosphorylates fructose to fructose-6P. The introduction of a hexokinase may furthermore be beneficial when lignocelluloses hydrolysates are used as a source of carbon, as this also allows for the phosphorylation and degradation of mannose, a hexose present in hemicelluloses. By establishing a glycolytic pathway, one additional mol of ATP is generated per mol of glucose, when compared with the ED pathway.

6.3.2.3 By-Products and Optimized Carbon Yield

P. aeruginosa produces several intracellular and extracellular by-products, which are undesirable regarding carbon yield and overall efficiency of the process. These by-products include the extracellular polymers alginate and lipopolysaccharide as well as the intracellular polyhydroxyalkanoate (PHA) synthesis (Figure 6.3). To identify potential targets for strain engineering, the interrelation of metabolism, rhamnolipid biosynthesis, and by-product formation has to be considered. Extracellular by-products alginate and lipopolysaccharide originate from D-glucose-6P, which is generated during gluconeogenesis. D-Glucose-6P is then converted into GDP mannuronic acid, the direct precursor for alginate synthesis via the genes *algACD*. However, all three are not suitable targets for deletion to alter alginate biosynthesis, since *algC* is also required during the generation of dTDP-L-rhamnose, a precursor for rhamnolipid synthesis, where it catalyzes the conversion of D-glucose-6P into D-glucose-1P (Figure 6.3). D-Fructose-6P is converted into UDP-N-acetyl-D-glucosamine, a direct precursor for lipopolysaccharide biosynthesis, via the genes *glmSMU* (Figure 6.3). UDP-*N*-acetyl-D-glucosamine is then used in lipopolysaccharide biosynthesis by several enzymatic conversions

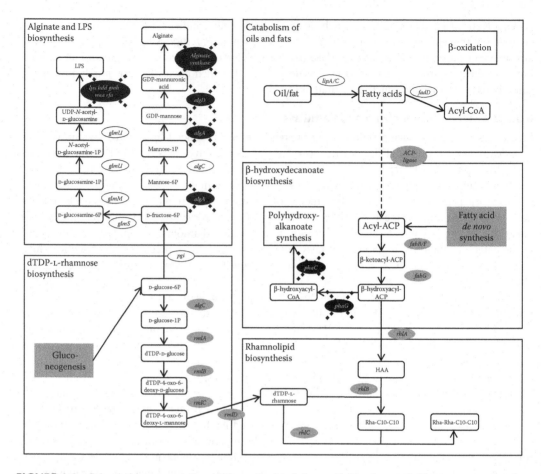

FIGURE 6.3 Schematic representation of biosynthesis of rhamnolipids, rhamnolipid precursor molecules, and potentially relevant by-products associated with the production of rhamnolipids (Rehm et al. 2001; Soberón-Chávez et al. 2005a,b). Precursors, intermediates and products are represented by boxes. Arrows indicate enzymatic conversion, and respective genes are shown in ellipses. Potential targets for strain engineering are indicated by filled ellipses, black crossed-out elements indicate potential points for deletion, gray elements indicate points for insertion or upregulation.

(mediated by the genes *lpx, hdd, gmh, waa,* and *rfa*). These genes may be suitable targets for strain engineering. Further investigations are required to see whether *glmSMU* are also suitable targets for deletion, as UDP-*N*-acetyl-D-glucosamine is also an intermediate in amino sugar and nucleotide sugar metabolism. The synthesis of intracellular PHAs in *P. aeruginosa* consumes β-hydroxyacyl-ACP, which is converted into β-hydroxyacyl-CoA and introduced into PHA synthesis by the gene *phaGC* (Figure 6.3). As hydroxyacyl-ACP is required for HAA synthesis, a component of rhamnolipids (Figure 6.3), the synthesis of PHAs draws precursors away from rhamnolipid biosynthesis and may therefore be another potential target for strain engineering.

6.3.2.4 Influencing Regulatory Mechanisms

In *P. aeruginosa*, the biosynthesis of rhamnolipids is controlled by a complex quorum-sensing regulatory network (Henkel et al. 2013; Soberón-Chávez et al. 2005b). Components of this regulatory network are potential targets for manipulation, for example, by substituting affected promoters to achieve desired expression levels or by modifying the action of regulators due to altered affinity to the regulatory sequence. However, it should be noted that the auto-enhancing effect of quorum sensing, which upregulates genes for rhamnolipid biosynthesis (Schmidberger et al. 2013), may be lost during this process. With respect to rhamnolipid biosynthesis, recombinant expressions of relevant genes in *P. putida* are not coupled to the complex quorum-sensing network, which may allow for easier process control strategies and product formation independent of cell growth.

6.3.2.5 Availability of Precursor Molecules

During the biosynthesis of rhamnolipids, the sugar moiety, dTDP-L-rhamnose, as well as the hydrophobic precursor HAA are synthesized *de novo* (Rehm et al. 2001). Therefore, in addition to the rhamnosyl transferase RhlB that links both units together, the production of these precursor molecules is a potential target to enhance rhamnolipid biosynthesis. However, it should be noted that such an approach usually requires further investigation of metabolic flux distributions to identify potential bottlenecks (Müller and Hausmann 2011). There are many genes involved in gluconeogenesis and fatty acid *de novo* synthesis that could be targeted for overexpression. The activated rhamnose is produced from gluconeogenesis by the *rmlBDAC* operon (Cabrera-Valladares et al. 2006). Up to reaching the metabolic capacity of gluconeogenesis, the formation of dTDP-L-rhamnose could possibly be enhanced by overexpressing the *rmlBDAC* operon, which then converts D-glucose-6P into dTDP-L-rhamnose. It should be noted that one genetic modification does not necessarily lead to enhanced rhamnolipid or precursor production, as altering a metabolic pathway may potentially shift the metabolic flux distributions, resulting in the development of another bottleneck.

6.4 CONCLUSION AND OUTLOOK

Environmental concerns and the demand for sustainable production of chemicals have become a major issue in recent years. The advantages assigned to rhamnolipids have been demonstrated very conclusively and the potential for rhamnolipids replacing conventional surfactants in numerous applications has been demonstrated. However, in contrast to this situation, the actual application of rhamnolipids is still rather restricted. In this chapter, it was shown that current yields still have room for significant improvements. Potential targets for this include the use of low-cost substrates as well as strains engineered for rhamnolipid production. Consequently, processes known today are running far below their theoretical productivity. Also, the effective downstream processing of rhamnolipids is a neglected topic that shows an equally relevant potential for optimization. In this context, it still remains an open question whether rhamnolipids eventually will be produced and applied on larger scales.

REFERENCES

Abalos, A., A. Pinazo, M. R. Infante, M. Casals, F. García, and A. Manresa. 2001. Physicochemical and antimicrobial properties of new rhamnolipids produced by *Pseudomonas aeruginosa* AT10 from soybean oil refinery wastes. *Langmuir* 17 (5):1367–1371.

Andrä, J., J. Rademann, J. Howe, M. H. J. Koch, H. Heine, U. Zähringer, and K. Brandenburg. 2006. Endotoxin-like properties of a rhamnolipid exotoxin from *Burkholderia (Pseudomonas) plantarii*: Immune cell stimulation and biophysical characterization. *Biological Chemistry* 387 (3):301–10.

Arino, S., R. Marchal, and J. P. Vandecasteele. 1996. Identification and production of a rhamnolipidic biosurfactant by a *Pseudomonas* species. *Applied Microbiology and Biotechnology* 45 (1–2):162–168.

Babu, P. S., A. N. Vaidya, A. S. Bal, R. Kapur, A. Juwarkar, and P. Khanna. 1996. Kinetics of biosurfactant production by *Pseudomonas aeruginosa* strain BS2 from industrial wastes. *Biotechnology Letters* 18 (3): 263–268.

Benincasa, M., J. Contiero, M. A. Manresa, and I. O. Moraes. 2002. Rhamnolipid production by *Pseudomonas aeruginosa* LBI growing on soapstock as the sole carbon source. *Journal of Food Engineering* 54 (4): 283–288.

Bergström, S., H. Theorell, and H. Davide. 1946. On a metabolic product of *Ps. pyocyanea*, pyolipic acid, active against *Mycobact. tuberculosis. Arkiv för Kemi, Mineralogi och Geologi.* 23 A (13):1–12.

Burger, M., L. Glaser, and R. M. Burton. 1963. Enzymatic synthesis of a rhamnolipid by extracts of *Pseudomonas aeruginosa. Federation Proceedings* 21 (2):82.

Cabrera-Valladares, N., A. P. Richardson, C. Olvera, L. G. Trevino, E. Deziel, F. Lepine, and G. Soberon-Chavez. 2006. Monorhamnolipids and 3-(3-hydroxyalkanoyloxy)alkanoic acids (HAAs) production using *Escherichia coli* as a heterologous host. *Applied Microbiology and Biotechnology* 73 (1):187–194.

Cha, M., N. Lee, M. Kim, and S. Lee. 2008. Heterologous production of *Pseudomonas aeruginosa* EMS1 biosurfactant in *Pseudomonas putida*. *Bioresource Technology* 99 (7):2192–2199.

Chen, S. Y., W. B. Lu, Y. H. Wei, W. M. Chen, and J. S. Chang. 2007. Improved production of biosurfactant with newly isolated *Pseudomonas aeruginosa* S2. *Biotechnology Progress* 23 (3):661–666.

Davey, M. E., N. C. Caiazza, and G. A. O'Toole. 2003. Rhamnolipid surfactant production affects biofilm architecture in *Pseudomonas aeruginosa* PAO1. *Journal of Bacteriology* 185 (3):1027–1036.

De Lima, C. J. B., F. P. Franca, E. F. C. Servulo, A. A. Resende, and V. L. Cardoso. 2007. Enhancement of rhamnolipid production in residual soybean oil by an isolated strain of *Pseudomonas aeruginosa*. *Applied Biochemistry and Biotechnology* 137:463–470.

de Sousa, J. R., J. A. da Costa Correia, J. G. L. de Almeida, S. Rodrigues, O. D. L. Pessoa, V. M. M. Melo, and L. R. B. Goncalves. 2011. Evaluation of a co-product of biodiesel production as carbon source in the production of biosurfactant by *P. aeruginosa* MSIC02. *Process Biochemistry* 46 (9):1831–1839.

Develter, D., B. Walcarius, and S. Fleurackers. 2007. Biosurfactants in cleaning and cosmetics. In *Third International Conference on Renewable Resources & Biorefineries, 4–6 June 2007*. Ghent, Belgium.

Déziel, E., F. Lépine, S. Milot, and R. Villemur. 2000. Mass spectrometry monitoring of rhamnolipids from a growing culture of *Pseudomonas aeruginosa* strain 57RP. *Biochimica et Biophysica Acta—Molecular and Cell Biology of Lipids* 1485 (2–3):145–152.

Déziel, E., F. Lépine, S. Milot, and R. Villemur. 2003. *rhlA* is required for the production of a novel biosurfactant promoting swarming motility in *Pseudomonas aeruginosa*: 3-(3-hydroxyalkanoyloxy)alkanoic acids (HAAs), the precursors of rhamnolipids. *Microbiology-Sgm* 149:2005–2013.

DSMZ. German Collection of Microorganisms and Cell Cultures. Strains degrading/utilizing natural or xenobiotic compounds. http://www.dsmz.de/ [cited September 2013].

Dubeau, D., E. Déziel, D. E. Woods, and F. Lépine. 2009. *Burkholderia thailandensis* harbors two identical *rhl* gene clusters responsible for the biosynthesis of rhamnolipids. *BMC Microbiology* 9 (263).

Dubey, K., and A. Juwarkar. 2001. Distillery and curd whey wastes as viable alternative sources for biosurfactant production. *World Journal of Microbiology & Biotechnology* 17 (1):61–69.

Dunlap, C. E., and L. H. Chiang. 1980. Cellulose degradation—A common link. In *Utilization and Recycle of Agricultural Wastes and Residues*, M. L. Shuler (Ed.). Florida: CRC Press.

EBB. European Biodiesel Board: The EU biodiesel industry 2010. http://www.ebb-eu.org/ (cited September 2013).

Edwards, J. R., and J. A. Hayashi. 1965. Structure of a rhamnolipid from *Pseudomonas aeruginosa*. *Archives of Biochemistry and Biophysics* 111 (2):415.

Gartshore, J., Y. C. Lim, and D. G. Cooper. 2000. Quantitative analysis of biosurfactants using Fourier transform infrared (FT-IR) spectroscopy. *Biotechnology Letters* 22 (2):169–172.

Giani, C., D. Wullbrandt, R. Rothert, and J. Meiwes. 1997. Patent Number US005658793A. *Pseudomonas aeruginosa* and its use in a process for the biotechnological preparation of L-Rhamnose. Germany: Hoechst Aktiengesellschaft, Frankfurt am Main.

Guerra-Santos, L., O. Kappeli, and A. Fiechter. 1984. *Pseudomonas aeruginosa* biosurfactant production in continuous culture with glucose as carbon source. *Applied and Environmental Microbiology* 48 (2):301–305.

Haba, E., M. J. Espuny, M. Busquets, and A. Manresa. 2000. Screening and production of rhamnolipids by *Pseudomonas aeruginosa* 47T2 NCIB 40044 from waste frying oils. *Journal of Applied Microbiology* 88:379–387.

Hauser, G., and M. L. Karnovsky. 1957. Rhamnose and rhamnolipide biosynthesis by *Pseudomonas aeruginosa*. *Journal of Biological Chemistry* 224 (1):91–105.

Hauser, G., and M. L. Karnovsky. 1958. Studies on the biosynthesis of L-rhamnose. *Journal of Biological Chemistry* 233 (2):287–291.

Hembach, T. 1994. Untersuchungen zur mikrobiellen Umsetzung von Maiskeimöl zu Rhamnolipid. In *PhD Thesis*. Stuttgart, Germany: Universität Hohenheim.

Henkel, M., M. M. Müller, J. H. Kügler, R. B. Lovaglio, J. Contiero, C. Syldatk, and R. Hausmann. 2012. Rhamnolipids as biosurfactants from renewable resources: Concepts for next-generation rhamnolipid production. *Process Biochemistry* 47 (8):1207–1219.

Henkel, M., A. Schmidberger, C. Kühnert, J. Beuker, T. Bernard, T. Schwartz, C. Syldatk, and R. Hausmann. 2013. Kinetic modeling of the time course of N-butyryl-homoserine lactone concentration during batch cultivations of *Pseudomonas aeruginosa* PAO1. *Applied Microbiology and Biotechnology* 97 (17):7607–7616.

Heyd, M., A. Kohnert, T. H. Tan, M. Nusser, F. Kirschhofer, G. Brenner-Weiss, M. Franzreb, and S. Berensmeier. 2008. Development and trends of biosurfactant analysis and purification using rhamnolipids as an example. *Analytical and Bioanalytical Chemistry* 391 (5):1579–1590.

Hörmann, B., M. M. Müller, C. Syldatk, and R. Hausmann. 2010. Rhamnolipid production by *Burkholderia plantarii* DSM 9509(T). *European Journal of Lipid Science and Technology* 112 (6):674–680.

Jacquemin, L., R. Zeitoun, C. Sablayrolles, P.-Y. Pontalier, and L. Rigal. 2012. Evaluation of the technical and environmental performances of extraction and purification processes of arabinoxylans from wheat straw and bran. *Process Biochemistry* 47 (3):373–380.

Jarvis, F. G., and M. J. Johnson. 1949. A glyco-lipide produced by *Pseudomonas aeruginosa. Journal of the American Chemical Society* 71:4124–4126.

Kanehisa, M., and S. Goto. 2000. KEGG: Kyoto Encyclopedia of Genes and Genomes. *Nucleic Acids Research* 28 (1):27–30.

Kim, B. S., J. Y. Lee, and B. K. Hwang. 2000. *In vivo* control and *in vitro* antifungal activity of rhamnolipid B, a glycolipid antibiotic, against *Phytophthora capsici* and *Colletotrichum orbiculare. Pest Management Science* 56 (12):1029–1035.

Koch, A. K., J. Reiser, O. Kappeli, and A. Fiechter. 1988. Genetic construction of lactose-utilizing strains of *Pseudomonas aeruginosa* and their application in biosurfactant production. *Bio-Technology* 6 (11):1335–1339.

Lee, K. M., S. Hawang, S. D. Ha, J. Jang, D. Lim, and J. Kong. 2004. Rhamnolipid production in batch and fed-batch fermentation using *Pseudomoas aeruginosa* BYK-2 KCTC 18012P. *Biotechnology and Bioprocess Engineering* 9:267–273.

Makkar, R. S., and S. S. Cameotra. 1997. Utilization of molasses for biosurfactant production by two *Bacillus* strains at thermophilic conditions. *Journal of the American Oil Chemists Society* 74 (7):887–889.

Makkar, R. S., S. S. Cameotra, and I. M. Banat. 2011. Advances in utilization of renewable substrates for biosurfactant production. *AMB Express* 1 (5).

Maneerat, S. 2005. Production of biosurfactants using substrates from renewable-resources. *Songklanakarin Journal of Science and Technology* 27 (3):675–683.

Manresa, M. A., J. Bastida, M. E. Mercade, M. Robert, C. Deandres, M. J. Espuny, and J. Guinea. 1991. Kinetic studies on surfactant production by *Pseudomonas aeruginosa* 44T1. *Journal of Industrial Microbiology* 8 (2):133–136.

Mata-Sandoval, J., J. Karns, and A. Torrents. 1999. High-performance liquid chromatography method for the characterization of rhamnolipid mixtures produced by *Pseudomonas aeruginosa* UG2 on corn oil. *Journal of Chromatography A* 864:211–220.

Meijnen, J. P., J. H. de Winde, and H. J. Ruijssenaars. 2008. Engineering *Pseudomonas putida* S12 for efficient utilization of D-xylose and L-arabinose. *Applied and Environmental Microbiology* 74 (16):5031–5037.

Meijnen, J. P., J. H. de Winde, and H. J. Ruijssenaars. 2009. Establishment of oxidative D-xylose metabolism in *Pseudomonas putida* S12. *Applied and Environmental Microbiology* 75 (9):2784–2791.

Mercade, M. E., M. A. Manresa, M. Robert, M. J. Espuny, C. Deandres, and J. Guinea. 1993. Olive oil mill effluent (OOME)—New substrate for biosurfactant production. *Bioresource Technology* 43 (1):1–6.

Miller, E. N., L. R. Jarboe, P. C. Turner, P. Pharkya, L. P. Yomano, S. W. York, D. Nunn, K. T. Shanmugam, and L. O. Ingram. 2009. Furfural inhibits growth by limiting sulfur assimilation in ethanologenic *Escherichia coli* strain LY180. *Applied and Environmental Microbiology* 75 (19):6132–6141.

Mohan, P. K., G. Nakhla, and E. K. Yanful. 2006. Biokinetics of biodegradation of surfactants under aerobic, anoxic and anaerobic conditions. *Water Research* 40 (3):533–540.

Müller, M. M., and R. Hausmann. 2011. Regulatory and metabolic network of rhamnolipid biosynthesis: Traditional and advanced engineering towards biotechnological production. *Applied Microbiology and Biotechnology.* 91 (2):251–264.

Müller, M. M., B. Hörmann, M. Kugel, C. Syldatk, and R. Hausmann. 2011. Evaluation of rhamnolipid production capacity of *Pseudomonas aeruginosa* PAO1 in comparison to the rhamnolipid over-producer strains DSM 7108 and DSM 2874. *Applied Microbiology and Biotechnology* 89:585–592.

Müller, M. M., B. Hörmann, C. Syldatk, and R. Hausmann. 2010. *Pseudomonas aeruginosa* PAO1 as a model for rhamnolipid production in bioreactor systems. *Applied Microbiology and Biotechnology* 87 (1):167–174.

Mulligan, C. N., and B. F. Gibbs. 1993. Factors influencing the economics of biosurfactants. In *Biosurfactants: Production—Properties—Applications*, N. Kosaric (Ed.). New York: Marcel Dekker Inc.

Nielsen, C. J., D. M. Ferrin, and M. E. Stanghellini. 2006. Efficacy of biosurfactants in the management of *Phytophthora capsici* on pepper in recirculating hydroponic systems. *Canadian Journal of Plant Pathology-Revue Canadienne de Phytopathologie* 28 (3):450–460.

Nitschke, M., S. G. Costa, and J. Contiero. 2010. Structure and applications of a rhamnolipid surfactant produced in soybean oil waste. *Applied Biochemistry and Biotechnology* 160 (7):2066–2074.

Nitschke, M., S. G. Costa, R. Haddad, L. A. G. Goncalves, M. N. Eberlin, and J. Contiero. 2005. Oil wastes as unconventional substrates for rhamnolipid biosurfactant production by *Pseudomonas aeruginosa* LBI. *Biotechnology Progress* 21 (5):1562–1566.

Novaol. Raw Glycerin Safety Data Sheet. http://www.novaol.it/ [cited October 2013].

Ochsner, U. A., A. Fiechter, and J. Reiser. 1994a. Isolation, characterization, and expression in *Escherichia coli* of the *Pseudomonas aeruginosa rhlAB* genes encoding a rhamnosyltransferase involved in rhamnolipid biosurfactant synthesis. *The Journal of Biological Chemistry* 269 (0):1–9.

Ochsner, U. A., A. Koch, A. Fiechter, and J. Reiser. 1994b. Isolation and characterization of a regulatory gene affecting rhamnolipid biosurfactant synthesis in *Pseudomonas aeruginosa*. *Journal of Bacteriology* 176 (7):2044–2054.

Ochsner, U. A., J. Reiser, A. Fiechter, and B. Witholt. 1995. Production of *Pseudomonas aeruginosa* rhamnolipid biosurfactants in heterologous hosts. *Applied and Environmental Microbiology* 61 (9):3503–3506.

Patel, R. M., and A. J. Desai. 1997. Biosurfactant production by *Pseudomonas aeruginosa* GS3 from molasses. *Letters in Applied Microbiology* 25 (2):91–94.

Pearson, J. P., E. C. Pesci, and B. H. Iglewski. 1997. Roles of *Pseudomonas aeruginosa las* and *rhl* quorum-sensing systems in control of elastase and rhamnolipid biosynthesis genes. *Journal of Bacteriology* 179 (18):5756–5767.

Pesci, E. C., and B. H. Iglewski. 1997. The chain of command in *Pseudomonas* quorum sensing. *Trends in Microbiology* 5 (4):132–134.

Rahim, R., L. L. Burrows, M. A. Monteiro, M. B. Perry, and J. S. Lam. 2000. Involvement of the *rml* locus in core oligosaccharide and O polysaccharide assembly in *Pseudomonas aeruginosa*. *Microbiology-Sgm* 146:2803–2814.

Raza, Z. A., M. S. Khan, and Z. M. Khalid. 2007. Physicochemical and surface-active properties of biosurfactant produced using molasses by a *Pseudomonas aeruginosa* mutant. *Journal of Environmental Science and Health. Part A—Toxic/Hazardous Substances & Environmental Engineering* 42 (1):73–80.

Rehm, B. H. A., T. A. Mitsky, and A. Steinbüchel. 2001. Role of fatty acid de novo biosynthesis in polyhydroxyalkanoic acid (PHA) and rhamnolipid synthesis by *Pseudomonads*: Establishment of the transacylase (PhaG)-mediated pathway for PHA biosynthesis in *Escherichia coli*. *Applied and Environmental Microbiology* 67 (7):3102–3109.

Robert, M., M. E. Mercadé, M. P. Bosch, J. L. Parra, M. J. Espuny, A. Manresa, and J. Guinea. 1989. Effect of the carbon source on biosurfactant production by *Pseudomonas aeruginosa* 44T1. *Biotechnology Letters* 11 (12):871–874.

Santa Anna, L. M., G. V. Sebastian, E. P. Menezes, T. L. M. Alves, A. S. Santos, N. Pereira, and D. M. G. Freire. 2002. Production of biosurfactants from *Pseudomonas aeruginosa* PA1 isolated in oil environments. *Brazilian Journal of Chemical Engineering* 19 (2):159–166.

Schenk, T., I. Schuphan, and B. Schmidt. 1995. High-performance liquid-chromatographic determination of the rhamnolipids produced by *Pseudomonas aeruginosa*. *Journal of Chromatography A* 693 (1):7–13.

Schmidberger, A., M. Henkel, R. Hausmann, and T. Schwartz. 2013. Expression of genes involved in rhamnolipid synthesis in *Pseudomonas aeruginosa* PAO1 in a bioreactor cultivation. *Applied Microbiology and Biotechnology* 97 (13):5779–5791.

Siegmund, I. and F. Wagner. 1991. New method for detecting rhamnolipids excreted by *Pseudomonas* species during growth on mineral agar. *Biotechnology Techniques* 5 (4):265–268.

Soberón-Chávez, G., M. Aguirre-Ramírez, and R. Sánchez. 2005a. The *Pseudomonas aeruginosa* RhlA enzyme is involved in rhamnolipid and polyhydroxyalkanoate production. *Journal of Industrial Microbiology and Biotechnology* 32:675–677.

Soberón-Chávez, G., F. Lépine, and E. Déziel. 2005b. Production of rhamnolipids by *Pseudomonas aeruginosa*. *Applied Microbiology and Biotechnology* 68 (6):718–725.

Stover, C. K., X. Q. Pham, A. L. Erwin, S. D. Mizoguchi, P. Warrener, M. J. Hickey, F. S. Brinkman. et al. 2000. Complete genome sequence of *Pseudomonas aeruginosa* PAO1, an opportunistic pathogen. *Nature* 406 (6799):959–64.

Syldatk, C., S. Lang, U. Matulovic, and F. Wagner. 1985. Production of four interfacial active rhamnolipids from n-alkanes or glycerol by resting cells of *Pseudomonas* species DSM 2874. *Zeitschrift für Naturforschung* 40 (1–2):61–67.

Syldatk, C., S. Lang, F. Wagner, V. Wray, and L. Witte. 1985. Chemical and physical characterization of four interfacial-active rhamnolipids from Pseudomonas spec. DSM 2874 grown on n-alkanes. *Zeitschrift für Naturforschung* 40 (1–2):51–60.

Taherzadeh, M. J., and K. Karimi. 2007. Acid-based hydrolysis processes for ethanol from lignocellulosic materials: A review. *Bioresources* 2 (3):472–499.

Van Bogaert, I. N. A., K. Saerens, C. De Muynck, D. Develter, W. Soetaert, and E. J. Vandamme. 2007. Microbial production and application of sophorolipids. *Applied Microbiology and Biotechnology* 76 (1):23–34.

Van Bogaert, I. N. A., J. X. Zhang, and W. Soetaert. 2011. Microbial synthesis of sophorolipids. *Process Biochemistry* 46 (4):821–833.

Varnier, A. L., L. Sanchez, P. Vatsa, L. Boudesocque, A. Garcia-Brugger, F. Rabenoelina, A. Sorokin. et al. 2009. Bacterial rhamnolipids are novel MAMPs conferring resistance to *Botrytis cinerea* in grapevine. *Plant Cell and Environment* 32 (2):178–193.

Walter, V., C. Syldatk, and R. Hausmann. 2010. Microbial production of rhamnolipid biosurfactants. In *Encyclopedia of Industrial Biotechnology*, M. C. Flickinger (Ed.). Weinheim, Germany: Wiley-VCH Verlag GmbH & Co. KGaA.

Walter, Vanessa. 2009. New approaches for the economic production of rhamnolipid biosurfactants from renewable resources. In *PhD thesis*. Karlsruhe, Germany: Karlsruhe Institute of Technology (KIT).

Zhu, K., and C. O. Rock. 2008. RhlA converts beta-hydroxyacyl-acyl carrier protein intermediates in fatty acid synthesis to the beta-hydroxydecanoyl-beta-hydroxydecanoate component of rhamnolipids in *Pseudomonas aeruginosa*. *Journal of Bacteriology* 190 (9):3147–3154.

Zhu, Y., J. J. Gan, G. L. Zhang, B. Yao, W. J. Zhu, and Q. Meng. 2007. Reuse of waste frying oil for production of rhamnolipids using *Pseudomonas aeruginosa* zju.u1M. *Journal of Zhejiang University-Science A* 8 (9):1514–1520.

7 Utilization of Palm Sludge for Biosurfactant Production

Parveen Jamal, Wan Mohd Fazli Wan Nawawi, and Zahangir Alam

CONTENTS

7.1 Introduction 101
7.2 From Microbial Isolation to Media Screening 102
    7.2.1 Sludge Palm Oil as a Substrate 102
    7.2.2 Microbial Isolation and Screening 103
    7.2.3 Selection of Critical Medium Components 103
    7.2.4 Significant Medium Components 104
    7.2.5 Nonsignificant Medium Components 105
7.3 Optimization Study 105
    7.3.1 Univariate Optimization 106
    7.3.2 Multivariate Optimization 106
7.4 Media Optimization for Biosurfactant Production 107
    7.4.1 Co-Substrate Determination 107
    7.4.2 Univariate Optimization of Critical Media Requirement 108
    7.4.3 Multivariate Optimization of Critical Media Requirement 108
7.5 Preliminary Identification of Biosurfactant 112
7.6 Conclusion 113
References 113

7.1 INTRODUCTION

Demand for surfactant chemicals for household cleaning products, personal care sectors, agriculture, food, pharmaceutical, and environmental industries is steadily increasing. According to a 2013 Acmite Market Intelligence report [1], the world markets of surfactants reached US$26.8 billion in 2012, experiencing a 10% increase since 2010. These figures are predicted to increase by 3.8% annually in the coming years and, by 2016, the market is expected to reach US$31.1 billion. However, due to the potential hazard of synthetic surfactants toward human health and increasing consumer demand for chemical products that are both effective and environmentally compatible, it is natural to turn to the microbial world to fulfill this demand by means of biosurfactant utilization. Microbial-derived surfactants are produced on living surfaces mostly microbial cell surfaces, or excreted extracellularly, and contain hydrophilic and hydrophobic moieties capable of reducing surface tension and interfacial tension between individual molecules at the surface and interface. Such properties exhibit excellent detergency, emulsifying, foaming, and dispersing traits, which can be applied in various industries. They are also commercially promising alternatives to chemically synthesized surfactants due to their inherent biodegradability, lower toxicity, better foaming properties, and greater stability toward temperature and pH [2].

However, large-scale production of biosurfactants is still at its infancy due to expensive raw material, low production yield, and high purification cost. Selection of inexpensive and nutrient-rich raw materials is crucial to the economics of the process because it highly influences the overall production cost. Recently, several renewable substrates, especially from oily-based agroindustrial wastes, have been extensively studied for microbial surfactant production as it confers cost-free or low-cost feed stocks [3]. Mercade et al. [4] reported the use of olive oil mill effluents for rhamnolipid production by *Pseudomonas* sp. Soap stick oil has been used for rhamnolipid production with *P. aeruginosa* [5]. Mulligan and Cooper used water collected during drying of fuel grade peat [6]. Raza et al. [7] evaluated waste frying oil from canola, corn, and soybean as a substrate for rhamnolipid production by *Pseudomonas aeruginosa* mutant EBN-8. Several studies with water-immiscible raw material such as plant-derived oils and oil wastes have shown that they can act as effective and cheap raw materials for biosurfactant production. Biosurfactant products obtained by using water-soluble carbon sources such as glycerol, glucose, mannitol, and ethanol are reported to be inferior to those obtained with water-immiscible substrate such as *n*-alkanes and olive oil [8,9]. Banat [10] observed little biosurfactant production when cells were grown on readily available carbon sources. The production of biosurfactant was triggered only when all the soluble carbon was consumed and when a water-immiscible hydrocarbon was available. Rapeseed oil [11], canola oil, babassu oil, and corn oil [12,13] are plant-derived oils that have been used as raw material for biosurfactant production. Similarly, vegetable oils such as sunflower and soybean oils [14,15] were used for the production of rhamnolipid, sophorolipid, and mannosylerythritol lipid biosurfactants by various microorganisms.

Despite ongoing research using unconventional sources, selection of appropriate waste substrate is still a challenge. Researchers are facing the problem of finding a waste with the right balance of carbohydrates and lipids to support optimal growth of microorganisms and maximum production of biosurfactant [16]. Search for new strains for high productivity is also a challenge for the widespread application of microbial surfactants. In addition, improvement of process conditions through statistical optimization can be implemented as an effective approache to increase the production yield of biosurfactants. Enhanced product yield, closer conformance of the process output to target requirement and reduced process variability, and development time and cost can be realized by the application of statistical experimental design techniques in bioprocess development and optimization [17].

Therefore, the aim of this chapter is to demonstrate investigation and results on an inexpensive raw material derived from palm oil refinery waste for biosurfactant production by potential isolated strain as well as enhancing the development process through a series of optimization studies of nutritional requirement for maximum biosurfactant production. For this, lab research was conducted by the authors.

7.2 FROM MICROBIAL ISOLATION TO MEDIA SCREENING

7.2.1 SLUDGE PALM OIL AS A SUBSTRATE

Utilization of palm oil derivative for biosurfactant production has gained some interest recently. Thaniyavarn et al. [18] reported the production of biosurfactant by *P. aeruginosa* A41 using palm oil as a carbon source. In a separate study, Oliveira et al. [19] utilize palm oil as the sole carbon source for biosurfactant production by *P. aeruginosa* FR. Similarly, Marsudi et al. [20] reported direct utilization of palm oil for the simultaneous production of polyhydroxyalkanoates and rhamnolipids using the *P. aeruginosa* strain. However, only a very limited report concerning the usage of waste from palm oil industry for biosurfactant production has so far been documented. This fact triggers our interest to seek some potential application of waste from palm oil mill effluent (POME) in the biosurfactant field.

Given that Malaysia is one of the largest palm oil producers in the world, the management of the ever-increasing organic waste resulting from palm oil mill discharge has been one of the most-worrying environmental issues in the country. To alleviate this problem, a practical and economically viable approach is required. Extraction of crude and refined palm oil from fresh fruit requires steam for sterilization and substantial amounts of water for dilution, which later gets discharged in the form of POME. Sludge palm oil (SPO) is the floating residual oil that is separated during the initial stage of POME discharge. Owing to its low-grade quality, this residual oil cannot be used for a wide variety of application. However, the presence of high organic nutrient and mineral originating from the fibrous element of palm oil fruit, makes it possible to use waste from palm oil industry as a substrate for bacterial growth.

In this study, SPO was utilized for biosurfactant production as it is a cheaper substrate and produces a better environment through the waste management process. Samples of SPO were collected from West Palm Oil Mill, Carey Island, Selangor, Malaysia. It was orange in color and semi-solid in texture at room temperature. A separate study of the same samples by Hayyan et al. [21] showed that the moisture content of SPO was 1.2% and the solid part was attributed with an oil-rich fibrous residue that resulted from crude palm oil processing. The percentage of free fatty acid content was in a range of 21%–25%, and consisted mainly of palmitic acid (42.8%), oleic acid (39.6%), and linoleic acid (9.9%). The samples were stored at 4°C and brought to room temperature before use.

7.2.2 MICROBIAL ISOLATION AND SCREENING

In this research, 21 strains were isolated from mixed cultures originating from on-site isolation at various hydrocarbon-contaminated regions of West Palm Oil Mill, Carey Island, Selangor, Malaysia. Each isolate was subcultured three times prior to further experiments to ensure pure single isolate for each colony was obtained.

During microbial screening, seed cultures were prepared by growing primary inoculum of each bacterial colony until the optical density at 600 nm wavelength (OD_{600}) reached 1.85–1.87 absorbance units (AU), equivalent to approximately 4.57×10^7 cells/mL. It was then used to inoculate the production media at 2% (v/v). Initial production media that was used for liquid state fermentation had the following composition (g/L): glucose (5), meat extract (0.5), K_2HPO_4 (2.2), KH_2PO_4 (0.14), $NaNO_3$ (0.5), NH_4NO_3 (3.3), $CaCl_2$ (0.04), NaCl (0.04), $FeSO_4$ (0.2), $MgSO_4$ (0.6), and 2% (v/v) SPO.

The ability for biosurfactant production of each isolate was measured qualitatively using a drop collapse method [22] and quantitatively using surface tension measurement. All strains, which showed a large diameter of collapsed drop and a low surface tension value (<40 mN/m), were selected for further studies. During the centrifugation of fermentation broth, only a few strains showed the accumulation of microbial biomass in the form of a pellet, which indicated their efficiency to grow in the supplied production medium. The microbial growth in production medium was used as a parameter during initial screening as most biosurfactants are considered as secondary metabolites and are growth associated. Their growth on water-immiscible aliphatic hydrocarbon has been associated with the production of surfactant. In our study, *Klebsiella pneumoniae* WMF02 was identified as the most promising biosurfactant producer in the presence of palm sludge oil. Further details regarding this microbial screening can be found in authors' previous study [23].

7.2.3 SELECTION OF CRITICAL MEDIUM COMPONENTS

Selection of critical medium components was done in this research after the potential biosurfactant-producing strain was determined from the previous experiment. The production medium mentioned earlier was a mixture of SPO, and inorganic and organic salts that supported growth of the microorganisms. Because nutrient requirements vary among the organisms, not all the salts present in the media were necessary for their survival, growth, and biosurfactant production. To determine significant media constituents, experiments were designed in which the medium was kept deficient in one or the other components.

Statistical analysis using Plackett–Burman design is suitable for screening the effect of a large numbers of parameters in an experiment and can be used to determine the factors that give significant contributions [24]. This approach was used to design and analyze the experiment during

TABLE 7.1

Percentage Contribution and Effect of Each Nutritional Constituent

Constituents	Source for	Contribution (%)	Effect	Significance
Sucrose	Carbon	2.05	−3.72	Yes
Glucose	Carbon	9.94	−8.19	Yes
Meat extract	Nitrogen	0.01	−0.26	No
K_2HPO_4	K^+ and PO_4^{3-}	34.14	+15.19	Yes
KH_2PO_4	K^+ and PO_4^{3-}	0.26	−1.31	No
$NaNO_3$	Nitrogen	9.50	−8.01	Yes
NH_4NO_3	Nitrogen	0.45	+1.75	No
$CaCl_2$	Ca^{2+}	0.18	+1.10	No
NaCl	Na^+ and Cl^-	0.04	−0.52	No
$FeSO_4$	Fe^{2+}	34.65	−15.30	Yes
$MgSO_4$	Mg^{2+}	8.78	−7.70	Yes

critical nutritional requirement screening. A new carbon source, sucrose (raw table sugar), was also introduced as one of the variables during the screening process, to evaluate its contribution as compared with glucose, which is expensive. The selected raw material, SPO, was fixed at a concentration of 2% (v/v). Fermentation was carried out for 30 h and the surface tension of the cell-free broth was measured using Du Nuoy ring methodology. The time needed to achieve the maximum concentration of biosurfactant is correlated indirectly with the surface tension reduction. Thus, surface tension was used as a main response due to the fact that there was a big difference between iterations. Lowering of surface tension is directly correlated with higher biosurfactant concentration.

The end results of our media screening study are indicated in Table 7.1. Interested readers may find further explanations about the Plackett–Burman response and analysis in the previous published report [23]. Six components such as K_2HPO_4, $FeSO_4$, $NaNO_3$, $MgSO_4$, glucose, and sucrose were found to be the most significant for biosurfactant production in this study based on anlaysis of variance and half normal probability plot.

By analyzing the response obtained under the experimental Plackett–Burman design, the effect of each nutritional component can be calculated using the standard equation:

$$\text{Effect} = 2\left[\sum R(H) - \sum R(L)\right]/N \tag{7.1}$$

where, R(H) denotes all responses when component was at high levels, R(L) denotes all responses when component was at a low level, and N is total number of iterations. Positive effect explains that if a higher concentration is used, a better response could be achieved, while a negative effect means lower concentrations are favored for better results.

7.2.4 SIGNIFICANT MEDIUM COMPONENTS

It is evident from Table 7.1 that K_2HPO_4 shows the highest level of significance with contribution percentage of 34.1%. This salt is a source of K^+ and PO_4^{3-} and can also act as a buffer for media.

Both carbon sources glucose and sucrose evaluated in the screening study showed significant effect in negative terms. Therefore, lower amounts of either of these sources will give lower surface tension and higher biosurfactant production. It seems counterintuitive at first as most literature showed that glucose was the main contributor during bacterial metabolism of biosurfactant production [8,9]. However, it needs to be mentioned that in this study, glucose and sucrose were used only as co-substrate in the medium. These simple sugars were only used to feed the bacteria during initial phase of growth. After the populations matured, a complex carbon source, SPO was utilized as the main substrate in the production medium. The contribution of both co-substrates as co-carbon source was minimal. This would introduce the probability of using sucrose alone as co-carbon source compared with more expensive glucose in later studies.

Another critical component required for the growth and biosurfactant production is a source of Fe^{2+} ions from $FeSO_4$ in the medium. The contribution of $FeSO_4$ toward total effects was found to be 34.65%. According to other researchers, excess iron in media resulted in acidification, which caused biosurfactant precipitation and loss in cell viability [25]. This is also in agreement with our studies, which showed a better result at a lower concentration of $FeSO_4$.

The concentration level of $MgSO_4$, which acts as a source of Mg^{2+} ions in the medium, was also found to influence biosurfactant production with 8.78% contribution. This salt was found to possess a negative effect that signifies its effectiveness at lower concentrations in experimental design. Similar study on evaluating the important nutrient requirement for marine bacterium also reported that a lower concentration of $MgSO_4$ influenced the biosurfactant production [26].

$NaNO_3$, as a nitrogen source, was also found to have large contribution (9.50%) at a lower concentration. Hommel et al. [27] explain this nitrogen-limiting phenomenon by suggesting that it is the absolute quantity of nitrogen and not its relative concentration that appears to be important for optimum biomass yield. Besides growth, nitrogen played an important role in structural lipopeptide biosurfactant as a peptide part. This justified its significance in the biosurfactant synthesis, particularly for the lipopeptide type. Pruthi and Cameotra [28] also found that sodium nitrate gave the best emulsifying activities during biosurfactant production from *Pseudomonas putida* compared with ammonium sulfate and ammonium hydroxide when used as nitrogen source.

7.2.5 NONSIGNIFICANT MEDIUM COMPONENTS

KH_2PO_4, a source of K^+ and PO_4^{3-}, was found to be statistically insignificant, contributing only 0.26% to the total effect compared with the K_2HPO_4 salt. This might have been caused by major involvement of more potassium ions during biosurfactant production. The addition of hydrogen atoms also made the KH_2PO_4 more acidic compared with K_2HPO_4, which in turn affected the total broth pH during fermentation.

NaCl, a source of Na^+ and Cl^- ions in the medium showed no significant value (0.04% contribution). This was due to the important role of K^+ ions in the medium that masks the significance of Na^+ ions. $CaCl_2$, the source of calcium in the mineral salts medium, was also found to be insignificant with only 0.18% contribution. The insignificant nature of $CaCl_2$ was due to the non-involvement of Ca^{2+} in any important biochemical reaction and the presence of a more vital Mg^{2+} ion in the production medium. Both NaCl and $CaCl_2$ results were coherent with a screening study of significant medium components in enhanced biosurfactant production by a marine bacterium [26].

Meat extract as a complex organic nitrogen source does not show any significant contribution in this study. This suggests the tendency of the selected strain in favoring inorganic nitrogen source for biosurfactant production. Mulligan and Gibbs [29] in their study on biosurfactant production by *B. subtilis* showed the superiority of NH_4NO_3 as a nitrogen source over NH_4Cl or $NaNO_3$. However, in the current study NH_4NO_3 contributed less compared with another nitrogen source, $NaNO_3$, which is a significant variable.

7.3 OPTIMIZATION STUDY

Bioprocess development is the primary step toward commercialization of all biotechnological products and every profit-making biotechnology industry, including biosurfactant production. Such development requires an efficient and economical bioprocess foundation in terms of media formulation plus optimal process conditions from the upstream to downstream processing level. Any attempt to increase the yield of a biosurfactant will induce the maximum or the optimum productivity for the whole process and simultaneously reduce the overall cost, time, and labor [26].

7.3.1 UNIVARIATE OPTIMIZATION

The classical method of medium optimization involves changing one variable at a time, keeping the others at fixed levels. This approach is usually termed as the one-factor-at-a-time (OFAT) technique. Prominent textbooks and academic articles currently favor multivariate optimization such as fractional factorial designs and response surface methodology (RSM) where multiple factors are changed at once. Despite these criticisms, some researchers have articulated a role for OFAT and showed that they can be more effective than fractional factorials under certain conditions: number of runs is limited, primary goal is to attain improvements in the system, and experimental error is not large compared with factor effects, which must be additive and independent of each other [30]. The reality that OFAT is inferior in certain situations does not eliminate the possibility that this traditional technique has a useful place in the experimental toolbox. For example, some situations are limited by time and resource pressure, and only free experimentation, such as OFAT, is possible. Abouseoud et al. [8] had used the OFAT method exclusively to evaluate different carbon and nitrogen sources in the production of biosurfactant by *Pseudomonas fluorescens*.

Other situations demand some immediate improvement to the running condition. For example, using OFAT, it is also possible to learn something after each experimental run and it does not require the entire set of runs to be completed. Additional and more complete experiments can be run afterwards to tune the system. Mutalik et al. [31] had used the OFAT method during their preliminary screening experiments to identify critical medium components for biosurfactant production from *Rhodococcus* spp. MTCC 2574 before subjecting them to further experiments for obtaining a result of statistical multivariate optimization. Similarly, Sivapathasekaran et al. [32] optimized nutritional medium for biosurfactant production from *Bacillus circulans* using RSM only after identifying the critical components in modified marine medium using the OFAT technique.

Many experiments are conducted on processes and factors where little is known. It is, therefore, impossible to determine the variable ranges for the experiment with a reasonable degree of confidence. The only way to determine the possible ranges is to experiment on the system, and the OFAT framework can determine the maximum and minimum settings [33]. For example, in one study unrelated to biosurfactant, Bari et al. [34] utilized the OFAT approach to determine the possible optimum level of parameters for citric acid production before going to multivariate statistical experiments, as not much information was available in the literature on the production of citric acid using lignocellulosic substrates.

7.3.2 MULTIVARIATE OPTIMIZATION

The univariate optimization method of changing one variable at a time, while keeping the others at fixed levels, is laborious, time consuming, and does not guarantee the determination of the optimal conditions for metabolite production. Multivariate optimization usually involves a designed experiment of planned series of tests in which purposeful changes are made to the input variables of a process or a system so that the reasons for changes in the output response may be observed and identified [35]. Czitrom [36] claims a designed experiment that, performed with the help of statistical software, is a more effective way to determine the impact of two or more factors on a response compared with OFAT experiments because

1. Fewer resources (experiments, time, material, etc.) are required for the amount of information obtained. This can be of major importance in industry, where experiments can be very expensive and time consuming.
2. More precise effect of each factor can be estimated. This can be done by using more observations to estimate an effect. For example, for full and fractional factorial designs, all the observations are used to estimate the effect of each factor and each interaction, while typically only two of the observations in OFAT experiments are used to estimate the effect of each factor.

3. The interaction between factors can be estimated systematically. Interactions are not estimable from OFAT experiments. Researchers who are not using designed experiments often perform a hit-and-miss scattershot sequence of experiments from which it may be possible to estimate interactions, but they usually do not estimate them.
4. There is experimental information in a larger region of the factor space. This improves the prediction of the response in the factor space by reducing the variability of the estimates of the response in the factor space, and makes process optimization more efficient because the optimal solution is searched for over the entire factor space.

Currently, attention is being laid on the routine utilization of statistical experimental design in the improvement of the existing process or development of a new process or product. Appearance of computer facilities for acquiring and analyzing mass amount of data has made the routine use of statistical optimization and analysis techniques a reality [37]. Software such as Design Expert, STATISTICA, Minitab, and Matlab are among the most popular statistical utilities used by researchers and engineers in designing and analyzing their experiments more efficiently.

Optimization strategy, based on RSM, has been used by various researchers. This multivariate optimization method is a collection of statistical techniques that uses design of experiment (DOE) for building models, evaluating the effects of factors and searching for the optimum conditions [38]. Statistical experimental designs such as central composite design [31], Box-Benhken [39], and Taguchi [40] designs have been used to optimize media constituents and increase product yields. Prior to optimization, initial screening using a statistical approach was often employed to identify critically influential variables. Experimental design such as Plackett–Burman [26] and fractional factorial design [38] are often helpful in reducing the numbers of factors to be considered in optimization experiments. Although DOE provides a powerful means to achieve breakthrough improvements in product quality and process efficiency, it requires planning, discipline, and some knowledge of statistics. To have successful experimental design, the objective of the experiment should be well defined. The focus of the study may be to screen out the factors that are not critical to the process, or it may be to optimize a few critical factors. A well-defined objective leads the experimenter to the correct DOE. The experiment also should be replicated in order to dampen uncontrollable variation. Replication improves the chance of detecting a statistically significant effect in the midst of natural process variation. Run order in which the experiment is conducted should be randomized to avoid influence by uncontrolled variables such as tool wear and changing of ambient temperature. These changes, which often are time related, can significantly influence the response [41].

7.4 MEDIA OPTIMIZATION FOR BIOSURFACTANT PRODUCTION

7.4.1 Co-Substrate Determination

During earlier Plackett–Burman screening studies, both glucose and sucrose were identified among the critical media components. It is assumed that both simple sugars play a vital role in feeding the microorganisms during the initial phase of growth. After the population matured, a complex carbon source, SPO was utilized as the main substrate in the production medium.

Isolated *K. pneumoniae* WMF02, which was used in this study, was known as both glucose and sucrose fermenter. Thompson et al. [42] reported that the growth of *K. pneumoniae* is unique among bacterial species in its ability to metabolize not only simple sugar like glucose but also has the ability to breakdown disaccharides such as sucrose and some isomeric fructose. In this study, post-analysis showed that the contribution of both co-substrates as co-carbon source was minimal relative to others critical media components. This would introduce the possibility of using either sucrose only or glucose only as co-carbon sources. Figure 7.1 shows the results obtained when the production media consists of glucose only and sucrose only as their co-carbon sources.

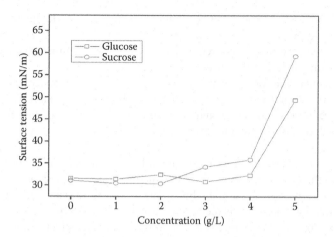

FIGURE 7.1   Effect of glucose and sucrose on the surface-reducing capabilities by *Klebsiella pneumoniae* WMF02 in shake flask culture.

It was observed that there is not much difference in reduction of surface activity when using either co-substrate at a lower concentration (below 5 g/L). Sepahy et al. [43] have drawn the same conclusion when they studied the effect of glucose and sucrose on biosurfactant production by isolates from Iranian oil. They observed the level of surface and interfacial tension reduction in glucose was similar when sucrose medium was used. Owing to the high cost associated with pure commercial glucose, raw table sugar (sucrose) was selected as a co-carbon source for further study.

7.4.2   Univariate Optimization of Critical Media Requirement

In this study, due to limited information available in the literature on the production of biosurfactant by *Klebsiella* sp. genus, OFAT optimization was used to determine the possible optimum level of significant variables obtained from previous Plackett–Burman results. Selected co-substrate was then subjected to OFAT optimization along with other nutritional components, that is, SPO, $MgSO_4$, $FeSO_4$, $NaNO_3$, and K_2HPO_4. Fermentation was carried out in 100 mL Erlenmeyer flasks with a 50 mL working volume with 2% (v/v) primary inocula for 30 h at 37°C and 180 rpm. Reduction of surface tension was used in an indirect method to determine the amount of biosurfactant after fermentation. Lowering of surface tension is directly correlated with higher biosurfactant concentration.

Table 7.2 shows the result for OFAT optimization which suggests optimal concentration of nutrients at (g/L): sucrose (5), $MgSO_4$ (0.4), $FeSO_4$ (0.3), $NaNO_3$ (2), K_2HPO_4 (4), and 4% (v/v) of SPO. This step greatly reduced the surface tension of nonoptimized control from 36.2 to 25.70 mN/m.

7.4.3   Multivariate Optimization of Critical Media Requirement

Statistical optimization was used as a viable approach to optimize the nutritional constituent for biosurfactant production in various literature. Mukherjee et al. [26] reported 84.7% increase in biosurfactant yield over nonoptimized medium after statistical screening of nutritional parameters for marine bacterium using the Plackett–Burman design. Joshi et al. [39] determined significant components for lichensyin production by *Bacillus lincheniformis* K51 using the Plackett–Burman design, followed by full factorial central composite design to optimize the concentration level of significant variables. With the optimization procedure, the relative biosurfactant yield expressed as the critical micelle dilution reported to be ten times higher than that obtained in the nonoptimized reference medium. Mutalik et al. [31] obtained 3.4-fold increase of crude biosurfactant yield after using response surface optimization for the production of biosurfactant from *Rhodococcus* spp. MTCC 2574.

TABLE 7.2

Tabulated Data from OFAT Media Optimization Study

Nutrient	Concentration				
SPO	0%	2%	4%	6%	8%
	58.69	32.10	28.7	28.7	29.22
Sucrose	0 g/L	5 g/L	10 g/L	15 g/L	20 g/L
	28.05	26.6	28.33	28.59	28.55
MgSO$_4$	0 g/L	0.2 g/L	0.4 g/L	0.6 g/L	0.8 g/L
	28.4	27.96	27.16	27.72	32.16
FeSO$_4$	0.05 g/L	0.1 g/L	0.2 g/L	0.3 g/L	0.4 g/L
	34.61	26.41	26.00	25.90	26.17
NaNO$_3$	0 g/L	1 g/L	2 g/L	3 g/L	4 g/L
	29.04	28.59	26.09	27.55	27.8
K$_2$HPO$_4$	0 g/L	2 g/L	4 g/L	8 g/L	10 g/L
	45.22	26.74	25.70	26.25	27.84

Source: Adapted from P. Jamal et al. *Aust. J. Basic Appl. Sci.*, 6(1): 100–108, 2012.

In this study, SPO, sucrose, MgSO$_4$, and K$_2$HPO$_4$ were further optimized using an experimental design derived from multivariate statistical software to see the possible effect of the interaction between variables and optimal concentration of media constituents for maximum biosurfactant production. Optimal levels obtained in the OFAT optimization study were used as a central level for each independent variable in design matrix. FeSO$_4$ and NaNO$_3$ concentration were fixed at 0.3 g/L and 2.0 g/L, respectively, to reduce the number of total experiments designed by the software. This selection was based on the consistency of triplicate reading of surface tension during OFAT study of FeSO$_4$ and NaNO$_3$.

A face-centered central composite design (FCCCD) from statistical software package Design-Expert (version 7.1.6, Stat-Ease, Minneapolis, USA) was used to design and analyze the experiment. The FCCCD employed had four independent variables, namely, concentration of SPO (A), concentration of sucrose (B), concentration of MgSO$_4$ (C), and concentration of K$_2$HPO$_4$ (D). Each of the independent variables was studied at three levels with 30 experiments (data not shown).

Throughout this media optimization experiment, the critical micelle at ten times dilution (CMD^{-1}) was used as a response. Surface tension can also be used as response, provided the difference and variation of each response is high. However, during this optimization study, there was not much variation in surface tension. The reason was, once the surface reached a saturation point due to surfactant, the value of surface tension became constant; thus, addition of more surfactant did not lead to further surface tension reduction. Two solutions having similar surface tension reading may contain different concentrations of surfactant. Hence, the critical micelle dilution at ten times dilutions (CMD^{-1}) was preferred as the response in this optimization study. Lower CMD^{-1} value correlates with higher production of biosurfactant.

On the basis of the parameter estimate and statistical analysis of the experimental results, an empirical relationship between response and tested variable in coded units was obtained through the following regression equation:

$$Y = 33.71 - 8.86A - 0.55B + 3.88C - 2.22D + 5.49A^2 + 2.40B^2 + 0.77C^2 + 2.60D^2$$
$$+ 4.65AB - 1.46AC + 2.96AD + 2.23BC + 0.63BD - 0.87CD \qquad (7.2)$$

where Y is the predicted response (CMD^{-1}) and A, B, C, and D are the coded values for SPO, sucrose, MgSO$_4$, and K$_2$HPO$_4$, respectively.

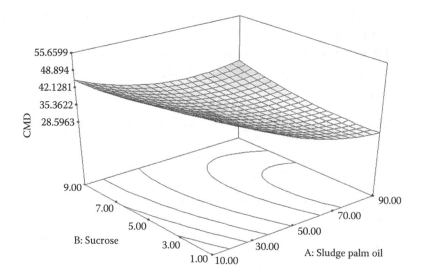

FIGURE 7.2 3D response surface for interaction between sucrose and sludge palm oil.

Response surface plots are graphical representations of the regression equation to determine the optimum values of the variables. The main goal of the response surface is to hunt efficiently for the optimum values of the variables in order to maximize the response. Each contour curve represents an infinite number of combinations of two test variables with another set of two maintained at their respective zero level. Figure 7.2 shows the interaction between SPO and sucrose. As illustrated, lower CMD^{-1} value, which corresponds to a higher amount of biosurfactant, can be obtained by increasing the concentration of SPO while decreasing the amount of co-substrate, sucrose.

In this study, the upper limit of SPO used was at 90 g/L. This is because an excess of oily substance from SPO will reduce the oxygen level in fermentation broth and consequently affect the growth of microorganisms [45]. Furthermore, beyond this concentration, the recovery and separation process after fermentation will prove to be difficult.

An elliptical response plot was obtained via second-order quadratic equation of interaction between SPO concentration and K_2HPO_4 concentration (Figure 7.3). The result showed that the

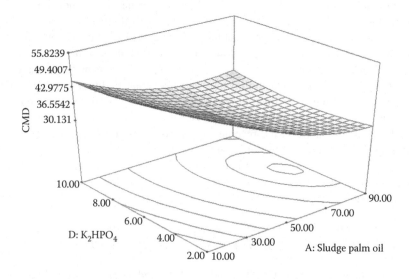

FIGURE 7.3 3D response surface for interaction between K_2HPO_4 and sludge palm oil.

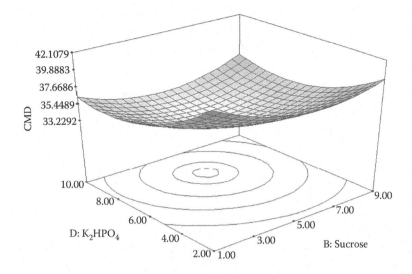

FIGURE 7.4 3D response surface for interaction between K$_2$HPO$_4$ and sucrose.

response (CMD^{-1}) was affected considerably by varying the concentration of SPO and K$_2$HPO$_4$. The model suggested that maximum biosurfactant production was achieved when SPO and K$_2$HPO$_4$ concentration level was about 80 g/L and 6 g/L, respectively.

Figure 7.4 shows the elliptical response surface plot of CMD^{-1} response as a function of K$_2$HPO$_4$ and sucrose concentration. The greater reduction of CMD^{-1} was predicted at higher concentration of K$_2$HPO$_4$ while maintaining the concentration of sucrose around 3–4 g/L. Both Figures 7.3 and 7.4 show that optimal concentration of K$_2$HPO$_4$ at 6–7 g/L is required for optimal biosurfactant production. There was no consensus from the literature on whether K$_2$HPO$_4$ is limiting or non-limiting mineral salt. For example, Huszcza and Burczyk [46] reported its usage as high as 9.5 g/L for biosurfactant production by *Bacillus coagulans* while Mukherjee et al. [26] reported its usage as low as 1.1 g/L as optimal concentration after the optimization process of nutritional parameters.

In this study, media optimization results were in agreement with previous Plackett–Burman analysis that revealed K$_2$HPO$_4$ indeed had a positive main effect toward biosurfactant production by selected strain. Figure 7.5 shows the interaction between MgSO$_4$ with SPO.

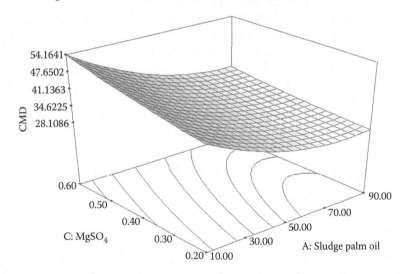

FIGURE 7.5 3D response surface for interaction between MgSO$_4$ and sludge palm oil.

TABLE 7.3

Validation of Developed Quadratic Model and Optimum Medium Constituents

Run	SPO	Sucrose	MgSO$_4$	K$_2$HPO$_4$	Response, CMD^{-1}		Error %
					Predicted	Experimental	
1	80.00	2.00	0.20	6.00	28.96	30.69	5.63
2	85.00	2.00	0.20	7.00	29.07	29.54	1.59
3	85.00	4.00	0.20	6.00	28.16	30.33	7.17
4	80.00	4.00	0.10	6.00	27.97	30.47	8.22
5	80.00	3.00	0.15	6.00	28.41	30.83	7.86

Lower CMD^{-1} value was obtained at the lower level of MgSO$_4$ (below 0.2 g/L) in the optimization study. However, during validation studies, reduction of MgSO$_4$ from 0.2 g/L to 0.1 g/L shows less difference in term of CMD^{-1} reduction (Table 7.3, run 3 and 4). This shows that the effect of multiple metal cations and trace minerals, when used together for biosurfactant production, is greater than their utility when used individually. MgSO$_4$, which was a limiting phenomenon in this experiment, is in agreement with other studies conducted by Mukherjee et al. [26] during statistical screening of nutritional parameters by marine bacterium. They found that optimal concentration of MgSO$_4$ at 0.3 g/L could yield more biosurfactant compared with when MgSO$_4$ was used at high concentration. In a separate study, Makkar and Cameotra [47] used a different concentration of MgSO$_4$ and examined their effect on bacterial growth and biosurfactant yield. The maximal yield of biosurfactant obtained at a low concentration of Mg^{2+} (2.43 mM) and a concentration of Mg^{2+} higher than 9.74 mM inhibited the biosurfactant production by almost 80%.

In order to validate the reliability of developed regression model, numerical optimization based on desirability function was carried out. A series of experiments were conducted to determine the optimum CMD^{-1} reduction when the variables were set to a different concentration as suggested by the Design Expert software. The concentration of independent variables along with predicted and observed response is shown in Table 7.3.

All experimental values tested during the validation study have percentages of error lower than 10% when compared with predicted response. Ten percent difference is acceptable because the experiment deals with living organisms that bring some uncertainty during the study. This indicates that media optimization by FCCCD was a capable and reliable way to optimize nutritional requirement for biosurfactant production. For example, the validation study lowered the CMD^{-1} value (29.54 mN/m) below 2% difference compared with predicted value. It was obtained at concentration of 85 g/L SPO, 2 g/L sucrose, 0.2 g/L MgSO$_4$, and 7 g/L K$_2$HPO$_4$ in addition of fixed concentration of 0.3 g/L FeSO$_4$ and 2 g/L NaNO$_3$. These optimal medium compositions greatly reduced the surface tension of nonoptimized control from 36.2 to 27 mN/m. These values correspond to an improvement of biosurfactant yield from 20 to 85 g/L, which account, for 75% increase from nonoptimized culture.

7.5 PRELIMINARY IDENTIFICATION OF BIOSURFACTANT

Chloroform: Methanol (2:1) extraction was carried out to separate extracellular biosurfactant from other constituents, proteins, polysaccharides, cell debris, sugars, and so on. Acidification of the sample was done (pH 2) prior to extraction to enhance the extraction yield of biosurfactants. At low pH, biosurfactants will be present in their protonated form and, hence, are less soluble in aqueous solution [48]. Three biochemical tests for detecting glycolipids, lipopeptide, and phospholipid were carried out in order to determine the type of biosurfactant produced in this study.

CTAB/methylene-blue agar is a semiquantitative assay for the detection of extracellular glycolipids or other anionic surfactants. Blue agar plates composed of 15 g/L agar bacteriological no. 1, 0.2 g/L N-Cetyl-N,N,N-trimethylammonium bromide, and 0.005 g/L methylene blue were used to detect extracellular glycolipid production. Formation of dark blue halos around the colonies indicates rhamnolipid biosurfactant production [49]. If glycolipid biosurfactants are secreted by the microbes growing on the plate, they form dark blue halos. In this study, a negative result was obtained when the strain was cultivated on the light blue mineral salts plate containing cationic surfactant cetyltrimethylammonium bromide.

A biuret test was carried out in order to detect lipopeptide biosurfactant in the sample [50]. Two milliliters of crude extract solution were first heated at 70°C before being mixed with 1 M NaOH solution. Drops of 1% $CuSO_4$ were added slowly and observation on color change was recorded. A positive result is indicated by a violet or pink ring, due to the reaction of peptide bond proteins or short-chain polypeptides, respectively. In this study, a negative result was obtained (no color change to violet) when crude biosurfactant extract was dissolved in biuret reagent.

A phosphate test was carried out for phospholipid detection. Ten drops of 6 M HNO_3 were added to 2 mL of crude extract solution followed by heating at 70°C. Drops of 5% ammonium molybdate were added slowly and observation was recorded. Formation of yellow color may be followed by slow formation of a fine yellow precipitate indicating the presence of phospholipid biosurfactant [51]. In this study, a colorless phosphate assay solution that changed to yellow indicated a positive result for phospholipid. In addition, formation of fine yellow precipitate was also observed after a few minutes. The result of preliminary identification revealed that the biosurfactant produced by *K. pneumoniae* WMF02 in this study was of a phospholipid type.

7.6 CONCLUSION

In this study, SPO was introduced as a novel substrate for biosurfactant production as a solution to provide an effective way to manage palm industry waste with less cost and high production of value-added end product. Utilization of complex hydrocarbons of SPO requires microorganisms that are able to solubilize the hydrophobic molecules for nutrient uptake during fermentation while subsequently producing high amounts of biosurfactants.

The selection of microbial strains for biosurfactant production from SPO via liquid state bioconversion was done based on the growth rate, drop collapse method, and surface tension activity. Of the 21 strains studied, *K. pneumoniae* WMF02 was selected as the potential biosurfactant producer. Sucrose, glucose, $NaNO_3$, $FeSO_4$, $MgSO_4$, and K_2HPO_4 were identified by the Plackett–Burman design as important parameters for improving biosurfactant production. K_2HPO_4 showed positive effect, whereas sucrose, glucose, $NaNO_3$, $FeSO_4$, and $MgSO_4$ had a negative main effect.

We also reported statistical optimization of nutritional requirement as an effective approache to increase the production yield of biosurfactants. Improved product yield, closer conformance of the process output to target requirement and reduced process variability and development time and cost can be realized by the application of statistical experimental design methodology during experimentation. In this study, media optimization with OFAT and FCCCD statistical design revealed an optimal concentration of nutrients, which were found at (g/L) sucrose (2), $MgSO_4$ (0.2), $FeSO_4$ (0.3), $NaNO_3$ (2), K_2HPO_4 (7), and SPO (85). This step greatly reduced the surface tension by about 75% from nonoptimized control at 36.2–27 mN/m (CMD^{-1} = 30 mN/m with biosurfactant yield of 85 g/L).

REFERENCES

1. Acmite Market Intelligence, *Market Report: World Surfactant Market,* Retrieved September 8, 2013. http://www.acmite.com/market-reports/chemicals/global-surfactant-market.html

2. J. D. Desai, and I. M. Banat, Microbial production of surfactants and their commercial potential. *Microbiol. Mol. Biol. R.*, *61*(1): 47–64, 1997.

3. S. Joshi, C. Bharucha, S. Jha, S. Yadav, S. A. Nerurkar, and A. J. Desai, Biosurfactant production using molasses and whey under thermophilic conditions. *Bioresource Technol.*, *99*: 195–199, 2008.

4. M. E. Mercade, M. A. Manresa, M. Robert, M. J. Espuny, C. Andres, and J. Guinea, Olive oil mill effluent (OOME): New substrate for biosurfactant production. *Bioresource Technol.*, *43*: 1–6, 1993.

5. M. F. Mercade, and M. A. Manresa, The use of agroindustrial byproducts for biosurfactant production. *J. Am. Oil Chem. Soc.*, *71*: 61–64, 1994.

6. C. N. Mulligan, and D. G. Cooper, Pressate from dewatering as substrate for bacterial growth. *Appl. Environ. Biotechnol.*, *50*: 160–162, 1985.

7. Z. A. Raza, M. S. Khan, Z. M. Khalid, and A. Rehman, Production kinetics and tensioactive characteristics of biosurfactant from a *Pseudomonas aeruginosa* mutant grown on waste frying oils. *Biotechnol. Lett.*, *28*: 1623–1631, 2006.

8. M. Abouseoud, R. Maachi, A. Amrane, S. Boudergua, and A. Nabi, Evaluation of different carbon and nitrogen sources in production of biosurfactant by *Pseudomonas fluorescens*. *Desalination*, *223*: 143–151, 2008.

9. M. Robert, M. E. Mercade, M. P. Bosh, J. L. Parra, M. J. Espuny, M. A. Manresa, and J. Guinea, Effect of the carbon source on biosurfactant production by *Pseudomonas aeruginosa* 44T1. *Biotechnol. Lett.*, *11*: 871–874, 1989.

10. I. M. Banat, Characterization of biosurfactants and their use in pollution removal—state of the art. *Acta Biotechnol.*, *15*: 251–267, 1995.

11. K. Trummler, F. Effenberger, and C. Syldatk, An integrated microbial/enzymatic process for production of rhamnolipids and L-(+)-rhamnose from rapeseed oil with *Pseudomonas* sp. DSM 2874. *Eur. J. Lipid Sci. Technol.*, *105*: 563–571, 2003.

12. M. H. Vance-Harrop, N. B. Gusmao, and G. M. Campos-Takaki, New bioemulsifier produced by *Candida lipolytica* using D-Glucose and Babassu oil as carbon sources. *Braz. J. Microbiol.*, *34*: 120–123, 2003.

13. G. Pekin, E. Vardar-Sukan, and N. Kosaric, Production of sophorolipids from *Candida bombicola* ATCC 22214 using Turkish corn oil and honey. *Eng. Life Sci.*, *5*: 357–362, 2005.

14. K. S. M. Rahman, T. J. Rahman, S. McClean, R. Marchant, and I. M. Banat, Rhamnolipid biosurfactant production by strains of *Pseudomonas aeruginosa* using low-cost raw materials. *Biotechnol. Progr.*, *18*: 1277–1281, 2002.

15. H. S. Kim, J. W. Jeon, B. H. Kim, C. Y. Ahn, H. M. Oh, and B. D. Yoon, Extracellular production of a glycolipid biosurfactant, mannosylerythritol lipid, by *Candida* sp. SY16 using fed batch fermentation. *Appl. Microbiol. Biotechnol.*, *70*: 391–395, 2006.

16. R. S. Makkar, and S. S. Cameotra, Biosurfactant production by microorganisms on unconventional carbon sources. *J. Surfactants Deterg.*, *2*(2): 237–241, 1999.

17. R. Sen, and T. Swaminathan, Application of response-surface methodology to evaluate the optimum environmental conditions for the enhanced production of surfactin. *Appl. Microbiol. Biotechnol.*, *47*: 358–363, 1997.

18. J. Thaniyavarn, A. Chongchin, N. Wanitsuksombut, S. Thaniyavarn, P. Pinphanichakam, N. Leepipatpiboon, M. Morikawa, and S. Kanaya, Biosurfactant production by *Pseudomonas aeruginosa* A41 using palm oil as carbon source. *J. Gen. Appl. Microbiol.*, *52*: 215–222, 2006.

19. F. J. S. Oliveira, L. Vasquez, and F. P. de Franca, Biosurfactant production by *Pseudomonas aeruginosa* FR using palm oil. *Appl. Biochem. Biotechnol.*, *129*: 129–132, 2006.

20. S. Marsudi, H. Unno, and K. Hori, Palm oil utilization for the simultaneous production of polyhydroxyalkanoates and rhamnolipids by *Pseudomonas aeruginosa*. *Appl. Microbiol. Biotechnol.*, *78*: 955–961, 2008.

21. A. Hayyan, M. Z. Alam, M. E. S. Mirghani, N. A. Kabbashi, N. I. N. M. Hakimi, Y. M. Siran, and S. Tahiruddin, Production of biodiesel from sludge palm oil by esterification process. *J. Energy Power Eng.*, *4*(1): 11–17, 2008.

22. D. K. Jain, D. L. Collins-Thompson, H. Lee, and J. T. Trevors, A drop-collapsing test for screening biosurfactant-producing microorganisms. *J. Microbiol. Meth.*, *13*: 271–279, 1991.

23. W. M. F. W. Nawawi, P. Jamal, and M. Z. Alam, Utilization of sludge palm oil as a novel substrate for biosurfactant production. *Bioresource Technol.*, *101* (23): 9241–9247, 2010.

24. R. L. Plackett, and J. P. Burman, The design of optimum multifactorial experiments. *Biometrica*, *33*: 305–325, 1946.

25. Y. H. Wei, and I. M. Chu, Enhancement of surfactin production in iron-enriched media by *Bacillus subtilis* ATCC 21332. *Enzyme Microb. Technol.*, *22*: 724–728, 1998.

26. S. Mukherjee, P. Das, C. Sivapathasekaran, and R. Sen, Enhanced production of biosurfactant by a marine bacterium on statistical screening of nutritional parameters. *Biochem. Eng. J., 42*: 254–260, 2008.

27. R. K. Hommel, O. Stuwer, W. Stuber, D. Haferburg, and H. P. Kleber, Production of water-soluble surface-active exolipids by *Torulopsis apicola. Appl. Microbiol. Biotechnol., 26*: 199–205, 1987.

28. V. Pruthi, and S. S. Cameotra, Effect of nutrients on optimal production of biosurfactants by *Pseudomonas putida*—A Gujarat oil field isolate. *J. Surfactants Deterg., 6*: 65–68, 2003.

29. C. N. Mulligan, and B. F. Gibbs, Factors influencing the economics of biosurfactants, in *Biosurfactant: Production, Properties, Applications*, Vol. 48, Surfactant Science Series (N. Kosaric, ed.), Marcel Dekker, New York, 1993, pp. 329–371.

30. C. Daniel, One-at-a-time plans. *J. Am. Statist. Assoc., 68*: 353–360, 1973.

31. S. R. Mutalik, B. K. Vaidya, R. M. Joshi, K. M. Desai, and S. N. Nene, Use of response surface optimization for the production of biosurfactant from *Rhodococcus* spp. MTCC 2574. *Bioresource Technol., 99*: 7875–7880, 2008.

32. C. Sivapathasekaran, S. Mukherjee, and R. Sen, Optimization of a marine medium for augmented biosurfactant production. *Int. J. Chem. React. Eng.*, 8: A92 m 2010.

33. M. Friedman, and L. J. Savage, Planning experiments seeking maxima, in *Selected Techniques of Statistical Analysis* (C. Eisenhart, M. W. Hastay, and W. A. Wallis, eds.), McGraw-Hill, New York, 1947, pp. 365–372.

34. M. N. Bari, M. Z. Alam, S. A. Muyibi, P. Jamal, and A. Al-Mamun, Improvement of production of citric acid from oil palm empty fruit bunches: Optimization of media by statistical experimental designs. *Bioresource Technol., 100*: 3113–3120, 2009.

35. D. C. Montgomery, *Design and Analysis of Experiments*, 3rd ed., Wiley, New York, 1991.

36. V. Czitrom, One-factor-at-a-time versus designed experiments. *Am. Stat., 53*(2): 126–131, 1999.

37. M. Winkler, Optimization and time-profiling in fermentation processes, in *Progress in Industrial Microbiology*, Vol. 25, (M. E. Bushell, ed.), Elsevier, Amsterdam, 1988, pp. 91–150.

38. L. Rodrigues, J. Teixeira, R. Oliveira, and H. Van der Mei, Response surface optimization of the medium components for the production of biosurfactants by probiotic bacteria. *Process Biochem., 41*: 1–10, 2006.

39. S. Joshi, S. Yadav, A. Nerurkar, and A. J. Desai, Statistical optimization of medium components for the production of biosurfactant by *Bacillus licheniformis* K51. *J. Microbiol. Biotechnol., 17*(2): 313–319, 2007.

40. Y. H. Wei, C. C. Lai, and J. S. Chang, Using Taguchi experimental design methods to optimize trace element composition for enhanced surfactin production by *Bacillus subtilitis* ATCC 21332. *Process Biochem., 42*: 40–45, 2007.

41. M. J. Anderson, and S. L. Kraber, Keys to successful designed experiments. Paper presented at the Quality Management Division of American Society for Quality Conference, 1999.

42. J. Thompson, S. A. Robrish, S. Immel, F. W. Lichtenthaler, B. G. Hall, and A. Pikis, Metabolism of sucrose and its five linkage-isomeric α-D-glucosyl-D-fructoses by Klebsiella pneumoniae. *J. Biol. Chem.* 276: 37,415–37,425, 2001.

43. A. A. Sepahy, M. M. Assadi, V. Saggadian, and A. Noohi, Production of biosurfactant from Iranian oil fields by isolated *Bacilli. Int. J. Environ. Sci. Technol., 1*(4): 287–293, 2005.

44. P. Jamal, W. M. F. W. Nawawi, and M. Z. Alam, Optimum medium components for biosurfactant production by *Klebsiella pneumoniae* WMF02 utilizing sludge palm oil as a substrate. *Aust. J. Basic Appl. Sci., 6*(1): 100–108, 2012.

45. S. M. Sauid, and V. P. S. Murthy, Effect of palm oil on oxygen transfer in a stirred tank bioreactor. *J. Appl. Sci., 10*(21): 2745–2747, 2010.

46. E. Huszcza, and B. Burczyk, Biosurfactant production by *Bacillus coagulans. J. Surfactants Deterg., 6*(1): 61–64, 2003.

47. R. S. Makkar, and S. S. Cameotra, Effects of various nutritional supplements on biosurfactant production by strain of *Bacillus subtlitis* at 45°C. *J. Surfactants Deterg., 5*(1): 11–17, 2002.

48. M. Heyd, A. Kohnert, T. H. Tan, M. Nusser, F. Kirschhofer, G. Brenner-Weiss, M. Franzreb, and S. Berensmeier, Development and trends of biosurfactant analysis and purification using rhamnolipids as an example. *Anal. Bioanal. Chem., 391*: 1579–1590, 2008.

49. I. Siegmund, and F. Wagner, New method for detecting rhamnolipids excreted by *Pseudomonas* species during growth on mineral agar. *Biotechnol. Tech., 5*: 265–268, 1991.

50. C. Feigner, F. Besson, and G. Michel, Studies on lipopeptide biosynthesis by *Bacillus subtilis*: isolation and characterization of iturin, surfactin mutants. *FEMS Microbiol. Lett., 127*: 11–15, 1995.

51. G. C. Okpokwasili, and A. A. Ibiene, Enhancement of recovery of residual oil using a biosurfactant slug. *Afr. J. Biotechnol., 5*(5): 453–456, 2006.

8 Bioreactors for the Production of Biosurfactants

Janina Beuker, Christoph Syldatk, and Rudolf Hausmann

CONTENTS

8.1 Introduction 117
8.2 Fermentation Strategies 117
    8.2.1 Batch Cultivation 117
    8.2.2 Fed-Batch Cultivation 118
    8.2.3 Continuous Cultivation 118
8.3 Obstacles of Biosurfactant Production in a Bioreactor 118
    8.3.1 Dispersing 118
    8.3.2 Foaming 119
        8.3.2.1 Foam Disruption 119
        8.3.2.2 Foam Avoidance 121
8.4 Conclusion 125
References 126

8.1 INTRODUCTION

Biosurfactants are a heterogeneous group of microbial metabolites and, therefore, exhibit vastly different physical properties. Especially with respect to foaming properties, reduction of surface tension, and emulsification capacities, this has consequences for handling biosurfactant fermentations in bioreactors. For example, the biosurfactants rhamnolipid, surfactin, and mannosylerythritol lipids can be grouped into high-foaming biosurfactants, whereas sophorolipids are low-foaming biosurfactants (Rau et al. 2005; Hirata et al. 2009; Müller et al. 2010).

Owing to this diversity, several bioreactors have been reported for biosurfactant production reaching from "standard" constructions like stirred tank bioreactors (Müller et al. 2010) to "exotic" constructions like the rotating disc bioreactor (Chtioui et al. 2012). This chapter presents different approaches to encounter the obstacles regarding biosurfactant production from the perspective of equipment used for cultivation without providing an economic evaluation.

8.2 FERMENTATION STRATEGIES

Regarding fermentation strategies in biosurfactant production, no consistent methods have been applied. Batch as well as fed batch and continuous processes have been reported.

8.2.1 BATCH CULTIVATION

Batch cultivations have been reported in many biosurfactant production processes (Davis et al. 1999, 2001; Yeh et al. 2006; Müller et al. 2010). The production of rhamnolipid with *Pseudomonas aeruginosa* PAO1 as a model, for instance, is realized by a batch fermentation process (Müller et al. 2010). In batch fermentation strategies, no regulation of the amount of substrates in the bioreactor or growth rate over time is possible. However, due to complex regulation in rhamnolipid production coupled with the quorum-sensing system, applied fed-batch strategies in rhamnolipid production are based on heuristic approaches leading to a production comparable to the batch processes. Next to batch fermentation, sequential batch processes are also used in biosurfactant production (Pornsunthorntawee et al. 2009).

8.2.2 Fed-Batch Cultivation

When using fed-batch strategies, concentrations of substrates and the growth rate can be controlled by feeding. This may lead to effective process strategies achieving high amounts of biosurfactants by fermentation. One example is the production of sophorolipids (Davila et al. 1992; Pekin et al. 2005; Van Bogaert et al. 2007). Using fed-batch cultivation, the production process can be divided into two phases (Davila et al. 1992). In the first phase biomass is produced, whereas sophorolipid production is relatively low. Upon nitrate depletion, biomass production stops and an extensive production of sophorolipids starts by utilizing the fed substrates glucose and rapeseed ethyl esters. Product inhibition is circumvented by cultivation parameter leading to a low solubility of sophorolipids in the aqueous media. These fermentation procedures lead to a production of 320 g/L sophorolipids with a weight yield of up to 65% with respect to the carbon source. The highest yield of sophorolipids reported in a fed-batch process was 422 g/L utilizing a two-stage process (Daniel et al. 1998).

8.2.3 Continuous Cultivation

A third fermentation strategy is continuous fermentation. Two examples are the production of surfactin (Lin et al. 1994; Noah et al. 2002; Chen et al. 2006) and rhamnolipids (Guerra-Santos et al. 1984; Reiling et al. 1986; Gruber 1991). Different systems were established exhibiting quite low dilution rates of 0.065 h^{-1} in the case of rhamnolipids and 0.1–0.2 h^{-1} in the case of surfactin for an optimal biosurfactant production.

8.3 OBSTACLES OF BIOSURFACTANT PRODUCTION IN A BIOREACTOR

The production of biosurfactants involves several problematic elements. Often a hydrophobic carbon source is advantageous regarding induction of the production regulation of biosurfactants (Henkel et al. 2012). Furthermore, other two phase systems are described to be advantageous using immiscible aqueous phases such as a polymer and salt in the water (Drouin and Cooper 1992). In this case, *Bacillus subtilis* and surfactin tend to concentrate in different phases reducing the end product inhibition and, therefore, leading to a higher overall biosurfactant production. However, these two-phase systems have to be dispersed properly to ensure proper biosurfactant production.

Another obstacle evolves during the production of high foaming biosurfactants, namely the foaming itself. If a hydrophobic carbon source is chosen, it might have antifoaming activities in the beginning depending on the type of source as well as on the media (Vardar-Sukan 1988; Rau et al. 2005). However, the hydrophobic carbon source is utilized during the production process. Additionally, cleavage products such as fatty acids have an amphiphilic character that could lead to an increased foaming activity (Kanicky et al. 2000). Therefore, a foam control is strongly needed.

8.3.1 Dispersing

Owing to their amphiphilic character, biosurfactants are capable of dispersing hydrophobic substances in the culture broth (Zhang and Miller 1992). Therefore, dispersing issues of the hydrophobic phase arise especially in the very beginning of the fermentation process. In many biosurfactant production processes stirred tank reactors are equipped with rushton turbines, which are radial impellors. These impellers lack the capacity for axial mixing. This could lead to dispersing issues of the hydrophobic carbon source due to phase separation on top of the fermentation media. Additionally, the rushton turbines lack the capacity to efficiently mix the foamy fermentation broth at the end of the production process (Abdel-Mawgoud et al. 2011). In this context, a combination of an axial propeller and a radial impeller housed in a draft tube is proposed by Abdel-Mawgoud et al. (2011) to enhance the dispersion of the hydrophobic substrate by forced vertical circulation as described by Walas (1997). Additionally, a foam stirrer may be used in this setup also to efficiently mix the foamy fermentation broth. These stirrers consist of large blades fostering foam agitation.

8.3.2 FOAMING

Generally, foams are dispersions of a gas in a liquid. Structure-wise, two different categories are distinguishable (Pugh 1996). Kugelschaum or sphere foam mainly occurs near to the liquid surface consisting of small nearly round bubbles separated by thick films. After foam ripening, polyeder-schaum may occur composed of thin bubbles separating films, called lamella and plateau borders at the junctions of two lamellas (Pelton 2002). The ripening of foam involves different mechanisms. The water reduction by drainage is caused by gravitational forces and capillary effects leading to thinner lamellas and dryer foam. Furthermore, differences in Laplace pressure lead to gas diffusion from small bubbles to bigger ones and therefore to a coarser foam (Colin 2012).

Regarding unstable kinetics, transient foams with a lifetime of seconds can be distinguished from metastable, permanent foam with a lifetime of hours or days (Ghildyal et al. 1988). Metastable, permanent foams are preferably formed in a liquid containing amphiphilic molecules, for example, biosurfactants. In a hydrophilic surrounding, the hydrophobic moisture of the surfactant arranges outside of the liquid phase due to hydrophobic–hydrophilic repulsions. This leads to an alignment of surfactant molecules on the liquid borders influencing the stability of its foam (Pelton 2002). Several different forces occur between two foam bubbles stabilizing or destabilizing the foam. Stabilizing effects of amphiphilic molecules are, for example, steric stabilization (Pelton 2002) and the Gibbs–Marangori-Elasticity (Vardar-Sukan 1998). Steric stabilization occurs if, for instance, an electrically charged surfactant is used. In this case, the surfaces of adjacent bubbles repulse each other, hindering liquid drainage that leads to bubble stabilization. If local thinning of bubbles occurs, the concentration of amphiphilic molecules reduces at this area. This gradient results in a movement of amphiphilic molecules accompanied by part of the bulk liquid to the thinned area to balance surface tension forces. This balancing is called the Gibbs–Marangori-Elasticity. However, destabilizing effects like van der Waals forces affect adjacent bubbles leading to thinning of the lamella (Pelton 2002). If the repulsing forces are too low to separate adjacent bubbles, coalescence occurs, whereas the quantity of coalescence increases with dryer foam.

Foaming issues arise especially during the production of high-foaming biosurfactants in aerated and agitated processes using hydrophilic carbon sources. One strategy to encounter foam formation is foam disruption during the biosurfactant production process. A second method is the strategic avoidance of foam formation using specific fermentation strategies and constructions. This part of the chapter will provide an overview of the different strategies.

8.3.2.1 Foam Disruption

Foam disruption is the principle of destroying foam during fermentation. Various strategies may be applied.

8.3.2.1.1 Antifoaming Agents

An often-applied strategy to control foaming is the usage of chemical antifoaming agents. Extensive overviews concerning antifoaming agents and their method of action are given by Pelton (2002), Pugh (1996), and Junker (2007). The composition of many antifoaming agents is complex and partly unknown, but usually contains surfactants, oil, and hydrophobic particles in the range of 1–10 µm (Pelton 2002). Several antifoaming agents grouped by their ingredients are summarized by Junker (2007). The antifoaming agents can be divided into two classes based on their mode of action. "Defoamers" are added during fermentation to disrupt foam and "antifoamers" are added prior to fermentation to prevent foam. Nevertheless, so far, the way of action is complex and not fully understood as shown by Pugh (1996), summarizing the different approaches to explain foam-breaking actions of antifoaming agents. Even though the usage of antifoaming agents seems to be an easy and convenient method to control the foam, these agents also exhibit some disadvantages. The usage of antifoam agents has an impact on oxygen transfer rates (Vardar-Sukan 1998). Additionally, antifoam agents are rather expensive in comparison to the low-priced product. Furthermore, down-stream processing of biosurfactants becomes extremely difficult due to the similar physiological and chemical behavior of antifoam and biosurfactants. These issues lead to higher overall costs for the production of biosurfactants.

8.3.2.1.2 Mechanical Foam Disruption

As an alternative to chemical antifoaming agents, mechanical possibilities for foam disruption have been developed. The main principles of mechanical foam disruption are shear stress, as well as pressure interaction inside of the foam breaker and collision of the condensed foam against the pri-mary foam and the bioreactor walls (Zlokarnik 1999). Several mechanical foam-disrupting devices are described in the literature. A mechanical foam breaker may be mounted to the agitator shaft (Deshpande and Barigou 1999). Also, separate foam breakers are described either in a stirring tank reactor or in a bubble column (Andou et al. 1997; Takesono et al. 2003; Müller et al. 2010). Separate foam breakers are supplementary constructional systems provided with its own powerful gear. However, Furchner and Mersmann (1990) developed an equation (Equation 8.1) for minimum velocity, to be applied for breaking the primary foam in a rotor stator defoamer, stating that the critical peripheral velocity for foam breaking is more or less independent of the defoamer type and its diameter.

Minimum velocity for foam breaking for primary foam in a rotor stator defoamer (Furchner and Mersmann 1990):

$$w = 7.33 * We^{1/2} * \left(\frac{2 * \Delta\sigma * l}{d_b^2 * \rho_L} \right)^{1/2} * \varphi^{-1/4} \tag{8.1}$$

Notation:

w	[m/s]	Minimum velocity for foam breaking
We	$0.0209 * r_1 * \rho_L * w^2 * d_B/(\Delta\sigma * l)$	Weber number of a plateau border
σ	[N/m]	Surface tension
l	[m]	Characteristic length of acceleration
d_B	[m]	Bubble diameter
ρ	[kg/m^3]	Density
φ_L	[m^3/m^3]	Liquid hold-up of foams
r_1	[m]	Radius of curvature of a plateau border

However, high amounts of secondary foam accumulated in the bioreactor may result in flood-ing of the foam breaker, which cannot be avoided by higher rotor speeds (Furchner and Mersmann 1990).

Nevertheless, due to an additional power unit and additional maintenance, mechanical foam disruption devices may lead to increased production costs for the respective biosurfactants (Norris and Ribbons 1970; Hoeks et al. 1997).

8.3.2.2 Foam Avoidance

A second strategy to control foam issues is to avoid foam formation in the beginning. However, antifoam agents could be applied and therefore prevent foaming from the start. Antifoaming agents have been discussed in the previous paragraph. Next to chemical antifoaming agents, optimized fermentation strategies could also be used to prevent or control foaming.

8.3.2.2.1 Rotating Disc Bioreactor

An exotic strategy to avoid foam formation that is presented in the literature is the rotating disk contractor/bioreactor (Chtioui et al. 2012), which is shown in Figure 8.1. This bioreactor circumvents bubble formation by avoiding gas bubbles in the liquid. The aeration is realized by surface contact of the flowing fermentation broth as well as the rotating discs. In this article, different airflow rates and amounts of discs in the bioreactor have been compared. The microorganism *Bacillus subtilis* grows in this kind of bioreactor planctonic as well as in biofilms on the rotating discs. Increasing the airflow rate from 100 to 200 L h^{-1} reduced the amount of biofilm and increased the amount of planctonic biomass leading to a negligible impact of the biofilm on the total amount of biomass. By increasing the number of discs in the bioreactor from 7 to 14 discs, the amount of biofilm as well as the amount of planctonic biomass increased. However, regarding biosurfactant production increasing the aeration rate from 100 L h^{-1} to 200 L h^{-1}, this resulted in a higher surfactin per biomass yield (71 mg g^{-1} compared to 52 mg g^{-1}) and lower fengicin per biomass yield (175 mg g^{-1} compared to 236 mg g^{-1}). Increasing the number of discs from 7 to 14 discs resulted in higher surfactin per biomass yield (68 mg g^{-1} compared to 52 mg g^{-1}) and a higher fengicin per biomass yield (268 mg g^{-1} compared to 236 mg g^{-1}). These dependencies gave the possibility to influence the production of lipopeptides by bioreactor design. Nevertheless, oxygen transfer limitations were reported in all bioreactor setups leading to low overall microbial growth. However, biosurfactant production was realized without any foaming issues.

8.3.2.2.2 Bubble-Free Membrane Bioreactor

Another strategy to realize aeration without bubble formation is the usage of a bubble-free membrane bioreactor. Such bioreactors were initially used in animal cell culture (Schneider et al. 1995) or in wastewater treatment (Brindle and Stephenson 1996). Giving the possibility to transfer oxygen bubble-free into a liquid, this strategy was also adopted for the production of biosurfactants (Coutte et al. 2010). The concept of this strategy is based on the diffusion of oxygen through a hollow fiber membrane into the surrounding liquid without bubble formation (Ahmed and Semmens 1996). The membranes consist of hydrophobic materials such as Teflon (polytetraflouruethylene), polypropylene, polyethersulfone, or polyethylene allowing an oxygen transfer due to their porosity.

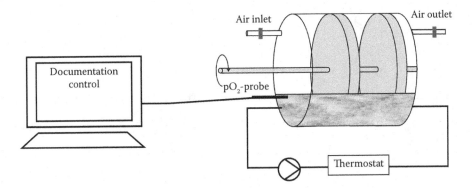

FIGURE 8.1 Schematic representation of the rotating disc bioreactor. (Described by Chtioui, O. et al. 2012. *Process Biochemistry* 47 (12):2020–2024.)

Despite contact with the surrounding liquid, the membranes and their pores are not penetrated by the liquid surrounding and stay dry in case of the usage of gases (Ahmed et al. 2004). Using this principle, Coutte et al. (2010) developed three different bubbleless bioreactors for the production of surfactin and fengycin by *Bacillus subtilis*, differing in membrane material and design. The different bioreactors are shown in Figure 8.2. Two different materials (polyethersulfone [PES], pore size 0.65 µm, area 2.5 m^2 and polypropylene [PP], pore size 0.2 µm, area 1.9 m^2) were tested as an external aeration module. In the first case, the culture flowed inside the hollow fibers at 0.021 m s^{-1} with an air pressure on the outside of 0.5 bar, whereas in the second case the culture flowed outside the hollow fibers at 0.11 m s^{-1} and a freely vented airflow inside the fibers. Additionally, a submerged polypropylene module was evaluated (pore size 0.05 µm, area 0.375 m^2). These three reactor concepts were tested and compared with a standard aerated foaming bioreactor. The values of this comparison are summarized in Table 8.1. The initial oxygen volumetric mass transfer coefficient (k_La) of the foaming bioreactor with a value of 27.29 h^{-1} was comparable to the k_La values of the different membrane reactors with a k_La of 39.36 h^{-1} for the PES external module, 8.99 h^{-1} for the PP external, and 28.24 h^{-1} for the PP internal module. However, one main characteristic of the bubbleless bioreactors was the adsorption of biomass and surfactin onto the membranes. From the overall surfactin produced up to 83% adsorbed on the membranes in case of the PES external module (19–32% in case of the PP hollow fibers). Evaluating the biomass on the hollow fibers aeration modules, 23–24% of the total biomass produced was adsorbed onto the external large hollow fiber PES and PP membranes, whereas in the case of the smaller submerged PP module, about 4% of the biomass was adsorbed. These adsorptions influenced the oxygen transfer. The adsorption of the biomass and surfactin decreased the k_La of the PES external module to 4.07 h^{-1}, of the PP external module to 0.91 h^{-1} and of the internal PP module to 1.75 h^{-1}.

Depending on the type of surfaces and the pore size, the amount of adsorbed surfactin and biomass differed as well as their impact on the k_La. For example, regarding the submerged PP module, the adsorbed surfactin had the biggest influence on the k_La with a relative k_La decrease of 91%, whereas the relative k_La decrease for the external PP module by surfactin was just 15%.

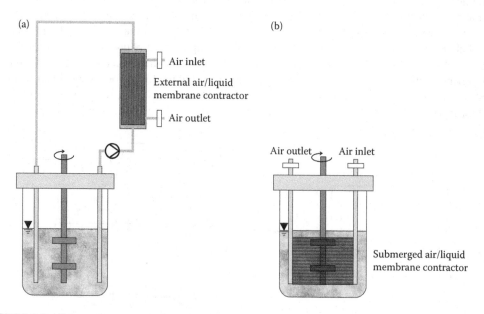

FIGURE 8.2 Schematic representation of the bubbleless membrane reactors, (a) bioreactor with external hollow fiber module, (b) bioreactor with submerged hollow fiber module. (Described by Coutte, F. et al. 2010. *Applied Microbiology and Biotechnology* 87 (2):499–507.)

TABLE 8.1
Parameter of the Bubbleless Bioreactors Compared to a Standard Aerated Foaming Bioreactor Summarized according to Coutte et al. (2010)

Parameter \ Bioreactor	External PES	External PP	Submerged PP	Standard Bioreactor
Type of aeration	External module	External module	Submerged module	Sparger
Material	Polyethersulfone	Polypropylene	Polypropylene	—
Pore size	0.65 μm	0.2 μm	0.05 μm	—
Area	2.5 m²	1.9 m²	0.375 m²	—
Adsorption surfactin	83%	19%	32%	—
	604 from 727 mg	130 from 690 mg	150 from 469 mg	
Adsorption biomass	23%	24%	4%	—
	2.1 from 9.3 g	1.2 from 5 g	0.3 from 8.4	
Initial $k_L a$	39.36 h⁻¹	8.99 h⁻¹	28.24 h⁻¹	27.29 h⁻¹
Decreased $k_L a$	4.07 h⁻¹	0.91 h⁻¹	1.75 h⁻¹	—

Another noticeable difference of the membrane reactors compared among each other and to the foaming reactor was the ratio of produced surfactin to fengicin. This ratio may be dependent on the oxygen transfer, whereas with lower oxygen transfer rates a larger amount of fengicin was produced. Another example for a bubble-free bioreactor is described by Pinzon et al. (2013). In this case, cells were immobilized outside of the fiber and the medium was circulating inside of the fibers. Additionally, nitrate instead of oxygen was used as electron acceptor. However, when using this bioreactor concept satisfactory microbial growth as well as biosurfactant production could be achieved without bubble formation.

8.3.2.2.3 Solid State Cultivation

Another fermentation strategy to avoid foaming is solid state cultivation. This topic is discussed in the section "Solid state cultivation of microorganisms for biosurfactant production."

8.3.2.2.4 Foam Fractionation

In situ product removal by foam fractionation leads to a reduced amount of surface active components in the culture broth and is therefore, also a method to reach a controllable foam level. Foam fractionation is based on the physicochemical properties of biosurfactants. Owing to their surface, activity, biosurfactants adsorb on the surface of air bubbles. The adsorption is dependent on different parameters, for example, on the type of biosurfactant and in case of rhamnolipid on the pH and the ionic strength of the solution (Helvaci et al. 2004; Özdemir et al. 2004). In the foam fractionation method, arising foam leaves the bioreactor and is collected in an external foam collector and collapses either by time, shear stress, or the usage of acid (Davis et al. 2001; Yeh et al. 2006; Winterburn et al. 2011). The collapsed foam is called "foamate." Foam fractionation as an *in situ* product removal method to separate and concentrate biosurfactants has already been described in several publications mainly for surfactin production. One possibility of such a strategy is coupling the air outlet with an external foam collector either cooled or non-cooled in a batch or a fed-batch fermentation (Cooper et al. 1981; Yeh et al. 2006; Guez et al. 2007; Chenikher et al. 2010). Yeh et al. (2006) additionally coupled the foam collector to a cell cycler and a surfactin precipitator. In this case, the liquid fermentation broth was pumped back into the bioreactor, whereas the surfactin was immediately precipitated. Next to using the gas outlet of a bioreactor, specific devices for foam fractionation were also constructed to achieve a satisfactory foam fractionation outcome (Davis et al. 2001; Winterburn et al. 2011) as shown in Figure 8.3. These fractionation columns could be used either in a non-integrated or in an integrated way. In a non-integrated method, the foam fractionation process is unconnected to the fermentation process and a defined volume of either fermentation broth or cell-free broth is poured into the foam column that is equipped with an own aeration device (Figure 8.3a). In integrated methods, foam fractionation is realized continuously and simultaneously to the fermentation process. In this case, the foam fractionation column is either coupled directly to the headplate of the bioreactor (Figure 8.3b) or integrated via an actively pumped

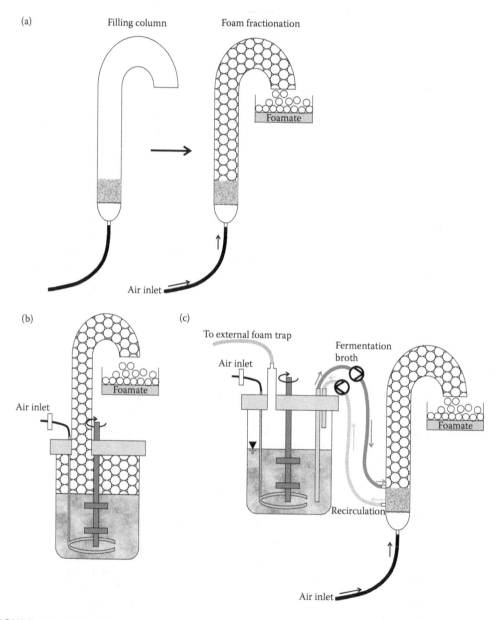

FIGURE 8.3 Schematic representation of foam fractionation columns: (a) non-integrated fractionation column, (b) integrated fractionation column in headplate of bioreactor, (c) integrated fractionation column with circulation loop. (Described by Davis, D. A., H. C. Lynch, and J. Varley. 2001. *Enzyme and Microbial Technology* 28 (4–5):346–354; Winterburn, J. B., A. B. Russell, and P. J. Martin. 2011. *Biochemical Engineering Journal* 54 (2):132–139.)

circulation loop (Figure 8.3c). Characteristic of the foam fractionation columns are their height, the volume air flow, as well as the porosity of the sintered glass disc if an external aeration applies (Gruber 1991). These parameters have an influence on the enrichment and the recovery of the biosurfactant. The enrichment factor "*E*" is the ratio of the biosurfactant concentration in the foamate (c_f) to its concentration in the fermentation broth (c_b) (Equation 8.2). However, the recovery "*R*" is the ratio of total biosurfactant amount in the foamate (n_f) to its overall amount in the fermentation broth and in the foamate (n_o) (Equation 8.3) (Davis et al. 2001).

Definition of the enrichment factor (Davis et al. 2001):

$$E = \frac{c_f}{c_b} \tag{8.2}$$

Notation:

E	—	Enrichment factor
c_f	g/L	Concentration in foamate
c_b	g/L	Concentration in fermentation broth

Definition of the recovery (Davis et al. 2001):

$$R = \frac{n_f}{n_o} * 100 = \frac{n_f}{n_f + n_b} * 100 \tag{8.3}$$

Notation:

R	%	Recovery
n_f	g	Total mass in foamate
n_b	g	Total mass in fermentation broth
n_o	g	Overall total mass

In foam fractionation, many different parameters have contrary effects on enrichment and recovery. Therefore, the settings have to be adjusted to the specific application. Increasing the resistance time of the foam, for instance, increases the enrichment, whereas it reduces the recovery. The drainage of water leads to an accumulation of surfactant molecules on the surface increasing the enrichment. However, some surfactant molecules also drain, thereby reducing the recovery of the biosurfactant (Gruber 1991). Nevertheless, an important drawback of foam fractionation is cells' accumulation in the foam. This leads to a continuous reduction of cells in the fermentation broth. However, the foam affinity for various cells is different (Shedlovsky 1948). For instance, the enrichment factor for *Bacillus subtilis* was reported to be 0.4 (Davis et al. 2001), whereas *P. aeruginosa* was reported to have an enrichment factor of 3.4 (Gruber 1991). To prevent cell accumulation in the foam, the latter has been immobilized in magnetic particles leading to retention of the cells (Heyd et al. 2009).

8.4   CONCLUSION

Overall, producing biosurfactants efficiently poses several obstacles, excessive foaming being the most prominent one. Several approaches are reviewed for that purpose. For low-foaming biosurfactants like sophorolipids using stirred tank bioreactors and sophisticated fermentation strategies leads to quite a high final concentration of biosurfactants from over 400 g/L. The main obstacle of other biosurfactants such as rhamnolipids or surfactin arises from their high-foaming activity. Approaches dealing with this difficulty are still in the stage of development which is also demonstrated by the diversity of the different methods presented in this chapter. To avoid foam formation, several systems are based on an alternative to aeration by spargers. These different approaches are often less efficient than aeration by spargers leading to oxygen limitation, which could influence growth as well as the product spectrum. Additionally, *in situ* product removal like foam fractionation methods could be utilized for foam control. The foam fractionation methods are mainly based on heuristic approaches due to the complexity and insufficient understanding of foam formation and coalescence of the bubbles originated from fermentation broth. Therefore, a reliable prediction of foam formation in the fermentation process is not possible. However, some approaches have been made to describe the lifetime of foam for primary and secondary foam (Furchner and Mersmann 1990). Nevertheless, these equations just apply to pure surfactant solution and are not applicable for impure surfactants. Furchner and Mersmann (1990) declare these impurities as an essential reason for difficulties in foam prediction.

To improve biosurfactant production, a further understanding of foam and its generation is urgently required to enhance the performance of foam fractionation and other fermentation methods.

REFERENCES

Abdel-Mawgoud, A. M., R. Hausmann, F. Lépine, M. M. Müller, and E. Dèziel. 2011. Rhamnolipids: Detection, analysis, biosynthesis, genetic regulation, and bioengineering of production. In *Biosurfactants: From Genes to Applications*, ed. G. Soberón-Chavez. Springer: Berlin/Heidelberg.

Ahmed, T. and M. J. Semmens. 1996. Use of transverse flow hollow fibers for bubbleless membrane aeration. *Water Research* 30 (2):440–446.

Ahmed, T., M. J. Semmens, and M. A. Voss. 2004. Oxygen transfer characteristics of hollow-fiber, composite membranes. *Advances in Environmental Research* 8 (3):637–646.

Andou, S., M. Yoshida, K. Yamagiwa, and A. Ohkawa. 1997. Performance characteristics of mechanical foam-breakers with rotating parts fitted to bubble column. *Journal of Chemical Technology and Biotechnology* 68 (1):94–100.

Brindle, K. and T.Stephenson. 1996. The application of membrane biological reactors for the treatment of wastewaters. *Biotechnology and Bioengineering* 49 (6):601–610.

Chen, C.-Y., S. C. Baker, and R. C. Darton. 2006. Continuous production of biosurfactant with foam fractionation. *Journal of Chemical Technology and Biotechnology* 81 (12):1915–1922.

Chenikher, S., J. S. Guez, F. Coutte, M. Pekpe, P. Jacques, and J. P. Cassar. 2010. Control of the specific growth rate of *Bacillus subtilis* for the production of biosurfactant lipopeptides in bioreactors with foam overflow. *Process Biochemistry* 45 (11):1800–1807.

Chtioui, O., K. Dimitrov, F. Gancel, P. Dhulster, and I. Nikov. 2012. Rotating discs bioreactor, a new tool for lipopeptides production. *Process Biochemistry* 47 (12):2020–2024.

Colin, A. 2012. Coalescence in foams. In *Foam Engineering: Fundamentals and Application*, ed. P. Stevenson. Wiley: Chinchester.

Cooper, D. G., C. R. Macdonald, S. J. Duff, and N. Kosaric. 1981. Enhanced production of surfactin from *Bacillus subtilis* by continuous product removal and metal cation additions. *Applied and Environmental Microbiology* 42 (3):408–412.

Coutte, F., D. Lecouturier, S. A. Yahia, V. Leclère, M. Béchet, P. Jacques, and P. Dhulster. 2010. Production of surfactin and fengycin by *Bacillus subtilis* in a bubbleless membrane bioreactor. *Applied Microbiology and Biotechnology* 87 (2):499–507.

Daniel, H.-J., M. Reuss, and C. Syldatk. 1998. Production of sophorolipids in high concentration from deproteinized whey and rapeseed oil in a two stage fed batch process using *Candida bombicola* ATCC 22214 and *Cryptococcus curvatus* ATCC 20509. *Biotechnology Letters* 20 (12):1153–1156.

Davila, A.-M., R. Marchal, and J.-P. Vandecasteele. 1992. Kinetics and balance of a fermentation free from product inhibition: Sophorose lipid production by *Candida bombicola*. *Applied Microbiology and Biotechnology* 38 (1):6–11.

Davis, D. A., H. C. Lynch, and J. Varley. 1999. The production of surfactin in batch culture by *Bacillus subtilis* ATCC 21332 is strongly influenced by the conditions of nitrogen metabolism. *Enzyme and Microbial Technology* 25 (3):322–329.

Davis, D. A., H. C. Lynch, and J. Varley. 2001. The application of foaming for the recovery of surfactin from *B. subtilis* ATCC 21332 cultures. *Enzyme and Microbial Technology* 28 (4–5):346–354.

Deshpande, N. S. and M. Barigou. 1999. Performance characteristics of novel mechanical foam breakers in a stirred tank reactor. *Journal of Chemical Technology and Biotechnology* 74 (10):979–987.

Drouin, C. M. and D. G. Cooper. 1992. Biosurfactants and aqueous two-phase fermentation. *Biotechnology and Bioengineering* 40 (1):86–90.

Furchner, B. and A. Mersmann. 1990. Foam breaking by high speed rotors. *Chemical Engineering and Technology* 13 (1):86–96.

Ghildyal, N. P., B. K. Lonsane, and N. G. Karanth. 1988. Foam control in submerged fermentation: State of the art. In *Advances in Applied Microbiology*, ed. A. I. Laskin. Elsevier Science: Amsterdam.

Gruber, T. 1991. *Verfahrenstechnische Aspekte der kontinuierlichen Produktion von Biotensiden am Beispiel der Rhamnolipide*. Universität Stuttgart: Stuttgart.

Guerra-Santos, L., O. Käppeli, and A. Fiechter. 1984. *Pseudomonas aeruginosa* biosurfactant production in continuous culture with glucose as carbon source. *Applied and Environmental Microbiology* 48 (2):301–305.

Guez, J.-S., S. Chenikher, J. Ph. Cassar, and P. Jacques. 2007. Setting up and modelling of overflowing fed-batch cultures of *Bacillus subtilis* for the production and continuous removal of lipopeptides. *Journal of Biotechnology* 131 (1):67–75.

Helvaci, S. S., S. Peker, and G. Özdemir. 2004. Effect of electrolytes on the surface behavior of rhamnolipids R1 and R2. *Colloids and Surfaces B: Biointerfaces* 35 (3):225–233.

Henkel, M., M. M. Müller, J. H. Kügler, R. B. Lovaglio, J. Contiero, C. Syldatk, and R. Hausmann. 2012. Rhamnolipids as biosurfactants from renewable resources: Concepts for next-generation rhamnolipid production. *Process Biochemistry* 47 (8):1207–1219.

Heyd, M., P. Weigold, M. Franzreb, and S. Berensmeier. 2009. Influence of different magnetites on properties of magnetic *Pseudomonas aeruginosa* immobilizates used for biosurfactant production. *Biotechnology Progress* 25 (6):1620–1629.

Hirata, Y., M. Ryu, Y. Oda, K. Igarashi, A. Nagatsuka, T. Furuta, and M. Sugiura. 2009. Novel characteristics of sophorolipids, yeast glycolipid biosurfactants, as biodegradable low-foaming surfactants. *Journal of Bioscience and Bioengineering* 108 (2):142–146.

Hoeks, F. W. J. M. M., C. V. Wees-Tangerman, K. Ch. A. M. Luyben, K. Gasser, S. Schmid, and H. M. Mommers. 1997. Stirring as foam disruption (SAFD) technique in fermentation processes. *The Canadian Journal of Chemical Engineering* 75 (6):1018–1029.

Junker, B. 2007. Foam and its mitigation in fermentation systems. *Biotechnology Progress* 23 (4):767–784.

Kanicky, J. R., A. F. Poniatowski, N. R. Mehta, and D. O. Shah. 2000. Cooperativity among molecules at interfaces in relation to various technological processes: Effect of chain length on the pKa of fatty acid salt solutions. *Langmuir* 16 (1):172–177.

Lin, S.-C., K. S. Carswell, M. M. Sharma, and G. Georgiou. 1994. Continuous production of the lipopeptide biosurfactant of *Bacillus licheniformis* JF-2. *Applied Microbiology and Biotechnology* 41 (3):281–285.

Müller, M. M., B. Hörmann, C. Syldatk, and R. Hausmann. 2010. *Pseudomonas aeruginosa* PAO1 as a model for rhamnolipid production in bioreactor systems. *Applied Microbiology and Biotechnology* 87 (1):167–174.

Noah, K. S., S. L. Fox, D. F. Bruhn, D. N. Thompson, and G. A. Bala. 2002. Development of continuous surfactin production from potato process effluent by *Bacillus subtilis* in an airlift reactor. *Applied Biochemistry and Biotechnology* 98–100:803–813.

Norris, J. R. and D. W. Ribbons. 1970. *Methods in Microbiology*. Elsevier Science: Amsterdam.

Özdemir, G., S. Peker, and S. S. Helvaci. 2004. Effect of pH on the surface and interfacial behavior of rhamnolipids R1 and R2. *Colloids and Surfaces A: Physicochemical and Engineering Aspects* 234 (1):135–143.

Pekin, G., F. Vardar-Sukan, and N. Kosaric. 2005. Production of sophorolipids from *Candida bombicola* ATCC 22214 using Turkish corn oil and honey. *Engineering in Life Sciences* 5 (4):357–362.

Pelton, R. 2002. A review of antifoam mechanisms in fermentation. *Journal of Industrial Microbiology and Biotechnology* 29 (4):149–154.

Pinzon, N. M., A. G. Cook, and L.-K. Ju. 2013. Continuous rhamnolipid production using denitrifying *Pseudomonas aeruginosa* cells in hollow-fiber bioreactor. *Biotechnology Progress* 29 (2):352–358.

Pornsunthorntawee, O., S. Maksung, O. Huayyai, R. Rujiravanit, and S. Chavadej. 2009. Biosurfactant production by *Pseudomonas aeruginosa* SP4 using sequencing batch reactors: Effects of oil loading rate and cycle time. *Bioresource Technology* 100 (2):812–818.

Pugh, R. J. 1996. Foaming, foam films, antifoaming and defoaming. *Advances in Colloid and Interface Science* 64:67–142.

Rau, U., L. A. Nguyen, H. Roeper, H. Koch, and S. Lang. 2005. Fed-batch bioreactor production of man-nosylerythritol lipids secreted by *Pseudozyma aphidis*. *Applied Microbiology and Biotechnology* 68 (5):607–613.

Reiling, H. E., U. Thanei-Wyss, Luis Guerra-Santos, R. Hirt, Othmar Käppeli, and A. Fiechter. 1986. Pilot plant production of rhamnolipid biosurfactant by *Pseudomonas aeruginosa*. *Applied and Environmental Microbiology* 51 (5):985–989.

Schneider, M., F. Reymond, I. W. Marison, and U. Von Stockar. 1995. Bubble-free oxygenation by means of hydrophobic porous membranes. *Enzyme and Microbial Technology* 17 (9):839–847.

Shedlovsky, L. 1948. A Review of fractionation of mixtures by foam formation. *Annals of the New York Academy of Sciences* 49 (2):279–294.

Takesono, S., M. Onodera, M. Yoshida, K. Yamagiwa, and A. Ohkawa. 2003. Performance characteris-tics of mechanical foam-breakers fitted to a stirred-tank reactor. *Journal of Chemical Technology and Biotechnology* 78 (1):48–55.

Van Bogaert, I. N. A., K. Saerens, C. De Muynck, D. Develter, W. Soetaert, and E. J. Vandamme. 2007. Microbial production and application of sophorolipids. *Applied Microbiology and Biotechnology* 76 (1):23–34.

Vardar-Sukan, F. 1988. Efficiency of natural oils as antifoaming agents in bioprocesses. *Journal of Chemical Technology and Biotechnology* 43 (1):39–47.

Vardar-Sukan, F. 1998. Foaming: Consequences, prevention and destruction. *Biotechnology Advances* 16 (5):913–948.

Walas, S. M. 1997. Chemical reactors. In *Perry's Chemical Engineers' Handbook*, eds. R. H. Perry and D. W. Green. New York: McGraw-Hill.

Winterburn, J. B., A. B. Russell, and P. J. Martin. 2011. Integrated recirculating foam fractionation for the con-tinuous recovery of biosurfactant from fermenters. *Biochemical Engineering Journal* 54 (2):132–139.

Yeh, M.-S., Y.-H. Wei, and J.-S. Chang. 2006. Bioreactor design for enhanced carrier-assisted surfactin produc-tion with *Bacillus subtilis*. *Process Biochemistry* 41 (8):1799–1805.

Zhang, Y. and R. M. Miller. 1992. Enhanced octadecane dispersion and biodegradation by a *Pseudomonas rhamnolipid* surfactant (biosurfactant). *Applied and Environmental Microbiology* 58 (10):3276–3282.

Zlokarnik, M. 1999. *Rührtechnik: Theorie und Praxis*. Springer-Verlag GmbH: Berlin/Heidelberg.

9 Purification of Biosurfactants

Andreas Weber and Tim Zeiner

CONTENTS

9.1 Introduction 129
9.2 Influence of Microbial Production on Downstream Processing........................... 131
9.3 Downstream Processing of Biosurfactants... 137
    9.3.1 Product Isolation by Phase Separation 139
    9.3.2 Foam Separation... 139
    9.3.3 Cell Separation 140
    9.3.4 Ultrafiltration 141
    9.3.5 Extraction... 142
    9.3.6 Precipitation/Crystallization.. 143
    9.3.7 Chromatography 145
9.4 Conclusion 146
References.. 148

9.1 INTRODUCTION

Surfactants—with a worldwide production of approximately 19 million tons in the year 2008—are an important class of products in chemical industry.[1] According to the amphiphilic structure of surfactants, there are many applications in the detergent, cosmetic, food, environmental, and biomedical industries. Apart from soaps that are mainly based on vegetable oil, the production of conventional surfactants is predominantly based on petroleum. Owing to depleting fossil resources and the increasing importance of environmental concerns such as biodegradability, safety, and sustainability from the consumers, biosurfactants based on renewable resources have a large market potential[2,3] and hence, has gained more attention in scientific and industrial research (Figure 9.1). An increasing number of publications deal with biosurfactants in general, whereas about half of them address purification of biosurfactants.

Possible fields of application are in cosmetic, detergent, personal care, and pharmaceutical sectors. Moreover, they can be utilized as auxiliary material in food, textile, paint, oil, pulp, and paper industries.[4]

In contrast to chemically synthesized surfactants, biosurfactants are produced by various microorganisms and are not classified according to their charge, but according to their chemical composition and microbial origin. The hydrophilic portion of biosurfactants can be built of carbohydrates, amino acids, phosphate, or cyclic peptides, resulting in the main classes of biosurfactants: glycolipids, lipopeptides, phospholipids, and polymeric biosurfactants (Table 9.1). The hydrophobic portion is usually based on long-chain fatty acids or fatty acid derivatives.

Considering the necessary amounts and prices of possible products, biosurfactants are a typical representative for a so-called white (or industrial) biotechnological product. Products of white biotechnology need to be competitive compared to conventional chemically synthesized products with respect to performance and price. The performance of biotechnological products is usually comparable or even better,[5] but prices are often too high, avoiding deep market penetration due to uneconomic production.[6] To overcome economic limitations, selective conversion of cheap raw

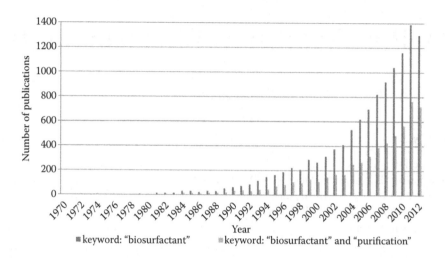

FIGURE 9.1 Number of publications. Search conducted with Google scholar.

materials and high productivities are required in upstream processing. For downstream processing, high yields and capacities are imperative. These requirements differ in their extent compared to processes of red biotechnology (e.g., antibodies, vaccines), since short time to market of small amounts of highly pure products (e.g., enantio-pure, free from viruses, sterile) is essential. Additionally, product prices and capacities of red and white biotechnological products differ in orders of magnitude, ranging from 100,000 to 1 €/kg and 100 kg/a to 100,000,000 kg/a, respectively. Nevertheless, the challenges in downstream processing of biotechnological products remain, being the complex multicomponent cultivation broth, containing components with unknown thermodynamic data as unknown phase behavior. Furthermore, product concentrations are comparably small and large aqueous streams need to be processed. In consequence, process requirements in white biotechnology with respect to upstream and downstream processing lie in between red biotechnology and conventional chemical engineering.

The same situation is expected to be present when considering the cost allotment of upstream and downstream processing. Usually it is stated that 60–80% of the manufacturing costs of biotechnological products are allocated to downstream processing,[2] without considering the nature of the process. This is mainly due to the complexity of the mixtures (reaction medium, substrates, cells, extracellular components) as well as the often small concentration of the target product and high-purity requirements of pharmaceutical products. Whether this holds true for products of white biotechnology is doubtful due to the aforementioned differences. In case of biosurfactants, cost analysis of sophorolipid (SL) production was performed by Deshpande et al.[7] A production scale of 2000 t/a and a fermentation medium were assumed and associated costs were calculated, resulting

TABLE 9.1
Overview of Biosurfactants

Biosurfactant Class	Example	Microorganism
Glycolipids	Sophorolipids	*Candida bombicola*
	Trehaloselipids	*Rhodococcus erythropolis*
	Rhamnolipids	*Pseudomonas aeruginosa*
	Mannosylerythritol lipids	*Pzeudozyma antarctica*
Lipopeptides	Surfactin	*Bacillus subtilis*
Phospholipids	Phosphatidylethanolamine	*Acinetobacter* sp.
Polymeric biosurfactants	Emulsan	*Acinetobacter calcoaceticus*

in a production cost of 3.6 $/kg sophorolipid, based on the assumption that raw material costs make up 20% of total production cost. Furthermore, another method was applied, considering the cost of fermentation and processing on a daily basis (12,000 $/day), using experience with similar processes. This approach results in SL production cost of 2.7 $/kg sophorolipid, in which raw material costs make up 26%. Apart from the outdated prices, the disadvantage of both approaches is that they are not traceable and representative, since they are based on expert knowledge. Additionally, downstream processing costs are not calculated separately and only estimated as a given percentage, without considering purity requirements or amounts of initial impurities.

Cooper et al.[8] used a similar approach, stating that 35% of total cost is due to raw materials and estimate production cost of 2.75 $/kg sophorolipid, based on soy bean oil and a substrate yield of 0.35 g/g. Mulligan et al.[9] analyzed the factors that influence the economic potential of biosurfactants stating that inexpensive raw materials, high product yields, and reduced total batch time play a major role in an economic production. They estimated 10–30% of the total production cost was for raw materials.

The interaction of upstream and downstream processing is an important factor in process design and takes place both ways. The choice of substrates (as possible source of impurities), contamination risk, time of harvest, and the operation mode have a major impact on the purification effort, whereas losses in the downstream process increase necessary cultivation capacities so that concomitant analysis of both process sections is essential for a competitive production. In general, the higher the purification effort, the larger the required number of downstream processing steps. Figure 9.2a depicts the decrease of overall downstream process recovery with an increasing number of process steps, idealized in terms of constant recoveries in each step (80%, 90%, 95%, and 99%, respectively). Figure 9.2b shows the influence of decreased downstream process recovery on the upstream section in terms of additional capacity and substrate consumption to compensate the losses.

Systematic investigation of downstream processing unit operations is scarce in the literature. The same applies for process synthesis and analysis for the production of biosurfactants. Some reviews list possible unit operations and respective sources[2,10]; however, purification procedures are usually only conducted and described in connection with quantification of microbial production. Operational parameters and their influence on purity, yield, or necessary utilities are rarely investigated.

Upstream costs determine whether a production may actually be profitable. How much downstream processing costs are theoretically allowed for a given competitive price of the biosurfactant, microbial production of sophorolipids—one of the best described biosurfactants—is investigated first in terms of productivity, yield, and utilized substrates to enlighten the influence of those parameters on the production cost of upstream processing.

In the following, possible downstream processing unit operations for biosurfactants and their evaluation with respect to feasibility, purification performance, energy efficiency, costs, and sustainability are presented.

9.2 INFLUENCE OF MICROBIAL PRODUCTION ON DOWNSTREAM PROCESSING

Sophorolipids, first described by Gorin et al.,[11] have several advantages with respect to an economic production, being produced in high concentrations of up to 400 g/L by wild-type nonpathogenic yeast *Candida bombicola*.[12,13] They consist of the glucose disaccharide sophorose as hydrophilic portion and a fatty acid chain of 16–18 carbon atoms, which constitutes the hydrophobic part of the surfactant. Usually, a mixture of over 20 sophorolipid congeners[14] is produced, whereas two major groups can be discriminated. The carboxyl group of the fatty acid can either be in the free form—acidic sophorolipids—or esterified with the sophorose—lactonic sophorolipids.[15] Additionally, both forms can be de-, mono-, or diacetylated and the length and saturation of the fatty acid chain may differ.[16] At higher concentrations, sophorolipids self-organize to large aggregates leading to a viscous sophorolipid-rich phase, which is heavier than water.[17]

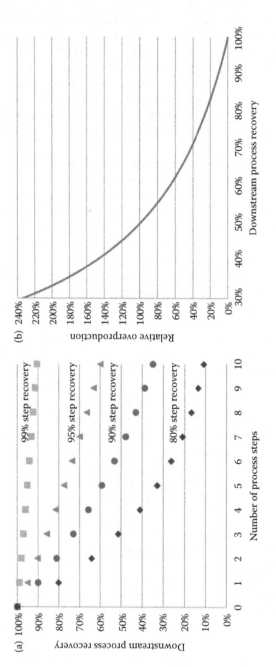

FIGURE 9.2 (a) Influence of number of process steps and respective step recovery on overall downstream process recovery. (b) Influence of overall downstream process recovery on the relative overproduction in upstream processing to compensate the losses.

Because of these structural differences, the two main groups of sophorolipids exhibit different chemical and physical properties and thus, find application in various industrial fields.[2,10] Acidic sophorolipids show better foam production and higher solubility in water. Lactonic sophorolipids are poorly water-soluble and are responsible for phase formation at higher concentrations. At lower temperatures lactonic sophorolipids crystallize, either in saturated aqueous phase[18] or in the product phase.

Possible fields of application with lower purity requirements are environmental remediation, agriculture, and microbial-enhanced oil recovery, whereas increasing purity is required in detergent, cosmetic, and pharmaceutical industry.

Reduction in the production cost of upstream processing section is generally done by selecting cheap raw materials, optimizing cultivation conditions, and improving productivity by metabolic engineering. All those measures are investigated in the case of sophorolipid production and are presented in the following section. Microbial production of sophorolipids and corresponding influences on productivity and product range are described in numerous publications. The review article of van Bogaert et al.[15] provides an overview on the relationships, and hence the same will be discussed here only briefly. Of particular interest is the fact that renewable resources such as vegetable oil and glucose represent optimal substrates in terms of productivity. Furthermore, it is possible to use waste streams from biodiesel[19] or food production, for example, deproteinized whey[19] or fatty acid residues.[20] Makkar et al.[21] summarize, in their review, the application of different renewable resources either waste streams or agricultural products such as vegetable oil. But a systematic analysis and comparison on the basis of key numbers of the different substrates is missing.

What controls the internal esterification to the lactone is unknown. It is reported that by feeding alkanes the proportion of the lactone form rises to 85%, which leads to the formation of white crystals.[22–24] In comparison to the otherwise present oily product phase, separation with greater purity is possible with a crystalline product. However, alkanes are a relatively expensive substrate. A patent of Kaneka Corporation claims the preferential production of lactonic sophorolipids based on limited oxygen supply of 0% during the cultivation.[25] Other possible factors affecting lactonization are the oil and yeast extract substrate concentration.[16,26]

Analyzing reported cultivation data in the literature with respect to utilized carbon source, achieved yield, and productivity, it is possible to identify the cultivation conditions and substrates that show the most promising performance. The price of the substrate (Table 9.2) and, more importantly, the product yield on substrate determine substrate cost, which plays an important role in operating cost. Table 9.3 summarizes selected literature data of *C. bombicola* cultivations to produce sophorolipids with different carbon sources, ranging from glucose, vegetable oils, and industrial

TABLE 9.2
Substrates for Sophorolipid Production and Respective Prices

Substrate	Price (€/t)
Glucose	576[104]
Rapeseed oil	920[105]
Sunflower oil	970[105]
Corn oil	500[106]
Animal fat	248[7]
Waste frying oil	554[106]
Deproteinized whey	250 (Assumption)
Biodiesel co-product stream	271[20]
Soy molasses	165[107]
Oleic acid	1323[108]
Fatty acid methylester	1580 (Assumption)
Hexadecane	1500 (Assumption)

TABLE 9.3
Sophorolipid Production with *Candida bombicola* ATCC 22214[a]

Method Reference	Substrates	Substrate (g/L)	SL Titer (g/L)	Yield ($g_{SL}/g_{Subs.}$)	Cultivation Time (h)	Space–Time Yield [$g_{SL}/(lh)$]	Substrate Cost/Sales Ratio ($\text{€}_{substrate}/\text{€}_{SL}$)
Batch[29]	Glucose	100	37.5	0.357	192	0.186	0.54
Resting cell[17]	Glucose, sunflower oil	200	120	0.600	200	0.600	0.42
Fed-batch[30]	Glucose, rapeseed oil	440	300	0.682	125	2.400	0.36
Fed-batch[14]	Corn oil, gluc., honey	666	400	0.601	436	0.917	0.42
Batch[7]	Glucose, animal fat	200	120	0.600	68	1.765	0.24
Fed-batch[36]	Glucose, frying oil	144	50	0.347	528	0.095	0.55
2 stage fed-batch[13]	Whey, rapeseed oil	510	422	0.827	410	1.029	0.32
Fed-batch[20]	BCS, FAME	206	60	0.291	168	0.357	0.31
Fed-batch[21]	Glucose, FA residues	180	120	0.667	240	0.500	0.21
Fed-batch[31]	Soy molasse, oleic acid	640	75	0.117	168	0.446	1.71
Fed-batch[27]	Glucose, oleic acid	206	180	0.874	200	0.900	0.35
Fed-batch[33]	Glucose, FAME	488	317	0.650	190	1.668	0.49
Feed cont. fed-batch[34]	Glucose, FAME	540	367	0.680	144	2.549	0.50
Conti.[35]	Glucose, oleic acid	98	44.1	0.450	D = 0.09	3.969	0.68
Fed-batch, crystalline[23]	Glucose, hexadecane	130	18	0.138	80	0.225	1.90
Batch, crystalline[25]	Glucose, heptadecane	15	5.9	0.391	264	0.022	0.75
Fed-batch, GMO[40]	Glucose, rapeseed oil	137.5	5	0.041	456	0.011	5.45

[a] Substrate cost/sales ratio is based on a selling price of sophorolipids of 3000 €/t.

carbon-rich waste streams, for example, tallow, whey, and molasses. Even though prices for renewable resources show large volatility due to global market fluctuations and increasing demand in the energy sector, a comparison of present substrate prices is possible, keeping in mind that those prices are subject to change. Prices for waste streams are also uncertain, due to different local availability and unknown savings by avoidance of waste treatment. Consequently, prices of waste streams are used as reported in respective literature, even though they might be outdated, since no or few other reliable information are available.

The substrate cost-to-sales (SCS) ratio (9.1) is a measure of the theoretical portion of substrate cost relative to possible revenues, similar to the fraction of revenue for feedstock introduced by Landucci et al.[27]

$$SCS = \frac{\sum_i C_{Si} \times w_{Si}}{Y_{P/S} \times C_{SL}} \qquad (9.1)$$

The cost of substrates is calculated according to their respective prices (C_{Si}) and their portion of total substrate addition (w_{Si}), and is related to the achieved substrate yield ($Y_{P/S}$) and sophorolipid price (C_{SL}). Here, a selling price of 3000 €/t of sophorolipid is assumed. Recovery losses are, at that stage, not included due to unknown procedures and impurities. The lower the SCS value, the better the cultivation/substrate combination. If the SCS is close to or of larger unity, an economic production for the desired product price is infeasible. The SCS can also be used to determine a necessary selling price for a certain cultivation/substrate combination by setting a desired SCS.

The space–time yield (or productivity) dictates the size and number of fermenters to produce a given annual amount of product, dictating investment cost and associated fixed capital costs.

Most studies were carried out in batch experiments; however, the reported high concentrations can be achieved only by a fed-batch operation, feeding of hydrophilic as well as hydrophobic carbon source and a long culture period of up to 500 h. Cells are grown in a nitrogen-rich medium for approximately 24 h. Sophorolipid production starts after nitrogen depletion in the stationary phase. It is also possible to produce sophorolipids solely on glucose,[28] whereas productivity and substrate yield are limited, 0.19 g/lh and 0.36 $g_{SL}/g_{Substrate}$, respectively (Table 9.3). As a consequence, an SCS of 0.54 $€_{Substrate}/€_{SL}$ is calculated for the production of sophorolipids on glucose. With combinations of renewable resources as glucose, honey, and vegetable oils substrate yields of about 0.6 $g_{SL}/g_{Substrate}$ are achievable, resulting in SCS of approximately 0.36–0.42 $€_{Substrate}/€_{SL}$ (Table 9.3). Productivity ranges from 0.6 to 2.4 g_{SL}/lh, whereas the outstanding productivity of 2.4 is claimed to be due to the control of substrate ratio of 3 (w/w, glucose/rapeseed oil).[29]

Feeding of waste streams is an attractive measure to decrease substrate costs, as long as high substrate yields are maintained. Selected SCS of waste stream processes range between 0.21 and 1.70 $€_{Substrate}/€_{SL}$ (Table 9.3), indicating the dominating importance of substrate yield over substrate cost (Figure 9.3).

According to the process of Solaiman et al.[30] soy molasses—the cheapest investigated substrate—and oleic acid, a comparably expensive substrate, are fed and result in a low substrate yield of only 0.12 g/g and consequently an SCS of 1.70 $€_{Substrate}/€_{SL}$. However, it is possible to reduce the SCS by utilizing cheap waste streams while maintaining high substrate yields, as it is reported by Deshpande et al.[7] and Felse et al.[20] resulting in an SCS of 0.24 and 0.21, respectively. The low SCS of 0.21 $€_{Substrate}/€_{SL}$ is achieved by feeding glucose and tallow-based fatty acid residues with a substrate yield of 0.67 $g_{SL}/g_{Substrate}$. Considering additionally the productivity, the process of Deshpande et al. is superior compared to Felse et al.; 1.77 to 0.50 g/lh. Lactose of whey is not consumed by *C. bombicola*, therefore a two-stage process was developed, where in the first-stage lactose is utilized by *Cryptococcus curvatus* to single-cell-oil, which then is used as co-substrate for *C. bombicola* in the second stage.[31] Nevertheless, the majority of substrate for *C. bombicola* is rapeseed

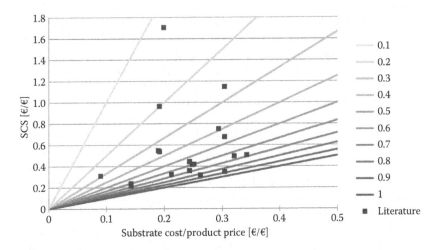

FIGURE 9.3 Dependency of SCS on substrate-cost-to-product price ratio and substrate yield (grayscaled). Literature data from Table 9.3.

oil,[12] so that the advantage of cheap whey substrate is small. The process, however, has a good SCS of 0.32 $\text{€}_{\text{Substrate}}/\text{€}_{\text{SL}}$ due to the high yield of 0.83 $g_{\text{SL}}/g_{\text{Substrate}}$.

Optimized cultivation operation, such as feed-rate control or continuous operation, and special substrates to influence product composition or productivity are also extensively investigated in the literature. By the use of oleic acid[26] or fatty acid methylesters[32,33] increased yield (0.65–0.87 $g_{\text{SL}}/g_{\text{Substrate}}$) and productivities of up to 2.55 g/lh can be achieved. Nevertheless, SCS are in the range of 0.35–0.50 $\text{€}_{\text{Substrate}}/\text{€}_{\text{SL}}$ due to higher substrate cost. Continuous production was also operated successfully by a two-step process that takes into account the nongrowth-coupled production of sophorolipids, resulting in the outstanding productivity of 3.97 g/lh.[34] The yield of 0.45 $g_{\text{SL}}/g_{\text{Substrate}}$ indicates the residual substrate content in the product of about 11 g/L oleic acid and 6 g/L glucose, as it is typical for continuous processes where complete substrate consumption is hard to achieve. Alkanes are also a possible substrate for *C. bombicola*, and Hommel et al.[22] and Glenns et al.[24] report that with this a crystalline product is already formed during cultivation without any further treatment. Nevertheless, due to the small yield of 0.13 $g_{\text{SL}}/g_{\text{Substrate}}$ and high cost of alkanes, an SCS of 1.90 $\text{€}_{\text{Substrate}}/\text{€}_{\text{SL}}$ is calculated for the process of Hommel et al.[22] Glenns et al.[24] achieve a lower SCS of 0.75 $\text{€}_{\text{Substrate}}/\text{€}_{\text{SL}}$ due to a higher yield of 0.39 $g_{\text{SL}}/g_{\text{Substrate}}$ but the experiments were conducted only in small-scale shake flasks.

Apart from direct economic consequences as substrate and investment cost, the choice of substrates and operation mode have also indirect consequences in terms of purification effort. Substrates may contain impurities and have by themselves different effects on the product as being an impurity. Hydrophobic substrates, such as oils or fatty acids, are solubilized by the surfactants; consequently, removal is laborious but necessary since they decrease the foaming ability.[35] Consequently, production solely on glucose might still be economically reasonable even though the SCS is larger (Table 9.3). If production is conducted additionally with hydrophobic substrates like vegetable oils and fatty acids to reduce the SCS, those are advantageously processed products with defined composition and high purity. In contrast, waste streams contain a broader spectrum of components with possible inhibitory effects or which are not convertible and accumulate in the raw product, as it happens with lactose from whey.[36] In the case of waste frying oil, for example, it was shown that polymerized triglycerides, a common reaction product in waste frying oil, are metabolized and no accumulation takes place.[35] Continuous production has higher productivity, but purification effort may be increased due to larger amounts of residual substrate[34] (see above). Fed-batch production enables (almost) complete consumption of substrate and reduces downstream processing effort. Nevertheless, the price for complete conversion probably is increased cultivation time and consequently reduced productivity.

Another method of improving microbial production is metabolic engineering that alters product properties, increases productivity and substrate yield by shutting down alternative metabolic pathways. The biosynthetic pathway of sophorolipid production by *C. bombicola* is investigated by the group of Soetaert, identifying associated genes and establishing a transformation and selection system.[37,38] On the basis of this knowledge, an acetyltransferase-deficient mutant was developed,[39] which produces only unacetylated sophorolipids, resulting in a more hydrophilic product. Consequently, having a product mixture of solely unacetylated sophorolipids the formation of large amounts of acetate after alkaline hydrolysis is avoided. Nevertheless, it needs to be noted that the reported productivity of the mutant is significantly decreased (Table 9.3). After 19 days only 5 g/L is produced with a substrate yield of 0.04 $g_{\text{SL}}/g_{\text{Substrate}}$,[39,40] which makes economic production probably infeasible (SCS = 5.45 $\text{€}_{\text{Substrate}}/\text{€}_{\text{SL}}$). Another successful example of metabolic engineering of *C. bombicola* was shown by Schaffer et al.[41] who sequenced the genome and identified five genes responsible for sophorolipid biosynthesis: a cytochrome P450 mono-oxygenase (SBG1), two glucosyltransferases (SBG2 and SBG5), an acetyltransferase (SBG3), and an ABC-transporter (SBG4). Using deletion mutants, each missing the respective gene, it was possible to show that the enzymes encoded by SBG1, SBG2, and SBG4 are essential for sophorolipid synthesis. In contrast, the mutants that were deficient of the genes SBG3 or SBG5 showed sophorolipid production. More interestingly, it was possible to construct mutants, which overexpress the respective enzymes.

Those mutants show increased productivity of 2.5–6 mg/L/h/OD600 compared with the control (2 mg/L/h/OD600). Unfortunately, no information on the OD600 and substrate yield is given to compare the performance of the mutants with conventional wild types in the literature (Table 9.3).

9.3 DOWNSTREAM PROCESSING OF BIOSURFACTANTS

So far there are no reliable strategies for the modeling and optimization of downstream processing technologies for biosurfactants. The components to be separated are characterized by a broad concentration range depending on the product, high viscosity, and unknown phase behavior due to the amphiphilic properties. For downstream processing of various biosurfactants, different unit operations have been described, such as washing, filtration, foam separation, solvent extraction, ultrafiltration, precipitation, crystallization, adsorption, or chromatography. However, each unit operation has different operating windows and is therefore restricted, depending on the characteristics of the considered biosurfactant and process requirement. Biosurfactants such as sophoro-, rhamno-, and mannosylerythritol lipids, which are produced in high concentration (>100 g/L), pose a different purification task as low-concentrated biosurfactants as surfactin (about 9 g/L)[42] or trehalose lipids (~4 g/L).[43] Moreover, different applications and associated purities need to be considered. In the case of detergents, for example, it is not necessarily required to obtain only one particular form of a surfactant, in this case even a range of products might be desirable. In contrast, in medical or cosmetic use purity and conformity requirements would be higher. For both applications, it is mandatory to remove any cells, debris, and proteins due to possible allergic reactions. Any use of hydrocarbons (extraction, chromatography, and adsorption) is problematic if the product will be used in food industry, since complete removal without trace amounts in the product is difficult. In the case of oil recovery and soil remediation such increased purity requirements are not necessary and a raw product can be applied. Corresponding to the different purity requirements of the applications, the numerous unit operations have different operation windows in terms of product/contaminant concentration, yield, and purity.

In the reviews of Desai et al.[10] and Mukherjee et al.[2] several references for the aforementioned unit operations for biosurfactant downstream processing are listed, but systematic investigation—where key numbers, such as recovery and purity, are estimated and the influence of operational parameters is analyzed—is scarce in the literature. Often purification procedures are only conducted to quantify microbial production and efficiency or, large-scale feasibility plays only a subordinate role.

Therefore, unit operations described in the literature are analyzed and evaluated—if possible—in terms of concentration, purity, and recovery. The results are summarized in Table 9.4 and described in the following paragraphs. An increase in concentration is evaluated by the concentration factor (CTF; 9.2), based on the mass fraction. When purity is considered, only the amount of impurities is important (9.3) and improvements are evaluated by the purification factor (PF; 9.4) and clearance factor (CF; 9.5).[44]

$$CTF = \frac{w_{out,i}}{w_{in,i}} \tag{9.2}$$

$$x_i = \frac{\text{Mass of product}_i}{\text{Mass of product}_i + \text{Mass of impurities}} \tag{9.3}$$

$$PF = \frac{x_{out,i}}{x_{in,i}} \tag{9.4}$$

$$CF = \frac{x_{out,i}}{x_{in,i}} \left/ \frac{1 - x_{out,i}}{1 - x_{in,i}} \right. \tag{9.5}$$

The definition of impurities has a large impact on the values for the respective purity and consequently PF and CF. If water is considered as an impurity is the most influential decision, due to its large content in microbial cultivation suspension. Since aqueous and solid formulations are both possible in later application, water is not considered as an impurity.

Calculated values of CTF, PF, and CF for the unit operations (Table 9.4) are not necessarily comparable quantitatively, since raw products differ in composition and different analytic procedures are applied. Measurements and calculations of reported purities and recoveries may be based on different assumptions; consequently, comparison of calculated key numbers should be done carefully.

Possibilities for in situ product removal are directly referred to in the respective paragraphs of the unit operations.

TABLE 9.4
Key Numbers of Unit Operations for Biosurfactant Downstream Processing

Biosurfactant Ref.	Conc. before	Conc. after	CTF	Purity before	Purity after	CF	PF	Recovery
Phase Formation								
SL[46]	12.7%	52.7%	4.17	89.0%	93.0%	1.6	1.0	95%
MEL[47]	9.0%	87.0%	9.7	–	87.0%	–	–	93%
Foam Separation								
Surfactin[50]	–	–	2.9	–	–	–	–	90%
Surfactin[52]	–	–	55.0	–	–	–	–	92%
RL[55]	0.04%	–	4.0	–	–	–	–	97%
Cell Separation: No respective information								
Ultrafiltration								
Surfactin[64]	–	5.2%	160.0	–	52.6%	–	–	–
RL[65]	0.3%	1.0%	3.9	–	–	–	–	–
Surfactin[67]	–	–	–	–	–	–	–	95%
Surfactin[66]	0.1%	16.6%	166.0	–	70.0%	–	–	–
Surfactin[71]	0.2%	–	–	55.0%	85.0%	4.6	87%	
Extraction								
Surfactin[75]	–	–	–	55.0%	60.0%	1.2	1.1	99%
MEL[47]	–	–	–	–	75.0%	–	–	–
MEL[47]	–	–	–	75.0%	91.0%	3.4	1.2	80%
MEL[47]	–	–	–	91.0%	100.0%	–	1.1	8%
Adsorption								
RL[95]	0.2%	–	–	–	60.0%	–	–	–
Surfactin[98]	–	–	–	76.0%	88.0%	2.3	1.2	95%
Precipitation								
RL[86]	–	–	–	–	–	–	–	90–99%
RL[88]	–	–	–	–	90.0%	–	–	72%
Crystallization								
SL[26]	–	–	–	–	–	–	–	26%
Chromatography								
RL[95]	–	–	–	60.0%	90.0%	6.0	1.5	90%
Surfactin[98]	–	–	–	76.0%	80.0%	1.3	1.1	98%

9.3.1 PRODUCT ISOLATION BY PHASE SEPARATION

The formation of a product-rich phase due to exceeding the solubility limit is often described in the literature for high productivity cultivations of sophoro-, rhamno-, and mannosylerythritol lipids. The product phase is heavier than water and can easily be separated from the cultivation suspension. Cells remain dominantly in the aqueous phase, but product phase is still contaminated. Centrifugation for phase separation is also an option, but the disadvantage is the increased contamination with cells, since they are heavier than the product phase, but transfer is hindered due to the high viscosity of the product phase. The product phase can be washed with water after phase separation to remove hydrophilic impurities, which are still dissolved, and cells.[26]

The product phase of sophorolipids is a brownish viscous liquid with a sophorolipid content of about 50 w%. Hydrophobic components are solubilized, and also hydrophilic molecules are present due to the water content, ranging from 30 to 50 w%. A systematic study of sophorolipid solubility in water was conducted in the temperature range of 20–70°C.[45] Solubility of sophorolipid in water decreases with increasing temperature from 0.8 w% at 70°C to 1.6 w% at 30°C, which is in the range of the results published by Daniel et al.[36] but contradictory to the findings of Hu et al.[18] where higher solubility was observed at elevated temperatures. An explanation might be the different composition of the crude product with 95% lactonic sophorolipids,[18] which is due to the substrate hexadecane, compared to a lactone content of about 80%. The corresponding sophorolipid concentration in the product phase decreases from 52.7 w% at 70°C to 49.3 w% at 30°C. Contrary to the expected composition due to the higher water solubility of acidic sophorolipids, there is no higher concentration of acidic forms in the aqueous phase compared to lactonic sophorolipids. The ratio of lactonic to acidic sophorolipids in the product phase is approximately 4.5 to 1 and only approximately 2.5 to 1 in the aqueous phase. On the basis of a phase separation at 70°C, the recovery of the step is 95% (Table 9.4). The sense of differentiation between concentration-based (CTF) and purity-based (PF or CF) evaluation becomes evident by comparing the CTF of 4.17 and the CF of 1.64. The increase in concentration is described by the CTF, whereas change in purity cannot be reflected by it. The opposite is the case for CF, which can describe improvement of purity in contrast to concentration.

In the case of mannosylerythritol lipids (MELs), aggregated beads are formed during cultivation, which are hard to isolate.[46] A homogenous MEL-rich phase can be induced by heating the culture suspension to 110°C for 10 min, which can easily be separated from the aqueous cell-containing top-phase. Recovery is about 93% and purity 87 w%, whereas residual fatty acids and oil make up the rest.

Phase separation is a very attractive means for product isolation, since the volume of the product stream can be reduced by approximately 75%, which is beneficial for all following purification steps. Purity and recovery are high, nevertheless additional processing of the aqueous phase needs to be considered, depending on loss and value of the product. In contrast to MEL, phase formation of a sophorolipid-rich phase could also be used for in situ product removal, since no heating is necessary.

9.3.2 FOAM SEPARATION

Foaming during cultivation is usually a problem in biosurfactant production. The working volume is decreased and cells and substrates are removed with the foam from the broth. In rhamnolipid production, typically only about 40% working volume is used[47] to overcome exhaust filter clogging by foam. However, foaming can be used for in situ product removal as it is reported for surfactin[48–51] to reduce product inhibition or rhamnolipids.[52–54] As sophorolipids are rather low-foaming biosurfactants, foam fractionation is no option, but cultivation is much easier since no foam breakers or other devices are necessary.

Usually there is a trade-off between recovery and enrichment as it is described in the review about foam separation by Winterburn et al.[55] If the foam is only collected, high recovery but low enrichment is achieved.[48] Enrichment in the foam can be increased if a foam column is used and foam fractionation is done.[56] Davis et al.[49] and Chen et al.[51] increased enrichment to 34 and 55, respectively; at a slightly decreased recovery of about 90% and 92%, respectively. The same is observed for rhamno-lipids, where high recovery of 97% is at the expense of a reduced enrichment of 4.[54]

No information is given with respect to the purity of the product after foam separation and it needs to be considered that cells and other components are also present. Cooper et al.[42] conducted further purification by centrifugation to remove biomass and surfactin precipitation at pH 2. After extraction with dichloromethane and evaporation, a solid product is produced, which is further purified by recrystallization. Nevertheless, foam fractionation is an appropriate unit operation for product isolation of diluted low-concentrated biosurfactants.

9.3.3 Cell Separation

The removal of cells and cell debris from the product is important if it will be used in washing, cos-metic, or medical applications. Several unit operations can be applied, as settling, filtration, immo-bilization, centrifugation, and flotation. To which extent cells are present depends on the process and when cell separation is conducted. Either the complete cultivation suspension is processed, or pre-purified streams after phase separation or foam fractionation are treated.

Settling for cell separation is already described in the paragraph on phase formation, but Palme[57] showed that ultrasonic enhanced agglomeration is an attractive option to improve cell separation in sophorolipid production. Owing to ultrasonic waves, cells agglomerate and the increase in particle size results in increased sedimentation velocity. Cell separation efficiency of 93–99% was achieved with product-free cultivation suspension in 10 L/d and 50 L/d scale. Unfortunately, the presence of oil decreases separation efficiency drastically to 20% at a concentration of 37 g/L. Sophorolipids are not affected by ultrasound and do not decrease cell separation. Ultrasonic enhanced agglomera-tion was successfully applied in a fed-batch cultivation to remove biomass prior to in situ product removal, which was done by sedimentation of the product phase. About 23% of sophorolipids could be separated with this external loop, but due to the aforementioned negative effect of oil on cell separation, oil concentration was kept low and resulted in low productivity and substrate yield of 0.34 g/lh and 0.15 $g_{sophorolipid}/g_{substrate}$.

At a cell size of about 2 μm, filtration can be used to remove biomass from a product stream. Cross-flow filtration is advisable since the tangential flow along the filter minimizes and controls filter cake formation. Filtration was applied for retention of *C. bombicola* cells with hollow fiber-modules with a pore diameter of 0.1 μm.[57] A cultivation suspension with a biomass concentration of 10 g/L and a sophorolipid content of 1 g/L was clarified to a permeate biomass concentration of 0.5 g/L without any sophorolipid retention. No flux information is given. Anyway, it needs to be considered that this sophorolipid concentration is low and is only present in the aqueous phase after phase separation of the sophorolipid-rich phase. Filtration of the sophorolipid-rich phase is also pos-sible, but high viscosity needs to be considered with respect to flux and pressure drop.

Filtration, using a pore diameter not larger than 0.2 μm, was also applied for cell retention of *Pseudomonas aeruginosa*,[58] but then retention of rhamnolipids was also observed due to micelle formation.

Immobilization is a known measure to fix cells in a reactor or ease separation due to increased particle size or additional properties. Drawback of immobilization is increased transport resistance, both for substrates and products.

Adsorption of *P. aeruginosa* cells on organic and inorganic supports was investigated by Siemann et al.,[59] but was not durable under cultivation conditions and cells were removed from the supports. In contrast, immobilization in alginate particles is stable and can be improved if magnetic particles are used, as it is shown by Heyd et al.[60] *P. aeruginosa* cells are immobilized in magnetic particles to prevent loss during foam fractionation for continuous rhamnolipid production.[53]

Chtioui et al.[61] immobilized cells of *Bacillus subtilis* on polypropylene, foamed with powder-activated carbon, to produce surfactin in a rotating disk bioreactor. Foaming is avoided since aera-tion is done bubble-less in the gas phase.[62]

Since cell-free products are mandatory for many applications or particles clog devices downstream, cell separation is an important step. Even if cells are immobilized, retention is never complete and so some kind of separation will be necessary, which can be achieved either by filtration or centrifugation.

9.3.4 Ultrafiltration

Ultrafiltration can be used for the purification of biosurfactants, making use of micelle formation above the critical micelle concentration (CMC). The micellar aggregates are retained by the membrane, whereas molecular impurities pass. Nevertheless, single biosurfactant molecules pass the membrane, resulting in a loss of product in the order of the CMC.

Ultrafiltration was first investigated by Mulligan et al.[63] for the purification of surfactin and rhamnolipids. It was used to concentrate both biosurfactants from collapsed foam after foam fractionation and centrifugation, so pre-concentration was already conducted and the feed was cell free. Different membranes with molecular weight cut-offs (MWCO) ranging from 10 to 300 kDa were investigated, showing that surfactin retention was approximately 98% if MWCO was at least 50 kDa. If 160-fold concentration was conducted, retention decreased to 90% and the filtration rate declined due to increased viscosity. Retention of 92% of rhamnolipids was only achievable if a membrane with MWCO of 10 kDa was used, indicating the smaller size of rhamnolipid micelles. Gruber et al. report diameters of rhamnolipid micelles of about 80–200 nm and if proteins are present, aggregates are formed, which are in the range of 1500 nm.[58] An ultrafiltration process for rhamnolipid concentration is patented by Buchholz et al.[64] claiming that membranes with an MWCO of 100–200 kDa can be used, if the feed solution is acidified to a pH of 3. For a 100 kDa membrane, rhamnolipid retention of 99% and a permeate flux of 50 L/m²h are reported, but no information on purity is given.

A systematic investigation of the influence of operation parameters trans-membrane pressure, feed pH, and feed concentration on ultrafiltration of surfactin was conducted by Sen et al.[65] The presence of surfactin and other macromolecules after cultivation decreases permeate flux significantly, from app. 850 L/m²h to approximately 200 L/m²h at the optimal pressure of 196.2 kPa. An increase of pH to 8.5 resulted in an improved permeate flux of approximately 250 L/m²h. Diluting the feed solution (0.1–1 g/L) resulted in increasing flux and indicated that concentration polarization becomes significant at surfactin concentration of approximately 0.5 g/L. Retention of 98%, purity of 70%, and a concentration factor of approximately 166 were achieved.

On the basis of the work of Mulligan et al., a two-step ultrafiltration process has been developed using the micelle formation of surfactin and subsequent break-down by methanol addition.[66] In the first step, a cell-free cultivation supernatant is filtered through a 30 kDa MWCO membrane corresponding to the micelle size, resulting in retention of approximately 99%. The retentate contains the micellar surfactants and large molecules, usually extracellular proteins. Addition of methanol to the retentate to a concentration of 50 v% resulted in complete break-down of the micelles so that single surfactin molecules pass the membrane in a second filtration step with a recovery of approximately 95%.

Isa et al.[67] applied the same two-step concept, comparing centrifugal and stirred devices and membrane material, polyethersulfone, and regenerated cellulose. Polyethersulfone membranes were better in the second ultrafiltration step due to smaller rejection of surfactin, but for both devices and materials high recovery and purity of approximately 95% were achieved. Nevertheless, long-term stability toward methanolic solutions is essential for membrane selection.[68]

Instead of methanol, Jauregi et al.[69] used ethanol for micelle destabilization in the second ultrafiltration step, but instead of 50 v% a higher concentration of 75 v% was necessary. Furthermore, surfactin micelle size was determined via dynamic light scattering to a range of 5–105 nm, depending on concentration. Contradictive results among different publications regarding micelle size are due to several influential parameters, as biosurfactant concentration, pH, temperature, ionic strength, and impurities.

Chen et al. compared different membrane materials and MWCO for surfactin rejection and permeate flux. Membranes with an MWCO from 30 to 100 kDa had a rejection of about 90%, whereas flux varied between 40 and 90 L/m²h.[70] An MWCO of 300 kDa resulted in a decreased rejection of only 40%, but an increased flux of 530 L/m²h. Additionally, the ultrafiltration process was extended by acid precipitation. Surfactin was precipitated by HCl addition (pH 4) and precipitate was dissolved in NaOH (pH 11), which resulted in a raw product with a purity of 55% and a recovery of 97%. Ultrafiltration of this raw product to remove salts increased purity to 75%. Subsequent ethanol addition (33 v%) for micelle destabilization and macromolecule removal resulted in a purity of about 85% and recovery of 87%. Furthermore, it was shown that nanofiltration membranes with an MWCO of 1 kDa are suitable for retention of single surfactin molecules, but permeate flux is very low (5 L/m²h).[70]

A hybrid salting-out/ultrafiltration process was investigated by Chen et al.[71] where surfactin was salted-out with ammonium sulfate (23% w/v) prior to ultrafiltration. Apart from large amount of used salt, no significant improvement of recovery (81%) and purity (78%) was achieved.

Apart from low-concentrated solutions of surfactin, ultrafiltration was also used for concentration of dissolved sophorolipids.[57] Crossflow ultrafiltration with polysulfon-ceramic membranes with an MWCO of 5 kDa was used for sophorolipid retention. A feed with a sophorolipid content of 10.2 g/L was concentrated to 16 g/L in the retentate, whereas permeate concentration was about 2.75 g/L, resulting in a retention of about 73%. No flux information is given. Nevertheless, in the case of the highly concentrated and viscous, product-phase application is unlikely.

In general, ultrafiltration is an attractive option for the purification of low-concentrated biosurfactants from hydrophilic impurities such as proteins, glucose, and salts. But no separation from hydrophobic impurities as oil or fatty acids is possible, since those are solubilized in the micelles. Additionally, fouling is a problem of filtration biomolecules and needs to be considered in process development. Instead of small-scale dead-end filtration equipment, cross-flow operation should be further investigated to overcome permeate flux limitation and enable continuous operation.

9.3.5 EXTRACTION

Extraction is the most common purification method in laboratory scale. However, there is no systematic study of extraction with respect to a comparison of different solvents and determination of process-relevant data such as partition coefficients. Furthermore, extraction is usually applied for quantitative recovery of biosurfactants, using large amounts of solvent in repetitive steps and no efficient counter-current operation to reduce solvent amount is investigated. Selectivity of the solvent toward the impurities and the biosurfactant is crucial. Polar solvents like methyl-tert-butyl ether (MTBE),[72] ethylacetate,[73,74] or alcohols[26,75] dissolve hydrophobic impurities and biosurfactants so that only separation from water, hydrophilic contaminants, and biomass is possible. Different mixtures of the aforementioned solvents are also applied, ethylacetate-isopropanol (4:1 v/v) for sophorolipids[76] and chloroform–methanol (2:1 v/v) for trehaloselipids.[77] After evaporation of the solvent, the product is extremely viscous and adhesive.

Nonpolar solvents such as n-pentane[23] or n-hexane[8] dissolve only hydrophobic impurities, whereas the product remains in the aqueous phase so that selective purification with high recovery is possible. Although solubility of n-hexane in water or sophorolipid-rich product phase is very poor, small amounts of solvent are found and need to be removed by subsequent evaporation.

To remove hydrophilic as well as hydrophobic impurities, multistep extractions with polar and nonpolar solvents are necessary, as it is described for sophorolipid purification by many groups.[8,78] First, the cultivation suspension is extracted twice with equal volumes of ethylacetate, followed by solvent evaporation. The crude product is then washed three times with equal volumes of n-hexane. Palme[57] extracted the cultivation suspension two times with MTBE (1:1; v/v) to isolate sophorolipids. After evaporation and dissolving the residue in methanol/water (3:1; v/v, pH 3.5), the mixture was extracted two times with n-hexane (3:1:1).

In case of rhamnolipids, first *n*-hexane extraction to remove residual substrate and subsequent acidification and ethylacetate extraction is described.[79]

Chen et al.[74] used ethylacetate and hexane for surfactin extraction. Ethylacetate extraction of surfactin had a recovery of 99% and resulted in purity of about 60%. Hexane is not as suitable for surfactin purification as for sophorolipids, since significant amounts of surfactin are solved.

Rau et al.[46] conducted a multistep extraction procedure for purification of MELs. In the first step, the culture suspension was extracted three times with MTBE (75 w% purity), second the dried extract was mixed three times with cyclohexane (91 w% purity; 80% recovery). The last step consists of dissolving the dried extract in an *n*-hexane–methanol–water (1:6:3) mixture and extract it again three times with *n*-hexane. After drying, the residual contains 100% MEL, whereas recovery is only 8%.

In general, it needs to be considered that due to the amphiphilic properties of biosurfactants, emulsification may happen or long phase separation times are necessary. Stable micro-emulsions or three-phase systems may form, either preventing or impeding phase separation. These phenomena depend on the properties and concentration of biosurfactant, applied solvent, and impurities and need to be tested experimentally.

Furthermore, organic solvents are usually toxic; hence, no residues may be present in the product and additional removal of residual organic solvent by vacuum evaporation is necessary. In general, toxicity of solvents should be considered during process development; ethylacetate and MTBE are better alternatives than chloroform as polar solvent and *n*-heptane may be safer than *n*-hexane as a nonpolar solvent.

In addition to the technical apparatus required for carrying out the extraction, additional equipment is needed for solvent regeneration, due to the high boiling temperature difference between *n*-hexane and hydrophobic impurities; thermal recovery, like evaporation, is easily applicable. Nevertheless, energy consumption for solvent evaporation increases operating costs.

A solvent-free alternative are aqueous-two-phase-systems (ATPS). An ATPS based on polyethylene-glycol (PEG) and dextran was investigated for surfactin extraction and in situ product removal during batch cultivation.[80] Batch cultivation experiments revealed that cells partition to the lower dextran-rich phase and surfactin dissolves preferably in the upper PEG-rich phase. Nevertheless, reextraction of surfactin needs to be considered for further purification and possible PEG recycling.

9.3.6 PRECIPITATION/CRYSTALLIZATION

Transferring biosurfactants into a solid state, either by precipitation—amorphous state—or crystallization—crystalline structure—is an often described way of isolation and purification. Acidifying the culture supernatant to a pH of 2–3 and subsequent cooling results in the cases of rhamnolipids[81] and surfactin[82] in precipitation. Low pH neutralizes negative charges on the molecules, reducing solubility in aqueous solution.[83]

Precipitation of rhamnolipids is described after acidification and heating of the emulsified culture suspension.[84] For this, the cultivation suspension is acidified (pH < 4) and heated to about 100°C. After cooling to 20°C, it is centrifuged and rhamnolipids can be obtained in a lower phase with a recovery of 90–99%. No information concerning purity is given. Phase formation instead of precipitation of solids—which is usually described for rhamnolipids—is probably due to residual oil in the cultivation suspension, which is emulsified by the rhamnolipids. If the rhamnolipid-rich phase is hydrolyzed to obtain L-rhamnose, two phases are formed: a lipophilic phase and a hydrophilic phase.[85] Combined with an ethylacetate extraction, precipitated rhamnolipids are purified with a recovery of 72% and a purity of 90%.[86]

Surfactin is also precipitated by the acid addition (pH 2) of either cell-free culture supernatant[87] or collected foam.[42] Collected precipitate can be re-suspended either in distilled water (pH 7)[88] or dichloromethane. If a solvent is used, the organic phase is separated and evaporated, and further purification can be done by dissolving the extract in distilled water (pH 7)—filtration and precipitation by acidification is repeated.[42]

Another possibility of precipitation is salting-out due to the addition of ammonium sulfate, as it is described for emulsan[89] or aluminum sulfate for rhamnolipids.[90]

Instead of precipitation of the biosurfactants, removal of impurities by precipitation is also possible. Fleurackers et al. describe a method by which fatty acids can be precipitated with $CaCl_2$.[35] Fatty acids form insoluble soaps with Ca_2^+ ions, which can be separated by centrifugation.

Crystallization of lactonic congeners of sophorolipids can be initiated by washing with cold water or direct cooling.[18,29] Sophorolipids crystallize in rod shape, ranging in length from 5 to 120 μm, and thickness of up to 40 μm (Figure 9.4). Crystal size depends on many parameters, such as temperature, cooling rate, duration, and so on.

If a significant proportion of acidic sophorolipids are present in the product mixture, achievable recovery is limited. Hu et al.[18] investigated crystallization of sophorolipids in aqueous buffers and ethanol. The effect of pH in the range of 4.0–8.0 at 25°C in aqueous buffer was investigated and resulted in sophorolipid solubility of 0.017–0.025 g_{SL}/g_{buffer}, which is far less compared to ethanol. Within the range of 10–45°C increased solubility was measured from 0.12 to 0.62 $g_{SL}/g_{ethanol}$, showing that aqueous buffers are better solvents for crystallization. Nevertheless, solubility of sophorolipids in water is affected due to the phase change from the liquid product-rich phase to the solid crystalline state, as it was shown by Weber et al.[45] Crystallization takes place at 20°C and a significant increase of total sophorolipid solubility to approximately 6.7 w% is observed in the aqueous phase. The same explanation for the divergent results in phase separation might be the different composition of the crude product with 95% lactonic sophorolipids,[18] which is due to the substrate hexadecane, compared to a lactone content of about 80%. Nevertheless, a systematic study in terms of the influence of operational parameters like temperature, cooling rate, crystallization mode, and so on, on crystal shape and size or yield is missing up to now. Only in the patent of Hommel et al.[22] properties of sophorolipid crystals, directly produced during cultivation, are reported to be of tetrahedral structure and having a melting temperature of 93°C.

An example for limited recovery of crystallization is the process described by Yanagisawa et al.[25]: after cultivation, approximately 300 g of oily product phase is separated and common extractions with n-hexane and ethylacetate are conducted. Afterwards the residue is mixed with water and cooling crystallization is applied, resulting in 39 g of solid product. Since no data are available with respect to the purity of the product phase before crystallization, only a rough calculation of sophorolipid recovery in the crystallization step is possible. Considering a water content of approximately 50% of the product phase, 150 g of sophorolipid enters crystallization yielding 39 g of solid product. On the basis of these data, recovery of crystallization is only 26%.

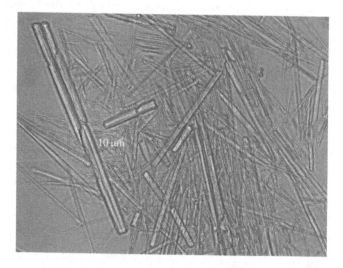

FIGURE 9.4 Microscopic image of crystalline sophorolipids. Magnification: ×630.

Crystallization of surfactin is also described, but the procedure is not optimized and takes two months.[82]

The advantage of crystallization is the high purity of the product and generation of solid particulate matter. If solid–liquid separation by sedimentation, centrifugation, or filtration is easy, it depends on many factors such as crystal size, shape, and homogeneity. In case of precipitation, purity can be decreased due to co-precipitation of proteins or fatty acids. Furthermore, residual hydrophobic substrate has a large influence on precipitation and crystallization behavior as it is seen with rhamno and sophorolipids. Owing losses to the mother liquor because of high solubility and additional water/solvent utilization, recovery is limited and additional purification steps for the spent mother liquor may be necessary.

9.3.7 CHROMATOGRAPHY

Preparative chromatography can be used for product purification for medical or cosmetic purposes with high-purity requirements and is especially suited for the isolation of specific congeners, if requested. Separation is due to a different degree of interaction of product and impurities with the stationary and mobile phase, in which interaction is based on hydrophobicity, size, charge, or affinity.

Adsorption of biosurfactants is often conducted with hydrophobic polymeric XAD resins, which interact with the hydrophobic moiety of the surfactant. Elution is then carried out with organic solvents. For the adsorption of rhamnolipids, XAD-2 resin is used and desorption is conducted with ethylacetate, due to polarity differences between the biosurfactant and co-adsorbing fatty acids.[91] Regeneration of the resin is carried out with methanol and water. Disadvantageous is the low loading capacity of adsorbents. To adsorb about 1 g of rhamnolipid, approximately 20 g of XAD-2 resin were needed. Instead of ethylacetate, isopropanol can also be used for elution of rhamnolipids.[92]

For continuous production of rhamnolipids in the pilot plant scale, two chromatographic operations were used in a four-step downstream process: adsorption and ion-exchange chromatography.[93] Adsorption on XAD-2 resin was used for product isolation after biomass removal, followed by ion-exchange chromatography for purification due to the carboxylic acid group. For adsorption, a bed volume of 10.4 L was equilibrated with 0.1 M potassium phosphate buffer (pH 6.1) and loaded with cultivation supernatant with a capacity of 10–20 L/h. After the breakthrough of biosurfactants, loading was stopped and the resin was washed with 2- to 3-bed volumes of distilled water. Elution was conducted with methanol and fractions were concentrated by the evaporation of methanol. Regeneration of the resin was done with 2- to 3-bed volumes of 1 N NaOH in methanol. Capacity of XAD-2 was calculated to 60 $g_{rhamnolipid}/kg_{resin}$ and concentrated fractions had a purity of 60% and recovery was 75%.[93]

DEAE-sepharose CL 6B (bed volume of 6.2 L) was used for ion-exchange chromatography to increase purity of concentrated rhamnolipid fractions. Equilibration was done with 10 mM trishydrochloride buffer (pH 8) containing 10% ethanol. Concentrated adsorbate from previous adsorption step was diluted 10-fold in buffer and loaded with a capacity of 6 L/h. Afterwards, washing was conducted with 2- to 3-bed volumes of 0.1 M NaCl in equilibration buffer, followed by elution with the same buffer containing 0.8 M NaCl. Finally, regeneration was done with 2- to 3-bed volumes of 2 M NaCl in buffer. This procedure results in a capacity of 50 $g_{rhamnolipid}/l_{gel}$, recovery of 90%, and 90% purity of the ion-exchange step. Adsorption was applied second time after ion-exchange chromatography and product was finally lyophilized. The overall downstream process of biomass removal, adsorption, ion-exchange, and second adsorption had a recovery of 60% and resulted in a purity of 90%.[93]

Wood-based activated carbon is another adsorption material, which was used by Dubey et al. to purify rhamnolipids from collapsed foam.[94] Using acetone for elution resulted in recovery of 89%.

Amberlite XAD resins were also used for the adsorption of glycolipids of *Tsukamurella* sp., where selective separation of glycolipid and oil was achieved by binding to XAD-16 and elution with methanol.[95] Nevertheless, adsorption capacity was only 2 $g_{glycolipid}/kg_{resin}$.

Adsorption of MELs on different XAD resins (Amberlite XAD-4, XAD-16, and XAD-7HP) was analyzed by Rau et al.[46] but no specific binding and elution of either MEL or hydrophobic impurities was achieved, all components adsorbed.

The disadvantage of low capacity can be avoided by operating the adsorption in flow-through mode. Thereby, primarily low-concentrated impurities adsorb while the product passes the packed bed. This method is used for the removal of proteins, polysaccharides, and peptides from a pre-purified (see paragraph on ultrafiltration) cultivation supernatant on slightly polar XAD-7 resin to obtain surfactin in a purity of 88% and a recovery of 95%.[96]

Besides hydrophobic interaction, normal and reversed phase and ion-exchange chromatography are applied for biosurfactant purification. Normal phase chromatography using silica can be used to separate lipids and pigments from rhamnolipids,[52,97] sophorolipids,[73,76] and surfactin. Biosurfactants are loaded on the silica in chloroform or ethylacetate and eluted by different mixtures of chloroform/methanol. Reversed phase chromatography (e.g., RP18) is also possible, corresponding to the different stationary phase, elution of rhamnolipids,[52,98] sophorolipids,[14,29] and surfactin[99,100] is done by mixtures of acetonitrile and water.

Walter investigated centrifugal partition chromatography with ethylacetate/water (1:1) and n-hexane/ethylacetate/methanol/water (5:1:3:5) to purify crude extracts of rhamnolipid from hydrophobic impurities.[79] With solvent-system 1 no separation was achieved, since all components were solved in ethylacetate (see paragraph on extraction). Addition of n-hexane to the nonpolar phase and methanol to the polar phase allowed separation of rhamnolipid.

Magnetic nanoparticles are used for ion-exchange purification of rhamnolipids by Abadi et al.[101] avoiding packed beds of column chromatography, claiming a final purity of 90%.

The strongly basic ion-exchange resin AG1-X4 was used for surfactin purification, having a comparatively large capacity of 1.73 $g_{surfactin}/g_{resin}$ at 0.2 g/L initial concentration.[96]

In summary, adsorptive chromatography can be used for isolation and purification of low-concentrated biosurfactants, if an adequate resin/solvent combination can be found. Since packed-bed processes are prone to clogging due to particles or polysaccharides, as described by Reiling et al.,[93] solid–liquid separation must be carefully controlled. Furthermore, due to the batch-wise operation of loading and elution, at least two adsorption columns are needed. Considering the low capacity of the resins, necessary amounts of resins, size of apparatus, and consequently investment cost will be large. The same probably holds true for operation costs, since large amounts of buffers and solvents for equilibration, washing, elution, and regeneration of the resins are necessary. Consequently, chromatography may only be suited for small-scale purification of valuable products, such as pharmaceuticals or cosmetics, where specific congeners are required.

9.4   CONCLUSION

Influential parameters of upstream and downstream processing of biosurfactants are analyzed and discussed in this contribution. Choice of substrate and operation mode have dominating influence on the overall process performance since these determine yield and productivity of microbial production. Furthermore, purification effort to separate the impurities and nonconverted substrate from the product is defined by those decisions and govern associated downstream processing costs.

On the example of sophorolipids, it is shown that different cultivation options and substrates are already investigated to a large extent and it is shown which influence they have with respect to yield and productivity. Resulting SCS ratios indicate which combinations are promising for economic production of biosurfactants. The use of cheap waste streams is a potent measure to decrease production cost, but still high substrate yield needs to be achieved, as can be seen, for example, at the processes of Felse et al.[20] and Deshpande et al.[7] (c.f. Table 9.3). Otherwise, advantage of low substrate price is overruled by increased substrate demand, as can be seen in the example of Solaiman et al.[30] Presented relationships of sophorolipid production hold true for other biosurfactants like rhamnolipids or surfactin, where same circumstances and requirements are faced with similar strategies.

Apart from substrate cost and resulting yield, associated purification effort and purity requirements need to be considered for process development. Investigated unit operations for biosurfactant purification cover all areas of downstream processing technology. For every purification task, from low to high concentrated cultivation suspensions and from high to reduced purity requirements, respective purification of biosurfactants is achievable (Table 9.4). For high productivity processes like sophoro- or rhamnolipids, product isolation by phase formation and separation is most promising. If the product shall be applied in oil recovery or environmental remediation, additional steps are probably not necessary (Figure 9.5). Further purification for large-scale industrial products such as detergents can be conducted by solvent extraction or crystallization, but additional efforts are necessary for solvent regeneration or recovery improvement.

In situ product removal is especially attractive for low-concentrated products such as surfactin to overcome product inhibition and increase productivity. Foam fractionation is very valuable, since a problem in the cultivation process is actually utilized for separation. No measures against foaming need to be taken and a utility-free separation process is used for product isolation.

Filtration processes play a major role in biosurfactant purification, as well for cell separation as concentration and purification of diluted product streams. Chromatography, using different mechanisms, is the only unit operation enabling separation of specific congeners of a biosurfactant mixture, as they are required in medical or cosmetic application (Figure 9.5). Anyhow solvent and buffer consumption is large and results in increased operation cost.

Avoiding the use of organic solvents is desirable, but if hydrophobic substrates are fed—usually mandatory to achieve high productivity—it is hard to fulfill. Unit operations like extraction and adsorption/chromatography, which are capable of separation of hydrophobic components, are based on organic solvents. Careful feeding control during continuous cultivation or timing of harvesting in fed-batch cultivation is necessary to achieve minimal residual substrate and purification effort.

Described processes are often only on a small scale and/or for analytical purposes and are consequently not systematically investigated and optimized with respect to recovery and purity in industrial application. Hence, there is large potential for significant improvement. Small-scale experiments need to be conducted to prove feasibility of unit operation and behavior of complex cultivation suspension during separation. Studies of operational parameters and their influence on the separation can also be conducted in small-scale or pilot-plant-scale to prove scale-up ability. Experiments need to be evaluated systematically in terms of amounts of used utility, purity improvement, and achieved recovery to improve comparability and enable identification of the most suitable unit operations for a purification task.

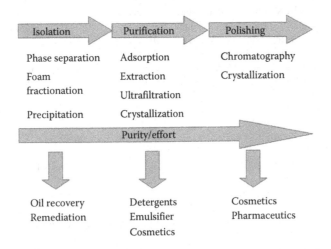

FIGURE 9.5   Scheme for biosurfactant downstream processing.

Computer-aided process modeling, simulation, and analysis is one option, as it is done in the chemical industry for past decades. In biotechnology, these tools are also readily used, but application is often limited due to missing thermodynamic models to describe multicomponent mixtures of polyfunctional biomolecules. Therefore, rather simple equilibrium-based models can be used and necessary experiments for model parameter estimation need to be conducted. Model-based process development enables simulation and optimization of downstream processing to design an economically competitive production process of biosurfactants.

REFERENCES

1. Wagner G. *Waschmittel: Chemie, Umwelt, Nachhaltigkeit.* 4th ed. Weinheim: Wiley-VCH; 2010.
2. Mukherjee S, Das P, Sen R. Towards commercial production of microbial surfactants. *Trends in Biotechnology*; **24**(11):509–15, 2006.
3. Banat IM, Franzetti A, Gandolfi I et al. Microbial biosurfactants production, applications and future potential. *Applied Microbiology and Biotechnology*; **87**(2):427–44, 2010.
4. Banat IM, Makkar RS, Cameotra SS. Potential commercial applications of microbial surfactants. *Applied Microbiology and Biotechnology*; **53**(5):495–508, 2000.
5. Lang S. Biological amphiphiles (microbial biosurfactants). *Current Opinion in Colloid and Interface Science*; **7**(1-2):12–20, 2002.
6. Lynd LR, Wyman CE, Gerngross TU. Biocommodity engineering. *Biotechnol Progress*; **15**(5):777–93, 1999.
7. Deshpande M, Daniels L. Evaluation of sophorolipid biosurfactant production by *Candida bombicola* using animal fat. *Bioresources Technology*; **54**(2):143–50, 1995.
8. Cooper DG, Paddock DA. Production of a biosurfactant from *Torulopsis bombicola*. *Applied and Environmental Microbiology*; **47**(1):173–6, 1984.
9. Mulligan CN, Gibbs BF. Factors influencing the economics of biosurfactants. In: Kosaric N, editor. *Biosurfactants: Production, Properties, Application.* New York: Marcel Dekker, Inc.; 1993. p. 329–71 (Surfactant Science Series).
10. Desai JD, Banat IM. Microbial production of surfactants and their commercial potential. *Microbiology and Molecular Biology Reviews*; **61**(1):47–64, 1997.
11. Gorin PAJ, Spencer JFT, Tulloch AP. Hydroxy fatty acid glycosides of sophorose from *Torulopsis magnoliae*. *Canadian Journal of Chemistry*; **39**(4):846–55, 1961.
12. Daniel HJ, Reuss M, Syldatk C. Production of sophorolipids in high concentration from deproteinized whey and rapeseed oil in a two stage fed batch process using *Candida bombicola* ATCC 22214 and *Cryptococcus curvatus* ATCC 20509. *Biotechnology Letters*; **20**(12):1153–6, 1998.
13. Pekin G, Vardar-Sukan F, Kosaric N. Production of Sophorolipids from *Candida bombicola* ATCC 22214 using Turkish corn oil and honey. *Engineering in Life Science*; **5**(4):357–62, 2005.
14. Davila AM, Marchal R, Monin N, Vandecasteele JP. Identification and determination of individual sophorolipids in fermentation products by gradient elution high-performance liquid-chromatography with evaporative light-scattering detection. *Journal of Chromatography*; **648**(1):139–49, 1993.
15. van Bogaert INA, Saerens K, Muynck C, Develter D, Soetaert W, Vandamme EJ. Microbial production and application of sophorolipids. *Applied Microbiology and Biotechnology*; **76**(1):23–34, 2007.
16. Casas JA, Garcia-Ochoa F. Sophorolipid production by *Candida bombicola*: Medium composition and culture methods. *Journal of Bioscience and Bioengineering*; **88**(5):488–94, 1999.
17. Zhou S, Xu C, Wang J et al. *Supramolecular Assemblies of a Naturally Derived Sophorolipid*; 2004. Available from: URL:http://pubs.acs.org/doi/abs/10.1021/la048590s.
18. Hu YM, Ju LK. Purification of lactonic sophorolipids by crystallization. *Journal of Biotechnology*; **87**(3):263–72, 2001.
19. Ashby RD, Nunez A, Solaiman DKY, Foglia TA. Sophorolipid biosynthesis from a biodiesel co-product stream. *Journal of American Oil Chemists' Society*; **82**(9):625–30, 2005.
20. Felse PA, Shah V, Chan J, Rao KJ, Gross RA. Sophorolipid biosynthesis by *Candida bombicola* from industrial fatty acid residues. *Enzyme and Microbial Technology*; **40**(2):316–23, 2007.
21. Makkar R, Cameotra S, Banat I. Advances in utilization of renewable substrates for biosurfactant production. *AMB Express*; **1**(1):1–19, 2011.
22. Hommel R, Kirste S, Weber L, Kleber HP. Verfahren zur Herstellung mikrokristalliner Glycolipide; DD 298 273 A5, 1992.

23. Cavalero DA, Cooper DG. The effect of medium composition on the structure and physical state of sophorolipids produced by *Candida bombicola* ATCC 22214. *Journal of Biotechnology*; **103**(1):31–41, 2003.

24. Glenns RN, Cooper DG. Effect of substrate on sophorolipid properties. *JAOCS, Journal of the American Oil Chemists' Society*; **83**(2):137–45, 2006.

25. Yanagisawa S, Kawano S, Yasohara Y. Method for producing sophorose lipid; EP2351847A1, 2009.

26. Rau U, Manzke C, Wagner F. Influence of substrate supply on the production of sophorose lipids by *Canidida bombicola* ATCC 22214. *Biotechnology Letters*; **18**(2):149–54, 1996.

27. Landucci R, Goodman B, Wyman C. Methodology for evaluating the economics of biologically producing chemicals and materials from alternative feedstocks. *Appl Biochem Biotechnol*; **45–46**(1):677–96, 1994.

28. Hommel RK, Weber L, Weiss A, Himmelreich U, Rilke O, Kleber H. Production of sophorose lipid by *Candida (Torulopsis) apicola* grown on glucose. *Journal of Biotechnology*; **33**(2):147–55, 1994.

29. Rau U, Hammen S, Heckmann R, Wray V, Lang S. Sophorolipids: A source for novel compounds. *Industrial Crops and Products*; **13**(2):85–92, 2001.

30. Solaiman D, Ashby R, Zerkowski J, Foglia T. Simplified soy molasses-based medium for reduced-cost production of sophorolipids by *Candida bombicola*. *Biotechnology Letters*; **29**(9):1341–7, 2007.

31. Daniel HJ, Otto RT, Binder M, Reuss M, Syldatk C. Production of sophorolipids from whey: Development of a two-stage process with *Cryptococcus curvatus* ATCC 20509 and *Candida bombicola* ATCC 22214 using deproteinized whey concentrates as substrates. *Applied Microbiology and Biotechnology*; **51**(1):40–5, 1999.

32. Davila AM, Marchal R, Vandecasteele JP. Kinetics and balance of a fermentation free from product inhibition—Sophorose lipid production by *Candida bombicola*. *Applied Microbiology and Biotechnology*; **38**(1):6–11, 1992.

33. Marchal RLJSC; Institut Francais du Petrole. Method of production sophorosides by fermentation with fed batch supply of fatty acid esters or oils; US 5 616 479, 1997.

34. Fiehler K. *Verfahrensetnwicklung zur zweistufigen, kontinuierlichen Produktion von Sophoroselipiden mit Candida bombicola*. Braunschweig: Technische Univeristät Carolo-Wilhelmina; 1997.

35. Fleurackers SJJ. On the use of waste frying oil in the synthesis of sophorolipids. *European Journal of Lipid Science and Technology*; **108**(1):5–12, 2006.

36. Daniel HJ, Otto RT, Reuss M, Syldatk C. Sophorolipid production with high yields on whey concentrate and rapeseed oil without consumption of lactose. *Biotechnology Letters*; **20**(8):805–7, 1998.

37. van Bogaert INA, Maeseneire SL de, Schamphelaire W de, Develter D, Soetaert W, Vandamme EJ. Cloning, characterization and functionality of the orotidine-5′-phosphate decarboxylase gene (URA3) of the glycolipid-producing yeast *Candida bombicola*. *Yeast*; **24**(3):201–8, 2007.

38. van Bogaert INA, Maeseneire SL de, Develter D, Soetaert W, Vandamme EJ. Development of a transformation and selection system for the glycolipid-producing yeast *Candida bombicola*. *Yeast*; **25**(4):273–8, 2008.

39. Saerens KSLSW. One-step production of unacetylated sophorolipids by an acetyltransferase negative *Candida bombicola*. *Biotechnology and Bioengineering*; **108**(12):2923–31, 2011.

40. Saerens K, van Bogaert I, Soetaert W. Producing unacetylated sophorolipids by fermentation; WO2012080116A1, 2012.

41. Schaffer S, Wessel M. Zelllen, Nukleinsäuren, Enzyme, und deren Verwendung sowie Verfahren zur Herstellung von Sophorosenlipiden; WO20111061032A2, 2011.

42. Cooper DG, Macdonald CR, Duff SJB, Kosaric N. Enhanced production of surfactin from *Bacillus subtilis* by continuous product removal and metal cation additions. *Applied and Environmental Microbiology*; **42**(3):408–12, 1981.

43. Rosenberg E, Ron EZ. High- and low-molecular-mass microbial surfactants. *Applied Microbiology and Biotechnology*; **52**(2):154–62, 1999.

44. Asenjo JA. *Separation Processes in Biotechnology*. New York: Marcel Dekker Inc.; 1990.

45. Weber A, May A, Zeiner T, Górak A. Downstream processing of biosurfactants. *Chemical Engineering Transactions*; **27**:115–20, 2012.

46. Rau U, Anh Nguyen L, Roeper H, Koch H, Lang S. Downstream processing of mannosylerythritol lipids produced by *Pseudozyma aphidis*. *European Journal of Lipid Science and Technology*; **107**(6):373–80, 2005.

47. Trummler K, Effenberger F, Syldatk C. An integrated microbial/enzymatic process for production of rhamnolipids and L-(+)-rhamnose from rapeseed oil with *Pseudomonas* sp. DSM 2874. *European Journal of Lipid Science and Technology*; **105**(10):563–71, 2003.

48. Mulligan C, Chow T, Gibbs B. Enhanced biosurfactant production by a mutant *Bacillus subtilis* strain. *Applied Microbiology and Biotechnology*; **31**(5–6):486–9, 1989.
49. Davis DA, Lynch HC, Varley J. The application of foaming for the recovery of surfactin from *B. subtilis* ATCC 21332 cultures. *Enzyme and Microbial Technology*; **28**(4-5):346–54, 2001.
50. Noah KS, Fox SL, Bruhn DF, Thompson DN, Bala GA. Development of continuous surfactin production from potato process effluent by *Bacillus subtilis* in an airlift reactor. *Applied Biochemistry and Biotechnology*; **98**:803–13, 2002.
51. Chen C, Baker SC, Darton RC. Batch production of biosurfactant with foam fractionation. *Journal of Chemical Technology and Biotechnology*; **81**(12):1923–31, 2006.
52. Heyd M, Kohnert A, Tan TH et al. Development and trends of biosurfactant analysis and purification using rhamnolipids as an example. *Analytical and Bioanalytical Chemistry*; **391**(5):1579–90, 2008.
53. Heyd M, Franzreb M, Berensmeier S. Continuous rhamnolipid production with integrated product removal by foam fractionation and magnetic separation of immobilized *Pseudomonas aeruginosa*. *Biotechnology Progress*; **27**(3):706–16, 2011.
54. Sarachat T, Pornsunthorntawee O, Chavadej S, Rujiravanit R. Purification and concentration of a rhamnolipid biosurfactant produced by *Pseudomonas aeruginosa* SP4 using foam fractionation. *Bioresource Technology*; **101**(1):324–30, 2010.
55. Winterburn JB, Martin PJ. Foam mitigation and exploitation in biosurfactant production. *Biotechnology Letters*; **34**(2):187–95, 2012.
56. Rujirawanich V, Chuyingsakultip N, Triroj M, Malakul P, Chavadej S. Recovery of surfactant from an aqueous solution using continuous multistage foam fractionation: Influence of design parameters. *Chemical Engineering and Processing: Process Intensification*; **52**(0):41–6, 2012.
57. Palme OH. *Sophoroselipide von Candida bombicola: Eine neue Molekülstruktur sowie Anwendung von Ultraschall-Zellseparations-Technik als innovative Trennmethode zur Erleichterung des Downstream Processing*. Braunschweig: Technische Univerität Carolo-Wilhelmina; 2010.
58. Gruber T, Chmiel H, Käppeli O, Sticher P, Fiechter A. Integrated process for continous rhamnolipid biosynthesis. In: Kosaric N, editor. *Biosurfactants: Production, Properties, Application*. Marcel Dekker, Inc.; 1993. p. 157–73 (Surfactant Science Series).
59. Siemann M, Wagner F. Prospects and limits for the production of biosurfactants using immobilized biocatalysts. In: Kosaric N, editor. *Biosurfactants: Production, Properties, Application*. Marcel Dekker, Inc.; 1993. p. 99–134 (Surfactant Science Series).
60. Heyd M, Weigold P, Franzreb M, Berensmeier S. Influence of different magnetites on properties of magnetic *Pseudomonas aeruginosa* immobilizates used for biosurfactant production. *Biotechnology Progress*; **25**(6):1620–9, 2009.
61. Chtioui O, Dimitrov K, Gancel F, Nikov I. Biosurfactants production by immobilized cells of *Bacillus subtilis* ATCC 21332 and their recovery by pertraction. *Bioprocess Engineering—SFGP2009*; **45**(11):1795–9, 2010.
62. Chtioui O, Dimitrov K, Gancel F, Dhulster P, Nikov I. Rotating discs bioreactor, a new tool for lipopeptides production. *Process Biochemistry*; **47**(12):2020–4, 2012.
63. Mulligan CN, Gibbs BF. Recovery of biosurfactants by ultrafiltration. *Journal of Chemical Technology and Biotechnology*; **47**(1):23–9, 1990.
64. Buchholz R, Fricke U, Mixich J, Hoechst AG. Verfahren zur Herstellung gereinigter Glycolipide durch Membrantrennverfahren; DE 4030264 A1, 1992.
65. Sen R, Swaminathan T. Characterization of concentration and purification parameters and operating conditions for the small-scale recovery of surfactin. *Process Biochemistry*; **40**(9):2953–8, 2005.
66. Lin S, Jiang H. Recovery and purification of the lipopeptide biosurfactant of *Bacillus subtilis* by ultrafiltration. *Biotechnology Techniques*; **11**(6):413–6, 1997.
67. Isa MHM, Coraglia DE, Frazier RA, Jauregi P. Recovery and purification of surfactin from fermentation broth by a two-step ultrafiltration process. *Journal of Membrane Science*; **296**(1-2):51–7, 2007.
68. Isa MHM, Frazier RA, Jauregi P. A further study of the recovery and purification of surfactin from fermentation broth by membrane filtration. *Separation and Purification Technology*; **64**(2):176–82, 2008.
69. Jauregi P, Coutte F, Catiau L, Lecouturier D, Jacques P. Micelle size characterization of lipopeptides produced by *B. subtilis* and their recovery by the two-step ultrafiltration process. *Separation and Purification Technology*; **104**:175–82, 2013.
70. Chen HL, Chen YS, Juang RS. Separation of surfactin from fermentation broths by acid precipitation and two-stage dead-end ultrafiltration processes. *Journal of Membrane Science*; **299**(1-2):114–21, 2007.
71. Chen H, Chen Y, Juang R. Recovery of surfactin from fermentation broths by a hybrid salting-out and membrane filtration process. *Separation and Purification Technology*; **59**(3):244–52, 2008.

72. Kuyukina MS, Ivshina IB, Philp JC, Christofi N, Dunbar SA, Ritchkova MI. Recovery of *Rhodococcus biosurfactants* using methyl tertiary-butyl ether extraction. *Journal of Microbiological Methods*; 46(2):149–56, 2001.

73. Asmer HJ, Lang S, Wagner F, Wray V. Microbial-production, structure elucidation and bioconversion of sophorose lipids. *Journal of American Oil Chemists' Society*; 65(9):1460–6 1988.

74. Chen HL, Juang RS. Recovery and separation of surfactin from pretreated fermentation broths by physical and chemical extraction. *Biochemical Engineering Journal*; 38(1):39–46, 2008.

75. Hammen S. Chemoenzymatische Modifikation von nativen und hydrolysierten Sophoroselipiden: Braunschweig, Technischen Universität; 2003.

76. Daverey APK. Kinetics of growth and enhanced sophorolipids production by *Candida bombicola* using a low-cost fermentative medium. *Applied Biochemistry and Biotechnology*; 160(7):2090–101, 2010.

77. Ristau E, Wagner F. Formation of novel anionic trehalosetetraesters from *Rhodococcus erythropolis* under growth limiting conditions. *Biotechnology Letters*; 5(2):95–100, 1983.

78. Ito S, Inoue S. Sophorolipids from *Torulopsis bombicola*: Possible relation to alkane uptake. *Applied and Environmental Microbiology*; 6(43):1278–83, 1982.

79. Walter V. *New Approaches for the Economic Production of Rhamnolipid Biosurfactants from Renewable Resources*. Karlsruhe: TH Karlsruhe; 2009.

80. C. M. Drouin DGC. Biosurfactants and aqueous two-phase fermentation; (1):86–90, 1992.

81. Müller M, Hörmann B, Syldatk C, Hausmann R. *Pseudomonas aeruginosa* PAO1 as a model for rhamnolipid production in bioreactor systems. *Applied Microbiology and Biotechnology*; 87(1):167–174, 2010.

82. Arima K, Kakinuma A, Tamura G. Surfactin, A crystalline peptidelipid surfactant produced by *Bacillus subtilis*: Isolation, characterization and its inhibition of fibrin clot formation. *Biochemical and Biophysical Research Communication*; 31(3):488–94, 1968.

83. Abdel-Mawgoud A, Hausmann R, Lépine F, Müller M, Déziel E. Rhamnolipids: Detection, analysis, biosynthesis, genetic regulation, and bioengineering of production. In: *Biosurfactants*. Berlin: Springer; 2011. p. 13–55 (Microbiology Monographs). Available from: URL:http://dx.doi.org/10.1007/978-3-642-14490-5_2.

84. Mixich J, Rothert R, Wullbrandt D, Hoechst AG. Verfahren zur quantitativen Aufreinigung von Glycolipiden; DE 4237334 A1, 1994.

85. Mixich J, Rapp K, Vogel M. Process for producing rhamnose from rhamnolipids; WO92/05182 1992.

86. Wei YH, Chou CL, Chang JS. Rhamnolipid production by indigenous *Pseudomonas aeruginosa* J4 originating from petrochemical wastewater. *Biochemical Engineering Journal*; 27(2):146–54, 2005.

87. Sen R. Response surface optimization of the critical media components for the production of surfactin. *Journal of Chemical Technology and Biotechnology*; 68(3):263–70, 1997.

88. Javaheri M, Jenneman GE, McInerney MJ, Knapp RM. Anaerobic production of a biosurfactant by *Bacillus licheniformis* JF-2. *Applied and Environmental Microbiology*; 50(3):698–700, 1985.

89. Rosenberg E, Zuckerberg A, Rubinovitz C, Gutnick DL. Emulsifier of Arthrobacter RAG-1: Isolation and emulsifying properties. *Applied and Environmental Microbiology*; 37(3):402–8, 1979.

90. Schenk T, Schuphan I, Schmidt B. High-performance liquid chromatographic determination of the rhamnolipids produced by *Pseudomonas aeruginosa*. *Journal of Chromatography A*; 693(1):7–13, 1995.

91. Matulovic U. *Verfahrensentwicklung zur Herstellung grenzflächenaktiver Rhamnolipide mit immobilisierten Zellen von Pseudomonas spec. DSM 2874*. Braunschweig: Braunschweig, T. U.; 1987.

92. Wittgens A, Tiso T, Arndt T et al. Growth independent rhamnolipid production from glucose using the non-pathogenic *Pseudomonas putida* KT2440. *Microbial Cell Factories*; 10(1):1–18 2011.

93. Reiling HE, Thaneiwyss U, Guerrasantos LH, Hirt R, Kappeli O, Fiechter A. Pilot-plant production of rhamnolipid biosurfactant by *Pseudomonas aeruginosa*. *Applied and Environmental Microbiology*; 51(5):985–9, 1986.

94. Dubey KV, Juwarkar AA, Singh SK. Adsorption-desorption process using wood-based activated carbon for recovery of biosurfactant from fermented distillery wastewater. *Biotechnology Progress*; 21(3):860–867, 2005.

95. Langer O. HPLC-gestützte Analyse, Optimierung der mikrobiellen Produktion und enzymatische Modifizierung von Glycolipiden aus Tsukamurella spec. nov: Braunschweig, Technischen Universität.

96. Chen H, Lee Y, Wei Y, Juang R. Purification of surfactin in pretreated fermentation broths by adsorptive removal of impurities. *Biochemical Engineering Journal*; 40(3):452–9, 2008.

97. Monteiro SA, Sassaki GL, de Souza LM et al. Molecular and structural characterization of the biosurfactant produced by *Pseudomonas aeruginosa* DAUPE 614. *Chemistry and Physics of Lipids*; 147(1):1–13, 2007.

98. Mata-Sandoval JC, Karns J, Torrents A. High-performance liquid chromatography method for the characterization of rhamnolipid mixtures produced by *Pseudomonas aeruginosa* UG2 on corn oil. *Journal of Chromatography A*; **864**(2):211–20, 1999.

99. Razafindralambo H, Paquot M, Hbid C, Jacques P, Destain J, Thonart P. Purification of antifungal lipopeptides by reversed-phase high-performance liquid chromatography. *First European Symposium on Fast Past Protein Liquid Chromatography of Biomolecules*; **639**(1):81–5, 1993.

100. Lin S, Chen Y, Lin Y. General approach for the development of high-performance liquid chromatography methods for biosurfactant analysis and purification. *Journal of Chromatography A*; **825**(2):149–59, 1998.

101. Abadi HAH, Rashedi H, Amoabediny G, Asadi M. Purification of rhamnolipid using colloidal magnetic nanoparticles. *African Journal of Biotechnology*; **8**(13):3097–106, 2009.

102. United States Department of Agriculture. U.S wholesale list price for glucose syrup. Available from: URL: http://www.ers.usda.gov/data-products/sugar-and-sweeteners-yearbook-tables.aspx#25442.

103. Fachagentur Nachwachsende Rohstoffe. Wholesale prices of plant oils [cited March 2013]. Available from: URL:http://mediathek.fnr.de/grafiken/daten-und-fakten/preise-und-kosten/pflanzenole-grosshandelspreise-interaktiv.html.

104. United States Department of Agriculture. USDA energy round-up [cited March 2013]. Available from: URL: http://www.ams.usda.gov/mnreports/lswagenergy.pdf.

105. Choi J, Lee SY. Factors affecting the economics of polyhydroxyalkanoate production by bacterial fermentation. *Appl Microbiology and Biotechnology*; **51**(1):13–21, 1999.

106. ICIS. Outlook '13: Prospects up for refined glycerine, oleic fatty acids [cited March 2013]. Available from: URL: http://www.icis.com/Articles/2013/01/04/9625559/outlook-13-prospects-up-for-refined-glycerine-oleic-fatty-acids.html.

10 Cost Analysis of Biosurfactant Production from a Scientist's Perspective

Gunaseelan Dhanarajan and Ramkrishna Sen

CONTENTS

10.1 Introduction 153
10.2 Biosurfactant Yields and Cost Parameters 154
    10.2.1 Method and Parameters of Cost Estimation 155
10.3 Costing of Biosurfactant Production: A Case Study 155
    10.3.1 Capital Cost Estimation 155
    10.3.2 Operating Cost Estimation 156
        10.3.2.1 Raw Materials' Cost 157
        10.3.2.2 Labor Cost 157
        10.3.2.3 Consumables and Utilities Costs 157
    10.3.3 Cost of Sophorolipid Based on the Above Data 157
10.4 Biosurfactant Production Cost: R&D versus Commercial 158
10.5 Strategies for Feasible Commercial Biosurfactant Production 159
References 160

10.1 INTRODUCTION

Interest and use of chemical surfactants have grown during the last decade, mainly due to their application in various process industries including food, detergent, oil recovery, and environmental bioremediation. The annual global production of synthetic surfactants was 13 million metric tons in the year 2008, and the total surfactant market is expected to generate revenues of more than USD 41 billion by 2018 (Ashby et al. 2013). However, increasing awareness about the negative environmental impact of chemical surfactants has called for the use of ecologically safe surfactants. Microbial surfactants are green amphiphilic molecules that are capable of lowering the surface and interfacial tensions of fluid phases significantly. In contrast to the chemically synthesized surfactants, biosurfactants are less toxic, highly biodegradable, and stable at extremes of pH, temperature, and salinity (Mukherjee et al. 2006). Furthermore, microbial surfactants possess diverse chemical structures that favor the use of these molecules in various fields such as energy, environment, and healthcare (Das et al. 2008a,b; Mukherjee et al. 2009; Sen 2008).

Recent survey by transparency research market has revealed that the worldwide biosurfactants' market volume is expected to be 476,512.2 tons by 2018. The global market value of biosurfactants was US$1735.5 million in 2011 and it is expected to grow at a rate of 3.5% annually and reach US$2210.5 million by the year 2018 (Specialty Surfactants Market and Bio Surfactants Market: Global Scenario, Raw Material And Consumption Trends, Industry Analysis, Size, Share and Forecast 2010–2018). Several big companies such as BASF-Cognis (Germany and the USA) and Ecover (Belgium) have entered into biosurfactant manufacturing and are the new entrants in the biosurfactants' market, while Jeneil Biosurfactants (the USA) has become a renowned company in this field. AGAE Technologies LLC (the USA) and Rhamnolipid Inc. (the USA) are known to commercially produce rhamnolipids, whereas Fraunhofer IGB (Germany) is involved in the industrial scale production of mannosylerythritol lipids. Other companies involved in biosurfactant manufacturing and business include Ecochem Ltd (Canada), Saraya (Japan), Sigma-Aldrich Co. (the USA), and Intobio (South Korea) (Ashby et al. 2013). Despite the increasing demand of ecologically acceptable microbial surfactants among consumers, lack of cost competitiveness of these molecules is still a major constraint for their successful commercialization. However, research endeavors have been directed toward enhancing the production of microbial surfactants manifold by optimizing media and process including downstream processing (Mukherjee et al. 2008, 2009; Sen 1997; Sen and Swaminathan 1997, 2004, 2005; Sivapathasekaran et al. 2010a, 2011; Sivapathasekaran and Sen 2013a,b). Thus, this chapter attempts to provide an insight into the analysis of costing of biosurfactant production by considering the process and economic parameters and also illustrating the cost analysis with examples based on published literature information and suggesting strategies that could make biosurfactant manufacturing a commercially viable business proposition.

10.2 BIOSURFACTANT YIELDS AND COST PARAMETERS

A variety of microorganisms, mainly bacteria and yeast, synthesize biosurfactants extracellularly or as part of the cell membrane mainly to increase the substrate availability and to transport nutrients. These molecules are majorly classified into glycolipids, lipopeptides, polymeric, and particulate surfactants, based on their molecular weight, structural composition, and microbial source of origin (Mukherjee et al. 2006). Even though microbial surfactants possess diverse structures and better chemical properties than the synthetic equivalents, they could not overcome the chemical surfactants in cost and production capacity. Efforts of researchers working toward improving biosurfactant production technologies have helped to increase product yield to some extent and hopefully can bring about further developments. Table 10.1 summarizes the product yields of different biosurfactants available in the literature. Among the various microbial surfactants, glycolipids such as sophorolipids and rhamnolipids are the promising candidates for the mass production and successful commercialization owing to their superior physicochemical properties and high product yields. Sophorolipids are produced at a concentration of over 400 g/L in commercial scale and are now being used in sanitizer and detergent formulations (Van Bogaert et al. 2011). Another glycolipid biosurfactant rhamnolipid is also commercialized and produced at a maximum concentration of 112 g/L by *P. aeruginosa* DSM 7107 using soybean oil as substrate (Lang and Wullbrandt 1999). Since these molecules are produced in higher amounts, they can be priced less and are used in appropriate fields based on their purity. For instance, less pure glycolipids can be used for environmental applications such as bioremediation and oil recovery, whereas highly pure molecules can be used in cosmetics as emulsifiers.

In contrast, lipopeptide biosurfactants are produced in less quantity but have high value especially due to their exceptional biological properties. Lipopeptide-like surfactin is reported to possess antiadhesive (Das et al. 2009), antimicrobial (Das et al. 2008a,b), and cytotoxic activities (Sivapathasekaran et al. 2010b). Hence, these molecules are treated as low-volume high-value products that have the potential to find applications in cosmetic and pharmaceutical industries. However, great emphasis has been given by the researchers in this field to enhance the yield of lipopeptides using different production technologies in order to make them cost effective. By optimizing the medium components and culture conditions using advanced mathematical tools, a maximum lipopeptide concentration of 6.94 g/L was achieved at a production rate of 320 mg/L/h (Sivapathasekaran and Sen 2013a). In another attempt, the yield of surfactin was improved by mutating *B. subtilis* SD901 using a chemical mutagen *N*-methyl-*N*'-nitro-*N*-nitrosoguanidine to get a maximum of 50 g/L surfactin production (Yoneda et al. 2006).

TABLE 10.1

Yield and Volumetric Productivity of Various Microbial Surfactants

Biosurfactant	Microorganism	Substrate	Yield (g/L)	Productivity (mg/L/h)	Reference
Surfactin	*Bacillus subtilis* ATCC 21332	Glucose	6.45	190	Yeh et al. (2006)
Surfactin + Fengycin	*B. circulans*	Glucose	6.98	320	Sivapathasekaran and Sen (2013a)
Lipopeptide biosurfactant	*B. subtilis* C9	Glucose	7.0	97	Kim et al. (1997)
Sophorolipid	*Candida bombicola* ATCC 22214	Lactose + Rapeseed Oil	422	720	Daniel et al. (1998)
Sophorolipid	*C. bombicola* ATCC 22214	Glucose + Fat	120	1765	Deshpande and Daniels (1995)
Rhamnolipid	*P. aeruginosa* DSM 7107	Soybean oil	112	424	Lang and Wullbrandt (1999)
Rhamnolipid	*P. aeruginosa* UI 29791	Corn oil	46	240	Linhardt et al. (1989)
Trehaloselipid	*Rhodococcus erythropolis* DSM 43215	Mihagol L	32	200	Kim et al. (1990)
Trehaloselipid	*Rhodococcus wratislaviensis* BN38	Hexadecane	3.1	82	Tuleva et al. (2008)
Mannosylerythritol lipid + cellobiose lipid	*Ustilago maydis*	Glucose	23	–	Hewald et al. (2005)

10.2.1 METHOD AND PARAMETERS OF COST ESTIMATION

Design and economic evaluation of a process is essential before constructing and operating a plant and it requires combined knowledge from both scientific and engineering disciplines (Petrides 2003). The overall cost analysis for the production of a biological product involves the estimation of capital expenditure and operating expenditure, which is based on size and number of process equipments, supporting utilities, raw materials required, and other resources consumed. The costs associated with the purchase of equipments, building construction, and investment in consumable materials for initial few months come under capital costs, whereas the costs of raw materials, consumables, utilities, and waste treatment constitute overall operating cost (Petrides 2003). These parameters are discussed in detail in Section 10.3.

10.3 COSTING OF BIOSURFACTANT PRODUCTION: A CASE STUDY

10.3.1 CAPITAL COST ESTIMATION

Capital expenditure is the sum of all expenses involving equipment purchase, installation, wiring, insulation, buildings, and auxiliary facilities (Petrides 2003). The equipment purchase cost can be estimated based on the information from suppliers and published data, while the other expenses depend on the equipment cost. A process economic model was developed using modern simulation software for sophorolipids production by setting the annual production capacity of the plant as 90.7 million kg/year (Ashby et al. 2013). Equipments included in the sophorolipid production model were storage tanks, fermentor, decanter centrifuge, spray dryer, vacuum dryer, and molecular sieve dehydration units. The estimated total capital cost was USD 51.4 million in which one-third was associated with equipment purchase and the remaining was assumed to be installation cost. Equipments used for downstream processing accounted for majority of the capital costs (~76%), the cost of upstream processing equipments was approximately 21% of the overall capital costs, and the remaining 3% was facility-dependent cost (Ashby et al. 2013). The scale and type of equipments vary with the annual target and class of biosurfactant to be produced, and therefore, total capital cost will differ. For instance, surfactin production requires high-performance liquid chromatography purification in addition to solvent extraction, which was the only purification step in sophorolipid production. Another factor that influences equipment cost is the material used for construction. For example, cost of a vessel made of stainless steel is approximately 2.5–3 times more than that of a carbon steel vessel of the same size, whereas a vessel made of titanium costs around 15 times more than a carbon steel tank of the same size (Petrides 2003). The costs of land, building construction, and working capital were not included in the sophorolipid production model (Ashby et al. 2013). Estimation of building costs will be based on the process area required for setting up the equipments and the space required for their efficient operation and maintenance.

10.3.2 Operating Cost Estimation

Operating expenditure of biosurfactant production includes the cost of raw materials, consumables, utilities, labor, waste treatment and disposal, and facility-dependent items. Table 10.2 shows the cost range of these different items with respect to the total operating expenses. Unit production cost of biosurfactants can be estimated as the ratio of the annual operating cost to the annual rate of biosurfactant production. The unit manufacturing cost is inversely proportional to the market volume of biosurfactants. Glycolipid biosurfactants such as sophorolipids that are produced in large quantities cost around $2–$5/kg, whereas the production cost of lipopeptide biosurfactants such as surfactin and iturin is in the range of $10,000–$15,000/g, since they are produced in less quantities.

TABLE 10.2
Production Cost and Market Price of Biosurfactants

Biosurfactant	Substrate	Production Scale (m³)	Production Cost/ Market Price (USD)	Reference/Supplier
Sophorolipid	Glucose + high oleic sunflower oil	19,832	2.95/kg	Ashby et al. (2013)
	Glucose + oleic acid		2.54/kg	
Sophorolipid	Molasses + soybean oil	–	3.00/kg	Kosaric (1992)
Sophorolipid	Corn syrup + yellow grease	15×10^{-3}	2.69/kg	Deshpande and Daniels (1995)
Rhamnolipid	Soybean oil	100	5/kg	Lang and Wullbrandt (1999)
		20	20/kg	
RAG-1 Emulsan	Ethanol and phosphate	5×10^{-3}	50/kg	Rosenberg and Ron (1997)
Lipopeptide biosurfactant	Glucose + oil	–	23.43/mol	Youssef et al. (2007)
Rhamnolipid			150/g (90% pure)	AGAE Technologies, LLC, USA
			1500/g (95% pure)	
Rhamnolipid			1.62/g	Urumqi Unite Bio-Technology Co., Ltd., China
Surfactin			697/50 mg	Sigma-Aldrich Co. LLC, USA
Iturin			460/5 mg	Sigma-Aldrich Co. LLC, USA

10.3.2.1 Raw Materials' Cost

Generally, the cost of raw materials tends to be the major component for most of the fermentative products. It is about 10–80% of the overall operating cost. For high-volume low-value products such as sophorolipids and rhamnolipids, the cost of fermentation media accounts for a major part of raw materials' cost. In case of low-volume high-value products such as surfactin, major cost component is due to the materials used for product recovery and the requirement of maintaining asepsis till the final packaging step. In the sophorolipid production model, the cost of glucose and oleic acid used to produce 90.7 million kg sophorolipids was estimated to be $150 million, accounting for approximately 75% of the total operating cost (Ashby et al. 2013). This can be reduced by substituting the expensive substrates with industrial by-products and agro-based low-cost raw materials. In another study, the use of inexpensive substrates for biosurfactant production was found to enhance the yield by 1.5 times and reduce the medium cost by 60–80%. The costs of cheese whey and molasses-based fermentation media were found to be €2.6–3.8 per liter (Rodrigues et al. 2006). During biosurfactant production, use of complex substrates and addition of antifoam agents will result in additional purification steps, thus making downstream processing a bit more expensive. Therefore, different cultivation strategies have been applied to reduce foam formation (Kronemberger et al. 2008; Rangarajan and Sen 2013).

10.3.2.2 Labor Cost

Labor cost can be estimated based on the number of operators and shifts per year. Companies with highly automated equipments will have less number of laborers when compared with companies with less automated plants. Normally, a company involved in the production of fermentative products will assign at least one operator for each equipment or processing step. Labor expense varies with country, for example, the annual labor cost is $4000–$10,000 in developing nations and can exceed $50,000 in developed countries (Petrides 2003). For sophorolipids production, the labor cost was estimated as US$50 per hour including base rate, fringe benefits, and indirect costs (Ashby et al. 2013).

10.3.2.3 Consumables and Utilities Costs

Consumables include items that must be periodically replaced such as membranes, filter cloths, adsorption resins, chromatography columns, and so on. This may vary according to the downstream processing steps required for particular class of biosurfactant and its purity level. In this case study involving sophorolipid production, the annual facility-dependent operating cost, including replacement of molecular sieve fill materials, was estimated to be $257,000 (Ashby et al. 2013). Utilities' expenses during the operation of the plant are due to the requirements of power, pressured steam, chilled water, process water, and compressed air with other accessories including pipelines. Fermentation accounts for a major part of utilities' cost as it requires clean steam to sterilize the reactor and media. It also needs a lot of cooling and heating agents to maintain the temperature during the process. The cost of electricity is about $0.05 to $0.15/kWh, whereas the cost of chilled water and refrigerants is in the range of $0.002 to $0.1 per 1000 kcal of heat removed. The cost of steam varies from $15 to 1000 per 1000 kg depending on the purity level of water (Petrides 2003). Annual utilities' cost to produce 90.7 million kg sophorolipids was estimated to be approximately $22.07 million (10% of the total operating cost) in which steam accounted for $20.4 million (~93%) (Ashby et al. 2013). The annual usage of steam and chilled water was calculated as 1.1 and 6.2 million metric tons, respectively (Ashby et al. 2013).

10.3.3 Cost of Sophorolipid Based on the Above Data

On the basis of the process economic model developed by Superpro Designer process simulation software, the annual operating cost for the production of 90.7 million kg sophorolipid from glucose and high oleic sunflower oil was US$268 million with a unit production cost of US$2.95/kg.

However, the annual operating cost to produce the same amount of sophorolipid from glucose and oleic acid was US$230 million, which resulted in a lesser unit production cost of US$2.54/kg (Ashby et al. 2013). Even though the calculated cost of sophorolipid was more expensive than the chemically synthesized surfactants, wider application potential of these green molecules will make them more attractive in the surfactant market.

10.4 BIOSURFACTANT PRODUCTION COST: R&D VERSUS COMMERCIAL

To be competitive and find economic acceptance, the market price of biosurfactants should be equal or lower than that of chemical surfactants. The production cost and market price of various biosurfactants are given in Table 10.3. The cost of high-volume low-value biosurfactants is around $2–$3/kg, which is 20–30% more than the cost of chemical surfactants (de Gusmao et al. 2010). The cost of fermentative production of sophorolipid using molasses and soybean oil as substrates was determined to be Canadian $3/kg (Kosaric 1992). The cost for the production of sophorolipid at 200 m³ scale from corn syrup and yellow grease was estimated to be $2.69/kg based on 100 g/L product concentration (Deshpande and Daniels 1995). The above cost calculations were based on R&D data. However, in large scale (19,832 m³), the cost of sophorolipids production using glucose and high oleic sunflower oil was reported to be $2.95/kg and that using glucose and oleic acid was US$ 2.54/kg (Ashby et al. 2013). From the above-mentioned data, it is evident that selection of raw materials and scale of production influence the production cost of microbial surfactants. Hence, manufacturing biosurfactants at an economical scale using inexpensive and renewable substrates could make them competitive in the surfactant market. The production cost of another glycolipid biosurfactant rhamnolipid by *Pseudomonas aeruginosa* using soybean oil as substrate was estimated for different scales of fermentation. The cost was inversely proportional to the scale of production, as the production cost was $20/kg in 20 m³ scale and $5/kg in 100 m³ scale (Lang and Wullbrandt 1999). The current market price of rhamnolipids is in the range of $1.5–$1500/g depending on the level of purity and manufacturer. Urumqi Unite Bio-Technology Co., Ltd., China, sells low-grade rhamnolipids that can be used in detergent formulations and other environmental applications at a price of $1.62/g, while AGAE Technologies, LLC, USA, sells 95% pure rhamnolipids at $1500/g. Rhamnolipids with high level of purity can be used in food and cosmetic industries as well as pharmaceutical industries since they are known to posses antimicrobial properties (Abalos et al. 2001; Haba et al. 2003).

Lipopeptide biosurfactants that are produced in less quantities but have application potential in pharmaceutical industries cost around $20 to $130/mg. The costs of surfactin and iturin marketed by Sigma-Aldrich Co. LLC, USA, are 184.5/10 mg and $460/5 mg, respectively. The high

---

TABLE 10.3

Operating Costs Distribution

Item	% of Total Operating Cost
Raw materials	10–80
Consumables	1–50
Utilities	1–30
Labor	5–30
Facility maintenance	1–40
Miscellaneous	0–20

Source: Adapted from Petrides, D. 2003. *Bioseparations Science and Engineering.* New York: Oxford University Press.

manufacturing cost of the lipopeptide biosurfactants comparing to glycolipids is mainly due to their low yields and use of synthetic production medium containing expensive substrates. However, *in situ* fermentation of *Bacillus* strains in a limestone petroleum reservoir resulted in the economical production of lipopeptide. Seven moles of lipopeptide were recovered from the produced fluids of the reservoir and the production cost was estimated to be \$23.43/mol (Youssef et al. 2007). In addition to *in situ* production of biosurfactants in oil reservoirs, co-production of these molecules with other useful metabolites can lead to their successful commercialization. In another study, the production cost of a high molecular weight biosurfactant RAG-1 emulsan in 5 L fermentor by fed-batch mode was estimated to be approximately \$50/kg which is three times the cost of existing commercial emulsifiers (Rosenberg and Ron 1997). In order to make biosurfactant production competitive in the surfactant market, it is essential to use low-cost substrates and substantially increase production rates.

10.5 STRATEGIES FOR FEASIBLE COMMERCIAL BIOSURFACTANT PRODUCTION

Since the use of expensive substrates and low product yield are the main reasons for the high production cost of biosurfactants, implementation of the following strategies can facilitate the successful commercialization of these molecules:

- Although diverse microbes are capable of producing biosurfactants, only *Bacillus*, *Pseudomonas*, and *Candida* sp. are primarily focused. Hence, other hyper-producing genera have to be closely examined for large-scale biosurfactant production (Mukherjee et al. 2006). Microorganisms from contaminated soils, effluents, and wastewater sources can be isolated and screened for surfactant production since they are capable of utilizing industrial wastes.
- Apart from naturally occurring biosurfactant-producing strains, hyper-producing microorganisms can be engineered by genetic recombination and mutation. By doing so, not only can the product yield be enhanced but also the characteristics of biosurfactants can be improved (Shaligram and Singhal 2010).
- Systems biology is an interesting approach and can be applied to enhance biosurfactant production by increasing the metabolic fluxes toward the product and reducing the formation of other undesired metabolites. In addition to random and targeted genetic alteration, this knowledge-based genetic and metabolic engineering approach can greatly enhance biosurfactant production (Muller et al. 2012).
- The type and amount of biosurfactants produced depend on the medium components and environmental conditions. Hence, different statistical and mathematical tools can be used to optimize these variables in order to enhance the product yield and volumetric productivity (Sen 1997; Sen and Swaminathan 1997; Sivapathasekaran et al. 2010a).
- Since raw materials constitute 30–80% of the overall production cost of biosurfactants, use of industrial and agro-based wastes and low-cost renewable substrates can lead to significant reduction in the operating cost involved in the process (Mukherjee et al. 2006; Nitschke and Pastore 2003; Noah et al. 2005).
- Comparing the selection of raw materials, choice of purification steps is equally an important factor in establishing an economic process and, therefore, use of cost-effective downstream processes is a positive step toward successful commercialization of biosurfactants.
- Another interesting approach in making profitable biosurfactants is co-production of these molecules with other metabolites like industrially important enzymes (Ramnani et al. 2005) and polyhydroxyalkanotes (Hori et al. 2002). Furthermore, *in situ* production of biosurfactants in oil reservoirs, where they are used to enhance oil recovery, can make the process economically viable (Youssef et al. 2007).

REFERENCES

Abalos, A., A. Pinazo, M. R. Infante, M. Casals, F. Garcia, and A. Manresa. 2001. Physicochemical and anti-microbial properties of new rhamnolipids produced by *Pseudomonas aeruginosa* AT10 from soybean oil refinery wastes. *Langmuir* 17 (5):1367–1371.

Ashby, R. D., A. J. McAloon, D. K. Y. Solaiman, W. C. Yee, and M. Reed. 2013. A Process model for approximating the production costs of the fermentative synthesis of sophorolipids. *Journal of Surfactants and Detergents* 16 (5):683–691.

Daniel, H. J., M. Reuss, and C. Syldatk. 1998. Production of sophorolipids in high concentration from deproteinized whey and rapeseed oil in a two stage fed batch process using *Candida bombicola* ATCC 22214 and *Cryptococcus curvatus* ATCC 20509. *Biotechnology Letters* 20 (12):1153–1156.

Das, P., S. Mukherjee, and R. Sen. 2008a. Antimicrobial potential of a lipopeptide biosurfactant derived from a marine *Bacillus circulans*. *Journal of Applied Microbiology* 104 (6):1675–1684.

Das, P., S. Mukherjee, and R. Sen. 2008b. Improved bioavailability and biodegradation of a model polyaromatic hydrocarbon by a biosurfactant producing bacterium of marine origin. *Chemosphere* 72 (9):1229–1234.

Das, P., S. Mukherjee, and R. Sen. 2009. Antiadhesive action of a marine microbial surfactant. *Colloids and Surfaces B-Biointerfaces* 71 (2):183–186.

de Gusmao, C. A. B., R. D. Rufino, and L. A. Sarubbo. 2010. Laboratory production and characterization of a new biosurfactant from *Candida glabrata* UCP1002 cultivated in vegetable fat waste applied to the removal of hydrophobic contaminant. *World Journal of Microbiology and Biotechnology* 26 (9):1683–1692.

Deshpande, M., and L. Daniels. 1995. Evaluation of sophorolipid biosurfactant production by *Candida bombicola* using animal fat. *Bioresource Technology* 54 (2):143–150.

Haba, E., A. Pinazo, O. Jauregui, M. J. Espuny, M. R. Infante, and A. Manresa. 2003. Physicochemical characterization and antimicrobial properties of rhamnolipids produced by *Pseudomonas aeruginosa* 47T2 NCBIM 40044. *Biotechnology and Bioengineering* 81 (3):316–322.

Hewald, S., K. Josephs, and M. Bolker. 2005. Genetic analysis of biosurfactant production in *Ustilago maydis*. *Applied and Environmental Microbiology* 71 (6):3033–3040.

Hori, K., S. Marsudi, and H. Unno. 2002. Simultaneous production of polyhydroxyalkanoates and rhamnolipids by *Pseudomonas aeruginosa*. *Biotechnology and Bioengineering* 78 (6):699–707.

Kim, H. S., B. D. Yoon, C. H. Lee et al. 1997. Production and properties of a lipopeptide biosurfactant from *Bacillus subtilis* C9. *Journal of Fermentation and Bioengineering* 84 (1):41–46.

Kim, J. S., M. Powalla, S. Lang, F. Wagner, H. Lunsdorf, and V. Wray. 1990. Microbial glycolipid production under nitrogen limitation and resting cell conditions. *Journal of Biotechnology* 13 (4):257–266.

Kosaric, N. 1992. Biosurfactants in industry. *Pure and Applied Chemistry* 64 (11):1731–1737.

Kronemberger, F. D., L. M. M. S. Anna, A. C. L. B. Fernandes, R. R. de Menezes, C. P. Borges, and D. M. G. Freire. 2008. Oxygen-controlled biosurfactant production in a bench scale bioreactor. *Applied Biochemistry and Biotechnology* 147 (1–3):33–45.

Lang, S., and D. Wullbrandt. 1999. Rhamnose lipids—Biosynthesis, microbial production and application potential. *Applied Microbiology and Biotechnology* 51 (1):22–32.

Linhardt, R. J., R. Bakhit, L. Daniels, F. Mayerl, and W. Pickenhagen. 1989. Microbially produced rhamnolipid as a source of rhamnose. *Biotechnology and Bioengineering* 33 (3):365–368.

Mukherjee, S., P. Das, and R. Sen. 2006. Towards commercial production of microbial surfactants. *Trends in Biotechnology* 24 (11):509–515.

Mukherjee, S., P. Das, C. Sivapathasekaran, and R. Sen. 2008. Enhanced production of biosurfactant by a marine bacterium on statistical screening of nutritional parameters. *Biochemical Engineering Journal* 42 (3):254–260.

Mukherjee, S., P. Das, C. Sivapathasekaran, and R. Sen. 2009. Antimicrobial biosurfactants from marine *Bacillus circulans*: Extracellular synthesis and purification. *Letters in Applied Microbiology* 48 (3):281–288.

Muller, M. M., J. H. Kugler, M. Henkel et al. 2012. Rhamnolipids—Next generation surfactants? *Journal of Biotechnology* 162 (4):366–380.

Nitschke, M., and G. M. Pastore. 2003. Cassava flour wastewater as a substrate for biosurfactant production. *Applied Biochemistry and Biotechnology* 105:295–301.

Noah, K. S., D. F. Bruhn, and G. A. Bala. 2005. Surfactin production from potato process effluent by *Bacillus subtilis* in a chemostat. *Applied Biochemistry and Biotechnology* 121:465–473.

Petrides, D. 2003. Bioprocess design and economics. In *Bioseparations Science and Engineering*, edited by R. G. Harrison. New York: Oxford University Press.

Ramnani, P., S. S. Kumar, and R. Gupta. 2005. Concomitant production and downstream processing of alkaline protease and biosurfactant from *Bacillus licheniformis* RG1: Bioformulation as detergent additive. *Process Biochemistry* 40 (10):3352–3359.

Rangarajan, V., and R. Sen. 2013. An inexpensive strategy for facilitated recovery of metals and fermentation products by foam fractionation process. *Colloids and Surfaces B-Biointerfaces* 104:99–106.

Rodrigues, L. R., J. A. Teixeira, and R. Oliveira. 2006. Low-cost fermentative medium for biosurfactant production by probiotic bacteria. *Biochemical Engineering Journal* 32 (3):135–142.

Rosenberg, E., and E. Z. Ron. 1997. Bioemulsans: Microbial polymeric emulsifiers. *Current Opinion in Biotechnology* 8 (3):313–316.

Sen, R. 1997. Response surface optimization of the critical media components for the production of surfactin. *Journal of Chemical Technology and Biotechnology* 68 (3):263–270.

Sen, R. 2008. Biotechnology in petroleum recovery: The microbial EOR. *Progress in Energy and Combustion Science* 34 (6):714–724.

Sen, R., and T. Swaminathan. 1997. Application of response-surface methodology to evaluate the optimum environmental conditions for the enhanced production of surfactin. *Applied Microbiology and Biotechnology* 47 (4):358–363.

Sen, R., and T. Swaminathan. 2004. Response surface modeling and optimization to elucidate and analyze the effects of inoculum age and size on surfactin production. *Biochemical Engineering Journal* 21 (2):141–148.

Sen, R., and T. Swaminathan. 2005. Characterization of concentration and purification parameters and operating conditions for the small-scale recovery of surfactin. *Process Biochemistry* 40 (9):2953–2958.

Shaligram, N. S., and R. S. Singhal. 2010. Surfactin—A review on biosynthesis, fermentation, purification and applications. *Food Technology and Biotechnology* 48 (2):119–134.

Sivapathasekaran, C., P. Das, S. Mukherjee, J. Saravanakumar, M. Mandal, and R. Sen. 2010b. Marine bacterium derived lipopeptides: Characterization and cytotoxic activity against cancer cell lines. *International Journal of Peptide Research and Therapeutics* 16 (4):215–222.

Sivapathasekaran, C., S. Mukherjee, A. Ray, A. Gupta, and R. Sen. 2010a. Artificial neural network modeling and genetic algorithm based medium optimization for the improved production of marine biosurfactant. *Bioresource Technology* 101 (8):2884–2887.

Sivapathasekaran, C., S. Mukherjee, R. Sen, B. Bhattacharya, and R. Samanta. 2011. Single step concomitant concentration, purification and characterization of two families of lipopeptides of marine origin. *Bioprocess and Biosystems Engineering* 34 (3):339–346.

Sivapathasekaran, C., and R. Sen. 2013a. Performance evaluation of an ANN-GA aided experimental modeling and optimization procedure for enhanced synthesis of marine biosurfactant in a stirred tank reactor. *Journal of Chemical Technology and Biotechnology* 88 (5):794–799.

Sivapathasekaran, C., and R. Sen. 2013b. Performance evaluation of batch and unsteady state fed-batch reactor operations for the production of a marine microbial surfactant. *Journal of Chemical Technology and Biotechnology* 88 (4):719–726.

Specialty Surfactants Market and BioSurfactants Market: Global Scenario, Raw Material and Consumption Trends, Industry Analysis, Size, Share and Forecast 2010–2018. 2013. [cited July 2013]. Available from http://www.transparencymarketresearch.com/specialty-and-biosurfactants-market.html.

Tuleva, B., N. Christova, R. Cohen, G. Stoev, and I. Stoineva. 2008. Production and structural elucidation of trehalose tetraesters (biosurfactants) from a novel alkanothrophic *Rhodococcus wratislaviensis* strain. *Journal of Applied Microbiology* 104 (6):1703–1710.

Van Bogaert, I. N. A., J. X. Zhang, and W. Soetaert. 2011. Microbial synthesis of sophorolipids. *Process Biochemistry* 46 (4):821–833.

Yeh, M. S., Y. H. Wei, and J. S. Chang. 2006. Bioreactor design for enhanced carrier-assisted surfactin production with *Bacillus subtilis*. *Process Biochemistry* 41 (8):1799–1805.

Yoneda, T., Y. Miyota, K. Furuya, and T. Tsuzuki, inventors; K. K. Showa Denko, assignee. 2006. Preparing surface tension protein via genetically engineered *Bacillus* culture for use in detergents, emulsifiers, wetting agents, dispersants, solubilizing agents, antistatic agents, cosmetics, foods and medical preparations. United States patent US 7,011,969.

Youssef, N., D. R. Simpson, K. E. Duncan et al. 2007. In situ biosurfactant production by *Bacillus* strains injected into a limestone petroleum reservoir. *Applied and Environmental Microbiology* 73 (4):1239–1247.

Section II

Applications

11 Patents on Biosurfactants and Future Trends

E. Esin Hames, Fazilet Vardar-Sukan, and Naim Kosaric

CONTENTS

11.1   Introduction.. 165
11.2   Criteria for Search of Patents... 166
11.3   Patenting Activity 166
11.4   Country of Origin 167
11.5   Nature of Applicants... 167
11.6   Commercial Activity... 168
11.7   Producer Microorganisms... 169
11.8   Techniques 177
      11.8.1 New Microbial Producers... 177
            11.8.1.1 Screening Methods for Biosurfactant Producers 177
            11.8.1.2 Genetic Modifications 181
      11.8.2 Alternative Substrates 181
      11.8.3 Efficient Production Processes.. 183
      11.8.4 Separation and Purification Processes 183
11.9   Application Fields 183
      11.9.1 Biomedical.. 192
      11.9.2 Cosmetic... 198
      11.9.3 Agriculture 202
      11.9.4 Bioremediation 204
      11.9.5 Oil Recovery 210
      11.9.6 Detergent 213
      11.9.7 Chemical Production and Other Applications 215
11.10 Conclusion...220
Acknowledgments...221
References..222
      Patents 222

11.1 INTRODUCTION

Biosurfactants are a diverse group of surface-active amphipathic molecules that are produced by various bacteria, yeast, and filamentous microorganisms. Despite a long history of production of biosurfactant and bioemulsifier from microorganisms, interest in microbial biosurfactant and bioemulsifier has been steadily increasing due to their diversity, ecofriendly nature, promising developments in large-scale production, selectivity, effective function in extreme conditions, and wide range of applications (Saharan et al., 2011). Some hypotheses have emerged for physiological roles of biosurfactants due to their very different structures and surface properties. These compounds facilitate microbial growth on hydrophobic water-insoluble substrates by lowering surface tension at the phase boundary, emulsification, resistance to toxic compounds, enabling the microbial cells to adhere to the organic compounds, or increasing their water solubility, regulating attachment and detachment of microorganisms from the surfaces. Biosurfactants glide and swarm the motility of microorganisms and participate in the cellular physiological processes of signaling and differentiation, as well as in biofilm formation (Fracchia et al., 2012; Rosenberg, 2006; Van Hamme et al., 2006).

Their diverse physicochemical properties namely, emulsification, wetting, foaming, cleansing, phase separation, surface activity, and reduction in viscosity of heavy liquids such as crude oil, can be exploited for many industrial and domestic application purposes (Fracchia et al., 2012). In addition, certain biosurfactants exhibit antimicrobial, antitumor, and similar bioactivities.

By virtue of their multiple features, these molecules have the potential to be used in a variety of industries such as agriculture, cosmetic, pharmaceutical, food, pulp and paper, coal, textiles, ceramic processing, uranium ore-processing, oil recovery; as humectants, preservatives and detergents, in herbicides and pesticides formulations; and mechanical dewatering of peat as well as in environmental protection (Campos et al., 2013; Kosaric, 2001; Saharan et al., 2011; Shete et al., 2006). Owing to their diverse application, they have been employed in different sectors and many patents have been filed.

The first patent that the authors have been able to access relates to microbial biosurfactants filed in March 17, 1965, titled "Oil glycosides of sophorose and fatty acid esters thereof" (US 3312684, Publication date Apr. 4, 1967). This patent describes hydroxy fatty acids and their production as glycolipids by fermentation processes using *Torulopsis magnolia*, claiming to produce the sugar sophorose and hydroxy fatty acids, giving good yields, and allowing ready recovery. Although it has been over half a century since the issue of this first patent, and despite their extremely wide potential applications, biosurfactants are still not being fully exploited by the different sectors.

The aim of this chapter is to provide an overview of the spectrum of issued patents on microbial biosurfactants and related topics with respect to their distribution by years, origin, substrates, techniques, and applications to reemphasize the existing and underexploited know-how in the field.

11.2 CRITERIA FOR SEARCH OF PATENTS

This chapter covers patents issued after 1967 through searches in important international patent databases including EPO Espacenet—http://worldwide.espacenet.com/, EPO Global Patent Index—http://www.epo.org/searching/subscription/gpi.html, WIPO PatentScope—http://patentscope.wipo.int, GooglePatents—http://www.google.com/patents, MatheoPatents—http://www.matheo-software.com.

Patents related to plant-derived biosurfactants have not been evaluated. Different versions of the same patent in different languages were taken as one. The patents of the same inventors with similar titles with different patent numbers from the same patent offices were considered separately. Detailed information about the inventions including abstracts could not be reached for 23 patents. It will be useful to note that all the studies on biosurfactants are not limited to patent documents as patents represent only a portion of total inventions.

11.3 PATENTING ACTIVITY

Overall, 531 patents on microbial biosurfactants were issued until July 2013. If multiple applications are removed, the list is reduced to 500 patents included in this review (Figure 11.1). Although the first patent related to microbial biosurfactants was published in 1967, issued patents did not exceeded two-digit numbers until 1995. The number of issued patents after 2004 have shown an increasing trend due to the rising interest in biosurfactants by the pharmaceutical and cosmetic industries as a natural additive due to environmental concerns. Naturally, although patenting activity is a reflection of research interest on biosurfactants, there is a lag time of 3–4 years between the commencement of research projects and patent applications. The results of one group of inventors

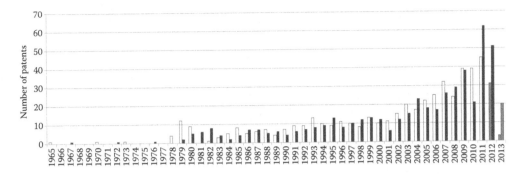

FIGURE 11.1 Number of patents applied for (□) and issued (■). 2012 applications and 2013 applications and publications are still in process (▩).

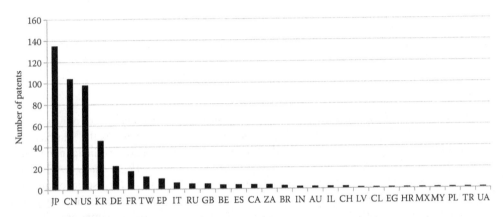

FIGURE 11.2 Number of patents by country of origin (as of July 2013). JP: Japan, CN: China, US: United States, KR: Korea, DE: Germany, FR: France, TW: Taiwan, EP: European Patent, IT: Italy, RU: Russia, BE: Belgium, GB: Great Britain, ES: Spain, CA: Canada, ZA: South Africa, BR: Brazil, IN: India, AU: Australia, IL: Israel, CH: Switzerland, LV: Latvia, CL: Chile, EG: Egypt, HR: Croatia, MX: Mexico, MY: Malaysia, PL: Poland, TR: Turkey, UA: Ukraine.

are responsible from the observed dramatic increase in the patenting activity in 2011. In the first 7 months of 2013, the upward trend in this field has been maintained with 20 patents.

11.4 COUNTRY OF ORIGIN

Based on priority numbers, it is seen that Japanese (136), Chinese (104), American (96), and Korean (46) inventors were at the forefront of patents related to biosurfactants. Germany (22), France (17), and Taiwan (11) are the followers. If the invention is protected with more than one country/location, the priority number of the patent and date were considered in Figure 11.2.

11.5 NATURE OF APPLICANTS

At least 80% of the patent applications were submitted by a company and/or a research institute/university, whereas small proportions of the patent applications were registered by individual inventors (%15.4). Out of 500 patents, 41.5% belong only to companies and 5.0% are owned jointly by inventors and companies. Companies are involved in half (50.1%) of all patents and nearly one-third (32.7%) are owned by academia and researchers. Joint ownership of academia and industry is very limited (3.6%), confirming the fact that collaborative research activity in the biosurfactants field is still very rare between universities and industrial companies (Figure 11.3).

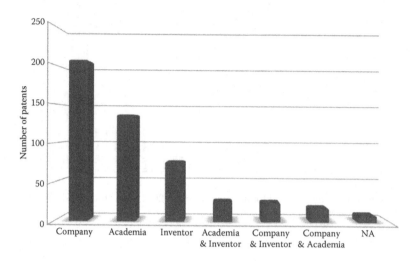

FIGURE 11.3   Distribution of patents with respect to types of applicants.

11.6   COMMERCIAL ACTIVITY

Although many biosurfactants and their production processes have been patented, only a few have been commercialized (Mukherjee et al., 2006; Reis et al., 2013). Emulsan was the first bioemulsifier/biosurfactant-based product manufactured on industrial scale and made available in the market. Emulsan is polyionic lipopolysaccharide produced by *Acinetobacter calcoaceticus* RAG-1 ATCC 31012. It was marketed by *Petroleum Fermentations* (the Netherlands) for use in cleaning oil-contaminated vessels, oil spills, and microbially enhanced oil recovery, besides facilitating pipeline transportation of heavy crude oil. This was followed by the production of rhamnolipids by Jeneil Biosurfactants Co. (Shete et al., 2006).

Currently, sophorolipids are produced by several companies in, for example, France, Japan, and Korea, with the material being used in products such as dishwasher formulations and Yashinomi vegetable wash. Examples of companies producing commercially available biosurfactant-based products are

- *Jeneil Biotech (Biosurfactants) Company* (www.jeneilbiotech.com): An American agroindustrial company produces rhamnolipids on a large scale (U.S. EPA approved fungicide Zonix™).
- *EcoChem Organics Company* (ecochem.com): A Canadian company, produces a dispersive agent (EC-601) of water-insoluble hydrocarbons containing rhamnolipids and bacterial preparations (EC 1800; EC2100 W) for cleaning up spills in sand, soil, and gravel as well as for applications in aqueous environments such as waste water treatment plants, lagoons, and storage tanks.
- *Pendragon Holdings Ltd.*: Produces an additive for fuels (PD5) based on a mixture of rhamnolipid biosurfactants and enzymes.
- *AGAE Technologies* (www.agaetech.com): Produces small quantities of highly purified rhamnolipids using *P. aeruginosa* strain NY3.
- *Saraya Co. Ltd.* (worldwide.saraya.com): In Japan, the company manufactures sophorolipids using Pseudozyma with palm oil as the main fermentation substrate.
- *Ecover* (www.ecover.com): A Belgium company, markets some products that contain "*Candida bombicola*/glucose/methyl rapeseedate ferment," cleansing agents.
- *MG Intobio* (http://mgintobio.en.makepolo.-com): A Korean company, it markets soaps containing sophorolipids specifically for acne treatment.

- *Soliance* (www.soliance.com): A French company, also produces sophorolipids from rapeseed fermentation for cosmetic applications in skin care through antibacterial and sebo regulator activity.
- *BioFuture Ltd.* (www.biofuture.ie): An Irish company, produces rhamnolipids (Biofuture) for bioremediation of hydrocarbon-contaminated soil.
- *Enzyme Technologies Inc.:* An American company, produces an undisclosed bacterial preparation (Petrosolv) for oil removal, recovery, and processing.
- *Sigma-Aldrich Co. LLC* (www.sigmaaldrich.com): An American company, produces surfactin with antifungal, antibacterial, and antitumor activities.

(Fracchia et al., 2012; Reis et al., 2013).

11.7   PRODUCER MICROORGANISMS

Although various species of bacteria, yeasts, and filamentous fungi are seen as producer organisms in several biosurfactants and bioemulsifier-related patents, species belonging to the *Acinetobacter, Bacillus, Pseudomonas, Torulopsis,* and *Candida* come to the forefront (Figure 11.4). Moreover, in many patents especially when the invention is focused on the formulation for certain applications and/or production methods, apparatus, or downstream process, the producer microorganism is not specified. In some cases, a bacterial consortium of unspecified origin was used.

Bacillus versus *Pseudomonas* species are the most common producers in patent publications (Tables 11.1 and 11.2). The members of rod-shaped genus *Bacillus*, chemo-organotrophic, aerobic, or facultatively anaerobic organisms forming resistant endospores have been used for many years (Slepecky and Hemphill, 2006). Surfactin discovered by Arima et al. (1968) from the culture broth of *Bacillus subtilis* is the most active biosurfactant (a cyclic lipopeptide) produced by the *Bacillus*

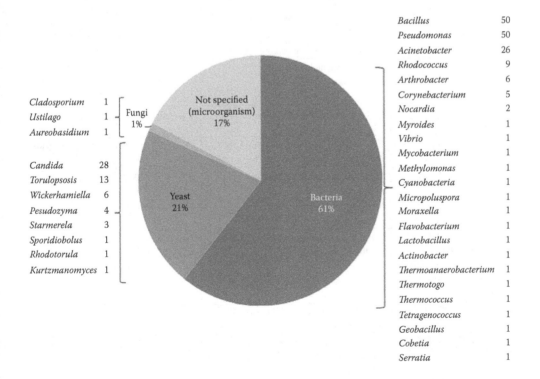

FIGURE 11.4   Distribution of biosurfactant producers by genus.

TABLE 11.1

Members of the Genus *Bacillus*

Organism	Short Title	Inventor and Date	Patent
Bacillus sp.	Surfactin	Arima et al. (1972)	US 3687926
Bacillus licheniformis JF-2 (ATCC No. 39307)	Biosurfactant and enhanced oil recovery	Mcinerney et al. (1985)	US 4522261
Bacillus sp. TY-8 (FERM P-13666) and *Bacillus* sp. TY-34 (FERM P-13665)	New microorganism assimilating petroleum and waste oil	Okura et al. (1995)	JP H078271
Bacillus subtilis, Bacillus licheniformis, Bacillus circulans	Bacterial preparation for agricultural use	Hibino and Minami (1998)	US 5733355
Bacillus (mutant)	Methods for producing polypeptides in surfactin	Sloma et al. (1998)	WO 9822598
	Methods for producing polypeptides in mutants	Sloma et al. (1999)	US 5958728
Bacillus Megateria-1 BD (Bacterial mixture)	Bacteria for regenerating crude oil contaminated earth and water	Arkadevitch et al. (1998)	DE 19652580
Bacillus subtilis	Novel strain to control plant diseases and corn rootworm	Heins et al. (1998)	WO 9850422
Bacillus megaterium-1BD	Bacterial composition for treatment of water and soil from pollution with mineral oil and process for producing thereof	Baburins et al. (2000)	LV 12348
Bacillus subtilis EB-162 (FERN P-17319)	EB-162 material and its production	Negishi et al. (2000)	JP 2000273100
Bacillus subtilis	Method for protecting or treating plants and fruit against insect pests, extract of lipopeptide and lipopeptide surfactin	Heins et al. (2001)	CZ 20011620
Bacillus sp.	Production process of surfactin	Yoneda et al. (2002)	WO 2002026961
	Production process of subtilin surfactin	Yoneda et al. (2004)	CN 1466626
Bacillus khr-10-mx (KCTC 8533P) and Mixture KHR-5-MX (KCTC 0078BP)	Biosurfactant and polymer for improvement of octane value	Fukunaga et al. (2003)	KR 20030044969
Bacillus sp.	Conversion of domestic garbage to compost	Shi et al. (2003b)	CN 1431314
Bacillus sp. LSC11 (KCTC 1038BP)	Biosurfactant producing microorganism *Bacillus* sp. LSC11	Choi et al. (2004a)	KR 20040046860
Bacillus sp. A8-8 (KACC 91018)	Biosurfactant and antifungal substance production	Choi et al. (2004b)	KR 20040055487
Bacillus subtilis	Method for preparing sodium surfactin and use thereof	Jun et al. (2004)	KR 20040055035
Myroides odoratus SM-1 (FERM P-19061), *Bacillus subtilis* SM-4 (FERM P-19059) and *Bacillus pumilus* SM-9 (FERM P-19060)	Novel microorganism and biosurfactant produced thereby	Kawai and Maneraato (2004)	JP 2004154090
Bacillus subtilis E2 (KCTC 10389BP)	Method for production and use of surfactin	Ahn et al. (2005)	KR 20050007670
Bacillus subtilis CH-1 (KCCM 10480)	Antithrombotic biosurfactant production for prevention, alleviation, and treatment of circulatory system diseases	Hong et al. (2005)	KR 20050014451

continued

TABLE 11.1 (continued)

Members of the Genus *Bacillus*

Organism	Short Title	Inventor and Date	Patent
Bacillus vallismortis TB40-3	Biosurfactant with oil degrading and antifungal property	Choi et al. (2005)	KR 20050109111
Bacillus subtilis 0017	Strain of bacterium *Bacillus subtilis* as producer of surfactin	Melent'ev et al. (2006)	RU 2270858
Bacillus subtilis BC1212	BC1212 from food and its production of surfactin	Kim et al. (2006c)	KR 20060045201
Bacillus subtilis strain E8 (1107)	Preparation method and usage for surfactin	Gong et al. (2007)	CN 101041846
Bacillus pumilus HY1 (KACC 91280P)	Production of kanjang and chungkookjang with surfactin	Yun et al. (2008)	KR 20080070127
Bacillus subtilis GWN-5(KCTC 11215BP)	Biosurfactant produced by this strain as a bioabsorbent	Ho and Su (2009)	KR 20090038546
Bacillus pumilus Y8A CCTCCM 209108	Lipopeptide-producing *Bacillus pumilus* and application	Biao et al. (2009)	CN 101560483
Bacillus licheniformis V9T14	Biosurfactant, uses, and products thereof	Ceri et al. (2010)	WO 2010067245
Bacillus subtilis BIT09S1, BIT09S2 and BIT09A2	Method for preparing *Bacillus subtilis* lipopeptid biosurfactant	Peng et al. (2010)	CN 101838621
Bacillus amyloliquefaciens BZ6-1 CGMCC No. 2892	Producing biosurfactant and application in oil sludge elution	Wuxing et al. (2011)	CN 101935633
Bacillus subtilis JK-1	Culture method of biosurfactant production	Kim and Joo (2011)	KR 20110016159
Bacillus subtilis, Bacillus licheniformis, Bacillus pumilus, Bacillus aeolius and *Bacillus subtillis* natto and mixtures	Method for improving the cleaning action of a detergent or cleaning agent	O'Connell et al. (2011)	US 2011201536
Bacillus subtilis	Lipopeptide biosurfactant preparation and application	Dawei (2012)	CN 102352227
Bacillus subtilis ACCC01430	Industrialized preparation method of a lipopeptide biosurfactant	Yujiang et al. (2012a)	CN 102373258
Bacillus spp. LEY11 (CGMCC No. 3767)	Preparing biosurfactant using high temperature *Bacillus* spp.	Zhiqiang et al. (2012)	CN 102399847
Bacillus subtilis ACCC01430	Lipopetide biosurfactant oil extraction agent	Yujiang et al. (2012b)	CN 102504789
Bacillus subtilis BS1	Remediation of soil PAH by bacteria and mycorrhiza	Xiang et al. (2012)	CN 102652957
Bacillus SCSGAB0092 CCTCC No. M2012339	Methylotrophic *Bacillus* for production of surfactins and iturin A	Qi et al. (2013)	CN 102994418
Bacillus sp.	Concurrent increased production of iturin A and surfactin	Hsieh (2013)	TW 201313898
Bacillus sp.	Biosurfactants composition, production, and use	Coutte et al. (2013)	WO 2013050700
Bacillus subtilis	Application of iturins sodium surfactin fermented liquid	Chen et al. (2013)	CN 103011978

species discovered so far (Ron and Rosenberg, 2001). Multiple biologic activity of surfactin such as insecticidal, antifungal, and antibacterial activity was also patented. A EB-162 is a biosurfactant obtained from a cultured material of a *B. subtilis* EB-162 (FERN P-17319) producing a surfactin analogue compound group, capable of exhibiting excellent surface tension-reducing activities, and useful as a biosystem surfactant (Negishi et al., 2000, JP 2000273100).

TABLE 11.2
Members of the Genus *Pseudomonas*

Organism	Short Title	Inventor and Date	Patent
Pseudomonas sp. DSM 2874	Process for the biotechnological production of rhamnolipids	Wagner et al. (1985)	EP 0153634
Pseudomonas sp.	Process for the production of rhamnolipids	Kaeppeli et al. (1986)	US 4628030
Pseudomonas aeruginosa DSM 4692 and *Rhodococcus erythropolis* DSM 4670	Microbial breakdown of petroleum hydrocarbons and mineral oil in contaminated soils, water, and aerosols	Not available (1990)	DE 3909324
Pseudomonas fluorescens 6519E01 ATCC 53860, *P. fluorescens* 6133D02 ATCC 53859, *Serratia plymuthica* 6109D01 ATCC 53858	Biological control of corn seed rot and seedling blight	Haefele et al. (1991)	US 4996049
Pseudomonas aeruginosa JAMM	Biosurfactants production using refined oils and fat	Manresa et al. (1991)	ES 2018637
Pseudomonas aeruginosa SB-1, SB-3, SB-30	Production of emulsifying agents and surfactants	Banerjee et al. (1991)	US 5013654
Pseudomonas aeruginosa ATCC 21996, ATCC 10145 or NCTC 10701	Rhamnolipid synthesis with glyceric acid ether lipid, to form intermediate for rhamnose used in synthesis of chiral cpds	Aretz and Hedtmann (1993)	DE 4127908
Pseudomonas aeruginosa	Use in the preparation of L-rhamnose	Giani et al. (1993)	EP 0575908
Pseudomonas aeruginosa TY-29 (FERM P-13667) *P. aeruginosa* TY-30 (FERM P-13668)	New microorganism assimilating petroleum and waste oil	Okura et al. (1995)	JP H078270
Pseudomonas cepacia ATCC 55487	To biodegrade hydrophobic organic compounds, for example, PCB	Rothmel (1996)	US 5516688
Pseudomonas aeruginosa 646011	Immunological activity of rhamnolipids	Piljac and Piljac (1996)	US 5514661
Pseudomonas spp.	Rhamnolipids for control of plant pathogen zoosporic fungi	Stanghellini et al. (1997)	WO 9725866
Pseudomonas or *Moraxella*	Cleaning porous surfaces with a washing liquid containing bacteria having an enzymatic activity	Dran and Debord (1997)	EP 0808671
Pseudomonas sp.	Novel *Pseudomonas* sp. that produces biosurfactant	Kim et al. (1998)	KR 0141065
Pseudomonas sp.	Biosynthesis of rhamnolipid–alginate polymer complexes	Eliseev et al. (1998)	DE 19654942
Pseudomonas sp.	Production of rhamnolipid by using ethanol	Matsufuji and Nakada (1998)	JP H1075796
Pseudomonas rubescens	Improvement in cellulosic fiber and composition	Masuoka et al. (1998)	JP H1096174
Pseudomonas aeruginosa (USB-CS1)	Oily emulsions mediated by a microbial tenso-active agent	Rocha et al. (1999)	US 5866376
Pseudomonas sp.	Rhamnolipid preparation and use in tertiary oil recovery	Yang et al. (2000)	CN 1275429
Pseudomonas putida HPLJS-1 (KCTC 0666BP)	*Pseudomonas putida* HPLJS-1 producing biosurfactant	Hwang et al. (2001)	KR 20010046256
Pseudomonas aeruginosa BYK-2	Simultaneous rhamnolipid and methyl rhamnolipid production	Ha et al. (2002)	KR 20020016707
Aerugino-pseudomonads	Biosurfactant of rhamnolipid and its application in artificial manure turned from household garbage	Shi et al. (2003a)	CN 1431036

continued

TABLE 11.2 (continued)
Members of the Genus *Pseudomonas*

Organism	Short Title	Inventor and Date	Patent
Pseudomonas stutzeri MEV-S1 B-8277	Strain used for treatment of soil, ground, and surface water from petroleum and products of its processing	Marchenko et al. (2004)	RU 2002122612
Pseudomonas fluorescens KPM-018P	Plant insect pest controlling agent and method for controlling plant insect pest	Mayama et al. (2005)	JP 2005102510
Not available	Rhamnolipid crude extract prepared by fermenting food and drink waste oil and application thereof	Meng and Zhang (2007)	CN 1908180
Pseudomonas chlororaphis NRRL B-30761	Processes for the production of rhamnolipids	Gunther et al. (2007)	US 7202063
Pseudomonas sp. G314 KACC 91263P	Novel *Pseudomonas* sp. G314-producing biosurfactant	Park and Shim (2008)	KR 20080017148
Pseudomonads SP27	Oil production method using biosurfactant	Tianbo (2008)	CN 101131086
Pseudomonas sp. VTS-1 CGMCC2200	Ind. preparation of rhamnolipid boil. fermentation liquor	Yumei et al. (2008)	CN 101177696
Pseudomonas aeruginosa	Method for enhancing yield of rhamnolipid produced by copper green pseudomonas	Xiaolan et al. (2008)	CN 101182560
Pseudomonas aeruginosa	Fermentation production technique for rhamnolipid biological surface activator	Xinzheng et al. (2008)	CN 101265488
Pseudomonas DSM7108 or DSM2874	Biosurfactants and production thereof	Leitermann et al. (2008)	WO 2008151615
Pseudomonas aeruginosa	Apparatus and method for producing strain of rhamnolipid	Zhang (2009)	TW 200909586
Pseudomonas aeruginosa	Method for producing biosurfactant	Chang et al. (2009)	TW 200909584
Pseudomonas aeruginosa	Method for preparing rhamnolipid by utilizing microorganism fermentation	Yu et al. (2009a)	CN 101613725
Pseudomonas sp. EP-3 KACC 91475P	Rhamnolipid-producing *Pseudomonas* sp. EP-3, and its use for control of aphids	Kim et al. (2010)	KR 100940231
Pseudomonas aeruginosa	Method for producing rhamnolipid and medium of the same	Wei et al. (2010)	TW 201009080
Pseudomonas aeruginosa	Preparation method and application of rhamnolipid	Meng and Zhang (2010)	CN 101845468
Pseudomonas aeruginosa	Biosurfactant preparation method and application thereof	Hongyan et al. (2010)	CN 101898100
Pseudomonas aeruginosa	Select rhamnolipid-producing bacterium and used medium	Wenjie et al. (2011a)	CN 101948787
Pseudomonas aeruginosa CGMCC No. 4002	Producing rhamnolipid with high yield and application	Wenjie et al. (2011b)	CN 101948786
Pseudomonas aeruginosa CGMCC No. 3034	Bacterium S2 for efficiently generating biosurfactant and fermentation culture medium thereof	Lixiang and Jie (2011)	CN 102250790
Pseudomonas aeruginosa NY3	Rhamnolipid biosurfactant and methods of use	Yin et al. (2011)	US 2011306569
Pseudomonas aeruginosa	Culture method and application thereof	Fulin et al. (2012)	CN 102409016
Pseudomonas fluorescens (PT)	Method of generating biosurfactant by using *Pseudomonas*	Zhuowei et al. (2012)	CN 102732573
Pseudomonas aeruginosa	Producing rhamnolipid by fermentation and separation	Liang et al. (2012)	CN 102796781
Pseudomonas aeruginosa	Method for pretreating crude oil using microorganism	Bin Mohd et al. (2012)	US 2012301940
Pseudomonas aeruginosa	Application of rhamnolipid as demulsifier	Meng et al. (2013b)	CN 102851059
Pseudomonas aeruginosa MZ01 CGMCC No. 6354	Crude oil-degrading bacterium for producing lipid biosurfactant and application	Guo et al. (2013)	CN 102978135

Species of *Pseudomonas* are well known for its metabolic versatility and capacity of utilizing a wide range of organic and inorganic compounds. Their ability to metabolize an extensive number of substrates, such as aliphatic and aromatic hydrocarbons, made them important for bioremediation (Moore et al., 2006).

The third most common prokaryotic species in biosurfactant production is *Acinetobacter* (Table 11.3). *Acinetobacter* are Gram-negative, aerobic, nonmotile, and oxidase negative coccobacillus. Because of their ability to utilize hydrocarbons, acinetobacters are capable of playing a role in the degradation of crude oil and may play an important role in a variety of commercially important industrial processes, as well as in the biodegradation of a wide range of environmental pollutants (Gutnick et al., 1991; Navon-Venezia et al., 1995; Towner, 2006).

Acinetobacters that degrade alkanes are frequently isolated from areas contaminated with petroleum. Most representatives of the genus *Acinetobacter* synthesize a high-molecular-weight biosurfactant that possesses emulsifying, but not surface-active properties (Rosenberg and Ron, 1999). The best-studied bioemulsans are produced by different species of *Acinetobacter* (Rosenberg and Ron, 1998). The derivatives of emulsan, α-emulsan, apo-0-emulsan, apo-α-emulsan, apo-psi-emulsan, proemulsans, and psi-emulsan are patented. The RAG-1 emulsan is a complex of an anionic heteropolysaccharide and protein (Rosenberg, 2006). With a patent issued in 1990, the biosurfactants produced by the novel *A. calcoaceticus* strain have emulsifying properties superior to the known biosurfactants produced by a known *A. calcoaceticus* strain RAG-1. In addition, the strain makes it possible to produce the biosurfactants on an industrial scale (Tanaka et al., 1990, EP 0401700). Alasan or E-KA53 produced by *Acinetobacter radioresistens* strain KA53 is a complex of an anionic polysaccharide and protein. Its emulsifying activity increases with preheating at increasing temperatures 60–90°C; resistance to strong alkali while retaining increased emulsifying activity; reduced viscosity that varies as a function of temperature treatment; and emulsifying activity that varies as a function of pH and magnesium ions (Rosenberg and Ron, 1996, WO 9620611).

Other prokaryotic producers are *Rhodococcus* and *Arthrobacter*. However, there are also patents involving other microorganisms belonging to different genera (Table 11.4). Among these, the *Corynebacteria salvinicum* strain SFC for materials having surfactant properties (Zajic et al., 1982, US 4355109), the genus *Micropolyspora* producing an oligosaccharide fatty acid ester and a partial ester of the tetraglucose (Ishigami et al., 1994, JP H06298784), an emulsifying agent from *Corynebacterium hydrocarboclastus* UWO 409 (Zajic and Knettig 1976a, CA 990668 and 1976b, US 3997398), and rhamnolipid from *Tetragenococcus koreensis* JS KCTC 3924 (Lee et al., 2007, KR 20070027151) have also been patented. Another producer, *Actinobacteria* produce halogenated emulsans that are cultured in halogenated compounds including fluorinated compounds such as fluorinated fatty acids and fluorinated fatty acid esters. Halogenated compounds that are assimilated by the microorganism are incorporated into the emulsan molecule (Yalpani, 2004, US 2004171128). Additionally, *Cobetia marina* (Dinamarca et al., 2012, WO 2012164508), *Methylomonas clara* (Kachholz and Schlingmann, 1984, EP 0121228), *Myroides odoratus* (Kawai and Maneraato 2007, JP 2007126469), and *Cyanobacteria* (Shilo and Fattom 1987, US 4693842) were all specified for biosurfactant/bioemulsifier production.

Biosurfactants are also synthesized by some yeast species including *Candida (Torulopsis) bombicola, Candida apicola,* and *Wickerhamiella domercqiae* (Table 11.5). Sophorolipids are glycolipid molecules formed by the combination of a sophorose sugar (dimeric glucose) and a lipid chain. Yeasts produce a number of emulsifiers. This is particularly interesting due to the food grade status of several yeasts. Among 65 patents using yeast, 9 do not specify species.

Chemically stable acid-type sophorose lipids can selectively be produced by one step, instead of a conventional sophorose lipid fermentation method by which the sophorose lipid is obtained as a mixture of a lactone type and the acid type. *Candida floricola* selectively produces the acid-type sophorose lipid to liquid culture (Konishi et al., 2008, JP 2008247845).

Recently, *Pseudozyma churashimaensis* isolated from sugarcane produced mannosyl erythritol lipid-A, which is suitable for uses requiring higher safety for living bodies, such as cosmetics, foods,

TABLE 11.3
Members of the Genus *Acinetobacter*

Microorganism	Title	Inventor and Date	Patent
Acinetobacter sp. ATCC 31012	Emulsans	Gutnick et al. (1980)	AU 5564380
		Gutnick et al. (1981)	ZA 8000912
Acinetobacter sp. ATCC 31012	Cleaning oil-contaminated vessels with alpha-emulsans	Gutnick and Rosenberg (1981)	US 4276094
Acinetobacter sp. ATCC 31012	Apo-beta-emulsans	Gutnick et al. (1982a)	US 4311829
Acinetobacter sp. ATCC 31012	Apo-alpha-emulsans	Gutnick et al. (1982b)	US 4311830
Acinetobacter sp. ATCC 31012	Apo-psi-emulsans	Gutnick et al. (1982c)	US 4311831
Acinetobacter sp. ATCC 31012	Proemulsans	Gutnick et al. (1982d)	US 4311832
Acinetobacter sp. ATCC 31012	Psi-emulsans	Gutnick et al. (1983a)	US 4380504
Acinetobacter sp. ATCC 31012	Polyanionic heteropolysaccharide biopolymers	Gutnick et al. (1983b)	US 4395353
Acinetobacter sp. ATCC 31012	Emulsans	Gutnick et al. (1983c)	US 4395354
Acinetobacter calcoaceticus	Cosmetic and pharmaceutical compositions containing bioemulsifiers	Hayes and Holzner (1986)	EP 0178443
Acinetobacter calcoaceticus	Personal care products containing bioemulsifiers	Hayes (1987)	EP 0242296
		Hayes (1991)	US 4999195
Acinetobacter calcoaceticus	Bathing agent	Osugi (1988)	JP S63156714
Acinetobacter calcoaceticus	Bioemulsifier production by *Acinetobacter calcoaceticus*	Gutnick et al. (1993)	CA 1316478
		Gutnick et al. (1989)	US 4883757
Acinetobacter calcoaceticus RAG-1	Novel *Acinetobacter calcoaceticus* and novel biosurfactant	Tanaka et al. (1990)	EP 0401700
Acinetobacter calcoaceticus	Recovery of oil from oil reservoirs	Sheehy (1992)	US 5083610
Acinetobacter calcoaceticus	Liquid detergent composition	Hwang et al. (1996)	KR 960007877
Acinetobacter radioresistens KA53	Novel bioemulsifiers	Rosenberg and Ron (1996)	WO 9620611
	Bioemulsifiers	Rosenberg and Ron (1998)	US 5840547
Acinetobacter calcoaceticus CL (KCTC 0081BP)	Microorganism producing biosurfactant	Hwang and Kim (1997)	KR 970006157
Acinetobacter sp. KRC-K4	Novel microorganism *Acinetobacter* sp. KRC-K4 and process for preparing bioemulsifier using the same	Park et al. (1999)	KR 0170107
Acinetobacter calcoaceticus	New lipopolysaccharide biosurfactant	Prosperi et al. (1999)	EP 0924221
Acinetobacter calcoaceticus	Compositions containing bioemulsifiers and preparation	Gutnick and Bach (2002)	WO 0248327
Acinetobacter sp.	Bioemulsifier production by *Acinetobacter* strains isolated from healthy human skin	Patil and Chopade (2005)	US 2005163739
		Patil and Chopade (2004)	US 2004138429
Rhodococcus baikoneurensis EN3 KCTC19082, *Acinetobacter johnsonii* EN67 KCTC12360, *Acinetobacter haemolyticus* EN96 KCTC12361	Novel microorganisms having oil biodegradability and method for bioremediation of oil-contaminated soil	Park et al. (2008)	US 2008020947
Acinetobacter calcoaceticus OM1 strain (KACC 91005 A)	Compositions and methods for degrading vehicle's waste oil or vessel's leaked oil	Chung et al. (2008)	KR 100866526
Acinetobacter BU03, *Bacillus subtilis* B-UM0	Biosurfactant and PAH-degrading strain-enhanced composting treatment of contaminated soil	Huanzhong and Zhenyong (2011)	CN 102125929
Acinetobacter spp.	Growth medium and method for producing biosurfactant	Chang et al. (2012)	TW 201209161

TABLE 11.4
Other Bacteria

Organism	Short Title	Inventor and Date	Patent Number
Corynebacterium hydrocarboclastus UWO 409	Emulsifying agents of microbiological origin	Zajic and Knettig (1976a)	CA 990668
		Zajic and Knettig (1976b)	US 3997398
Arthrobacter sp. ATCC 31012	Production of alpha—emulsans	Gutnick and Rosenberg (1980)	US 4230801
Arthrobacter sp. ATCC 31012	Production of alpha—emulsans	Gutnick et al. (1980)	US 4234689
Arthrobacter terregens, Arthrobacter xerosis, Bacillus megaterium, Corynebacterium lepus, Corynebacterium xerosis, Nocardia petrophilia, Pseudomonas asphaltenicus ASPH-AI, and *Vibrio ficheri*	Hydrocarbon extraction agents and microbiological processes for their production	Zajic and Gerson (1981)	CA 1114759
		Zajic and Gerson (1987)	US 4640767
Corynebacteria salvinicum SFC	Microbiological production of novel biosurfactants	Zajic et al. (1982)	US 4355109
Mycobacterium phlei	Method and installation for flooding petroleum wells and oil-sands	Wagner et al. (1982)	CA 1119794
Methylomonas clara ATCC 31 226	Lipotensides, process for their isolation and their use	Kachholz and Schlingmann (1984)	EP 0121228
Rhodococcus erythropolis DSM 43215, *Arthrobacter* DSM 2567 *Corynebacterium* DSM 2568	Trehalose-lipid-tetraesters	Wagner et al. (1987) Wagner et al. (1988)	CA 1226545 US 4720456
Rhodococcus erythropolis SD-74	Production of succinyl-trehalose lipid by microorganism	Tabuchi and Kayano (1987)	JP S6283896
Cyanobacteria J-1	Bioemulsifier composition and solution thereof	Shilo and Fattom (1987)	US 4693842
Arthrobacter sp. No. 38 (FERM BP-4435).	Peptide-surfactant, production process, and microorganisms	Imanaka and Sakurai (1994)	EP 0593058
	Biosurfactant cyclopeptide compound produced by culturing	Imanaka and Sakurai (1994)	US 5344913
Micropolyspora	Tetraglucose and its partial fatty acid ester	Ishigami et al. (1994b)	JP H06298784
Mixture of bacterial strains belonging to the *Pseudomonas, Flavobacterium, Arthrobacter, Corynebacterium, Moraxella, Nocardia* genera	Composition containing a surface-active compound and glycolipids and decontamination process for a porous medium polluted by hydrocarbons	Ducreux et al. (1997)	US 5654192
Lactobacillus	*Lactobacillus* therapies	Reid et al. (2000)	US 6051552
Rhodococcus globerulus (HPPDS-1) (KFCC-11168)	Bacterium with higher biosurfactant activity for biological soil remediation	Park and Park (2002)	KR 20020003460
Rhodococcus globerulus (KFCC-11171)	*Biosurfactant-secreting bacteria and its utilization method*	Park (2002)	KR 20020011251
Rhodococcus ruber Em (CGMCC No. 0868)	Erythro micrococcus Em and usage for generating biologic emulsifier as well as degrading PAH	Liu et al. (2004)	CN 1519312

continued

TABLE 11.4 (continued)
Other Bacteria

Organism	Short Title	Inventor and Date	Patent Number
Actinobacter	Halogenated emulsans	Yalpani (2004)	US 2004171128
Microbial consortium hyperthermophilic, barophilic, acidogenic, anaerobic bacterial strains (*Thermoanaerobacterium* sp., *Thermotoga* sp. and *Thermococcus* sp.)	A process for enhanced recovery of crude oil from oil wells using novel microbial consortium	Lal et al. (2005)	WO 2005005773
Rhodococcus erythropolis EK-1 IMB Ac-5017	Strain of *Rhodococcus erythropolis* EK-1 bacteria—producer of the surface-active substances	Pyroh et al. (2006)	UA 77345
Tetragenococcus koreensis JS KCTC 3924	Rhamnolipid-producing ability and method for use	Lee et al. (2007)	KR 20070027151
Myroides odoratus SM-1 (FERM P-19061)	Biosurfactant that new microorganism produces	Kawai and Maneraato (2007)	JP 2007126469
Geobacillus thermodenitrificans CGMCC-1228	Screening method and the uses thereof	Wang et al. (2009)	US 2009148881
Rhodococcus equi	Use thereof in petroleum microorganism yield increase	Xiaodan et al. (2011a)	CN 101935630
Cobetia marina MM1IDA2H-1, CECT No. 7764	Biosurfactant extract obtained from same	Dinamarca et al. (2012)	WO 2012164508

and agricultural livestock with more excellence in adaptability than conventional biosurfactants (Morita et al., 2011, JP 2011182660).

There are only five patents using eukaryotic microorganisms (Table 11.6), namely, filamentous fungi. In these, *Cladosporium resinae* (Jimenez and Morales 1993, ES 2039187), Ustilaginaceae (Morita et al., 2010, JP 2010200695), genus *Ustilago* (Omori et al., 2010, JP 2010158192), *Aureobasidium pullulans* (Kim et al., 2013, KR 101225110), and an unspecified fungus are mentioned.

11.8 TECHNIQUES

Although microbial biosurfactants exhibit a number of environmental advantages over their chemical equivalents, further studies are needed to develop economically competitive production processes. Relatively high substrate costs as well as production and downstream processing costs, low productivities, and intense foaming formation during the biosurfactant production are currently the major obstacles for the economic production of biosurfactants. Thus, the patents in the past 50 years may be classified under four main headings: *New microbial producers, Alternative substrates, Efficient production processes, and Separation and purification processes.*

11.8.1 NEW MICROBIAL PRODUCERS

11.8.1.1 Screening Methods for Biosurfactant Producers

There are four patents issued that are related to screening methods of biosurfactant producer microorganisms. In two of these, biosurfactant-producing microorganisms are selected according to colony morphotypes from culture of the petroleum-oxidizing microorganisms on solid medium in Petri dishes at room temperature. The method is simple in realization and allows screening of

TABLE 11.5
Yeast

Organism	Short Title	Inventor and Date	Patent
Torulopsis magnoliae	Oil glycosides of sophorose and fatty acid esters thereof	Theodore et al. (1967)	US 3312684
Torulopsis bombicola	Dehydrating purification of a fermentation product	Not available (1979)	GB 2002369
Torulopsis bombicola	Process for producing a hydroxyfatty acid ester	Not available (1979)	GB 2002756
		Inoue and Miyamoto (1980)	US 4201844
Torulopsis bombicola	Cosmetic composition	Tsutsumi et al. (1981)	US 4305961
Torulopsis bombicola	Powdered compressed cosmetic material	Kawano et al. (1981b)	US 4305931
Torulopsis bombicola	Stick-shaped cosmetic material	Kawano et al. (1981a)	US 4305929
Torulopsis bombicola	Skin-protecting cosmetic composition	Tsutsumi et al. (1982)	US 4309447
Torulopsis bombicola	Cosmetic composition for skin and hair treatment	Abe et al. (1981)	US 4297340
Yeast	Process for the preparation of rhamnose or fucose	Voelskow and Schlingmann (1984)	ZA 8305832
Torulopsis bombicola KSM-36 (FERM BP-799)	A novel microorganism isolated from cabbage leaves has sophorose-producing activity	Not available (1985)	GB 8516702
Torulopsis bombicola KSM-36 (FERM-P No. 7586)	Production of sophorose by microorganism	Inoue and Kimura (1986)	JP S6135793
Torulopsis bombicola KSM-36 (FERM BP-799)	Novel microorganism	Inoue and Kimura (1988)	US 4782025
Candida sp. b-1 (FERM P-5884)	Production of glycolipid	Kobayashi et al. (1988)	JP S6363389
Candida bombicola or *Candida apicola*	Process for the production of sophorosids by fermentation with continuous fatty acids ester or oil supply	Marchal et al. (1992)	CA 2075177
Candida bombicola or *Candida apicola*	Fed-batch production of sophorolipid(s)—including pre-culture step, giving prod. in acetylated acid form, useful as emulsifier, for example, in sec. oil recovery	Marchal et al. (1993)	FR 2691975
Candida bombicola or *Candida apicola*	Production de sophorolipides acétyles sous leur forme acide à partir d'un substrat consistant et une huile ou un ester	Marchal et al. (1993)	FR 2692593
Candida bombicola CBS 6009 or *Candida apicola* strain	Fermentative production of sophorolipid compsn.	Lemal et al. (1994)	DE 4319540
Candida bombicola ATCC 22214	Sophorolipid derivative	Ishigami et al. (1994a)	JP H06100581
Candida bombicola, Candida apicola, or *Candida magnoliae*	Production of sophorose glyco-lipid biosurfactants	Reus et al. (1995)	DE 19518768
Candida bombicola	Production of sophorolipids from sugars and oil	Garcia-Ochoa and Casas (1997a)	ES 2103687

continued

TABLE 11.5 (continued)
Yeast

Organism	Short Title	Inventor and Date	Patent
Candida bombicola	Production of sophorose by *Candida bombicola*	Garcia-Ochoa et al. (1997b)	ES 2103688
Candida bombicola or *Candida apicola*	Process for the production of sophorolipids by cyclic fermentation with fatty acid esters or oil supply	Marchal et al. (1998)	EP 0837140
Candida bombicola or *Candida apicola*	Process for the production of sophorolipids by cyclic fermentation with feed of fatty acid esters or oils	Marchal et al. (1999)	US 5879913
Yeast	New sophoroselipids, method for their production and use	Lang et al. (1999b)	WO 9924448
Candida sp.	Method for preparing glyceride-type biological surfactant	Chen et al. (2000)	CN 1242256
Torulopsis bombicola, *Torulopsis petrophilum*, and *Torulopsis apicola*	Red Tide-Preventing Agent	Kim (2002)	KR 20020003679
Candida sp.	Fermentation production of sophorose lipid	Furuta and Igarashi (2002)	JP 2002045195
Candida antarctica T-34	Method for treating soy sauce oil	Furubayashi et al. (2002)	JP 2002101847
Candida bombicola ATCC 22214	Cosmetics composition comprising sophorolipids	Han et al. (2004)	KR 20040033376
Candida bombicola	Spermicidal and virucidal properties of sophorolipids	Gross et al. (2004)	US 2004242501
Yeast	Antimicrobial properties of various forms of sophorolipids	Gross and Shah (2004)	AU 2003299557
Kurtzmanomyces sp., *Candida* sp., *Candida antarctica*	Method for producing mannosyl erythritol lipid	Tamai and Tamura (2004)	JP 2004254595
Pseudozyma sp. TM-453	Glycolipid and method for producing the same	Matsuura et al. (2005)	JP 2005104837
Wickerhamiella domercqiae var. sophorolipid. CGMCC No. 1576	Y2A for producing sophorose lipid and its uses	Song (2006)	CN 1807578
Wickerhamiella domercqiae	Preparation method of Y2A variation waufa glucolipid crude extract with antineoplastic activity	Song (2006)	CN 1839892
Candida bombicola ATCC 22214	Two-stage sophorolipids production with *Candida bombicola* ATCC 22214	Pekin (2006)	TR 200601906
Candida bombicola	Sophorolipids as protein inducers and inhibitors in fermentation medium	Gross et al. (2007)	WO 2007073371
Pseudozyma sp.	Method for producing biosurfactant	Morita et al. (2007a)	JP 2007209332
Pseudozyma parantarctica	Highly efficiently producing mannosylerythritol lipid	Morita et al. (2007c)	JP 2007252279
Yeast	Method for producing biosurfactant	Morita et al. (2007b)	JP 2007209333
Yeast	Application of sophorolipid in preparing antibiotic medicine	Song (2007)	CN 101019875
Candida floricola	Method for producing acid-type sophorose lipid	Konishi et al. (2008)	JP 2008247845

continued

TABLE 11.5 (continued)
Yeast

Organism	Short Title	Inventor and Date	Patent
Candida apicola, Candida bogoriensis, Candida bombicola, Candida gropengiesseri, Candida magnolia, Wickerhamiella domercqiae	A method for the production of short-chained glycolipids	Develter and Fleurackers (2008)	EP 1953237
Candida bombicola, Starmerella bombicola, Candida apicola, Rhodotorula bogoriensis, and *Wickerhamiella domericqiae*	Process for producing cellulase	Ju (2008)	US 2008241885
Candida sp.	Compounds of C-22 sophorolipid derivatives and the preparation method thereof	Kim and Shin (2010)	KR 20100022289
Candida and genus *Starmerela*	Method for producing biosurfactant	Morita et al. (2010)	JP 2010200695
Torulopsosis TJZKBA10326	Producing sophorolipid by cont. feeding and fermentation	Hua et al. (2010)	CN 101845469
Wickerhamiella domercqiae Y2A CGMCC3798	Sophorolipid fruit preservative and use thereof	Jing and Bingbing (2010)	CN 101886047
Candida bombicola	Sophorolipid transporter protein	Soetaert and Van Bogaert (2011)	WO 2011070113
Pseudozyma churashimaensis	New microorganism and method for producing sugar-type biosurfactant using the same	Morita et al. (2011a)	JP 2011182660
Pseudozyma tsukubaensis	New microorganism and method for producing sugar-type biosurfactant therewith	Morita et al. (2011b)	JP 2011182740
Candida bombicola ATCC 2214	Method for producing sophorose lipid	Yanagisawa et al. (2011)	EP 2351847
Torulopsis utilis TJZKBA 10326	Method for producing sophorose ester through fermenting waste molasses and waste glycerin	Yonghu et al. (2012)	CN 102329833
Candida bombicola	Purified ethyl ester sophorolipid for sepsis treatment	Falus et al. (2012)	US 2012142621
Yeast	Bioreactor and sophorolipid continuous production method	Guangru et al. (2012)	CN 102492605
Wickerhamiella domercqiae	Producing sophorolipid by fermentation of lignocellulose	Xin and Xiaojing (2012)	CN 102492753
Sporidiobolus salmonicolor AH3 CGMCC No. 4814	Yeast strain for producing biosurfactant and application	Yu et al. (2012)	CN 102766580

petroleum-oxidizing microorganisms as producers of extracellular and cell-bound biosurfactants (Karaseva et al., 2008, RU 2320715; Karaseva et al., 2007, RU 2006115033). The third invention comprises a simple operation by sprinkling a microorganism on oil that is spread on a solid culture medium, culturing the microorganism, and selecting the microorganism that formes a halo. The biosurfactant is dropped onto an oil film to measure the area of the formed emulsified circle. As a result, the activity of the biosurfactant is measured (Sakurai et al., 1993b, JP H05211892). According to the recently issued fourth invention, a new method is described. A polydiacetylene vesicle is used

TABLE 11.6
Filamentous Fungi

Organism	Short Title	Inventor and Date	Patent
Cladosporium resinae	Preparation of new biosurfactant	Jimenez and Morales (1993)	ES 2039187
Fungus	Method for preparing rhamnolipid	Yang (2007)	CN 1891831
Ustilago	Microorganism and method for producing saccharide-type biosurfactant	Omori et al. (2010)	JP 2010158192
Ustilaginaceae and microorganisms of genus *Candida* and genus *Starmerela*	Method for producing biosurfactant	Morita et al. (2010)	JP 2010200695
Aureobasidium pullulans	Novel biosurfactant produced by *Aureobasidium pullulans*	Kim et al. (2013)	KR 101225110

as a carrier, and biosurfactants produced by the strains are used for destroying molecular hydrogen bond structures of the polydiacetylene vesicle to promote polydiacetylen molecular conformation to be changed, so that optical signal changes are triggered. These are used for analyzing the performances of the biosurfactants (Huang et al., 2013, CN 103014121).

11.8.1.2 Genetic Modifications

There are 12 patents issued on the genetic modification of existing strains or new recombinant strains with improved capabilities for biosurfactant production. Most of these patents have used *Bacillus* strains (Table 11.7).

11.8.2 ALTERNATIVE SUBSTRATES

A number of researchers have focussed on the exploitation of various inexpensive agro-industrial residues and/or wastes as substrates, to reduce the production costs while contributing to the reduction of environmental impact generated by the discarded residues and related treatment costs (Nitschke et al., 2005; Saharan et al., 2011).

There are 50 patents describing the utilization of different substrates for the production of biosurfactants. These alternative substrates can be categorized under four groups as (Table 11.8)

1. Crude oil-based substrates
2. Oil-based substrates (vegetable/plant and animal oil)
3. Carbohydrate-based substrates/industrial wastes
4. Mixture and other substrates

Crude oil-based substrates include alkanes and/or polycyclic aromatic hydrocarbons, paraffinic hydrocarbons, polycyclic arylhydrocarbon, petroleum hydrocarbon, and mineral oil wastage. Oil-based substrates of vegetable/plant and animal origin include corn oil or chicken fat, fatty acid mixtures with a vegetable oil or waste edible oil, olive oil mills (olive waste water), rapeseed, soybean, sunflower, palm oil, or esters, waste grease, and oils. Carbohydrate-based substrates/industrial wastes include L-rhamnose, glucose, galactose, mannose, glucuronic acid, edible plants, ethanol, molasses, lignocellulose saccharification liquid, soybean flour or an extract, pulse flour, konjac gum finemeal, guar gum, and derivatives. In 13 patents, various mixtures and other substrates were used.

TABLE 11.7

Patents on Production of Biosurfactants by Genetically Modified New Strains

Organism/Details	Short Title	Inventor and Date	Patent
Bacillus subtilis ATCC 21332 (mutant; at least one mutation between Arg4 and HisA1 sites)	Enhanced production of biosurfactant through the use of a mutated *B. subtilis* strain	Mulligan et al. (1991)	US 5037758
B. subtilis (mutant)	Enhanced biosurfactant production by *B. subtilis* mutant	Mulligan and Chow (1992)	CA 2025812
Bacillus subtilis (mutant)	A mutant of *B. subtilis* and producing surfactin	Carrera et al. (1992)	EP 0463393
	Mutant and process for the production of surfactin	Carrera et al. (1994)	IT 1243392
	Mutant of *Bacillus subtilis*	Carrera et al. (1993)	US 5264363
srfA operon from *Bacillus subtilis*	Cloning and sequencing of chromosomal DNA which encodes the multienzymatic complex surfactin synthetase	Carrera et al. (1993)	EP 0540074
	Cloning and sequencing of chromosomal DNA which codes for the surfactin synthetase multienzyme complex	Grandi et al. (1995)	IT 1256366
	Cloning and sequencing of the srfA operon coding for the multienzyme surfactin synthetase complex	Grandi et al. (1995)	IT 1251680
Bacillus subtilis (mutant from the wild-type strain lodged as ATCC 55033)	Process for the production of surfactin	Carrera et al. (1995)	IT 1248979
Bacillus subtilis CB114 (KCTC 10792BP) (modified having a recombinant vector pDIA:sfp contains the sfp gene from *Bacillus subtilis* C9)	*Bacillus subtilis* CB114 producing a surfactin and having a genetic competency	Kim et al. (2006a)	KR 100642292
Bacillus subtilis (promoter Pspac replaces the promoter of *Bacillus subtilis* fmbR surfactin synthase gene)	Promotor replacement method for improving volume of production of *Bacillus subtilis* surfactin	Zhaoxin et al. (2009)	CN 101402959
Surfactin biosynthesis operon from *Bacillus subtilis* C9 (A transformed recombinant *E. coli* (KCTC 11248BP) having surface activity is produced by transforming with the recombinant vector pSRF536)	Amplification method of surfactin biosynthesis operon, re-construction plasmid, and transformed strain containing operon	Kim et al. (2009)	KR 20090066097
Yeast species	Yeast strains modified in their sophorolipid production	Soetaert et al. (2011)	WO 2011154523
Pseudomonas putida KT2440-rhlABRI (gene cluster rhlABRI related to the synthesis of the rhamnolipid biosurfactant from *Pseudomonas aeruginosa* BSFD5)	Bacterial strain for generating rhamnolipid biosurfactant and generated microbial inoculum thereof	Qing et al. (2011)	CN 101948793
Not available (designed polypeptide biosurfactants)	Designed biosurfactants, their manufacture, purification, and use	Middelberg et al. (2012)	WO 2012079125

continued

TABLE 11.7 (continued)
Patents on Production of Biosurfactants by Genetically Modified New Strains

Organism/Details	Short Title	Inventor and Date	Patent
Bacillus subtilis (expression system controlled by substitution of the promoter region)	Method for producing a modified strain of *B. subtilis*, method for producing surfactin and use of modified strain	Pastore and Moraes (2012)	WO 2012151647
Not available	Cells and nucleic acids for producing rhamnolipids, and also methods for producing rhamnolipids	Schaffer et al. (2012)	CA 2806430
Not available	Transformant for improving productivity of biosurfactin and method for producing surfactin using the same	Han et al. (2012)	KR 20120103187
Host cell comprising an rhlA gene or an ortholog under the control of a heterologous promoter	Means and methods for rhamnolipid production	Blank et al. (2013)	EP 2573172

11.8.3 EFFICIENT PRODUCTION PROCESSES

Table 11.9 summarizes the patents issued up to date on different production methods, reactor configurations, and related apparatus. It is evident that the number of studies in this area has been insufficient, explaining the fact that biosurfactants have been exploited at a very limited level at an industrial scale. There are 10 patents describing different methods of production, namely, continuous, fed-batch, and semicontinuous, two patents on immobilization applications, three describing various apparatus particularly for foam removal and defoaming, and two patents on simultaneous production and separation.

11.8.4 SEPARATION AND PURIFICATION PROCESSES

Similarly, there are only 10 patents issued, describing different approaches on downstream operations for the recovery of biosurfactants (Table 11.10).

11.9 APPLICATION FIELDS

Apart from their surfactant and emulsifier features, the bio-active properties of microbial biosurfactants further enhance their application fields. The antimicrobial and in certain cases insecticidal properties of biosurfactants are exploited to eliminate the unwanted microorganisms (bacteria, fungi, virus, parasites) in plants, animals, as well as in food and medical applications. Furthermore, the fact that biosurfactants exhibit inhibitory effects on tumor cells, enzymes, sperms, and various active substances, make them invaluable particularly over a wide spectrum of potential medical applications.

Biosurfactants can be widely exploited in agriculture for enhancement of biodegradation of pollutants to improve the quality of agriculture soil, for indirectly promoting plant growth based on their antimicrobial activities, and for increasing the plant microbe interaction. Several biosurfactants show promising results against plant pathogens (Sachdev and Cameotra, 2013).

There are more than 50 patents in the cosmetic sector applications exploiting the moisturization, softening, antiaging effects, dryness-improving, firmness, elasticity, skin-ameliorating effects of

TABLE 11.8

Substrates for the Production of Biosurfactants

Substrate	Biosurfactant/ Bioemulsifier	Microorganism	Inventor and Date	Patent Number
Crude Oil-Based Substrates				
Alkane and/or polycyclic aromatic hydrocarbons	Biosurfactant	*Sporidiobolus salmonicolor* AH3 CGMCC No. 4814	Yu et al. (2012)	CN 102766580
Crude oil	Biosurfactant	*Myroides odoratus* SM-1 (FERM P-19061)	Kawai and Maneraato (2004)	JP 2004154090
Paraffinic hydrocarbon	Emulsifying agents	*Corynebacterium hydrocarboclastus* UWO 409	Zajic and Knettig (1976)	US 3997398
Paraffin and polycyclic arylhydrocarbon	Emulsifier	*Rhodococcus ruber* Em (CGMCC No. 0868)	Liu et al. (2004)	CN 1519312
Petroleum hydrocarbon, mineral oil wastage	Glyceride-type biosurfactant	*Candida* sp.	Chen et al. (2000)	CN 1242256
Oil-Based Substrates (Vegetable/Plant and Animal Oil)				
Corn oil	Rhamnose from rhamnolipid	*Pseudomonas aeruginosa*	Daniels et al. (1988)	EP 0282942
Corn oil or chicken fat	Emulsifier	*Pseudomonas aeruginosa* SB-30	Banerjee et al. (1991)	US 5013654
Fatty acid substrates	Alpha-emulsans	*Arthrobacter* sp. ATCC 31012	Gutnick and Rosenberg (1980)	US 4230801
Fatty acid mix with a vegetable oil or waste edible oil	Sophorose lipid	*Candida* sp.	Furuta and Igarashi (2002)	JP 2002045195
Oil of C-22 fatty acid	C-22 sophorolipids	*Candida* sp.	Kim and Shin (2010)	KR 20100022289
Olive oil mills (olive waste water)	Biosurfactants	*Pseudomonas aeruginosa* JAMM	Manresa et al. (1991)	ES 2018637
Olive oil	Composite biosurfactant	*Pseudomonas aeruginosa, Bacillus subtilis Bacillus licheniformis Acinetobacter* sp.	Fu et al. (2003)	CN 1431312
Plant oil	Rhamnolipid	Aerugino-pseudomonads	Shi et al. (2003a)	CN 1431036
Rapeseed oil or esters	Rhamnolipid	*Pseudomonas aeruginosa* CGMCC No. 3034	Lixiang and Jie (2011)	CN 102250790
	Sophorolipid	*Bacillus subtilis* KCCM 10639, KCCM 10640, *Trichosporon loubieri* Y1-A (KCTC 10876BP), *Trichosporon cutaneum,* white-rot fungi	Yum (2008)	US 2008032383
Rapeseed, sunflower, palm and/or soya oil or esters	Sophorolipid	*Candida bombicola* or *C. apicola* (*C. bombicola* CBS 6009)	Lemal et al. (1994) Marchal et al. (1993)	DE 4319540 FR 2692593
Soybean oil	Rhamnolipid	*Pseudomonas* sp. EP-3 (KACC 91475P)	Kim et al. (2010)	KR 100940231
Soy sauce oil	Mannosyl erythritol lipid	*Candida antarctica* T-34	Furubayashi et al. (2002)	JP 2002101847

continued

TABLE 11.8 (continued)
Substrates for the Production of Biosurfactants

Substrate	Biosurfactant/ Bioemulsifier	Microorganism	Inventor and Date	Patent Number
Vegetable oil and fat	Mannosyl erythritol lipids	Microorganisms	Tamai and Tamura (2004)	JP 2004254595
	Mannosyl erythritol lipid	*Pseudozyma parantarctica*	Morita et al. (2007)	JP 2007252279
Waste grease	Rhamnolipid	*Pseudomonas aeruginosa*	Meng and Zhang (2010)	CN 101845468
Waste oil	Rhamnolipid	*Pseudomonas aeruginosa* CGMCC No. 4002	Wenjie et al. (2011)	CN 101948786
Carbohydrate-Based Substrates/Industrial Wastes				
Glucose	Mannosyl erythritol lipid	*Pseudozyma*	Morita et al. (2007)	JP 2007209332
Glucose and bean flour	Surfactin	*Bacillus subtilis* E2 (KCTC 10389BP)	Ahn et al. (2005)	KR 20050007670
Edible plant	Sugar-type biosurfactant mannosyl erythritol lipid-A	*Pseudozyma churashimaensis*	Morita et al. (2011)	JP 2011182660
Ethanol	Surfactin analogue compound group	*Bacillus subtilis* EB-162 (FERN P-17319)	Negishi et al. (2000)	JP 2000273100
	Alpha-emulsans	*Arthrobacter* sp. ATCC 31012	Gutnick et al. (1980)	US 4234689
	Biosurfactant	*Acinetobacter calcoaceticus* CL	Hwang et al. (1996)	KR 960007877
	Rhamnolipid	Not available	Mingxiu et al. (2009)	CN 101407831
	Rhamnolipid	*Pseudomonas* sp.	Matsufuji and Nakada (1998)	JP H1075796
L-rhamnose, glucose, galactose, mannose, glucuronic acid	Rhamnose-containing polysaccharide	*Pseudomonas paucimobilis* DSM 4429	Then et al. (1989)	EP 0339445
Lignocellulose saccharification liquid	Sophorolipid	*Wickerhamiella domercqiae*	Xin and Xiaojing (2012)	CN 102492753
Molasses	Lipopeptide biosurfactant	*Bacillus subtilis* ACCC01430	Yujiang et al. (2012b)	CN 102504789
Molasses, pulse flour, konjac gum finemeal, guar gum and gum breaking liquid of modified guar gum	Lipopeptid biosurfactant	*Bacillus subtilis* BIT09S1, BIT09S2, and BIT09A2	Peng et al. (2010)	CN 101838621
Soybean flour or an extract	Surfactin	*Bacillus* sp.	Showa et al. (2002)	WO 0226961
			Yoneda et al. (2004)	CN 1466626
Sugarcane cellulose particles	Iturin and surfactin	*Bacillus* sp.	Hsieh (2013)	TW 201313898

continued

TABLE 11.8 (continued)
Substrates for the Production of Biosurfactants

Substrate	Biosurfactant/ Bioemulsifier	Microorganism	Inventor and Date	Patent Number
		Mixture and Other Substrates		
>= 10C *n*-alkanes, fats or oils	Succinyl-trehalose lipid	*Rhodococcus erythropolis* SD-74	Tabuchi and Kayano (1987)	JP S6283896
Ambrette seed oil and waste molasses	Rhamnolipid	*Pseudomonas aeruginosa*	Yu et al. (2009a)	CN 101613725
Carbohydrate and hydrophobic hydrocarbon substrate	Short chained glycolipids	Microorganism	Develter and Fleurackers (2008)	EP 1953237
Carbohydrate, 10–24C satd. or unsatd. fatty acid ester, 10–20C satd. or unsatd. aliphatic hydrocarbon, aliphatic alcohol or acid, or their mix	Sophorolipid	*Candida bombicola* or *C. apicola*	Marchal et al. (1993)	FR 2691975
Crude oil and edible oil	Biosurfactant and antifungal substance	*Bacillus* sp. A8-8 (KACC 91018)	Choi et al. (2004)	KR 20040055487
Corn oil and honey	Sophorolipid	*Candida bombicola* ATCC 22214	Pekin (2006)	TR 200601906
Glucose and sunflower oil	Sophorolipid benzylamide	*Candida bombicola* ATCC 22214	Ishigami et al. (1994)	JP H06100581
Glucose and olive oil	Biosurfactant and rhamnose	*Pseudomonas putida* HPLJS-1 (KCTC 0666BP)	Hwang et al. (2001)	KR 20010046256
Glucose or sac or aviation fuel JP-8, automobile gasoline or sunflower oil	Biosurfactant	*Cladosporium resinae*	Jimenez and Morales (1993)	ES 2039187
Glycerin, succinate, mono-, di- and/or tri-saccharide	Sophoroselipids	Yeast	Lang et al. (1999)	WO 9924448
Oil and glucose	Sophorolipid	*Torulopsosis* sp. TJZKBA10326	Hua et al. (2010)	CN 101845469
Waste food and drink oil	Rhamnolipid	*Pseudomonas* sp.	Meng and Zhang (2007)	CN 1908180
Cryptococcus curvatus cell lysate	Sophorose glycolipid biosurfactants	*Candida bombicola, Candida apicola* or *Candida magnoliae*	Reus et al. (1995)	DE 19518768

biosurfactants on skin as well as on hair, describing special formulations comprising various biosurfactants (Table 11.16).

Similarly, biosurfactants have wide applications in bioremediation and oil recovery, facilitating the removal of oil-based compounds from soil or oil wells (Tables 11.18 and 11.19).

Although they are not currently economically competitive in comparison to their chemical equivalents, their ecofriendly nature is a significant advantage for the medical and cosmetic sectors,

TABLE 11.9

Details and Short Summaries of Patents on Processes, Reactors, Immobilization Applications, and Various Apparatus

Type	Short Summary	Inventor and Date	Patent Number
Processes and reactors	Processes and devices for the continuous production of biosurfactants using a production reactor, a preemulsification loop, two tangential ultrafiltration stages, and an extraction unit are disclosed	Mattei and Bertrand (1986)	FR 2578552
	A fed-batch production process using an excess supply of sugar and a continuous feed of at least one substrate is described where the residual concentration of the said substrate is maintained at a value of 18 g/L of initial reaction volume for said alimentation duration	Marchal et al. (1992)	CA 2075177
	A fed-batch production system where culture is fed continuously at 0.01–4 g/h/L of initial reactor volume such that the concentration of the residual substrate is at most 18 g/L during the feeding period is disclosed, leading to a product containing a high proportion of acetlylated acids	Marchal et al. (1993)	FR 2691975
	A two-step semicontinuous production process where a part of the broth is removed and a mineral medium containing nitrogen is added to the remaining broth	Marchal et al. (1998)	EP 0837140
	and subsequently a second fermentation cycle is carried out, reducing the duration of fermentation cycles to 30 h	Marchal et al. (1999)	US 5879913
	A semicontinuous production process exploiting a pH-stat fed-batch culture system where the fed medium or N-containing medium is supplied to adjust and stabilize the pH value to reduce the usage of alkaline solution achieving high yield glycolipids	Chang et al. (2009)	TW 200909584
	A preparation method for rhamnolipid fermentation fluid comprising extraction of *P. aeruginosa* from petroleum contaminated soil, its mutation by nitroso guanidine, subsequently fermentation in a two-stage process with advantages of high yielding, high efficiency and charging coefficient, simple technique short, and fermentation period is disclosed	Yumei et al. (2008)	CN 101177696
	A commercially viable process for production of biosurfactants and biopolymers by employing environmentally safe utilization of waste oil after recycling or during the early-stage recycling eliminating the use of strong solvents or acids is described	Ganti (2011)	US 2011151100
	A method using a three-stage bioreactor at low-cost with high efficiency, for bioremediation of oil-spoiled soil by extracting polluted water using microorganism which can decompose oil and secrete biosurfactant, and treating the extracted polluted water is revealed	Park (2003)	KR 20030066072

continued

TABLE 11.9 (continued)
Details and Short Summaries of Patents on Processes, Reactors, Immobilization Applications, and Various Apparatus

Type	Short Summary	Inventor and Date	Patent Number
	A sophorolipid continuous production method comprising centrifuging, recovering precipitated yeast by-product; mixing a centrifuged supernatant and a crude product, recovering a secondary purified product after settlement layering; disinfecting the secondary purified product by an autoclave and storing directly, or drying is described	Guangru et al. (2012)	CN 102492605
	An apparatus and a method for producing rhamnolipids using an irradiating mechanism for *Pseudomonas aeruginosa* and an actuating unit capable of detaining bacterial fluid to allow it to form the biological membrane are described in which, *Pseudomonas aeruginosa* can be facilitated to form the biological membrane environment so as to produce strains for biological surfactant improving the yield of surfactant	Zhang (2009)	TW 200909586
	Isolation and characterization of 13 strains capable of producing surface-active compounds are described where the most-promising strains producing a biosurfactant, polysaccharide in nature, were selected and the production medium was optimized, the surfactant was purified and its stability at different pHs and temperatures was studied	Hussein et al. (2006)	WO 2006136178
	A process for the production of a highly functional microbial surface-active substance (biosurfactant) by utilizing by-products of processes of producing the liquors, yielding a biosurfactant at lower cost in a shorter period of time is described	Omori et al. (2009)	JP 2009207493
	A method for producing a surfactin or its salt, comprising an extraction step by adding an organic solvent containing a branched alkyl alcohol into a liquid culture medium that contains a surfactin or its salt into a solution that is obtained by removing the insoluble fraction from the liquid culture medium is detailed	Izumida et al. (2012)	WO 2012043800
	High yield low-cost processes for generating nutrient-rich biomass, including SCP, and ECP containing biosurfactants; from waste oils, including oily sludge from ships, without added nutrients, and improved separation of oil from biomass are disclosed	Ganti and Ganti (2005)	US 2005227338
	A low-cost, efficient method for producing a glycolipid biosurfactant, in particular, lactonic sophorose lipids under oxygen limitations enabling their recovery in a solid form, as well as production of high purity acidic sophorose lipids by hydrolyzing high purity lactonic sophorose lipids leading to products that possess strong antibacterial and antifungal activities is revealed	Yanagisawa et al. (2011)	EP 2351847
	Sophorolipid production from soybean dark oil processing by-product is described	Kim (2005)	KR 20050076124
	Large-scale production method of glycolipid biosurfactant by fermentation is described	Kim et al. (2006b)	KR 100599248

continued

TABLE 11.9 (continued)

Details and Short Summaries of Patents on Processes, Reactors, Immobilization Applications, and Various Apparatus

Type	Short Summary	Inventor and Date	Patent Number
	A manufacturing method of phospholipids biosurfactant is described	Lee et al. (2013)	KR 20130000705
	A culture medium prescription for the industrial production of rhamnolipid fermentation liquor is described	Yanfang et al. (2008)	CN 101173238
	A method for sifting and producing generation agent of dual-rhamnolipid is described	Shulin et al. (2008)	CN 101173210
Chemical modification	An environmentally friendly method for producing a deacylated product of a glycolipid-type biosurfactant using a hydrolase and regioselectively cleaving of the acyl group of a glycolipid-type biosurfactant resulting in a product with high biodegradability, low toxicity, and improved water solubility is described	Kitagawa et al. (2008d)	JP 2008187902
	A method where a fatty acid or a fatty acid derivative is introduced into the glycolipid-type biosurfactant according to an esterification reaction or a transesterification reaction. A freeze-dried culture solution in which a lipid, etc., are added and a glycolipid biosurfactant-producing microorganism is cultured is dissolved in an organic solvent and reacted thereby, producing the triacyl derivative of the glycolipid-type biosurfactant with slight by-products inexpensively	Kitagawa et al. (2008b)	JP 2008043210
Immobilization	A method of producing biosurfactant is disclosed where biosurfactant-producing bacterial cells are mixed with a latex solution to form immobilized cells, increasing the yield of biosurfactant, reducing cost for down-streaming processing	Chang et al. (2006)	TW I257921
	The invention utilizes sterile solid carriers to increase the yield providing easy maintenance repetitive operation and reduced bioreactor volume, where the cell-free fermentation broth has a very significant emulsive activity	Chang et al. (2007)	TW I289605
De-foaming and foam removal apparatus	The invention describes a fermentation device which comprises a fermentation tank, a first-level foam backflow tank, a foam transfer tank, and a second-level foam backflow tank. The fermentation device can be used to control the large amount of foam generated in process of fermentation, reduce the wastage of fermentation bacteria cells, and fermentation products	Wen et al. (2011)	CN 102061282
	A bioreactor comprising two containers is disclosed. The first container contains a medium and mixed bacteria fixed on a solid carrier and the second is used to eliminate the foam generated by the fermentation and recycle the medium. The container containing an acid solution to precipitate and evaporate material directed from the foam eliminator	Chang et al. (2007)	TW I282366

continued

TABLE 11.9 (continued)

Details and Short Summaries of Patents on Processes, Reactors, Immobilization Applications, and Various Apparatus

Type	Short Summary	Inventor and Date	Patent Number
	A master fermentation cylinder and a slavery fermentation cylinder are combined to ferment in the process	Xinzheng et al. (2008)	CN 101265488
	A method for decreasing foam formation as well as maximizing expression of a biosurfactant in a microorganism, encompassing precipitation of the biosurfactant resulting in decreased foam formation	Heng and Bodo (2012)	US 2012252066
Simultaneous production/ separation	The invention describes an integrated system comprising a bioreactor for obtaining biological products on a large scale (biosurfactants, sugars, biofuels, and enzymes for the production of biofuels). Special membranes are used for, separating and recycling microbial cells, ensuring the sterilization of the culture media reducing foam formation, concentrating the products of interest and recycling streams	Melo et al. (2012)	WO 2012079138
	The invention describes an online production method of rhamnolipids using a process involving cellulose hydrolyzation with *Trichoderma reesei* ZM4-F3 followed by *Pseudomonas aeruginosa* BSZ-07 cultivation for online production of rhamnolipid biosurfactant and accelerating the hydrolyzation of cellulose. Eliminating the need for rhamnolipid purification reducing the production cost and improving the hydrolyzation efficiency of cellulose	Weimin and Qiuzhuo (2009)	CN 101538604

TABLE 11.10

Details and Short Summaries of Patents on Downstream Processes

Downstream Processes	Inventor and Date	Patent Number
The invention discloses a method for producing rhamnolipid by virtue of fermentation and separation of *Pseudomonas aeruginosa* involving a twice extraction and distillation on the pasty fermentation broth to obtain paste rhamnolipid with the purity of 80–90%. The method can meet the low-cost large-scale industrialized production requirements with more than 80% purity	Liang et al. (2012)	CN 102796781
A dehydrating purification process for a fermentation product, which comprises adding at least one polyhydric alcohol and distilling of water with heating under reduced pressure	Not available (1979) Inoue et al. (1980a)	GB 2002369 US 4197166
The patent provides a method for concentrating biosurfactants from a culture solution where the solution containing biosurfactant is subjected to ultrafiltration	Sakurai et al. (1993a)	JP H05211876
A method for obtaining a high concentration SL-hydrated product by liquid and solid removal from cultivation liquid without using an organic solvent. The method comprises a solid–liquid separation, subsequent acidification to insolubilize the sophorose again; a precipitation process and removal of the supernatant leading to a hydrated product of high concentration is provided	Furuta et al. (2003)	JP 2003009896

continued

TABLE 11.10 (continued)
Details and Short Summaries of Patents on Downstream Processes

Downstream Processes	Inventor and Date	Patent Number
A process of extracting rhamnolipids, involving addition of ammonium sulfate, centrifuging to eliminate protein precipitate; diluting with icy salt solution, eliminating protein precipitate to obtain supernatant, stilling in a fridge for residual grease to demulsify and agglomerate in the surface of the supernatant, skimming the grease; centrifuging and freeze drying is provided	Shi (2007)	CN 1974589
The production, in particular, to a post-extraction method of rhamnolipid, comprising pretreatment of fermentation broth, solid–liquid separation, extraction, dry milling with advantages such as easy realization of the transportation, storage, and packaging of products is detailed	Yu et al. (2009b)	CN 101613381
A method for separating and purifying rhamnolipid, which comprises dissolving the crude product in mixed organic solvent; putting the sample to a silica gel column; eluting with mixed organic solvent; separating pigments; and finally, using reduced-pressure distillation and concentration to obtain the purified rhamnolipid is provided	Jinfeng et al. (2010)	CN 101787057
No summary available	Witek-Krowiak and Witek (2011)	PL 390696
A two-step separation method of rhamnolipid by using an ultrafiltration membrane. The two-step ultrafiltration method yields more than 80% recovery, reducing the amount of organic solvent and energy consumption; thus reducing costs is disclosed	Qin and Guoliang (2012)	CN 102432643
A composition containing a basic physiologically active protein and sophorolipid attached to a composite having a particle size suitable to be absorbed in a living body is provided	Not available (2012)	JP 2012232963
An industrial production method of a rhamnolipid biosurfactant dry powder, comprising, coarse filtration; high-speed centrifugation; vacuum concentration; addition of an embedding medium; spray drying; and cyclone separation yielding a dry powder rhamnolipid which is convenient to store and transport is described	Shaojun et al. (2012)	CN 102766172
A method comprising centrifugation, membrane filtration, neutralization, and drying for purifying sodium surfactin yielding 75–88% purification so that the finished product of sodium surfactin can be widely applied in the field of cosmetics as a cosmetic additive is explained	Sun et al. (2013a)	CN 103059108
A method with beneficial effects to further improve the purity of the finished product of sodium surfactin by utilizing a dissolving agent and collecting the concentrated crystal in the upper organic phase, thus obtaining a purified product of sodium surfactin reaches 90–98%; enabling the sodium surfactin to be used as a food additive is disclosed	Sun et al. (2013b)	CN 103059107
A method to obtain pure sophorolipid useful as a cosmetic base and capable of being deodorized by the additional treatment, economically, by treating a crude sophorolipid derivative with a specific adsorbent and then hydrogenating and decoloring the treated product is disclosed	Ootani et al. (1980)	JP S554344
A method for purifying glycolipid methylester (I) comprising steps for the removal of impurities by use of a mixed-phase tower of anion exchange resin and cation exchange resin, a single-phase tower of anion exchange resin or a single-phase tower of cation exchange resin; or by direct-adding ion exchange resin into the reactants is described	Hong et al. (1993)	KR 930003489
An isolation and purification method of sophorose lipid is described	Kim et al. (1999)	KR 0181952

which is clearly reflected in the patents issued. The patents related to application were categorized under seven subtitles as

- Biomedical (Tables 11.11 through 11.15)
- Cosmetic (Table 11.16)
- Agriculture (Table 11.17)
- Bioremediation (Table 11.18)
- Oil recovery (Table 11.19)
- Detergent (Table 11.20)
- Chemical production and other applications (Table 11.21)

11.9.1 BIOMEDICAL

TABLE 11.11
Antimicrobial Activity against Pathogens

Type	Description of the Invention	Inventor and Date	Patent Number
Antimicrobial	The isolation of novel peptides whose sequences derive from microbial by-products including biosurfactants and the use of these novel peptides in the inhibition of infections caused by pathogens are described	Howard et al. (2002) Howard et al. (2004)	WO 0212271 US 6727223
	A sophorolipid which can inhibit *Candida albicans* and Gram-positive bacteria by virtue of high bacteriostasis effect or its composition is disclosed as an effective component for preparing antibiotic medicine with no toxic side effect	Song (2007)	CN 101019875
	Nucleic acid molecule encoding the peptide having antimicrobial activity and/or surfactant property and its production and uses are revealed	Lu (2008) Lu (2009)	TW 200823231 WO 2009044279
	An antimycotic composition containing a rhamnolipid is disclosed	Gandhi et al. (2007)	US 2007191292
Combined effect	Methods of using viscosin and analogs as biosurfactants and as antibacterial, antiviral, and antitrypanosomal therapeutic compounds, in particular, compositions inhibit the growth of the pathogens *Mycobacterium tuberculosis*, Herpes Simplex Virus 2, and/or *Trypanosoma cruzi,* are described	Burke et al. (1999)	US 5965524
	Various forms of sophorolipids as antibacterial, antiviral, and/or antispermidical agents are described	Gross and Shah (2004)	AU 2003299557
	A natural mixture of lactonic and nonlactonic sophorolipids compounds synthesis by fermentation of *Candida bombicola* as a spermicidal and/or antiviral agent is disclosed	Gross et al. (2004)	US 2004242501
Antiviral	A method for treating a herpes-related viral infection by administering an effective amount of at least one sophorolipid is described	Gross and Shah (2007)	WO 2007130738
	A pharmaceutical preparation comprising an active ingredient of at least one rhamnolipid for the treatment of dermatological diseases, for example, Papilloma virus infections is described	Piljac and Piljac (1993)	WO 9314767
	A physiologically active lipid comprising an organic acid-combined glycolipid composed of a monosaccharide, oligosaccharide, or polysaccharide, 3-7C dibasic acid and 6–24C fatty acid in a specific molar ratio with a high antiviral activity and reduced adverse effects useful as a direct remedy for viral diseases such as herpes is disclosed	Kayano and Funada (1990)	JP H0291024

TABLE 11.12
Antiadhesion Activity

Type	Description of the Invention	Inventor and Date	Patent Number
Oral and dental applications	The invention describes the inhibition of bacterial plaque formation of the oral cavity by dispersing biosurfactants (emulsan) in water as a preparation of toothpaste or mouthwash and contacting the aqueous dispersion with dental surfaces	Eigen and Simone (1988) Eigen and Simone (1986)	US 4737359 PT 82402
	A composition or its derivative to inhibit bacterial plaque formation of the oral cavity containing biosurfactants produced by microorganisms and further compounded with a nonionic and/or a cationic disinfectant is described	Morishima (2003)	JP 2003246717
	A composition for the oral cavity containing the biosurfactant selected from a glycolipid produced by the microorganism, containing different kinds of essential oil components selected from a thymol, anethole, eugenol, bisabolol, farnesol, and nerolidol and dicarboxylic acid salt that inhibits the formation of the pathogenic bacterial plaque, through bacteriostasis of the pathogenic bacteria without an effect on indigenous bacterial flora is provided	Morishima and Monoi (2005)	JP 2005298357
	A safe oral composition containing 0.01–50 mass% of one or more kinds of the reducing oligosaccharides selected from kojibiose, sophorose, laminaribiose, cellobiose, turanose, cellotriose, solatriose, cellotetraose, and cellopentaose, and one or more kinds of the linear or cyclic water-soluble salts of polyphosphoric acids and chemically inhibiting the formation of stain on a tooth plane, useful for bleaching is provided	Inoue and Uchiyama (2005)	JP 2005170867
Other	Methods to prevent urogenital infection in mammals and inhibition of microbial biofilm formation and displacement of adherent biofilm-forming bacteria from surfaces using *Lactobacillus* biosurfactants are disclosed	Reid et al. (2000)	US 6051552

TABLE 11.13
Antitumor Activity

Type	Description of the Invention	Inventor and Date	Patent Number
Antitumor or antimetastasis	A sophorosid biosurfactant produced by *Wickerhamiella domercqiae* var. sophorolipid CGMCC No. 1576 with its high restraining rate to tumor cell outside of body is provided	Song (2006)	CN 1807578
	An anticancer composition comprising surfactin inhibiting the cell proliferation, the proliferation of the human colon carcinoma cell by inducing pro-apoptotic activity and suppressing the differentiation and proliferation of LoVo cell by inhibiting cell cycle is disclosed	Hong et al. (2009)	KR 20090014428

continued

TABLE 11.13 (continued)
Antitumor Activity

Type	Description of the Invention	Inventor and Date	Patent Number
	A method of rapidly producing fermented soybeans and soy sauce by using the *Bacillus pumilus* HY1 strains as a starter is provided to obtain fermented soybeans and soy sauce with a high content of surfactin exhibiting cytotoxicity against breast cancer cells and intestinal cancer is described	Yun et al. (2008)	KR 20080070127
	Lipoic acid esters and spontaneously dispersible concentrates better co-surfactants for concentrate preparation. Then known compounds exhibiting antitumor and biosurfactant activities by forming micelles of radius 1.2–2.4 mm, which penetrate the plasma membrane of tumor cells easily by a diffusion process independent of cellular metabolism, are revealed	Eugster et al. (1994)	CH 683920
	Novel antitumor and biosurfactant esters of D, L alpha-liponic acid with vitamin D and vitamin E compounds with ergosterol and with special alcohols and spontaneously dispersible concentrates are disclosed	Eugster and Haldemann (1995)	DE 4400843
	A composition for the anti-metastasis containing surfactin is provided	Kim et al. (2013)	KR 20130012855
Combined effect	Treatment methods of the cancer or a microbial infection with therapeutic composition including syringopeptin, rhamnolipid, and pharmaceutically acceptable carrier are provided	Weimer (2008)	US 2008261891
	A process of producing extracellular mucopolysaccharide with biological effects of resisting tumor, raising immunity, resisting blood coagulation, resisting virus, etc. and wide application in food, medicine, papermaking, preservation, and other fields and produced by rhamnose lactobacillus as initial strain and glucose and lactose as carbon source and including the steps of compounding material, expanding culturing and fermentation, separation, washing, dewatering, and drying is described	Zhang (2006)	CN 1781950

TABLE 11.14
Activity on the Immune System

Type	Description of the Invention	Inventor and Date	Patent Number
Immunosuppressant/ anti-inflammator	A composition consisting of rhamnolipids from *Pseudomonas aeruginosa* 646011 for the treatment of various autoimmune diseases and for providing immunorestoration is disclosed	Piljac and Piljac (1996a) Piljac and Piljac (1995b)	US 5514661 US 5466675
	The immunosuppressant containing surfactin as an active ingredient suppressing antigen-presenting ability of macrophage or dendritic cells	Kim and Park (2011)	KR 20110013613

continued

TABLE 11.14 (continued)
Activity on the Immune System

Type	Description of the Invention	Inventor and Date	Patent Number
	A method for the microbial production of ethyl ester sophorolipid derivative with no acetylated groups for treating and preventing sepsis/septic shock acting primarily through decreasing inflammatory cytokines and eliciting other synergistic anti-inflammatory mechanisms and can be administered either intraperitoneally or intravenously, preferably within 48 h of sepsis inception is described	Falus et al. (2012)	US 2012142621
	A composition containing biosurfactant and anti-inflammatory agent is disclosed	Yamamoto et al. (2011c)	JP 2011001314
	A new amphipathic compound useful as a biodegradable surfactant or as an intermediate for synthesizing glycolipid-based biosurfactants, specifically bindable to L-secretin, therefore used as an anti-inflammatory agent is described	Uzawa and Usui (2000) Uzawa and Usui (2000)	JP 2000143686 JP 2000143687
	An agent, including a mannosyl-alditol lipid selected from the group consisting of mannosyl erythritol lipid (MEL), mannosyl mannitol lipid (MML), a triacyl derivative of the MEL, and a triacyl derivative of the MML, which can be used as an additive to a cosmetic, a quasi medicine, a medicine and a food and beverage for inhibiting allergic reaction is described	Kitagawa et al. (2011)	JP 2011105607
	A method for treating allergic rhinitis or sinusitis containing a formulation comprising an active ingredient of an effective amount of rhamnolipid is provided	Leighton (2010) Leighton (2011)	WO 2010080406 US 2011257115
Vaccine	The emulsan analog as an adjuvant in the immunization formulation, produced and secreted from *A. calcoaceticus* and transposon mutants of *A. calcoaceticus* RAG-1 cultured in the presence of varying fatty acid sources is disclosed	Kaplan et al. (2000)	WO 0051635

TABLE 11.15
Other Medical Patents of Biosurfactants

Type	Description of the Invention	Inventor and Date	Patent Number
Wound healing/ Cells and/or tissue regeneration	A composition containing one or more rhamnolipids as an active ingredient for wound healing with reduced fibrosis, treatment of burn shock, treatment and prevention of atherosclerosis, prevention and treatment of organ transplant rejection, treatment of depression and schizophrenia, and treatment of the signs of aging is described	Piljac and Piljac (1999) Piljac and Piljac (2007), (2008, 2007) Stipcevic et al. (2001)	WO 1999043334 US 7262171 EP 1889623 US 2007155678 WO 0110447
	Techniques for lung injury treatment including the step of administering a therapeutically effective amount of a sophorolipid to the patient are provided	Wadgaonkar et al. (2008)	WO 2008137891

continued

TABLE 11.15 (continued)
Other Medical Patents of Biosurfactants

Type	Description of the Invention	Inventor and Date	Patent Number
	The use of one or more rhamnmolipids where the composition of the mixture is useful in the treatment of combination radiation injuries/illnesses and for cell and tissue regenerations including radiation damages combined with burns, mechanical injuries, infections of digestive system, infections of lungs, neutropenia, sepsis, atherosclerosis, depression, schizophrenia, atopic eczema, and other illnesses in connection with radiation injuries are described	Piljac (2011)	WO 2011109200
Drug delivery vehicles	Methods for the production of liposomes, microbubbles that are common delivery vehicles, used for transferring a medicament, physiologically active substance or gene via encapsulation into a biological body or cell are provided	Ishigami et al. (1988) Ishigami et al. (1990) Imura et al. (2006) Kitamoto and Nakanishi (2009) Kitamoto and Nakanishi (2003)	JP S63182029 US 4902512 JP 2006028069 JP 2009203240 JP 2003040767
	Biosurfactants that can self-assemble or auto-aggregate into polymeric micellar structures and also have the ability to increase metabolic soluble proteins as well as their use in topically applied dermatologic products is described	Owen and Fan (2007)	WO 2007143006
	A method for the preparation of novel solid lipid nanoparticles consisting of a core and a shell, where the core is a biosurfactant forming a micelle or an aggregate to wrap active ingredients, or/and the biosurfactant and the active ingredients forming a synergic aggregate; and the shell is a solid lipid material provided for encapsulating the core for delivering protein medicaments	Yuan et al. (2011)	CN 102106821
	A method for the production of emulsan-alginate compositions for use as drug delivery vehicles, removal of protein toxins from food products or other products and solutions	Kaplan et al. (2006)	WO 2006028996
	A percutaneous absorption controller including a molecular assembly containing sophorolipid to be used as a vehicle or a huge micelle and at least one active ingredient (antioxidant, an antimicrobial agent, an antiinflammatory agent, a blood circulation promoter, a skin lightening ingredient, a rough skin inhibitor, an antiaging agent, a hair growth promoter, a humectant, a hormone agent, a pigment, a glycoside, a protein, a lipid, and a vitamin) is disclosed where the molecular assembly can be diffused through the skin	Imura et al. (2009)	JP 2009062288
Other	Surfactin useful for treating or preventing hypercholesterolemia and used in inhibiting loss of activity of various active substances is disclosed	Arima et al. (1972)	US 3687926

continued

TABLE 11.15 (continued)

Other Medical Patents of Biosurfactants

Type	Description of the Invention	Inventor and Date	Patent Number
	A sophorolipid composition to be used as antimicrobial agents, antifungal agents, biopesticides, for uses as drugs to treat HIV, septic shock, cancer, asthma, dermatological conditions, as spermicidal agents, as anti-inflammatory drugs, as ingredients in cosmetics and building blocks for monomers and polymers and self-assembled templates for further chemical elaboration is described	Gross and Schofield (2011)	WO 2011127101
	A topical compostion having rhamnolipid as an active ingredient, absorbed by the skin of a human or animal, absorbed by the blood stream, and distributed through the human or animal body for treating patients is described	Desantho (2011)	WO 2011056871
	An antithrombotic biosurfactant comprising a biosurfactant produced from *Bacillus subtilis* CH-1(KCCM 10480) is provided for the alleviation and treatment of circulatory system diseases including arterial disease, cardiovascular disease, and cerebral vasomotor disease	Hong et al. (2005)	KR 20050014451
	A biocompatible carbohydrate surfactant is disclosed which can be used as a lipase inhibitor of pharmaceuticals, health foods, and cosmetics that are effective for preventing, improving, and treating various lipase-related diseases	Fukuoka et al. (2010)	JP 2010248186
	An ameliorative substance containing a glucooligosaccharide composed of beta-glucoside bonds or a reduction product thereof as an active ingredient and capable of promoting proliferation of bifidus bacteria, lactic acid bacteria, etc., suppressing proliferation of *Clostridium perfringens*, useful for preventing adult diseases such as obesity is disclosed	Okada et al. (1991)	JP H03262460
	An application of rhamnolipid as oral medicine absorbent accelerant by 0.03–10 times is described.	Meng et al. (2013)	CN 103007287
	A pharmaceutical preparation comprising as an active ingredient at least one rhamnolipid or a pharmaceutically acceptable salt to be used as a carrier and/or diluent is provided	Piljac and Piljac (1995)	US 5455232
	A rhamnolipid biosurfactant which uses the rhamnolipid fermented liquor as main component, and containing ingredients of somatic stem cell, neutral lipid, polar lipid X, rhamnolipid, polysaccharide, and metal ion and anion, possessing good cooperative effect is described	Yang et al. (2000)	CN 1275429

11.9.2 Cosmetic

TABLE 11.16

Cosmetic Applications

Type	Description of the Invention	Inventor and Date	Patent Number
For skin	An isomer MEL-B with a vesicle-forming ability and with potential as a skin care agent, produced by *Pseudozyma tsukubaensis* newly isolated from leaves of *Perilla* is described	Morita et al. (2011)	JP 2011182740
	A formulation having excellent activating and antiaging effects on cells safe enough to be used for a long time obtained by formulating a biosurfactant, especially at least one kind selected from the group consisting of MEL, MML, a triacyl derivative of the MEL and a triacyl derivative of the mannosyl mannitol lipid as the active ingredient is disclosed	Suzuki et al. (2008) Suzuki et al. (2009)	JP 2008044855 EP 2055314
	A cosmetic aiming at preventing skin roughness/skin care which contains biosurfactants, in particular, MEL-A, MEL-B, and MEL-C is described	Masaru et al. (2008)	CN 101316574
	The oil-in-water-type emulsion cosmetic composition comprising a biosurfactant, in which a sugar alcohol is glycoside-bonded to the 1-hydroxy group of a mannose skeleton, more preferably a MEL, a phospholipid, a polyhydric alcohol, and water is described	Kitagawa and Yamamoto (2009)	JP 2009275017
	Several cosmetic compositions containing at least one or more biosurfactant such as MEL, (MEL-A, MEL-B or MEL-C), MML, glycolipid ester, sophorolipid, to prevent or improve skin problems (moisturizing, dryness-improving, firmness, elasticity, etc.) that a conventional oil-based cosmetic does not achieve a sufficient performance in terms of quality to last long on the skin despite various considerations is disclosed.	Tsutsumi et al. (1981) Not available and (1983) Kitagawa et al. (2009) Kitagawa and Inamori (2011) Kitagawa et al. (2008) Kitagawa et al. (2008)	US 4305961 GB 2033895 JP 2009149566 JP 2011148731 JP 2008044857 EP 1964546
	The oil-in-water-type emulsified composition comprising a biosurfactant MEL, exhibiting excellent safety and storage stability as a cosmetic with refreshing cool feeling and visually moist feeling is provided	Kitagawa and Inamori (2011c)	JP 2011173843
	A skin care preparation having long-lasting medicinal effects, imparting moisture and showing low irritation to the skin, highly effective in ameliorating skin troubles, comprising at least one or two biosurfactants (MEL) and at least one or two chosen active components is disclosed	Kitagawa and Inamori (2011d)	JP 2011132134
	A preparation of oil-in-water-type emulsified solid or a semi-solid fat, liquid oil improving moisture retention capacity and which is stable with time, comprising a hydrophilic polyglycerol fatty acid ester; a biosurfactant (MEL); liquid oil having a melting point of lower than 30°C; and an aqueous medium is disclosed	Kitagawa and Kondo (2011b) Kitagawa and Kondo (2011c) Kitagawa and Inamori (2011a)	JP 2011063558 JP 2011063559 JP 2011168527
	Skin-protecting cosmetic composition comprising a hydroxypropyl-etherified glycolipid ester obtained from sophorolipid produced by *Torulopsis bombicola* is disclosed	Tsutsumi et al. (1982)	US 4309447

continued

TABLE 11.16 (continued)
Cosmetic Applications

Type	Description of the Invention	Inventor and Date	Patent Number
	A skin care preparation with excellent stability with time, comprising a biosurfactant represented by general formula (R1 is an amino acid selected from L-leucine, L-isoleucine and L-valine; R2 is an alkyl; and R3 is hydrogen) and derived from a natural product and/or its salt and a chelating agent is described	Chikakura et al. (2003)	JP 2003201209
	Nontherapeutic cosmetic method to improve barrier function of skin and/or to hydrate the skin, comprising 2–60 wt.% of a sulfated polysaccharide with rhamnose is described	Potter et al. (2013)	FR 2982152
	A water-in-oil emulsion cosmetic containing a biosurfactant having a sugar alcohol bonded to the 1-position hydroxy group of a mannose skeleton through a glycoside bond, and at least two components comprising an oily component and an aqueous component and capable of continuously maintaining a moisturizing function close to the function of natural skin is disclosed	Kamikura et al. (2010)	JP 2010018560
	An oil-in-water-type emulsified skin cosmetic having excellent moisture-keeping properties, preventing the humectant component from being washed away by sweat or water, exhibiting high stability and excellent rough skin-ameliorating effects comprising a biosurfactant; a polyglycerol fatty acid ester having an HLB of 8 or lower; a water-soluble polymer; and glyceriol and/or diglycerol is disclosed	Kitagawa and Inamori (2011b) Kitagawa and Inamori (2011e)	JP 2011168548 JP 2011126790
	A gel-like and transparent composition on oil basis for tattooing comprising at least a hydrophilic component, a lipophilic component, and surfactin or its derivative, salts or solvates is described	Belter (2007)	DE 102005050123
	A nonsticky skin cosmetic having good ductility, and a superior feeling of use, containing a polyhydric alcohol-modified silicone, a biosurfactant, and silicone oil is described	Not available (2011)	JP 2011225453
	A sophorolipid used together with the surfactant as an adsorption suppressing agent of the surfactant, a rinse-enhancing agent of cleaning by the surfactant or a tense feeling suppressing agent after cleaning of the skin by the surfactant is disclosed	Ryu and Hirata (2009)	JP 2009275145
For skin and hair	A cosmetic composition comprising a glycolipid ester obtained from sophorolipid produced by *Torulopsis bombicola*, effective for skin and hair treatment is described	Abe et al. (1981)	US 4297340
	Different cosmetic and quasi-drug formulations containing at least a biosurfactant, a bleaching agent, an anti-aging agent, a moisture-retaining agent, an ultraviolet light absorber, antibacterial agent, a vegetable extract containing a flavonoid-based or non-flavonoid-based secondary metabolic product, acting as a collagen production-promoting action, a cell-activating action, an anti-aging action, a hair-growing action and a chapped skin-improving action ceramide is revealed.	Kitagawa et al. (2009a) Kitagawa et al. (2009b) Kitagawa et al. (2009c) Kitagawa et al. (2009d) Kitagawa et al. (2009e) Kitagawa et al. (2010e) Yamamoto et al. (2010) Yamamoto et al. (2011b)	JP 2009149566 JP 2009149567 JP 2009167157 JP 2009167158 JP 2009167159 JP 2010018559 JP 2010018558 JP 2011001313

continued

TABLE 11.16 (continued)
Cosmetic Applications

Type	Description of the Invention	Inventor and Date	Patent Number
	Personal care products for skin and hair comprising bioemulsifiers, produced by *Acinetobacter calcoaceticus*, are provided	Hayes (1987) Hayes (1991)	EP 0242296 US 4999195
	A conditioning shampoo composition containing biosurfactant (glycolipid, lipopeptide, lipoprotein, phospholipid, sugar esters, or polymer biosurfactant) is provided to maximize the formation of coacervate and ensure hair conditioning effect. Bioemulsifiers produced by *Acinetobacter calcoaceticus* containing soaps and shampoos is described for skin and hair	Park (2009) Hayes and Holzner (1986)	KR 20090117081 EP 0178443
	A composition mild to the skin, suitable for personal wash, shower gel, and shampoo formulations comprising a sophorolipid biosurfactant, an anionic surfactant and a foam boosting surfactant, with dispersed modified cellulose biopolymer as structurant is described	Cox et al. (2011)	WO 2011120776
	Multiple hair cosmetic preparations comprising a biosurfactant to protect hair and/or scalp from physical and chemical stimulations and provide moistness, smoothness, and flexibility with antidandruff, antipruritic, and deodorizing effects are described	Ishida et al. (1982) Mager et al. (1987) Kitagawa and Kondo (2011k) Kitagawa and Kondo (2011d) Kitagawa and Kondo (2011e) Kitagawa and Kondo (2011f) Kitagawa and Kondo (2011g) Kitagawa and Kondo (2011a) Kitagawa and Kondo (2011h) Kitagawa and Kondo (2011i) Kitagawa and Kondo (2011j)	US 4318901 EP 0209783 JP 2011046634 JP 2011026275 JP 2011026276 JP 2011026277 JP 2011026278 JP 2011026279 JP 2011026280 JP 2011026281 JP 2011026282
Cosmetics with medicinal applications	The use of sophorolipids in cosmetics and dermatology as an agent to stimulate the metabolism of dermal fibroblasts restructuring, repairing the skin, is described through a formulation comprising a major part of diacetyl lactones as agent for stimulating skin dermal fibroblast cell metabolism and more particularly as agent for stimulating collagen neosynthesis	Borzeix (2000) Borzeix (1998) Borzeix (1999)	US 6057302 EP 0850641 FR 2779057
	A cosmetics composition comprising sophorolipids that are produced from *Candida bombicola* (ATCC 22214) has excellent sterilization effect, on *Propionibacterium* acne, *Staphylococcus, Micrococcus, Corynebacterium, Propionibacterium,* and the like, as well as moisturization and softening effects on the skin is provided	Han et al. (2004)	KR 20040033376

continued

TABLE 11.16 (continued)
Cosmetic Applications

Type	Description of the Invention	Inventor and Date	Patent Number
	Cosmetic and dermatological composition comprising at least one unprocessed or acid sophorolipid or a sophorolipid associated with a monovalent or divalent salt have a free radical formation inhibiting activity, an elastase inhibiting activity, or an anti-inflammatory activity is disclosed	Hillion et al. (1995a) Hillion et al. (1995b) Hillion et al. (1998)	CA 2192595 WO 9534282 US 5756471
	Oligomeric acylated biosurfactants having low critical micelle concentrations to be used as dermatocosmetic compostion with the ability to increase metabolic soluble proteins and/or increase synthesis of extracellular skin matrix proteins and/or increase rates of cell turnover while at the same time exhibiting comparatively low toxicity—preferably, an LD50 of greater 200 ppm in 37-year-old female fibroblast cells are revealed	Oven and Fan (2009)	WO 2009148947
	An external skin preparation with moisturizing and dryness-improving effects inhibiting itchiness and eczema, comprising a biosurfactant, and at least one kind selected from di- and higher valent polyols, and at least one kind selected from the group consisting of compounds having an antiphlogistic action and compounds having an anti-inflammatory action is described	Kitagawa and Inamori (2011f)	JP 2011184322
	A novel and safe skin care composition with fibroblast cell-activating effect, collagen synthesis accelerating effect, skin roughness ameliorating effect, and skin protection effect against UV, containing a sophorolipid to Momordica from Grosvenori swingla extract is described	Ryu et al. (2007)	JP 2007106733
	Pharmaceutically acceptable salts of the acid form of the sophorolipid and of the ester of the deacetelated sophorolipid acid forming therapeutically active substances acting as an activator of macrophages, as fibrinolytic agent, as healing agent, as desquamation agent, and as depigmentation agent is described for the treatment of the skin	Maingault (1997) Maingault (1997)	FR 2735979 WO 9701343
	Sodium surfactin useful in cosmetics with improved collagen decomposition inhibiting activity and sebum and waste decomposing activity is disclosed	Jun et al. (2004)	KR 20040055035
	Cosmetic material comprising hydroxypropyl-etherified glycolipid ester produced by *Torulopsis bombicola* is disclosed for compressed powder cosmetic material for a stick-shaped cosmetic material	Kawano et al. (1981b) Kawano et al. (1981a)	US 4305931 US 4305929
	New reverse vesicles formed by a sugar-type biosurfactant compound being agitated in a liquid oily substance. To be used as a compounding agent for cosmetics or drugs having a wide range of functions is described	Fukuoka et al. (2011)	JP 2011074058
	New cosmetic compositions containing at least one sophorolipid, in conjunction with a lipolytic agent, the adenylate cyclase and the inhibiting agents of the enzyme phosphodiesterase is provided for reducing the subcutaneous fat overload via stimulating the leptin synthesis through adipocytes	Pellicier and Andre (2004a) Pellicier and Andre (2004b)	FR 2855752 WO 2004108063

11.9.3 Agriculture

TABLE 11.17

Agriculture and Food/Feed Applications

Type	Description of the Invention	Inventor and Date	Patent No.
Control, prevention and treatment of diseases	The supernatant of novel *Bacillus subtilis* (strain AQ713) containing an effective insecticidal, antifungal, and antibacterial, corn rootworm-active metabolite agents for the protecting or treating plants is described	Heins et al. (1998) Heins et al. (2001)	CZ 20011620 WO 9850422
	The control of plant diseases rhamnolipid B, glycolipid antibiotics as an effective component (natural fungicide) against *Phytophthora capsici, Cladosporium cucumerinum, Colletotrichum orbiculare, Magnaporthe grisea, Cercospora kikuchi, Cylindrocarpon destructans,* or *Rhizoctonia slain* is described	Kim (2001)	KR 20010036515
	A composition comprises a culture of emulsifier and surfactant producing bacteria and their respective genetic equivalents for biological control of corn seed rot and seedling blight is disclosed	Haefele et al. (1991)	US 4996049
	Rhamnolipids from *Pseudomonas aeruginosa* for inhibiting plant disease fungus, killing cockroaches and aphids to be used as an aid of pesticide, fertilizer, and feed and a dehydrating agent of oil-contained sludge; the obvious effect was reported	Meng and Zhang (2010)	CN 101845468
	Rhamnolipid biosurfactants produced by *Pseudomonas* spp. were demonstrated to rapidly kill zoospores of pathogenic microorganisms including *Pythium aphanidermatum, Plasmopara lactucae-radicis,* and *Phytophthora capsici*	Stanghellini et al. (1997)	WO 9725866
	A rhamnolipid compound from *Pseudomonas* sp. EP-3 (KACC 91475P) with surfactant activity and insecticidal activity to green peach aphid is described	Kim et al. (2010)	KR 100940231
	A composition for controlling scab disease in agricultural products is described	Saeki et al. (2008)	EP 1929869
	Nonpathogenic *Cobetia marina* for the production of biosurfactant and prevention of infectious pathologies in aquaculture and to prevent and treat diseases of birds or mammals and to be used as an additive in paint formulations is described	Ito and Kaneda (2010) Dinamarca et al. (2012)	JP 2010220516 WO 2012164508
	A preparation prepared by adopting *Bacillus amyloliquefaciens*, which is low in cost, reliable in technology, and good in effect with a wide range of applications for disease control is described	Jie et al. (2012)	CN 102765863
	The red tide-preventing agent with high treatment efficiency, comprising sophorolipid (*Torulopsis bombiocola, Torulopsis petrophilum*) or rhamnolipid (*Pseudomonas aeruginosa*) as glycolipid and sugarester and span as chemical glycolipid-type surfactants is disclosed	Kim (2002)	KR 20020003679
	Red tide inhibiting agent comprising yellow earth and biodegradable sophorolipid is disclosed	Kim (2004)	KR 20040083614

continued

TABLE 11.17 (continued)
Agriculture and Food/Feed Applications

Type	Description of the Invention	Inventor and Date	Patent No.
Pest control	Biosurfactants with pesticidal qualities, produced by cultivating a biosurfactant-producing microbe, or *in situ* in the environment of the pests by applying a carbon substrate to the pests' environment, to be used to control a variety of pests is described	Awada et al. (2005)	US 2005266036
	An attractant containing a biosurfactant as the active ingredient and method for nematode control is described	Matsuda et al. (2012)	WO 2012115225
	A method for controlling pests by modifying derivatives of sophorolipids and applying the modified sophorolipid derivatives to the plant pathogen or to an environment in which the pathogens may occur or are located in an amount such that the pathogens are substantially controlled is disclosed	Schofield et al. (2013)	US 2013085067
Food and feed applications	An aquatic animal feed and livestock feed prepared by two or more than two of sophorolipid, rhamnolipid, and trehalose, which improves feed fat digestion utilizing rate of fishes, shrimps, and crabs and the like, reduces bait coefficient, improves production performances of aquatic animals, improves body immune function, reduces morbidity and mortality of the animals is provided	Jinbo and Lili (2012a) Jinbo and Lili (2012b)	CN 102696895 CN 102696880
	The use of rhamnolipids can further be used to improve the properties of dough or batter stability, dough texture, volume, and shape, and for microbial conservation of bakery products is described	Van Haesendonck and Vanzeveren (2004); Vanzeveren (2005)	EP 1415538 MX PA05004797
	A sophorolipid for the production of bread from the wheat flour promoting dough formation during mixing, improving gas retention power during baking enlarging bread volume, and providing bread with good appearance and aging preventing properties is disclosed	Shigeta and Yamashita (1986)	JP S61205449
	A method to prevent fruit corrosion, keep the freshness of the fruit and prolong the preservation life of the fruits at room temperature by a sophorolipid produced by *Wickerhamiella domercqiae* Y2A CGMCC3798 is explained	Jing and Bingbing (2010)	CN 101886047
	A preservative that includes an extract from eucalyptus leaves as a bactericidal agent saccharides, disinfectants, water-soluble mineral substances, calcium phosphate compounds, surface-active substances, and plant hormones is provided	Futaki et al. (1996)	US 5536155
Soil amendment and fertilizier	A rhamnolipid from aerugino-pseudomonads to be used for preparing compost from lift garbage to improve efficiency and quality of compost through culturing by using plant oil as carbon source and fermenting to obtain fermented solution of rhamnolipid is revealed	Shi et al. (2003a)	CN 1431036
	A biological surfactant lipopetide to be used to improve the efficiency and quality of compost is provided	Shi et al. (2003b)	CN 1431314

continued

TABLE 11.17 (continued)
Agriculture and Food/Feed Applications

Type	Description of the Invention	Inventor and Date	Patent No.
	Bacterial preparation comprising material belonging to genus *Bacillus* and *Clostridium*, producing lipopeptides which decrease surface tension of water, and possess an ability to propagate in soil in the presence of vegetable cellulosic materials under anaerobic condition, and producing cellulases is described	Hibino and Minami (1998)	US 5733355
	A composite biological surfactant prepared from aeruginous pseudomonads, hay bacillus, *Bacillus licheniformis* and aplanobacillus to be used for compost of life garbage to speed up the assimilation action of microbes, shorten period, and improve fertility is described	Fu et al. (2003)	CN 1431312
	The application of iturins sodium surfactin fermented liquid in preparing fertilizer synergists is described	Chen et al. (2013)	CN 103011978
	A rhamnolipid auxin, comprising urea, monopotassium phosphate, ammonium molybdate, borax, magnesium sulfate, of manganese sulfate, zinc sulfate, and the balance of water improving the nutrient absorption efficiency of plants, and making the plants robust and have high yield is described	Lili et al. (2011)	CN 101948354
	A method for preparing microemulsion by mixing a pesticide, a liquid fertilizer, and bioemulsifier, which can be applied to disinsection, mite prevention, and sterilization of the plants, and can promote growth regulation of the plants and supply nutrition and microelements to the plants is described	Ren et al. (2012)	CN 102826916
	A method of cleaning foliage in agriculture involving a mineral or surface-active organic acid, a hydrotropic anionic alkaline alkylsulfonate, and a biological surfactant comprising liposaccharides having a strong enzymatic action is disclosed	Mercier and Lemarchand (2000)	FR 2787439

11.9.4 BIOREMEDIATION

TABLE 11.18
Bioremediation

Type	Description of the Invention	Inventor and Date	Patent Number
New strains	A *Cyanobacteria* sp. J-1 strain producing an emulsifying agent for hydrocarbons and oils in liquids such as water is reported	Shilo and Fattom (1987)	US 4693842
	Two different strains *of P. aeruginosa* TY-29 (FERM P-13667) and TY-30 (FERM P-13668) capable of producing a biosurfactant and decomposing and assimilating petroleum and waste oil are described	Okura et al. (1995a) Okura et al. (1995b)	JP H078270 JP H078271

continued

TABLE 11.18 (continued)
Bioremediation

Type	Description of the Invention	Inventor and Date	Patent Number
	Two strains (*Pseudomonas stutzeri* MEV-S1 B-8277 isolated from usual chernozem and *Pseudomonas alcaligenes* B-8278 isolated from wheat rhizosphere) that are capable of utilizing petroleum, mazut, diesel fuel, polycyclic aromatic hydrocarbons, and are resistant to heavy metal ions and producing biological surface-active substances, to be used for the treatment of soil, ground, and surface waters contaminated with petroleum and its processing products containing metals are reported	Marchenko et al. (2004b) Marchenko et al. (2004a)	RU 2002122612 RU 2002122613
	Novel strains of isolated and purified bacteria have been identified exhibiting the ability to produce a biosurfactant, degrading petroleum hydrocarbons including a variety of PAHs and binding heavy metal ions for removal from a soil or aquatic environment	Brigmon et al. (2005) Brigmon et al. (2006)	US 2005106702 WO 2006085848
	A strain, *Rhodococcus erythropolis* EK-1 IMB Ac-5017, which can be used for cleaning oil contaminations, and for the production of bioemulsifiers for the perfume-cosmetic industry is reported	Pyroh et al. (2006)	UA 77345
	A marine bacterium *P. aeruginosa* BYK-2, which simultaneously producing rhamnolipid and methyl rhamnolipid were patented for the remediation of an oil spill	Ha et al. (2002)	KR 20020016707
	A biosurfactant-producing strain, *Pseudomonas putida* HPLJS-1 (KCTC 0666BP), effectively eliminating oil in soil and oil-contaminated cleansing tanks is described	Hwang et al. (2001)	KR 20010046256
	A biosurfactant-producing strain, *Rhodococcus globerulus* (HPPDS-1) (KFCC-11168), showing superior oil collapsing property to hydrocarbons, over pH 5–7 under both aerobic and anaerobic condition with potential applications for soil recovery is reported	Park and Park (2002)	KR 20020003460
	The use of *Pseudomonas cepacia* ATCC 55487 strain producing a bioemulsifier for the remediation of medium contaminated with an organohalide (particularly PCBS) is reported	Rothmel (1996)	US 5516688
Microbial combinations	A microbial material, comprising at least one microorganism capable of degrading oil and toxic chemicals the group consisting of *Trichosporon loubieri* Y1-A (KCTC 10876BP), *Trichosporon cutaneum*, and white-rot fungi, *Bacillus subtilis* KCCM 10639 and *Bacillus subtilis* KCCM 10640, for degrading the oils and toxic chemicals is described. Rapeseed oils, lipophilic powders to increase the effective surface area for biodegradation, and finally a microbial nutrient is added to enhance sophorolipid production	Yum (2008)	US 2008032383
	A mixed culture hyphae of plural microorganisms, in a kind of symbiosis relation, secreting biosurfactants to decompose a chemical material (PCB) at high reaction rates is described	Kanehara et al. (1989)	JP S6468281

continued

TABLE 11.18 (continued)
Bioremediation

Type	Description of the Invention	Inventor and Date	Patent Number
	A composite bacterial agent comprises biosurfactant, organic acid, organic ester, organic ketone, biogas, and other metabolites produced by microbial metabolism and a biological (fermentation) method for treating flow-back fracturing fluid to obtain oil displacement active water, followed by the residual organic substance harmful to environment in the flow-back fluid being transformed into biological active water, enabling the reutilization technology of fracturing flow-back fluid is disclosed	Xiaodan et al. (2011b)	CN 101935615
	A microorganism-surfactant composite oil displacement system is described for improving the oil-water displacement efficiency of reinjection after hypotonic oil deposit	Chunning et al. (2012)	CN 102477855
A compound	A bioemulsifier with potential applications in bioremediation of petroleum polluted soil and for treatment of the oil-contained sewage, capable of decreasing the surface tension of aqueous solutions, exhibiting strong emulsifying power to ester substances, increasing the solubility of paraffin and polycyclic arylhydrocarbon in water, produced *by Rhodococcus ruber* Em (CGMCC No. 0868) is described	Lui et al. (2004)	CN 1519312
	A biosurfactant, surfactin, which can be used in various drilling and production processes such as oil-gas field fracturing, acidizing, de-plugging, controlling and water shut-off, oily water treatment, and environment renovation is defined	Peng et al. (2010)	CN 101838621
	The glycolipid biosurfactants produced by *Rhodococcus equi* are used to degrade the hydrocarbons in crude oil up to 60% and the solid paraffin degradation rate up to 30%; reducing the surface tension of the fermentation liquor to less than 30 mN/m	Xiaodan et al. (2011a)	CN 101935630
	A biological heavy metal adsorbing and flocculating agent containing chitosan, citric acid, ethanol, sophorolipid, and water with remarkable chelating effects on heavy metal substances is described	Qinglu and Fangming (2012)	CN 102764632
A biologically active mixture	The addition of concentrated nutrients to stimulate the growth of mineral oil-based hydrocarbons (MHC)-consuming micro-organisms is described for biologically cleaning almost any-sized area of soil contaminated with MHC, and especially the leftover pollutants associated with mineral oil-based hydrocarbons	Lauer and Kopp-Holtwiesche (1999)	CA 2303413
	A nutrient concentrate for the accelerated cultivation and growth of hydrocarbon-consuming microorganisms as well as a process for its use in bioremediation of polluted soils, waters, and articles using a diluted aqueous solution of the concentrated admixture is disclosed	Kopp-Holtwiesche et al. (1997)	US 5635392

continued

TABLE 11.18 (continued)
Bioremediation

Type	Description of the Invention	Inventor and Date	Patent Number
	The fermentation liquor of *Pseudomonas aeruginosa* CGMCC No. 4002, producing high yields of rhamnolipids using waste oil is utilized to treat heavy metals (lead) in wastewater, degrading the crude oil and fuel scavenge and oil extraction. The biosurfactant is reported to remarkably reduce the surface tension of a water solution, has strong emulsion capacity, effectively improve the solubility of alkane or crude oil in the water, promoting the degradation of the strain to the crude oil	Wenjie et al. (2011)	CN 101948786
	The use of glycolipids in their crude or modified form as surface-active agents for the decontamination of porous medium polluted by hydrocarbons is described	Ducreux et al. (1997)	US 5654192
	An aqueous preparation of *Pseudomonas* or *Moraxella*, containing biosurfactants and exhibiting enzymatic activity to hydrolyse hydrocarbons is reported as a cleaning method of porous surfaces	Dran and Debord (1997)	EP 0808671
	A spray-type preparation is described, containing an oil-digestive bacteria and an emulsifier (enzyme excreted from bacteria and a biosurfactant), water, a supplemental nutrient, and a porous material capable of decomposing leaked oil to eliminate the flammability and ignitability of the leaked oil in a short time without detrimental effects to the environment and traffic	Yoshida et al. (2003)	JP 2003183635
	A biological absorbent mainly consisting of *Penicillium simplicissimum* powder after being modified by rhamnolipid has been patented for the treatment of cadmium-containing wastewater	Xingzhong et al. (2011)	CN 102151551
Method	A method for the efficient bioremediation of crude-oil contaminated area using *Rhodococcus globerulus* (KFCC-11171) is described where hydrocarbons can be decomposed over wide range of pH 5–7 under both aerobic and anaerobic conditions	Park (2002)	KR 20020011251
	An *in situ* cleaning method for petroleum-based contaminants by direct introduction of a microbial strain acclimatized with the contaminant, and addition of saponin leading to increased amounts of biosurfactant excretion to the contaminated environmental system such as water and soil is described	Shimizu et al. (2003)	JP 2003320367
	An *in situ* microbial decontamination of soil contaminated with xenobiotics using a novel method is described	Bisa (1992)	EP 0475227
	A method is revealed, for the bioremediation of oil-contaminated soil using novel microorganisms specified as *Rhodococcus baikoneurensis* EN3 KCTC19082, *Acinetobacter johnsonii* EN67 KCTC12360, and *Acinetobacter haemolyticus* EN96 KCTC12361, having excellent biodegradability, purifying oil-contaminated soils	Park et al. (2008)	US 2008020947

continued

TABLE 11.18 (continued)
Bioremediation

Type	Description of the Invention	Inventor and Date	Patent Number
	A method, involving complete and uniform mixing, is provided for treating contaminated soil and ground water or converting the contaminants to harmless components in a shorter period, by the addition of a biosurfactant or oil-biodegrading accelerant, and/or an oil-degrading microorganism, nutrients for microorganism growth	Cheng et al. (2009)	TW 200909590
	A method for culturing *Sporidiobolus salmonicolor* AH3 CGMCC No. 4814 with alkane and polycyclic aromatic hydrocarbons as the carbon source is provided yielding an enhanced emulsifying ability of the produced biosurfactant for hydrophobic organic matters	Yu et al. (2012)	CN 102766580
	A treatment method for contaminated environmental materials containing hydrocarbons, heavy metals, or pesticides with rhamnolipids compositions and inhibiting microbial growth is described	Yin et al. (2011)	US 2011306569
	A remediation method of soil polycyclic aromatic hydrocarbon by combining surfactant producing bacteria (*B. subtilis* BS1) and mycorrhiza has been patented creating a novel efficient integrated remediation technology with improved efficiency, shortened remediation time for PAHs in soil	Xiang et al. (2012)	CN 102652957
	A combined remediation method exploiting oil degradation bacteria and rhamnolipids to effectively decompose, degrade, and absorb petroleum pollutants in the oil-polluted soil in alkaline environment is described, possessing high remediation speed, simplicity in operation, large treatment capacity	Jian et al. (2012)	CN 102771221
	A method for recovering petroleum-polluted waters using composite crude oil degrading bacteria by the help of simultaneously generated lipopeptide biosurfactant to reduce the interfacial tension has been patented	Zhenshan et al. (2010)	CN 101851027
	A method for bioremediation of oil-contaminated soils using a specific microorganism and biosurfactant agent is disclosed	Gwak et al. (2005)	KR 20050043506
	A combined bioremediation method of polycyclic aromatic hydrocarbon-contaminated soil, comprising the restoring steps of, planting lucerne as perennial leguminous plants in the polycyclic aromatic hydrocarbon-contaminated soil; inoculating a mixed bacterial inoculum comprising arbuscular mycorrhizal fungi and adding rhamnolipid as a biosurfactant, is described	Rui et al. (2011)	CN 101972772
	A method for removing sludge from a crude oil tank while recovering hydrocarbons from the sludge is described	Powell (2000)	US 6033901
	A simple low in cost, high in efficiency method for electrically repairing heavy metal As (Arsenic) contaminated soil by using rhamnolipid is described	Yu et al. (2013)	CN 102886374

continued

TABLE 11.18 (continued)
Bioremediation

Type	Description of the Invention	Inventor and Date	Patent Number
	A method for removing pollutants in contaminated soil is disclosed, including a system to induce growth and reproduce of at least one dominant synergistic flora for petroleum degradation	Lin (2012)	US 2012021499
	A method and an apparatus comprising a contaminated water inflow bioreactor, an emulsifier injection installed in rear of the bioreactor, a discharge part connected to the bioreactor to inject the treated water underground to effectively purify the contaminated soil is described	Kim et al. (2004)	KR 20040063515
	A method involving a di-rhamnolipid for removing phenol in wastewater by laccase catalysis yielding a high efficiency, simple operation, and low cost	Zhifeng et al. (2011)	CN 102020347
Process	A batch process for the microbial decontamination of soils contaminated with hydrocarbons is described where the surfactants are metered into the contaminated soils directly or into the removed contaminated soil layers, preventing the formation of toxic intermediates and improving the accessibility of mineral oils	Lindoerfer et al. (1992)	US 5128262
	A superior process for the remediation of contaminated soil using polycyclic aromatic hydrocarbon-degrading strain is detailed, targeting polycyclic aromatic hydrocarbons (phenanthrene), benzo(a)pyrene, claiming to improve the treatment process by 392%	Huanzhong and Zhenyong (2011)	CN 102125929
	An onsite or *in situ* process for cleaning soils, water, and aerosols containing petroleum hydrocarbons by continuous addition of the bacterial mixed culture of two bacterial strains *Rhodococcus erythropolis* DSM 4670 and *Pseudomonas aeruginosa* DSM 4692, with the addition of air, pure oxygen, or oxygen donors or in fixed bed, fluidized bed, or submerged processes in controllable reactors with gas introduction in the presence of growth-maintaining nutrient salts, is described	Not available (1990)	DE 3909324
	An apparatus for purification of contaminated soil permitting reliable removal and recovery of contaminants in injecting cleaning water into soil contaminated by contaminants based on organic compound is described	Ishikawa et al. (2005)	JP 2005238004
	A process where bitumen froth tailings are mixed with a growth media containing native hydrocarbon metabolizing microorganisms, to reduce the amount of asphaltenes as well as solids such as clays and sands is disclosed for the low temperature biological treatment of bitumen froth tailings produced from a tar sands treatment and bitumen froth extraction process	Huls et al. (2000) Duyvesteyn et al. (2000)	CA 2350927 WO 0029336

11.9.5 Oil Recovery

TABLE 11.19
Oil Recovery

Type	Description of the Invention	Inventor and Date	Patent Number
New strains	A novel mutant strain of *P. aeruginosa* strain SB-1 designated SB-3, having the property of growing on solid (C20+) paraffins but not on liquid alkanes, producing an emulsifier for reducing the viscosity of crude oil in secondary recovery methods as well as in oil spill management and the cleaning of oil-contaminated vessels and pipelines was reported, exhibiting an advantage in selective degradation of the solid paraffinic components in crude oil, reducing the viscosity of the oil for improving the recovery thereof from oil wells. A novel revertant strain of *P. aeruginosa* SB-3, designated SB-30, grows both on liquid and solid hydrocarbon substrates, but produces greater amounts of emulsifier	Banerjee et al. (1991)	US 5013654
	A new strain of *Myroides odoratus* SM-1 (FERM P-19061) isolated from seawater shows crude oil emulsification ability and produces a biosurfactant	Kawai and Maneraato (2007)	JP 2007126469
	A strain of *Geobacillus thermodenitrificans* CGMCC-1228 which can tolerate a high temperature, degrade alkanes, decrease viscosity, and increase fluidity of crude oil is reported to enhance the yield of oil and improving the transporting efficiency. The strain also has the ability of decreasing the surface tension of substances, ability of degrading oil and thus can be developed for the treatment and clearing of oil-contaminated water	Wang et al. (2009)	US 2009148881
Microbial combinations	A microbial consortium containing three hyperthermophilic, barophilic, acidogenic, anaerobic bacterial strains, unique in producing a variety of metabolic products mainly CO_2, methane, biosurfactant, volatile fatty acids, and alcohols in the presence of specially designed nutrient medium is reported to be suitable for enhanced oil recovery from oil reservoirs by *in situ* application where temperatures range from 70°C to 90°C.	Lal et al. (2005)	WO 2005005773
	A mixture of compounds containing alpha-emulsans—(a new class of unique extracellular microbial protein-associated lipopolysaccharides) produced by *Arthrobacter* sp. ATCC 31012 and apo-alpha-emulsans, obtained by deproteinization of alpha-emulsans, both of which strongly anionic, exhibiting high degree of specificity in the emulsification of hydrocarbon substrates with both aliphatic and cyclic components are disclosed as possessing efficient emulsifying characteristics and potential applications in cleaning oil-contaminated vessels, oil spill management, and enhanced oil recovery by chemical flooding	Gutnick and Rosenberg (1980) Gutnick et al. (1980)	US 4230801 US 4234689

continued

TABLE 11.19 (continued)
Oil Recovery

Type	Description of the Invention	Inventor and Date	Patent Number
Compound/ composition	Two classes of extracellular microbial protein-associated lipopolysaccharides produced by *Acinetobacter* sp. ATCC 31012 on various substrates and under varying conditions to be widely used in cleaning oil-contaminated vessels, oil spill management, and enhanced oil recovery by chemical flooding are disclosed. These classes have been named alpha-emulsans and beta-emulsans, both of which have substantially the same polymer backbone but differ from each other in certain important structural aspects	Gutnick et al. (1982)	US 4311830
	A residual polymeric material from a post-secondary oil recovery operation is reported to be consumed by microorganisms injected into the oil-bearing reservoir. The resulting metabolic products, including surfactant-acting substances, enhance additional oil production from the reservoir	Hitzman (1984)	US 4450908
	A complex biological oil displacement agent, which comprises a biosurfactant (rhamnolipid fermentation stock solution), is reported to have the advantages of high oil displacement efficiency, simple preparation process, low cost and is suitable for the popularization and application of the further improvement on a recovery ratio in high-water-content oil fields	Jianjun et al. (2012)	CN 102492409
	A composition containing solution of biosurfactant KshAS-M is reported to ensure the redistribution of flows of draining water in formation and improvement of their washing properties	Simaev et al. (1999)	RU 2143553
	Surfactin-related compound lichenysin, an effective surfactant over a wide range of temperatures, pH, salt, and calcium concentrations, produced by *Bacillus licheniformis* strain JF-2 (ATCC No. 39307), was patented for the enhancement of oil recovery from subterranean formations	Mcinemey et al. (1985)	US 4522261
	An anti-agglomeration composition containing a combination of a surfactant and an alcohol cosurfactant, used to prevent hydrates from binding together and forming plugs in pipelines, particularly from off shore gas wells is described	Firoozabadi et al. (2010)	WO 2010111226
	A biological oil-eliminating agent consisting of a sophorolipid, rhamnose, tween-85, span-80 and 48, ethylene glycol butyl ether, capable of eliminating floating oil on the sea surface and a preparation method is described	Guangru et al. (2012)	CN 102690634
Method and composition	A method for improving the extraction rate of crude oil by using industrial wastewater and industrial waste gas using microorganism and generating the metabolites that are favorable for improving the extraction rate of the stratum crude oil is disclosed to be widely used in the petroleum exploitation process	Weidong et al. (2009)	CN 101503956
	A method for pretreating a crude oil prior to a crude oil distillation process comprising the step of propagating a culture of a hydrocarbon-utilizing and biosurfactant-producing microbial strain and treating the crude oil with the propagated culture is described	Bin Mohd et al. (2012)	US 2012301940

continued

TABLE 11.19 (continued)
Oil Recovery

Type	Description of the Invention	Inventor and Date	Patent Number
	A method for the utilization of viscous hydrocarbons through the formation of low-viscosity hydrocarbon-in-water emulsions, including chemically nonstabilized hydrocarbon-in-water emulsions; chemically stabilized hydrocarbon-in-water emulsions; and bioemulsifier-stabilized hydrocarbon-in-water emulsions, in which the hydrocarbon droplets dispersed in the continuous aqueous phase are substantially stabilized from coalescence by the presence of bioemulsifiers and in particular, microbial bioemulsifiers, surrounding the droplets at the hydrocarbon/water interface is disclosed to facilitate the utilization of the above-described hydrocarbon fuels as clean-burning fuels	Hayes et al. (1987)	WO 8702376
	A method for increasing the recovery ratio of petroleum by using composite three-element emulsified liquid to displace oil is described	Liao et al. (2003)	CN 1420255
	A method for remediating polycyclic aromatic hydrocarbon-polluted soil by jointly enhancing plants through edible fungus residue and biosurfactant is disclosed to effectively promote PAHs organic pollutants to desorb from soil particles, and improve the bioavailability and degradation rate of the organic pollutants without secondary pollution; and the cost is greatly reduced	Xiangui et al. (2010)	CN 101780465
	Methods and compositions are provided to facilitate the transporation and combustion of highly viscous hydrocarbons by forming reduced viscosity hydrocarbon-in-water emulsions, and in particular, *bioemulsifier-stabilized hydrocarbon-in-water emulsions*	Hayes et al. (1985) Hayes et al. (1986) Hayes et al. (1989) Hayes et al. (1988) Hayes et al. (1990) Hayes et al. (1985) Hayes et al. (2000)	ZA 8408499 US 4618348 US 4821757 US 4793826 US 4943390 WO 8501889 US RE36983
	Methods and compositions are provided for the production and use of enzymes that degrade lipopolysaccharide bioemulsifiers, and in particular, emulsans. The enzymes may be used to demulsify bioemulsifier-stabilized hydrocarbon-in-water emulsions	Shoham et al. (1987) Shoham et al. (1989)	US 4704360 US 4818817
Process	A microbiological fermentation process using certain selected microorganisms is described for the production of hydrocarbon extraction agents for separating bitumen from tar sands	Zajic and Gerson (1981) Zajic and Gerson (1987)	CA 1114759 US 4640767
	A process, which allows working with viscous petroleum referred to as heavy and extra heavy crudes by adding an appropriate biosurfactant, is disclosed providing the formation of a stable crude/water emulsion even with salt present	Aburto et al. (2011)	US 2011139262
	A new process is disclosed for the preparation of a biosurfactant useful as microbial emulsifier for the recovery of oil	Deshpande et al. (2003)	IN 189459

11.9.6 DETERGENT

TABLE 11.20
Applications as Cleaning Agents, Disinfectants, and Related Methods

Type	Description of the Invention	Inventor and Date	Patent No.
Cleaning Agent	Detergent compositions comprising at least one glycolipid biosurfactant and at least one nonglycolipid surfactant are reported	Develter et al. (2004a) Develter et al. (2004b)	EP 1445302 US 2004152613
	A cleaning composition is provided, comprising at least one glycolipid biosurfactant and at least one solvent, to be employed in a spray dispenser for streak-free cleaning of hard surfaces, especially glass	Karsten et al. (2011)	DE 102009046169
	An adhesive composition in gel or paste form comprising biosurfactants has been patented for cleaning and/or fragrancing of a WC, by application to the inside of the WC ceramic	Schiedel et al. (2012)	DE 102011004771
	A rhamnolipid to be used with high efficiency, in cleaning membrane modules, as well as fruit and vegetable cleaning agents is disclosed	Meng and Zhang (2012)	CN 102399644
	A detergent comprising a wheat protein, maleic rosin-polyoxyethylene ether diester carboxylate sodium, a protein-based biosurfactant of thickener and chelating agent is described. The product is mild, good in wettability, and strong in skin affinity, and the washing is efficient, fast, simple, safe, and sanitary	Teng et al. (2011)	CN 102051270
	A combination of sophorolipid, which is a biosurfactant, an oxygenic bleaching agent, and a bleaching activator to be used in a dish washer and dryer, and providing better bleaching composition, is provided	Hirata et al. (2006b)	JP 2006274233
	A cleaning composition consisting of one or more biosurfactants and one or more enzymes. To the material to be used as a fabric cleaner, carpet cleaner, deodorizer, disinfectant, all purpose cleaner, kitchen counter cleaner, window cleaner, bar soap additive, septic and sewage cleaner, driveway and street cleaner, circuit board cleaner, wheel cleaner, toxic waste cleaner, clean-room cleaner, oil spill cleaner, soil treatment, or metal cleaner is provided	Nero (2009)	US 7556654
	A detergent/cleaning agent composition comprising at least one or more biosurfactant (sophorolipids, rhamnolipid, trehalose-lipid) for improving the cleaning action is disclosed	O'Connell et al. (2011) Hall et al. (1996) Hall et al. (1995) Hwang et al. (1996) Guerin et al. (1997) Furuta et al. (2004) Hirata et al. (2006a) Hirata et al. (2006c) Ito et al. (2009) Not available (2012) Hall et al. (1992a) Hall et al. (1992b) Cox et al. (2011)	US 2011201536 US 5520839 US 5417879 KR 960007877 FR 2740779 EP 1411111 JP 2006070231 JP 2006083238 JP 2009052006 EP 2410039 CA 2060698 EP 0499434 WO 2011120776

continued

TABLE 11.20 (continued)

Applications as Cleaning Agents, Disinfectants, and Related Methods

Type	Description of the Invention	Inventor and Date	Patent No.
		Parry et al. (2012a)	WO 2012010405
		Parry et al. (2012c)	WO 2012010406
		Parry et al. (2012b)	WO 2012010407
		Hirata et al. (2006a)	JP 2006070231
		Jung et al. (2004)	KR 20040020314
	The use of mixtures of glycolipids selected from rhamnose, glucose, sophorose, trehalose, and/or cellobiose lipids and surfactants which are anionic, nonionic and/or amphoteric or zwitterionic for the production of hand dish-washing detergents is disclosed	Hees and Fabry (1997)	DE 19600743
	A high-temperature foaming agent comprising the following components in parts by weight: 10–15 parts of lauryl sodium sulfate, 10–15 parts of acrylonitrile butadiene styrene (ABS) surfactant, 10–15 parts of alpha-olefin sodium sulfonate (AOS), 10–20 parts of white saponin, 5–10 parts of self-prepared biosurfactant, and the balance of water is disclosed	Junqi and Haifang (2011)	CN 102212356
	A biological oil cleaning agent, cleaning agent containing 1.85 to 2.75 g/L of coarse biological enzyme extract and 9.75 to 13.45 g/L of coarse biosurfactant product, is described for deplugging and oil displacement	Jianguo (2009)	CN 101451062
Disinfectant	A specific glycolipid (RCO is 7–20C aliphatic acyl group bonding to either of hydroxyl groups on mannose in the disaccharide via ester bond) produced by *Candida* sp. b-1 (FERM P-5884) is reported to be useful as a hygroscopic oil and fat component, a surfactant component, an antibacterial component, etc., for detergent and the like, on an industrial scale at a low cost	Kobayashi et al. (1988)	JP S6363389
	Rhamnolipid-based formulations are reported to create clean surface areas for medical procedures, chemical testing, during food preparation, as well as for daycare centers and hospitals by forming a bio-film preventing the growth of bacteria and fungus	Desanto (2008)	WO 2008013899
	Methods of treating the surface of an apparatus within an electrocoating operation using a biosurfactant to remove or prevent sessile microorganism growth on the apparatus are described	Contos et al. (2004) Bourdeau (2005)	WO 2004078222 MX PA05009254
	The methylotrophic *Bacillus* producing surfactins 1–7 and iturin A compounds 8–11, possessing advantages of fouling resistance, bacterium resistance, fungus resistance, and antiviral activity is described	Qi et al. (2013)	CN 102994418
	A germicidal composition, suitable for cleaning fruits, vegetables, skin, and hair, includes a mixture of fruit acids such as citric acid, glycollic acid, lactic acid, malic acid, and tartaric acid and a surfactant is disclosed and reported to be sufficient to kill 100% of *E. coli*, *Salmonella*, and *Shigella* in 30 s after application	Pierce and Heilman (1998)	CA 2267678

continued

TABLE 11.20 (continued)

Applications as Cleaning Agents, Disinfectants, and Related Methods

Type	Description of the Invention	Inventor and Date	Patent No.
	Method for removing and preventing the recurring formation of bio-films in water-transporting or water-containing systems by production of biosurfactants is reported	Koppe et al. (2007)	DE 102005062337
	A preparation containing a biosurfactant, promoter enzyme and *n*-acyl glutamate, dodecyl dimethyl benzyl ammonium bromide, fatty alcohol polyoxyethylene ether, ester succinate sulfonate, boric acid diglyceride, linear alkyl benzene sulfonate, sorbitan fatty acid, fatty acid diethanol amide, sunscreen, carbamide, sodium hypochlorite, is provided for applications in washing, sterilizing, and removing smell to protect clothing, plastic, rubber, office articles, household appliances, auto-inner ornament, and toilet in daily life	Jinrong et al. (2008)	CN 101126052
	A composition for microbial inactivation used in chirurgical instruments comprising a trisodium citrate composition, polymeric biguanides composition, a biosurfactant, and edetate disodium is disclosed	Rosito (2011)	WO 2011143726
Methods	A method for decontaminating porous medium, polluted with hydrocarbons, using an aqueous solution containing an anionic and/or nonionic surfactant, and a glycolipid is disclosed	Ducreux et al. (1995)	FR 2713655
	A process comprising sophorosides, for cleansing solid particles impregnated with hydrocarbons, for washing cuttings impregnated with a drilling fluid containing hydrocarbons is disclosed	Baviere et al. (1994)	US 5326407
	A water treatment system that uses rhamnolipids to prevent fouling and bio-film formation on the membrane and equipment has been patented	Desanto and Keer (2012)	US 2012255918
	Systems and methods involving biosurfactants and enzymes for cleaning materials by applying cleaning composition through a spray bottle is described	Nero (2006)	US 2006080785 US 2006084587

11.9.7 Chemical Production and Other Applications

TABLE 11.21

Chemical Production and Other Applications

Type	Description of the Invention	Inventor and Date	Patent Number
New compounds or combinations	The chemically modified sophorolipid composition, or isolated components of the chemically modified sophorolipid composition, to be used as inducers for protein production in filamentous fungi are reported	Huang (2013)	WO 2013003291
	A hydroxyalkyl-etherified glycolipid ester is defined	Inoue et al. (1980b)	US 4195177

continued

TABLE 11.21 (continued)
Chemical Production and Other Applications

Type	Description of the Invention	Inventor and Date	Patent Number
	A process for producing a glycolipid ester to obtain the high-purity, stable glycolipid surfactant is described	Inoue et al. (1980c)	US 4215213
	The functional polybutylene succinate composition comprising a polybutylene succinate resin as the principal ingredient, one or more kind of pigments and one or more kind of biosufactants selected from the group consisting of a sophorolipid, a rhamnolipid, and a surfactin produced by a hydrocarbon-assimilating microorganism is provided	Nakayama et al. (2003)	JP 2003253105
	A new amphipathic straight-chain surfactin exhibiting large surface-active action, having water soluble property or oil soluble property according to pH and useful as an extracting agent for biological substances, a biosurfactant and a surfactant, etc. is disclosed	Mohamado et al. (1996)	JP H0892279
	A method for producing a stable and widely used mixture of sophorolipid and polyhydric alcohol in high purity, high yield, by the methanolysis, and the methylation of low viscosity sophorolipid–alcohol mixture obtained from sophorolipid and a polyhydric alcohol is disclosed	Inoue et al. (1979)	JP S54109914
	A sophorolipid produced by a method involving the reaction of several compounds and a sophorolipid containing a carrier and at least one sophorolipid is described	Zerkowsk et al. (2007)	US 2007027106
	New glycolipids of specific formulae are provided	Wullbrandt et al. (1996)	EP 0745608
	A new derivative useful as a surfactant having high safety and sophorolipid action to vary the properties of various interfaces of biomembrane, etc., by adsorbing to and intruding into the interfaces is provided	Ishigami et al. (1995)	JP H07118284
	A fluorescent material of glucolipid type and useful for fluorescent probe material and research of biomembrane or liposome is described	Gama and Ishigami (1989)	JP H01283295
	New sophorose lipids to be used, for example, as surfactants, cosmetics, disinfecting agents, or pharmaceutical products are described	Lang et al. (1999a)	WO 9924448
	A composition obtained by adding a mixture of alkylglycoside-, rhamnolipid-, sophorolipid-, trehaloslipid-, alkylmannoside-, and alkanoyl-*N*-methylglucamide-based surfactants and other ionic or nonionic surfactants to water, various kinds of brine aqueous solutions, organic compound aqueous solution, etc. to improve flow and heat transfer in a cold storage tank, to facilitate cold storage transportation and to produce ice slurry capable of being transported by piping and pumps, is disclosed	Kitamoto et al. (2001)	JP 2001131538
	A pressure-reducing and injection-increasing agent containing a lipopeptide biological enzyme, for applications in injection water of oil field with beneficial effects such as decreasing the well head pressure and increasing the rate of water injection is provided	Yarong et al. (2012)	CN 102690640

continued

TABLE 11.21 (continued)
Chemical Production and Other Applications

Type	Description of the Invention	Inventor and Date	Patent Number
	A solid succinoyl trehalose lipid composition readily soluble in the aqueous solvent and capable of affording the colorless transparent aqueous solution and forming a stable emulsion with an aqueous ingredient and an oily ingredient, obtained by culturing a microorganism is disclosed	Miyazaki and Senba (2009)	JP 2009013160
	A biosurfactant-containing fire extinguishing agent which has both high fire-extinguishing performance and a high level of safety to the environment and the human body is described	Izumida et al. (2013)	WO 2013015241
	An antistatic agent, comprising a water-soluble electroconductive polymer and a hydrophilic site-bearing a biosurfactant with a molecular weight of 200–10,000, efficient in preventing film-thinning in chemically amplified resist, and forming an antistatic film and a coated product, are described	Okubo et al. (2006b) Okubo et al. (2006a)	JP 2006077236 WO 2006016670
	A composition for dyeing keratinous fiber providing good in dyeability without any crocking, deterioration, and damage by washing of the hair is provided	Tagami et al. (1994)	JP H06247833
	A safe and biodegradable blue coloring agent by gold nano fine particles prepared by adding monovalent inorganic salts to a red system gold hydrosol obtained by reducing chloroauric acid by spiculisporic acid a biosurfactant with a tribasic acid-type structure derived from an organism is disclosed	Choi et al. (2011)	JP 2011080137
	Compositions comprising a rhamnolipid component and one or more active agents, and related methods of use are provided	Gandhi and Skebba (2007)	WO 2007095258
	A new emulsifier or solubilizing agent using a self-assembly of a biosurfactant formed by mixing water or an aqueous medium, where the biosurfactant is used as an emulsion or solubilizing agent is described	Imura et al. (2007)	JP 2007181789
	A high-temperature foam-scrubbing agent with a very high thermal stability and high foamability is described for drainage gas recovery	Junqi et al. (2011)	CN 102212344
	A surface-active compound containing at least one rhamnolipid as surface-active substance is disclosed	Piljac (1994)	BE 1005825
	A composition comprising biosurfactant and polyhydric alcohol is provided	Yamamoto et al. (2011a)	JP 2011001312
Other processes/ methods containing biosurfactants	A nonsolvent homogeneous process for preparing a sucrose ester, biosurfactant containing rhamnolipid or modified rhamnolipid, or sophorolipid or modified sophorolipid, or their mixtures by mixing sucrose and fatty acid ester with soap body of fatty acid, adding biosurfactant and homogenous synthesis is provided	Li and Shi (1999)	CN 1232036
	A method with advantages of easy control, low production cost, short preparation cycle and feasibility of the industrialization for synthesizing $LiFePO_4$ with more ideal particle morphology: spherical particles, uniform grain size, and even dispersion by using a biosurfactant is described	Peng (2009)	CN 101555003

continued

TABLE 11.21 (continued)
Chemical Production and Other Applications

Type	Description of the Invention	Inventor and Date	Patent Number
	An environmentally friendly, safe, simple, convenient, and low-cost method comprising a reverse micelle solution containing a rhamnolipid biosurfactant, isooctane, and n-hexyl alcohol, involving stirring, centrifuging, phase splitting, and distilling for extracting and purifying a lignin peroxidase is described	Xingzhong et al. (2012)	CN 102533688
	A novel process of cellulose two-stage cohydrolysis involving *Trichoderma reesei* ZM4-F3 hydrolyzing sugar and combined with *Pseudomonas aeruginosa* BSZ-07 after 48 h to produce rhamnolipid biosurfactant to accelerate the zymohydrolysis of straws and to effectively solve the difficulty of mutual compatibility of the two organisms is provided	Weimin and Qiuzhuo (2009)	CN 101358226
	A simple and faster electrostatic spinning method of polyhydroxylated polymer where the diameter of the produced fiber is nano-level to sub-micron level with smooth surface and uniform distributed diameter; the quality of product improved is disclosed	Lixing et al. (2008)	CN 101275291
	A method for producing cellulase by adding sophorose lipid for inducing fermentation to induce the synthesis of cellulase, yielding a higher cellulase activity and lowering the production cost of cellulose, is described	Yonghu et al. (2011)	CN 102250859
	A method of degrading plastics using a recombinant microorganism, microorganism in the coexistence of a biosurfactant, and/or a plastic-degrading enzyme, containing a gene encoding a surface-active substance is described	Abe et al. (2004) Abe et al. (2005)	WO 2004038016 EP 1595949
	A composite material and a preparation method in the field of macromolecule modification, yielding a composite material high mechanical strength and adjustable performance is disclosed	Li et al. (2011)	CN 101974212
	A low-in-cost method for preparing a high belite cement retarder capable of improving rheological property of cement paste and quality of the cement in a wide temperature range is provided	Liwen et al. (2012)	CN 102584083
	A method for producing acid through promoting anaerobic fermentation of residual sludge by virtue of a rhamnolipid biosurfactant is described	Aijie et al. (2012)	
	A method for efficiently preparing natural abscisic acid through adding acetylcoenzyme A as precursor to synthesize the abscisic acid and simultaneously adding biosurfactant Tween-20 to improve the dissolved oxygen utilization rate and the acid yield, exhibiting higher substrate conversion rate and higher product synthesis velocity, improved yield and production efficiency, reduced energy consumption, and production cost, is disclosed	Hong et al. (2012)	CN 02796764
	Addition of a culture solution of a *Pseudomonas rubescens* containing a lipopeptide biosurfactant to reduce the amount of a cellulase and shorten the treatment time of the improvement of cellulosic fiber is reported	Masuoka et al. (1998)	JP H1096174

continued

TABLE 11.21 (continued)

Chemical Production and Other Applications

Type	Description of the Invention	Inventor and Date	Patent Number
	Improving degradation efficiency and shortening the degradation time of the straw by using the sophorolipids produced by *P. aeruginosa* are reported	Hongyan et al. (2010)	CN 101898100
	A method for producing cellulase using a culture comprising a sophorolipid producer and a cellulase producer with a substrate that is consumed by the sophorolipid producer is disclosed	Ju (2008)	US 2008241885
	An application of rhamnolipid as a printing and dyeing auxiliary to promote humidification, dispersion, retardant dyeing, and mobile dyeing; replacing various chemical surfactants and resisting alkaline and oxidants, simplifying and shortening the conventional pretreatment process consisting of three steps including stewing, refining, and bleaching is revealed	Meng and Zhang (2013)	CN 103061164
	An application of rhamnolipid serving as a protective agent of low-temperature or ultralow-temperature cell preservation, improving the cell viability and functions after cryopreservation and low-temperature preservation, is described	Meng and Jiang (2013)	CN 103027032
	A method for the production of rhamnose and 3-hydroxydecanoic acid by hydrolysis of rhamnolipid from the culture medium of *Pseudomonas* sp. is described	Daniels et al. (1988)	EP 0282942
	A method for the preparation of rhamnose from rhamnolipids is disclosed	Mixich et al. (1992) Mixich et al. (1996)	WO 9205182 US 5550227
	A process to isolate sophorose as a metabolic intermediate in the production of sophorolipids using *Candida bombicola* or as a degradation compound when the sophorolipids are consumed by the yeast is described. Sophorose is used as an inducer for cellulase synthesis by the fungus *Trichoderma reesei*	Garcia-Ochoa and Casas (1997)	ES 2103688
	Biosurfactant produced by *P. aeruginosa* (USB-CS1) is reported to stabilize emulsions of high viscosity crude oil	Rocha et al. (1999)	US 5866376
	The bioproduction of rhamnolipid-alginate polymer complexes by *Pseudomonas* sp. has been patented	Eliseev et al. (1998)	DE 19654942
	Methods of extinguishing fires and protecting objects using rhamnolipid-based formulations for bio-hazard coatings and fire extinguishing compositions and formulations are described	Desanto (2009)	US 2009126948
	Materials to form gold nanoparticles and a thick and stable nanocolloidal gold liquid with a high dispersion stabilizing action, by reducing a thick aqueous solution of chloroauric acid, using an alkali salt of spiculisporic acid (a safety biosurfactant of biological origin, having a tribasic acid-type structure) and an alkali salt of a lactone ring-opened substance as a reducing agent and a colloid protecting and stabilizing agent. In a wide temperature range without requiring concentration and the addition of a protective colloid agent are described	Sai et al. (2009)	JP 2009057627

continued

TABLE 11.21 (continued)
Chemical Production and Other Applications

Type	Description of the Invention	Inventor and Date	Patent Number
	A method of preparation of the emulsified composition comprises mixing water with the oily gel composition containing a glycolipid-type biosurfactant and a method of preparation of the emulsified composition containing a biosurfactant having a glycolipid structure as a surfactant, at least either of a polyhydric alcohol or water and an oily component is revealed	Miyazaki et al. (2009)	JP 2009079030

11.10 CONCLUSION

A burst in patents relating to all aspects within the application and production of biosurfactants is being witnessed worldwide. While there were just a few patents issued two decades ago, more than 500 patents have been registered since. This review represents a comprehensive and condensed documentation on patent literature worldwide. The present lead in the number of issued patents is attributed to Asia (Japan, China, Korea, and others), paralleled with the USA. All Europe, with the lead in Germany and France, follows in this trend. This fact by itself clearly points out where most of the developments and industrial/commercial production and application of biosurfactants will be in the future. In this respect, this chapter, representing a condensed summary of all aspects covered by the issuing patents, will serve as a most reliable source of information, of particular interest to commercial/industrial developments, in addition to wide specialties of scientific and applied R&D, University, and governmental researchers.

In an attempt to present this tremendous amount of information in a coherent and summary form, the text is being condensed with the help of systematic summary tables and figures. The diversified patents have been grouped relating to producing microorganisms, areas of application, methods for production and product recovery, and purification, properties for implementation, and apparatus/equipment for production and product recovery. The following conclusions can be drawn:

1. Patents on biosurfactants relate to many applications in a variety of industries such as agriculture, cosmetics, pharmaceuticals, food, pulp and paper, coal, textiles, ceramic processing, uranium ore processing, oil recovery, medical applications, humectants, preservatives and detergents, herbicides and pesticides, mechanical dewatering, and soil bioremediation.

2. Many microorganisms have been evaluated and tested for the production of biosurfactants, various bacteria being at the forefront (61%), followed by selected yeast species (21%). In addition to natural species, new genetically modified producers are gradually taking the lead. Most patents have been issued on the production, properties, and utilization of *Bacillus, Pseudomonas,* and *Acinetobacter,* followed by *Rhodococcus, Arthrobacter, Corynebacterium, Nocardia,* and so on. Surfactin (cyclic polypeptide) produced by *Bacillus* species is the subject of many patents dealing with its production, insecticidal, antifungal, and antibacterial activity, as well as an agent for oil extraction. Rhamnolipids produced by *Pseudomonas* species are the subject of many patents relating to their production and utilization in bioremediation by breaking down petroleum hydrocarbons and mineral oil in contaminated soil, water, and aerosols, and use in tertiary oil recovery. *Acinetobacters* are known to degrade hydrocarbons, producing emulsifying agents (emulsans), whose property is widely utilized in soil and water bioremediation, cosmetic preparations, pharmaceutical compositions, and personal care products. Emulsifying agents are also produced by other bacteria, patented for various uses such as in bioremediation,

handling oil/water emulsions, and as hydrocarbon extraction agents. Glycolipids (e.g., sophorolipids) produced by yeasts, such as *C. (Torulopsis) bombicola,* have been widely patented (65 patents relating to yeast biosurfactants). Owing to the food grade status of several yeasts, biosurfactants produced from yeast have found multiple applications in cosmetics, skin care products, medicinal applications, sepsis treatment, food preservative, and so on. Of particular interest for the future are patents relating to the production of biosurfactants by genetically modified new strains as summarized in Table 11.7. It is expected that this method for biosurfactant production will be considerably enlarged in the future, by developing unique and tailored producers for defined and specific applications.

3. Alternative substrates, such as crude oil-based substrates, vegetable/plant and animal oil substrates, carbohydrate/industrial waste substrates and mixtures, are considered. A number of patents relate to utilization of various inexpensive agro-industrial residues and/or wastes as substrates, to reduce the production costs while contributing to environmental impact. There are 50 patents utilizing the following substrates: crude oil-based substrates (alkanes and PAH, paraffins, petroleum hydrocarbons, and mineral oil wastage), vegetable/plant and animal oil substrates (corn oil or chicken fat, fatty acid mixtures, olive oil mill waste waters, waste grease, and oils), and various industrial waste substrates (carbohydrate based, edible plants, ethanol, molasses, lignocellulose saccharification liquid, oil extract, and industrial by-products).

4. Production, separation, and purification processes, including reactors, immobilization applications, and various apparatus, have extensively been subjects of numerous patents, as well as downstream processing, presented in this overview. Details, with short summaries of the contents of each patent, are presented in Tables 11.9 and 11.10.

5. Fields of biosurfactant applications are wide and diversified, and have been separately categorized and summarized for various applications (biomedical—Tables 11.11 through 11.15, cosmetic—Table 11.16, agriculture—Table 11.17, bioremediation—Table 11.18, oil recovery—Table 11.19, detergents—Table 11.20, and chemical production and other applications—Table 11.21). Patents represent agricultural applications in promoting plant growth based on their antimicrobial activities and increasing the plant–microbe interaction, showing promising results against plant pathogens, bioremediation, and oil recovery, facilitating removal of oil-based compounds from soil or oil wells. Medical applications include antiadhesion activity, antitumor activity, activity on immune system, wound healing, lung injury treatment, and so on. Cosmetic applications include skin care, emulsions, shampoo composition, dermatological compositions, and so on. Chemical production and other applications are summarized, including glycolipid surfactants, fluorescent material, improving flow and heat transfer in a cold storage tank, fire extinguishing agent, antistatic agent, high temperature foam-scrubbing agent, method for producing cellulase enzyme, method for degrading plastics, printing and dyeing auxiliary, and many other exotic industrial applications.

It can be concluded that biosurfactants do find a role and use in a variety of commercial/industrial processes. Therefore, there is no surprise that so many patents have been issued in this relatively new field in such a short recent period of time. This extensively categorized summary represents a unique and very valuable source of information for active researchers in the field and for new business ventures in this lucrative area. Further new discoveries in this field are expected to flourish, and this relatively long (extensively shortened) chapter should further boost this exciting, most profitable development.

ACKNOWLEDGMENTS

The authors would like to thank Mustafa Çakir from Ege University, EBILTEM-TTO Patent Office, Turkey, for his efforts toward a comprehensive search of patent databases and Eyüp Bilgi for his help in preparing graphics.

REFERENCES

Arima, K., Kakinuma, A., Tamura, G. Surfactin, a crystalline peptide-lipid surfactant produced by *Bacillus subtilis*: Isolation, characterization and its inhibition of fibrin clot formation. *Biochemical and Biophysical Research Communications,* 1968, 31:488–494.

Campos, J.M., Montenegro Stamford, T.L., Sarubbo, L.A., de Luna, J.M., Rufino, R.D., Banat, I.M. Microbial biosurfactants as additives for food industries. *Biotechnology Progress,* 2013, 29:1097–1108.

Fracchia, L., Cavallo, M., Martinotti, M.G., Banat, I.M. Biosurfactants and bioemulsifiers: Biomedical and related applications-present status and future potentials. In *Biomedical Science, Engineering and Technology*, ed. D. N. Ghista, InTech, Rijeka, 2012, pp. 325–370.

Gutnick, D.L., Allon, R. Levy, C. Petter, R., Minas, W. Applications of *Acinetobacter* as an industrial microorganism. In *The Biology of Acinetobacter,* ed. K. J. Towner, E. Bergogne-Bérézin, and C.A. Fewson, Plenum Press, New York, NY. 1991, pp. 411–441.

Kosaric, N. Biosurfactants and their application for oil bioremediation. *Food Technology and Biotechnology,* 2001, 39(4):295–304.

Moore, E.R.B., Tindall, B.J., Dos Santos, V.P.M., Pieper, D.H., Ramos, J-.L., Palleroni, N.J. Nonmedical: *Pseudomonas*. In *The Prokaryotes A Handbook on the Biology of Bacteria*, Third Edition, Volume 6, Ed M. Dworkin (editor-in-chief), eds. S. Falkow, E. Rosenberg, K.-H. Schleifer, E. Stackebrandt, Springer Science + Business Media, New York. 2006 pp. 646–703.

Mukherjee, S., Das, P., Sen, R. Towards commercial production of microbial surfactants. *TRENDS in Biotechnology,* 2006, 24(11):509–515.

Navon-Venezia, S., Zosim, Z., Gottlieb, A., Legmann, R., Carmeli, S., Ron, E.Z., Rosenberg, E. Alasan, a new bioemulsifier from *Acinetobacter radioresistens. Applied and Environmental Microbiology,* 1995, 61:3240–3244.

Nitschke, M., G.V.A.O. Costa Siddhartha, J. Contiero, Rhamnolipid surfactants: An update on the general aspects of these remarkable biomolecules. *Biotechnology Progress,* 2005, 21(6):1593–1600.

Reis, R.S., Pacheco, G.J., Pereira, A.G., Freire, D.M.G. Biosurfactants: Production and applications. In *Biodegradation—Life of Science*, eds. R. Chamy and F. Rosenkranz, InTech. 2013, pp. 31–61. Available from: http://www.intechopen.com/books/biodegradation-life-of-science/biosurfactants-production-and-applications

Ron, E.Z. and Rosenberg, E. Natural roles of biosurfactants. *Environmental Microbiology*, 2001, 3:229–236.

Rosenberg, E. Biosurfactants. In *The Prokaryotes A Handbook on the Biology of Bacteria*, Third Edition, Volume 1, Ed M. Dworkin (editor-in-chief), eds. S. Falkow, E. Rosenberg, K.-H. Schleifer, E. Stackebrandt, Springer Science + Business Media, New York. 2006, pp. 834–849.

Rosenberg, E. and Ron, E.Z., High- and low-molecular-mass microbial surfactants, *Applied Microbiology and Biotechnology,* 1999, 52(2):154–162.

Rosenberg, E. and Ron, E.Z. Surface active polymers from the genus *Acinetobacter*. In *Biopolymers from Renewable Resources*, ed. D.L. Kaplan, Springer-Verlag KG, Berlin, Germany. 1998, pp. 281–289.

Sachdev, D.P. and Cameotra, S.S. Biosurfactants in agriculture. *Applied Microbiology and Biotechnology,* 2013, 97:1005–1016.

Saharan, B.S., Sahu, R.K., Sharma, D. A review on biosurfactants: Fermentation, current developments and perspectives. *Genetic Engineering and Biotechnology Journal*, Volume 2011: GEBJ-29.

Shete, A.M., Wadhawa, G., Banat, I.M., Chopade, B.A. Mapping of patents on bioemulsifier and biosurfactant: A review. *Journal of Scientific and Industrial Research*, 2006, 65:91–115.

Slepecky, R.A. and Hemphill, H.E. The Genus *Bacillus*-Nonmedical. In *The Prokaryotes A Handbook on the Biology of Bacteria*, 3rd edn, Volume 4, Ed M. Dworkin (editor-in-chief), eds. S. Falkow, E. Rosenberg, K.-H. Schleifer, E. Stackebrandt, Springer Science + Business Media, New York. 2006, pp. 530–562.

Towner, K. The Genus *Acinetobacter.* In *The Prokaryotes A Handbook on the Biology of Bacteria*, 3rd edn, Volume 6 Ed. M. Dworkin (editor-in-chief), eds. S. Falkow, E. Rosenberg, K.-H. Schleifer, E. Stackebrandt, Springer Science + Business Media, New York. 2006, pp. 746–758.

Van Hamme J.D., Singh A., Ward O.P. Physiological aspects: Part 1 in a series of papers devoted to surfactants in microbiology and biotechnology. *BiotechnologyAdvances,* 2006, 24(6):604–620.

Patents (multiple applications in italics)

Abe Keietsu, Gomi Katsuya, Yamagata Yohei, Hasegawa Fumihiko, Maeda Hiroshi, Nakajima Tasuku, Machida Masayuki. Method of degrading plastic and process for producing useful substance using the same. 2005, EP Patent 1595949, Tohoku Techno Arch Co Ltd, Nat Inst of Advanced Ind Scien.

Abe Keietsu, Gomi Katsuya, Yamagata Yohei, Hasegawa Fumihiko, Maeda Hiroshi, Nakajima Tasuku, Machida Masayuki. Method of degrading plastic and process for producing useful substance using the same. 2004, WO Patent 2004038016, Tohoku Techno Arch Co Ltd, Nat Inst of Advanced Ind Scien, Abe Keietsu, Gomi Katsuya, Yamagata Yohei, Hasegawa Fumihiko, Maeda Hiroshi, Nakajima Tasuku, Machida Masayuki.

Abe Yoshiaki, Inoue Shigeo, Ishida Atsuo. Cosmetic composition for skin and hair treatment. 1981, US Patent 4297340, Kao Corp.

Aburto Anell Jorge Arturo, Zapata Rendon Beatriz, Mosqueira Mondragon Maria De Lourdes Araceli, Quej Ake Luis Manuel, Flores Oropeza Eugenio Alejandro, Vazquez Moreno Flavio Salvador, Bernal Huicochea Cesar, Clavel Lopez Juan De La Cruz, Ramirez De Santiago Mario. Process of preparing improved heavy and extra heavy crude oil emulsions by use of biosurfactants in water and product thereof 2011, US Patent 2011139262, Mexicano Inst Petrol.

Ahn Chi Yong, Jeon Jong Woon, Kim Hee Sik, Lee Hong Won, Oh Hee Mock, Park Chan Sun, Yoon Byung Dae. Microorganism *Bacillus subtilis* E2 highly producing surfactin which is lipoproteinic biosurfactant and method for producing the surfactin using the same. 2005, KR Patent 20050007670, Korea Res Inst of Bioscience.

Aijie Wang, Aijuan Zhou, Zechong Guo Jingwen Du. Method for producing acid through promoting anaerobic fermentation of residual sludge by virtue of rhamnolipid biosurfactant. 2012, CN Patent 102796764, Harbin Inst of Technology.

Aretz Werner and Hedtmann Udo. Inducing rhamno:lipid synthesis in *Pseudomonas aeruginosa*—with glyceric acid ether lipid, to form intermediate for rhamnose which is used in synthesis of chiral cpds.1993, DE Patent 4127908, Hoechst AG.

Arima Kei, Tamura Gakuzo, Kakinuma Atsushi. Surfactin. 1972, US Patent 3687926, Takeda Chemical Industries Ltd.

Arkadevitch Baburin Leonid, Cemyonovitch Mischkevitch Alex, Lwowitch Baranov Gennady. Bacterial mixture for regenerating crude oil contaminated earth and water. 1998, DE Patent 19652580, Arkadevitch Baburin Leonid, Cemyonovitch Mischkevitch Alex, Lwowitch Baranov Gennady.

Awada Salam M, Spendlove Rex S, Awada Mohamed. Microbial biosurfactants as agents for controlling pests. 2005, US Patent 2005266036, Agscitech.

Baburins Leonids, Muskevics Aleksandrs, Baranovs Genadijs. Bacterial composition for treatment of water and soil from pollution with mineral oil and process for producing thereof. 2000, LV Patent 12348, Plus K Sia M.

Banerjee Santimoy, Karns Jeffrey S, Chakrabarty Ananda M. Production of emulsifying agents and surfactants. 1991, US Patent 5013654, Univ Illinois.

Baviere Marc, Degouy Didier, Lecourtier Jacqueline. Process for washing solid particles comprising a sophoroside solution. 1994, US Patent 5326407, Inst Francais Du Petrole.

Belter Clemens. Cosmetic, gel-like and transparent composition on oil basis, useful for applying tattoo on the skin, comprises a hydrophilic component, a lipophilic component and surfactin. 2007, DE Patent 102005050123, Belter Clemens.

Biao Shen, Wei Ran, Juan Cao. Lipopeptide-producing *Bacillus pumilus* and application thereof. 2009, CN Patent 101560483, Univ Nanjing Agricultural.

Bin Mohd Yahya Ahmad Ramli, Mohamad Mohamad Nasir Bin, Binti Md Noh Nur Asshifa. Method for pre-treating crude oil using microorganism. 2012, US Patent 2012301940, Bin Mohd Yahya Ahmad Ramli, Mohamad Mohamad Nasir Bin, Binti Md Noh Nur Asshifa.

Bisa Karl Method and apparatus for the *in-situ* microbial treatment of contaminated soil. 1992, EP Patent 0475227, Xenex Ges Zur Biotechnischen S.

Blank Lars, Rosenau Frank, Wilhelm Susanne, Wittgens Andreas, Tiso Till. Means and methods for rhamnolipid production. 2013, EP Patent 2573172, Univ Duesseldorf H Heine, Univ Dortmund Tech.

Borzeix Concaix Frederique. No title available. 1999, FR Patent 2779057, Inst Francais Du Petrole.

Borzeix Frederique. Sophorolipids as stimulating agent of dermal fibroblast metabolism. 2000, US Patent 6057302, Institut Francais Du Petrole, Sophor S.A.

Borzeix Frederique. The use of a sophorolipid (I) in cosmetics and dermatology is new. 1999, EP Patent 0850641, Inst Francais Du Petrole, Sophor SA.

Borzeix Frederique. Use of sophorolipids as an agent to stimulate the metabolism of dermal fibroblasts, 1998, CA Patent 2223740 Sophor SA, Inst Francais du Petrole.

Bourdeau Michael J. Treating an electrocoat system with a biosurfactant. 2005, MX Patent PA05009254, Valspar Sourcing Inc.

Brigmon Robin L, Story Sandra, Altman Denis, Berry Christopher J. Surfactant biocatalyst for remediation of recalcitrant organics and heavy metals. 2006, WO Patent 2006085848, Westinghouse Savannah River Co, Brigmon Robin L, Story Sandra, Altman Denis, Berry Christopher J.

Brigmon Robin L, Story Sandra, Altman Denis, Berry Christopher J. Surfactant biocatalyst for remediation of recalcitrant organics and heavy metals. 2005, US Patent 2005106702.

Burke Jr Terrance, Chandrasekhar Bhaskar, Knight Martha. Analogs of viscosin and uses thereof. 1999, US Patent 5965524, Peptide Tech Corp.

Carrera Paolo, Cosmina Paola, De Ferra Francesca, Grandi Guido, Perego Marta, Rodriguez Francesco. Cloning and sequencing of that chromosomal DNA region of *Bacillus subtilis* comprising the SRFA operon which encodes the multienzymatic complex surfactin synthetase.1993, EP Patent 0540074, Eniricerche SpA.

Carrera Paolo, Cosmina Paola, Grandi Guido Process for the production of surfactin. 1995, IT Patent 1248979, Eniricerche SpA.

Carrera Paolo, Cosmina Paola, Grandi Guido. A mutant of Bacillus subtilis and a method of producing surfactin with the use of the mutant. 1992, EP Patent 0463393, Eniricerche SpA.

Carrera Paolo, Cosmina Paola, Grandi Guido. Mutant of Bacillus subtilis. 1993, US Patent 5264363, Eniricerche SpA.

Carrera Paolo, Cosmina Paola, Grandi Guido Mutant of Bacillus subtilis and process for the production of surfactin using this mutant. 1994, IT Patent 1243392, Eniricerche SpA.

Ceri Howard, Turner Raymond, Martinotti Maria Giovanna, Rivardo Fabrizio, Allegrone Gianna. Biosurfactant composition produced by a new *Bacillus licheniformis* strain, uses and products thereof. 2010, WO Patent 2010067245, Pan Eco S A, Univ Technologies Internat LP, Ceri Howard, Turner Raymond, Martinotti Maria Giovanna, Rivardo Fabrizio, Allegrone Gianna.

Chang Jo-Shu, Chen Shan-Yu, Wei Yu-Hong. Method for producing biosurfactant. 2009, TW Patent 200909584, Univ Nat Cheng Kung.

Chang Jo-Shu, Cheng Chieh-Lun, Chen Pei-Yu. Medium for *Acinetobacter* sp. and method for producing biosurfactant. 2012, TW Patent 201209161, Univ Nat Cheng Kung.

Chang Jo-Shu, Yeh Mao-Sung, Wei Yu-Hong. Fermentation method for producing biosurfactant. 2007, TW Patent I289605, Univ Nat Cheng Kung.

Chang Joshu, Yeh Mao-Sung, Wei Yu-Hong. Method of producing biosurfactant 2006, TW Patent I257921, Univ Nat Cheng Kung.

Chang Jo-Shu, Yeh Mao-Sung, Wei Yu-Hung. Bioreactor for producing biosurfactant and applications thereof. 2007, TW Patent I282366, Univ Nat Cheng Kung.

Chen Jian, Hua Zhaozhe, Lun Shiyi. Method for preparing glyceride type biological surfactant. 2000, CN Patent 1242256, Wuxi Light Industry Univ.

Chen Tianhe, Zhang Yiyan, Wang Yongguo, Ding Shiqi, Sun Wen, Ye Guogang, Gong Youchu, Xu Rumin. Application of iturins sodium surfactin fermented liquid. 2013, CN Patent 103011978, Anhui Province Kingorigin Biotechnology Co Ltd.

Cheng Sheng-Shung, Whang Liang-Ming, Liu Pao-Wen, Lin Ta-Chen, Tseng I-Cheng, Chang Tsung-Chain, Young Chiu-Chung, Chang Jo-Shu, Pan Po-Tseng, Liao Yi-Ting, Fan Yen-Chen. Bioremediation of oil-contaminated soil and/or groundwater. 2009, TW Patent 200909590, Univ Nat Cheng Kung.

Chikakura Yoshito, Yashiro Yoichi, Miyamoto Kunihiro, Kitahara Michio, Nakada Satoru. Skin care preparation. 2003, JP Patent 2003201209, Nonogawa Shoji YK.

Choi Yong Lark, Joo Woo Hong, Jung Youn Ju, Kwon Young Soo, Lee Sang Cheol, Yoo Ju Soon Microorganism *Bacillus* sp. A8–8 producing biosurfactant and antifungal substance. 2004, KR Patent 20040055487, Choi Yong Lark, Joo Woo Hong, Jung Youn Ju, Kwon Young Soo, Lee Sang Cheol, Yoo Ju Soon.

Choi Yong Lark, Jung Youn Ju, Kwon Young Soo, Lee Sang Cheol, Yoo Ju Soon Biosurfactant producing microorganism *Bacillus* sp. LSC11. 2004, KR Patent 20040046860, Choi Yong Lark, Jung Youn Ju, Kwon Young Soo, Lee Sang Cheol, Yoo Ju Soon.

Choi Yong Lark, Lee Sang Cheol, Lee Yong Suck, Kim Sun Hee, Yoo Ju Soon, Joo Woo Hong, Hwang Cheol Won. Novel strain *Bacillus vallismortis* TB40-3 producing biosurfactant with oil degrading activity and antifungal substance. 2005, KR Patent 20050109111, Choi Yong Lark.

Choi Young Kook, Cho Yakugun, Koike Takaki, Fukuoka Takao, Ishigami Yutaka, Hidaka Hisao. Blue coloring agent made of gold fine particle. 2011, JP Patent 2011080137, Ishigami Yutaka, Hyosung Corp, Cho Yakugun, Koike Takaki, Fukuoka Takao, Hidaka Hisao.

Chung Seon Yong, Oh Kyung Taek, Lee Ki Seok, Li Guang Chun. Compositions and methods for degrading vehicle's waste oil or vessel's leaked oil. 2008, KR Patent 100866526, Univ Nat Chonnam Ind Found.

Chunning Gao, Yue Luo, Pingcang Wu, Shaojin Yi, Wenhong Li, Chaohua Ren, Zhongyuan Xiang, Lei Liu, Huan Yang. Method for improving oil displacement efficiency of reinjection after hypotonic oil deposit high salt produced water disposal. 2012, CN Patent 102477855, Petrochina Co Ltd.

Contos Michael A, Bourdeau Michael J, Pillar Lonnie L. Treating an electrocoat system with a biosurfactant. 2004, WO Patent 2004078222, Valspar Sourcing Inc, Contos Michael A, Bourdeau Michael J, Pillar Lonnie L.

Coutte Francois, Jacques Philippe, Lecouturier Didier, Guez Jean-Sebastien, Dhulster Pascal, Leclere Valerie, Bechet Max. *Bacillus* sp. biosurfactants, composition including same, method for obtaining same, and use thereof. 2013, WO Patent 2013050700, Univ Lille 1 Sciences Et Technologies Ustl, Coutte Francois, Jacques Philippe, Lecouturier Didier, Guez Jean-Sebastien, Dhulster Pascal, Leclere Valerie, Bechet Max.

Cox Trevor Frederick, Crawford Robert John, Gregory Lee Garry, Hosking Sarah Louise, Kotsakis Panos. Mild to the skin, foaming detergent composition. 2011, WO Patent 2011120776 Unilever PLC, Unilever NV, Unilever Hindustan, Cox Trevor Frederick, Crawford Robert John, Gregory Lee Garry, Hosking Sarah Louise, Kotsakis Panos.

Daniels Lacy, Linhardt Robert J, Bryan Barbara Ann, Mayerl Friedrich, Pickenhagen Wilhelm. Method for producing rhamnose. 1988, EP Patent 282942, Univ Iowa Res Found.

Dawei Wang. Lipopeptide biosurfactant, preparation method thereof, and application thereof. 2012, CN Patent 102352227, China Nat Offshore Oil Corp, Cnooc Res Ct.

Desanto Keith, Keer Donald R. Use of rhamnolipids in the water treatment ndustry. 2012, US Patent 2012255918, Desanto Keith, Keer Donald R.

Desanto Keith. Rhamnolipid mechanism. 2011, WO Patent 2011056871, Desanto Keith.

Desanto Keith. Rhamnolipid-based formulations. 2008, WO Patent 2008013899, Aurora Advance Beauty Labs, Desanto Keith.

Desanto Keith. Use of rhamnolipid-based formulations for fire suppression and chemical and biological hazards. 2009, US Patent 2009126948, Desanto Keith.

Deshpande Megha, Mishra Kirti, Bal Anand Sureshchandra, Khanna Purushottaham, Juwarkar Asha, Babu Polukonda Sudhakar. A process for the preparation of a biosurfactant useful as microbial emulsifier for the recovery of oil. 2003, IN Patent 189459, Council Scient Ind Res.

Develter Dirk, Fleurackers Steve. A method for the production of short chained glycolipids. 2008, EP Patent 1953237, Ecover N V.

Develter Dirk, Renkin Mark, Jacobs Ilse. Detergent compositions. 2004, US Patent 2004152613.

Develter Dirk, Renkin Mark, Jacobs Ilse. Detergent compositions. 2004, EP Patent 1445302, Ecover Belgium.

Dinamarca Tapia Miguel Alejandro, Ojeda Herrera Juan Ricardo, Ibacache Quiroga Claudia Jimena. Strain of *Cobetia marina* and biosurfactant extract obtained from same. 2012, WO Patent 2012164508, Univ De Valparaiso, Dinamarca Tapia Miguel Alejandro, Ojeda Herrera Juan Ricardo, Ibacache Quiroga Claudia Jimena.

Dran Maurice, Debord Guy. Method for cleaning porous surfaces with a washing liquid containing bacteria having an enzymatic activity. 1997, EP Patent 0808671, Dran Maurice, Debord Guy.

Ducreux Jean, Ballerini Daniel, Aviere Marc, Bocard Christian, Monin Nicole. Composition containing a surface active compound and glycolipids and decontamination process for a porous medium polluted by hydrocarbons. 1997, US Patent 5654192, Inst Francais Du Petrole.

Ducreux Jean, Ballerini Daniel, Aviere Marc, Bocard Christian, Monin Nicole. Decontaminating porous medium, esp. soil, polluted with hydrocarbon(s). 1995, FR Patent 2713655, Inst Francais Du Petrole.

Duyvesteyn Willem P.C., Budden Julia R., Huls Bernardus J. Biochemical treatment of bitumen froth tailings. 2000, WO Patent 2000029336, BHP Minerals Int Inc.

Eigen Edward, Simone Alexander J. Control of dental plaque and caries using emulsan. 1988, US Patent 4737359, Colgate Palmolive Co.

Eigen Edward, Simone Alexander J. Process for preparing a composition for control of dental plaque and caries containing an emulsan. 1986, PT Patent 82402, Colgate Palmolive Co.

Eliseev Serguei, Sulga Aleksander, Karpenko Elena, Bolochovskaja Valentina. Bacterium of genus *Pseudomonas* useful for biosynthesis of rhamnolipid-alginate polymer complexes. 1998, DE Patent 19654942, MFH Marienfelde GmbH Unternehm.

Eugster Carl, Haldemann Walter. Antitumour and bio-surfactant ester(s) of D,L-alpha-liponic acid. 1995, DE Patent 4400843, Marigen SA.

Eugster Carl, Eugster Conrad Hans, Forni Guido, Haldemann Walter, Rivara Giorgio, Zina Giuseppe. New alpha-lipoic acid ester(s) with antitumour and bio-surfactant activity—formulated as spontaneously dispersible concentrate with other surfactants, opt. contg. pharmaceutical or cosmetic active ingredients. 1994, CH Patent 683920, Marigen SA.

Falus George, Nowakowski Maja, Bluth Martin, Aikens John. Purified ethyl ester sophorolipid for the treatment of sepsis. 2012, US Patent 2012142621, Biomedica Man Corp.

Firoozabadi Abbas, York Dalton, Xiaokai Li. A composition and method for inhibiting agglomeration of hydrates in pipelines. 2010, WO Patent 2010111226, Yale University Office Of Coop, Firoozabadi Abbas, York Dalton, Xiaokai Li.

Fu Haiyan, Zeng Guangming, Huang Guohe. Composite biosurfactant and its application in compost. 2003, CN Patent 1431312, Univ Hunan.

Fukunaga Toshio, Kim Hak Ro, Nozawa Mo. Liquid fuel additive composition comprising biosurfactant derived from mixture KHR-5-MX and polymer for inducing improvement of octane value. 2003, KR Patent 20030044969, Biosoft Co Ltd.

Fukuoka Tokuma, Morita Tomotake, Imura Tomohiro, Kitamoto Masaru. Reverse vesicle made of biosurfactant. 2011, JP Patent 2011074058, Nat Inst of Advanced Ind Scien.

Fukuoka Tokuma, Morita Tomotake, Imura Tomohiro, Kitamoto Masaru. Lipase Inhibitor. 2010, JP Patent 2010248186, Nat Inst of Advanced Ind Scien.

Fulin Chen, Yefang Sun, Wei Sun, Shuwen Xue. *Pseudomonas aeruginosa* strain, and culture method and application thereof. 2012, CN Patent 102409016, Xi An Rege Bio Technology Co Ltd, Univ Northwestern.

Furubayashi Makio, Nakahara Tadaatsu, Nomura Nobuhiko, Okada Kenji, Nakajima Toshiaki. Method for treating soy sauce oil. 2002, JP Patent 2002101847, Higashimaru Shoyu Co Ltd.

Furuta Taro, Igarashi Keisuke. Method for carrying out fermentation production of sophorose lipid. 2002, JP Patent 2002045195, Saraya KK.

Furuta Taro, Igarashi Keisuke, Hirata Yoshihiko. Low foaming detergent compositions. 2004, EP Patent 1411111, Saraya Co Ltd.

Furuta Taro, Igarashi Keisuke, Sakakibara Akiko, Ito Takahide. Method for purifying sophorose lipid. 2003, JP Patent 2003009896, Saraya KK, Iwata Kagaku Kogyo.

Futaki Kouji, Shigeno Keiko, Hoshi Keiko. Preservative of cut flowers. 1996, US Patent 5536155, Asahi Optical Co Ltd.

Gama Yasuo, Ishigami Yutaka. Rhamnolipid pyrenacyl ester. 1989, JP Patent H01283295, Agency Ind Science Techn.

Gandhi N. R, Skebba Victoria L Palmer. Rhamnolipid compositions and related methods of use. 2007, WO Patent 2007095258, Jeneil Biotech Inc., Gandhi N R, Skebba Victoria L Palmer.

Gandhi N. R, Skebba Victoria P, Takemoto Jon Y, Bensaci Mekki F. Antimycotic rhamnolipid compositions and related methods of use. 2007, US Patent 2007191292, Gandhi N R, Skebba Victoria P, Takemoto Jon Y, Bensaci Mekki F.

Ganti Satyanarayana. Green process for production of biosurfactants or biopolymers through waste oil utilization. 2011, US Patent 2011151100, Ganti Satyanarayana.

Ganti Satyanarayana, Ganti Dinkar. Production of biomass and single cell protein from industrial waste oils, oily sludge from ships and other sources of oily wastes. 2005, US Patent 2005227338, Ganti Satyanarayana & Ganti Dinkar.

Garcia-Ochoa Soria Felix, Casas De Pedro Jose Antonio. Process for the production of sophorolipids from sugars and oil using deenergized cells of *Candida bombicola*. 1997, ES Patent 2103687, Univ Madrid Complutense.

Garcia-Ochoa Soria Felix, Casas De Pedro Jose Antonio. Process for the production of sophorose by *Candida bombicola*. 1997, ES Patent 2103688, Univ Madrid Complutense.

Giani Carlo, Wullbrandt Dieter, Rothert Reinhardt, Meiwes Johannes. *Pseudomonas aeruginosa* and its use in the preparation of L-rhamnose. 1993, EP Patent 0575908, Hoechst AG.

Gong Guohong Liu. Preparation method and usage for novel lipopeptide type biosurfactant surfactin. 2007, CN Patent 10104184, Inst Plasma Physics Cas.

Grandi Guido, Cosmina Paola, Rodriguez Francesco, Perego Marta, De Ferra Francesca. Cloning and sequencing of the chromosomal DNA region of *Bacillus subtilis* including the srfA operon which codes for the surfactin synthetase multienzyme complex. 1995, IT Patent 1256366, Eniricerche SpA.

Grandi Guido, Cosmina Paola, Rodriguez Francesco, Perego Marta, De Ferra Francesca, Carrera Paolo. Cloning and sequencing of the srfA operon of *Bacillus subtilis* coding for the multienzyme surfactin synthetase complex. 1995, IT Patent 1251680, Eniricerche SpA.

Gross Richard A, Schofield Mark H. Sophorolipid analog compositions. 2011, WO Patent 2011127101, Politechnic Inst Univ New York, Gross Richard A, Schofield Mark H.

Gross Richard A, Shah Vishal. Anti-herpes virus properties of various forms of sophorolipids. 2007, WO Patent 2007130738, Univ Polytechnic, Gross Richard A, Shah Vishal.

Gross Richard A, Shah Vishal. Antimicrobial properties of various forms of sophorolipids. 2004, AU Patent 2003299557, Univ. Polytechnic, Gross Richard A, Shah Vishal, Nerud Frantisek, Madamwars Datta.

Gross Richard A, Shah Vishal, Doncel Gustavo F. Spermicidal and virucidal properties of various forms of sophorolipids. 2004, US Patent 20040242501.

Gross Richard A, Shah Vishal, Nerud Frantisek, Madamwars Datta. Sophorolipids as protein inducers and inhibitors in fermentation medium. 2007, WO Patent 2007073371, Univ Polytechnic, Gross Richard A, Shah Vishal, Nerud Frantisek, Madamwars Datta.

Guangru Li, Liang Qu, Ming Li, Yu Chen, Shengkang Liang, Dandan Song. Biological oil-eliminating agent capable of eliminating floating oil on sea surface and preparation method for biological oil-eliminating agent 2012, CN Patent 102690634, China Nat Offshore Oil Corp, Cnooc Energy Technology Co Ltd, Coes Ltd.

Guangru Li, Liang Qu, Yu Chen, Wenyu Lu, Xue Yang, Lingqing Zhu. Bioreactor and sophorolipid continuous production method by using the same. 2012, CN Patent 102492605, Coes Ltd.

Guerin Gilles, Guillou Veronique, Joubert Daniel. New detergent composition comprising enzyme. 1997, FR Patent 2740779, Rhone Poulenc Chimie.

Gunther Nereus W, Solaiman Daniel K Y, Fett William F. Processes for the production of rhamnolipids. 2007, US Patent 7202063, US Agriculture.

Guo Chuling, Zhang Hui, Dang Zhi, Yang Chen, Lu Guining. Crude oil degrading bacterium for producing lipid biosurfactant and application. 2013, CN Patent 102978135, Univ South China Tech.

Gutnick D, Rosenburg E, Belsky I, Zosim Z, Shabtai Y. Emulsans. 1981, ZA Patent 8000912 Biotech AG Emulsan.

Gutnick David L, Belsky Igal, Shabtai Yossef, Rosenberg Eugene, Zosim Zinaida, Emulsans. 1980, AU Patent 5564380, Biotech AG Emulsan.

Gutnick David L, Bach Horacio. Compositions containing bioemulsifiers and a method for their preparation. 2002, WO Patent 0248327, Univ Ramot, Gutnick David L, Bach Horacio R.

Gutnick David L, Rosenberg Eugene. Cleaning oil-contaminated vessels with α-emulsans. 1981, US 4276094, Biotech AG Emulsan.

Gutnick David L, Rosenberg Eugene. Production of alpha–emulsans. 1980, US Patent 4230801, Biotech AG Emulsan.

Gutnick David L, Nestaas Eirik, Rosenberg Eugene, Sar Nechemia. Bioemulsifier production by Acinetobacter calcoaceticus strains. 1993, CA Patent 1316478, Petroleum Fermentations.

Gutnick David L, Nestaas Eirik, Rosenberg Eugene, Sar Nechemia. Bioemulsifier production by Acinetobacter calcoaceticus strains. 1989, US Patent 4883757, Petroleum Fermentations.

Gutnick David L, Rosenberg Eugene, Belsky Igal, Zinaida Zosim. Apo-β-emulsans. 1982, US Patent 4311829, Petroleum Fermentations.

Gutnick David L, Rosenberg Eugene, Belsky Igal, Zinaida Zosim. Apo-ψ-emulsans. 1982, US Patent 4311831, Petroleum Fermentations.

Gutnick David L, Rosenberg Eugene, Belsky Igal, Zinaida Zosim. Emulsans. 1983, US Patent 4395354, Petroleum Fermentations.

Gutnick David L, Rosenberg Eugene, Belsky Igal, Zinaida Zosim. Polyanionic heteropolysaccharide biopolymers. 1983, US Patent 4395353, Petroleum Fermentations.

Gutnick David L, Rosenberg Eugene, Belsky Igal, Zinaida Zosim. Proemulsans. 1982, US Patent 4311832, Petroleum Fermentations.

Gutnick David L, Rosenberg Eugene, Belsky Igal, Zinaida Zosim. Ψ-emulsans. 1983, US Patent 4380504, Petroleum Fermentations.

Gutnick David L, Rosenberg Eugene, Belsky Igal, Zosim Zinaida. Apo-α-emulsans. 1982, US Patent 4311830 Petroleum Fermentations.

Gutnick David L, Rosenberg Eugene, Shabtai Yossef. Production of α-emulsans. 1980, US Patent 4234689, Biotech AG Emulsan.

Gwak Hyung Sub, Kim Myung Kyum, Kown Mi Jung, Lee Myung Jin, Lee Sung Taik, Park Byeong Deog. Method for bioremediation of oil-contaminated soils using a specific microorganism and biosurfactant agent. 2005, KR Patent 20050043506, Neo Pharm Co Ltd.

Ha Sun Deuk, Kim Bong Jo, Kim Hak Ju. Marine bacterium *Pseudomonas aeruginosa* BYK-2 producing rhamnolipid and methyl rhamnolipid simultaneously. 2002, KR Patent 20020016707, Kong Jae Yeol.

Haefele Douglas M, Lamptey Jonathan C, Marlow Joseph L. Biological control of corn seed rot and seedling blight. 1991, US Patent 4996049, Pioneer Hi Bred Int.

Hall Peter J, Haverkamp Johan, Van Kralingen Cornelis G, Schmidt Michael. Laundry detergent composition containing synergistic combination of sophorose lipid and nonionic surfactant. 1996, US Patent 5520839, Lever Brothers Co.

Hall Peter J, Haverkamp Johan, Van Kralingen Cornelis G, Schmidt Michael. Detergent compositions. 1992, EP Patent 0499434, Unilever PLC, Unilever NV.

Hall Peter J, Haverkamp Johan, Van Kralingen Cornelis G, Schmidt Michael. Detergent compositions. 1992, CA Patent 2060698, Unilever PLC.

Hall Peter J, Haverkamp Johan, Van Kralingen Cornelis G, Schmidt Michael. Synergistic dual-surfactant detergent composition containing sophoroselipid. 1995, US Patent 5417879, Lever Brothers Co.

Han Sang Gil, Kim Jeong Cheol, Kim Yun Seok. Cosmetics composition comprising sophorolipids. 2004, KR Patent 20040033376, LG Household & Health Care Ltd.

Han Sung Ok, Jung Ju, Yu Kyung Ok. Transformant for improving productivity of biosurfactin and method for producing surfactin using the same. 2012, KR Patent 20120103187, Univ Korea Res & Bus Found.

Hayes Michael E, Hrebenar Kevin R, Murphy Patricia L, Futch Jr Laurence E, Deal Iii James, F. Bolden Jr Paul L. Bioemulsifier-stabilized hydrocarbosols. 1988, US Patent 4793826, Petroleum Fermentations.

Hayes Michael E, Hrebenar Kevin R, Murphy Patricia L, Futch Jr Laurence E, Deal Iii James F, Bolden Jr Paul L. Bioemulsifier-stabilized hydrocarbosols. 1990, US Patent 4943390, Petroleum Fermentations.

Hayes Michael E, Hrebenar Kevin R, Murphy Patricia L, Futch Jr Laurence E, Deal Iii James.F, Bolden Jr Paul L. Bioemulsifier stabilized hydrocarbosols. 1989, US Patent 4821757, Petroleum Fermentations.

Hayes Michael E, Hrebenar Kevin R, Murphy Patricia L, Futch Jr Laurence E, Deal Iii James F. Combustion of viscous hydrocarbons. 1986, US Patent 4618348, Petroleum Fermentations.

Hayes Michael E, Hrebenar Kevin R, Murphy Patricia L, Futch Jr Laurence E, Deal Iii James F, Bolden Jr Paul L. Pre-atomized fuels and process for producing same. 2000, US Patent RE36983, Petroferm Inc.

Hayes Michael E. Personal care products containing bioemulsifiers. 1991, US Patent 4999195, Emulsan Biotechnologies Inc.

Hayes Michael Edward. Personal care products containing bioemulsifiers. 1987, EP Patent 0242296, Petroleum Fermentations.

Hayes Michael Edward, Deal James Frances Iii, Murphy Patricia Lord, Hrebenar Kevin Robert, Futch Laurence Ernest Jr, Bolden Paul Lester Jr. Bioemulsifier-stabilized hydrocarbosols. 1985, ZA Patent 8408499, Petroleum Fermentations.

Hayes Michael Edward, Hrebenar Kevin Robert, Murphy Patricia Lord, Futch Laurence Ernest Jr, Deal James Frances Iii, Bolden Paul Lester Jr. Bioemulsifier-stabilized hydrocarbosols. 1985, WO Patent 8501889, Petroleum Fermentations Inc.

Hayes Michael Edward, Hrebenar Kevin Robert, Deal James Francis Iii, Bolden Paul Lester Jr. Method for reducing emissions utilizing pre-atomized fuels. 1987, WO Patent 8702376, Petroferm Inc.

Hayes Michael Edwards, Holzner Günter. Cosmetic and pharmaceutical compositions containing bioemulsifiers. 1986, EP Patent 0178443, Firmenich & Cie.

Hees Udo, Fabry Bernd. Use of mixture of glyco-lipid and surfactant in hand dish-washing detergent. 1997, DE Patent 19600743, Henkel KGAA.

Heins Sherry D, Manker Denise C, Jimenez Desmond R, Mccoy Randy J, Marrone Pamela J, Orjala Jimmy E. Method for protecting or treating plants and fruit against insect pests, extract of lipopeptide and lipopeptide surfactin. 2001, CZ Patent 20011620, Agraquest.

Heins Sherry Darlene, Manker Denise Carol, Jimenez Desmond Rito, Mccoy Randy Jay, Orjala Jimmy Ensio. A novel strain of *Bacillus* for controlling plant diseases and corn rootworm. 1998, WO Patent 9850422, Agraquest Inc, Heins Sherry Darlene, Manker Denise Carol, Jimenez Desmond Rito, Mccoy Randy Jay, Orjala Jimmy Ensio.

Heng Meng H, Bodo Michael. Methods of foam control. 2012, US Patent 2012252066, Heng Meng H, Bodo Michael.

Hibino Susumu, Minami Zenrou. Bacterial Preparation for agricultural use. 1998, US Patent 5733355, Hibino Susumu, Nagase Biochemicasl Ltd, Risahru Kosan Ltd.

Hillion Gerard, Marchal Remy, Stoltz Corinne, Borzeix Frederique. Use of a sophorolipid to provide free radical formation inhibiting activity or elastase inhibiting activity. 1998, US Patent 5756471, Inst Francais Du Petrole, Sophor SA.

Hillion Gerard, Marchal Remy, Stoltz Corinne, Borzeix Frederique. Use of sophorolipids and cosmetic and dermatological compositions, 1995, CA Patent 2192595.

Hillion Gerard, Marchal Remy, Stoltz Corinne, Borzeix Frederique. Use of sophorolipids and cosmetic and dermatological compositions. 1995, WO Patent 9534282, Inst Francais Du Petrole, Sophor Sa, Hillion Gerard, Marchal Remy, Stoltz Corinne, Borzeix Frederique.

Hirata Yoshihiko, Igarashi Keisuke, Oda Tomoka. Biodegradable liquid detergent composition. 2006, JP Patent 2006070231, Saraya KK.

Hirata Yoshihiko, Igarashi Keisuke, Oda Tomoka. Bleaching agent composition. 2006, JP Patent 2006274233, Saraya KK.

Hirata Yoshihiko, Igarashi Keisuke, Oda Tomoka. Cleanser composition. 2006, JP Patent 2006083238, Saraya KK.

Hitzman Donald O. Enhanced oil recovery process using microorganisms. 1984, US Patent 4450908, Phillips Petroleum Co.

Ho Lee Chang, Su I Mun. *Bacillus subtilis* GWN-5 and biosurfactant produced by this strain as a bioabsorbent. 2009, KR Patent 20090038546, Lee Chang Ho.

Hong Jeong Hwa, Hur Sung Ho, Kim Chun Gyu, Lee Ho Jae, Park Hwa Jin. Microorganism *Bacillus subtilis* CH-1 and anti-thrombotic biosurfactant produced therefrom which is useful for prevention, alleviation and treatment of circulatory system diseases. 2005, KR Patent 20050014451, Kyongsangnam Do, Unifood Tech Co Ltd.

Hong Sung Youl, Kwon Moo Sik, Lee Choong Eun, Cho Jae Youl, Kim Seo Young. Anti-cancer composition comprising surfactin. 2009, KR Patent 20090014428, Univ Sungkyunkwan Found.

Hong Sung-Ki, Park Jong-Su, Park Won-Jae, Ju Jong-Dae. Process for purification of methyl ester of glycosieds. 1993, KR Patent 930003489, Pacific Chem Co Ltd.

Hong Tan, Jinyan Zhou, Juan Zhong, Jie Yang, Liang Xiao. Method for efficiently preparing natural abscisic acid. 2012, CN Patent 102399827, Chengdu Inst Biology Cas.

Hongyan Wang, Wei Zhao, Meiyun Geng, Huan Ren, Jie Meng, Tao Li, 2010Biosurfactant as well as preparation method and application thereof. 2010, CN Patent 101898100, Univ Northeast Agricultural.

Howard Jeffrey, Reid Gregor, Gan Bing Siang. Treatment of microbial infections with bacterial proteins and peptides. 2004, US Patent 6727223, Urex Biotech Inc.

Howard Jeffrey, Reid Gregor, Gan Bing Siang. Treatment of microbial infections with bacterial proteins and peptides. 2002, WO Patent 0212271, Howard Jeffrey, Reid Gregor, Gan Bing Siang.

Hsieh Chien-Yan. Method for concurrently increasing production of iturin A and surfactin of lipopeptides. 2013, TW Patent 201313898, Nat Kaohsiung Normal University.

Hua Guo, Yajun Hu, Ruiying Li, Xiao'ou Liu, Yun Liu, Yongle Wang, Yonghu Xu. Method for producing sophorolipid by continuous feeding and fermentation. 2010, CN Patent 101845469, Tianjin Sf Bio Ind Bio Tec Co Ltd.

Huang He, Han Yuwang, Lu Shengguo, Zhu Lingyan, Xu Qing. Method for screening strains producing biosurfactants. 2013, CN Patent 103014121, Univ Nanjing.

Huang Tom Tao. Chemically modified sophorolipids and uses thereof. 2013, WO Patent 2013003291, Huang Tom Tao.

Huanzhong Huang, Zhenyong Zhao. Biosurfactant and polycyclic aromatic hydrocarbon-degrading strain enhanced compositing treatment of contaminated soil. 2011, CN Patent 102125929, Hong Kong Baptist University.

Huls Bernardus J, Duyvesteyn Willem P C, Budden Julia R. Biochemical treatment of bitumen froth tailings. 2000, CA Patent 2350927, BHP Minerals Int Inc.

Hussein Hany Mohamed Ahmed, Gaballa Ahmed Abo El-Einin, Abdel-Fattah Yasser Refaat, Zinder Steven H. Microbial production of a biosurfactant exhibiting exellent emulsification and surface active properties. 2006, WO Patent 2006136178, Mubarak City For Scient Res & Hussein Hany Mohamed Ahmed.

Hwang Ae Min, Jung Uk Jin, Kim Hak Yeong, Kim Hui Sik, Lee Seung U, Lim Yeong Gyeong, Park Jin Hui. *Pseudomonas putida* HPLJS-1 (KCTC 0666 bp) producing biosurfactant 2001, KR Patent 20010046256, Samsung Everland Inc.

Hwang Kyun-Ah, Kim Yun-Suk. Microorganism producing biosurfactant. 1997, KR Patent 970006157, LG Chemical Co.

Hwang Kyung-Ah, Kim Yeu-Kyung, Lee Jung-Rae. Liquid detergent composition. 1996, KR Patent 960007877, LG Chemical Ltd.

Imanaka Tadayuki, Sakurai Shoji. Biosurfactant cyclopeptide compound produced by culturing a specific arthrobacter microorganism. 1994, US Patent 5344913, Nikko Bio Technica Co.

Imanaka Tadayuki, Sakurai Shoji. Peptide-surfactant, production process thereof and microorganisms producing the same. 2000, EP Patent 0593058, Nikko Bio Technica Co.

Imura Tomohiro, Hirata Yoshihiko, Igarashi Keisuke, Ryu Mizuyuki, Ichiyanagi Naoki. Percutaneous absorption controller containing sophorolipid and method for producing the same. 2009, JP Patent 2009062288, Nat Inst of Adv Ind & Technol, Saraya KK.

Imura Tomohiro, Kitamoto Masaru, Fukuoka Tokuma, Morita Tomotake. Emulsifier or solubilizing agent. 2007, JP Patent 2007181789, Nat Inst of Adv Ind & Technol.

Imura Tomohiro, Kitamoto Masaru, Yagishita Hiroshi, Negishi Hideyuki, Ikegami Toru. Method for producing liposome utilizing coacervate. 2006, JP Patent 2006028069, Nat Inst of Adv Ind & Technol.

Inoue Shigeo, Kimura Yoshiharu. Novel microorganism. 1988, US Patent 4782025, Kao Corp.

Inoue Shigeo and Kimura Yoshiharu. Production of sophorose by microorganism. 1986, JP Patent S6135793, Kao Corp.

Inoue Shigeo, Miyamoto Norioki. Process for producing a hydroxyfatty acid ester. 1980, US Patent 4201844, Kao Corp.

Inoue Shigeo, Kimura Yoshiharu, Kaneda Manzou. Preparation of glycolipid methyl ester. 1979, JP Patent S54109914, Kao Corp.

Inoue Shigeo, Kimura Yoshiharu, Kinta Manzo. Dehydrating purification process for a fermentation product. 1980, US Patent 4197166, Kao Corp.

Inoue Shigeo, Kimura Yoshiharu, Kinta Manzo. Hydroxyalkyl-etherified glycolipid ester. 1980, US Patent 4195177, Kao Corp.

Inoue Shigeo, Kimura Yoshiharu, Kinta Manzo. Process for producing a glycolipid ester. 1980, US Patent 4215213, Kao Corp.

Inoue Shimako and Uchiyama Akira. Oral composition. 2005, JP Patent 2005170867, Lion Corp.

Ishida Atsuo, Homma Itomi, Inoue Shigeo. Hair cosmetic composition. 1982, US Patent 4318901, Kao Corp.

Ishigami Yutaka, Gama Yasuo, Kitamoto Masaru. Sophorolipid derivative. 1994, JP Patent H06100581, Agency Ind Science Techn.

Ishigami Yutaka, Gama Yasuo, Nagahora Hitoshi, Hongu Tetsuhiko, Yamaguchi Muneo. Rhamnolipid liposomes. 1990, US Patent 4902512, Agency Ind Science Techn.

Ishigami Yutaka, Gama Yasuo, Nagahora Hitoshi, Motomiya Tatsuhiko, Yamaguchi Muneo. Liposome. 1988, JP Patent S63182029, Agency Ind Science Techn, Shinetsu Chemical Co.

Ishigami Yutaka, Gama Yasuo, Someya Junichiro, Takamori Yasuaki, Ito Takuji. Tetraglucose and its partial fatty acid ester. 1994b, JP Patent H06298784, Agency Ind Science Techn, Iwata Kagaku Kogyo.

Ishigami Yutaka, Sasaki Makoto, Gama Yasuo, Kawaguchi Yoshihiro, Matsukawa Kuranari. Sophorose derivative. 1995, JP Patent H07118284, Agency Ind Science Techn, Ishihara Chemical Co Ltd.

Ishikawa Yoji, Fujii Haruhiko, Takezaki Satoshi, Kiriyama Hisashi. Method and apparatus for purification of contaminated soil. 2005, JP Patent 2005238004, Ohbayashi Corp, Toho Gas KK.

Ito Hitoshi, Igarashi Keisuke, Hirata Yoshihiko. Enzyme-containing detergent composition. 2009, JP Patent 2009052006, Saraya KK.

Ito Shinji and Kaneda Koichi. Feed additive and feed. 2010, JP Patent 2010220516, Idemitsu Kosan Co.

Izumida Masashi, Kawasaki Hiroaki, Moroshima Tadashi. Method for producing surfactin and salt thereof. 2012, WO Patent 2012043800, Kaneka Corp, Izumida Masashi, Kawasaki Hiroaki, Moroshima Tadashi.

Izumida Masashi, Yanagisawa Satohiro, Ueda Yasuyoshi. Fire extinguishing agent and fire extinguishing method using same. 2013, WO Patent 2013015241, Kaneka Corp, Izumida Masashi, Yanagisawa Satohiro, Ueda Yasuyoshi.

Jian Jiang, Baoling Yang, Fang Fan, Bing Wang, Xiaohong Wen, Miao Liu. Method for multi-means combined remediation of oil pollution soil of alkaline lands. 2012, CN Patent 102771221, Univ Dalian Nationalities.

Jianguo Wang. Biological oil displacement agent and production method and use thereof. 2009, CN Patent 101451062, Beijing Flty Petroleum Science.

Jianjun Le, Xiaolin Wu, Lulu Bai, Qun Zhao, Yu Li, Zhaowei Hou, Ying Wang et al. Complex biological oil displacement agent and application thereof. 2012, CN Patent 102492409, Daqing Oilfield Co Ltd.

Jie Wang, Zhenyu Ma, Hongquan Wang, Xianmei Zeng. Preparation prepared by adopting *Bacillus amyloliquefaciens* for processing excrement in a livestock farm as well as preparation method and processing method thereof. 2012, CN Patent 102765863, Hubei Hongquan Biolog Science & Technology Co Ltd.

Jimenez Sanchez Alfonso and Morales Bruque Jose. Prepn. of new bio:surfactant. 1993, ES Patent 2039187, Univ Extremadura.

Jinbo Wang and Lili Qi. Biologic emulsifier for aquatic animal feed. 2012, CN Patent 102696895, Ningbo Inst Tech Zhejiang Univ.

Jinbo Wang and Lili Qi. Biologic emulsifier for livestock feed. 2012, CN Patent 102696880, Ningbo Inst Tech Zhejiang Univ.

Jinfeng Liu, Bozhong Mou, Dandan Wang, Shizhong Yang. Method for separating and purifying rhamnolipid. 2010, CN Patent 101787057, Univ East China Science & Tech.

Jing Chen and Bingbing Yuan. Sophorolipid fruit preservative and use thereof in fruit preservation. 2010, CN Patent 101886047, Shandong Inst Light Industry.

Jinrong Wang, Xu Li, Tianchen Li, Binhong Wang. Environmental protection remover. 2008, CN Patent 101126052, Jinrong Wang.

Ju Lu-Kwang. Process for producing cellulase. 2008, US Patent 2008241885, Univ Akron.

Jun Jong Un, Kim Hui Sik, Lee Jin Yeong. Method for preparing sodium surfactin and use thereof. 2004, KR Patent 20040055035, Enbioeng Co Ltd.

Jung Myeong Su, Kwon Seong Hwan, Lee Seung Ju, So Jae Choon, Yeo Pyeong Mo. Non-pollution and non-toxic dishwashing detergent composition. 2004, KR Patent 20040020314, So Jae Choon.

Junqi Wang and Haifang Gai. High temperature foaming agent. 2011, CN Patent 102212356, Univ Xian Shiyou.

Junqi Wang, Haifang Gai, Changwu Han. High-temperature foam scrubbing agent for drainage gas recovery. 2011, CN Patent 102212344, Univ Xian Shiyou.

Kachholz Traudel and Schlingmann Merten. Lipotensides, process for their isolation and their use. 1984, Patent EP 121228, Hoechst AG.

Kaeppeli Othmar and Guerra-Santos Luis. Process for the production of rhamnolipids. 1986, US 4628030, Petrotec Forschungs AG.

Kamikura Yoshiko, Kitagawa Masaru, Yamamoto Shuhei. Water-in-oil emulsion cosmetic composition containing biosurfactant. 2010, JP Patent 2010018560, Toyo Boseki.

Kanehara Kazuhide, Hashimoto Toshiyuki, Oana Takao, Fukuda Masao, Takagi Masamichi, Yano Keiji, Oishi Michio. Microbe and method for decomposition of materials with use of the same. 1989, JP Patent S6468281 Japan Ind Res Inst, Tokyo Electric Power Co.

Kaplan David L, Castro Guillermo R, Panilaitis Bruce. Emulsan-alginate microspheres and methods of use thereof. 2006, WO Patent 2006028996, Tufts College, Kaplan David L, Castro Guillermo R, Panilaitis Bruce.

Kaplan David L, Fuhrman Juliet, Gross Richard A. Emulsan and emulsan analogs immunization formulations and use. 2000, WO Patent 0051635, Tufts College, Univ Massachusetts Lowell, Kaplan David L, Fuhrman Juliet, Gross Richard A.

Karaseva Ehmma Viktorovna, Volchenko Nikita Nikolaevich, Samkov Andrej Aleksandrovich. Method for selection of petroleum-oxidizing microorganisms as producers of biosurfactants. 2007, RU Patent 2006115033, Gosudarstvennoe Obrazovatel'noe Uchrezhdenie Vysshego Professional'nogo Obrazovanija Kubanskij Gosu.

Karaseva Ehmma Viktorovna, Volchenko Nikita Nikolaevich, Samkov Andrej Aleksandrovich. Method for selection of petroleum-oxidizing microorganisms as producers of biosurfactants. 2008, RU Patent 2320715, G Obrazovatel Noe Uchrezhdenie.

Karsten Stefan, Dreja Michael, Noglich Juergen. Detergent. 2011, DE Patent 102009046169, Henkel AG & Co KGAA.

Kawai Fusako and Maneraato Spasil. Biosurfactant which new microorganism produces. 2007, JP Patent 2007126469, Japan Science & Tech Agency.

Kawai Fusako and Maneraato Spasil. Novel microorganism and biosurfactant produced thereby. 2004, JP Patent 2004154090, Japan Science & Tech Agency.

Kawano Junichi, Suzuki Toshiyuki, Inoue Shigeo, Hayashi Shizuo. Stick-shaped cosmetic material. 1981, US Patent 4305929, Kao Corp.

Kawano Junichi, Utsugi Toshiaki, Inoue Shigeo, Hayashi Shizuo. Powdered compressed cosmetic material. 1981, US Patent 4305931, Kao Corp.

Kayano Masako and Funada Tadashi. Physiologically active lipid. 1990, JP Patent H0291024, Nippon Oils & Fats Co Ltd.

Kim Beom Seok. Fungicide for control of plant diseases using rhamnolipid b as glycolipid antibiotic. 2001, KR Patent 20010036515, Korea Chungang Educational Fou.

Kim Eun Gi. Red tide inhibiting agent comprising yellow earth and biodegradable sophorolipid. 2004, KR Patent 20040083614, Univ Inha.

Kim Eun Gi. Red tide-preventing agent. 2002, KR Patent 20020003679, M & Bgreenus Co Ltd.

Kim Eun Ki and Shin Jae Dong. Compounds of C-22 sophorolipid derivatives and the preparation method thereof. 2010, KR Patent 20100022289, Inha Ind Partnership Inst.

Kim Eun Ki. Sophorolipid production from soybean dark oil processing by-product. 2005, KR Patent 20050076124, Univ Inha.

Kim Hee Sik, Oh Hee Mock, Yoon Byung Dae, Cho Dae Hyun, Kim Byung Hyuk, Lee Young Ki, Baek Kyung Hwa, Ahn Chi Yong. Amplification method of surfactin biosynthesis operon, re-construction plasmid and transformed strain containing operon. 2009, KR Patent 20090066097, Korea Res Inst of Bioscience.

Kim Hee Sik, Yoon Byung Dae, Lee Young Ki, Baek Kyung Hwa, Oh Hee Mock, Lee Hong Won, Park Chan Sun, Ahn Chi Yong Bacillus subtilis CB114 producing a surfactin and having a genetic competency. 2006, KR Patent 100642292, Korea Res Inst of Bioscience.

Kim Hee Sik, Yoon Byung Dae, Oh Hee Mock, Park Chan Sun, Ahn Chi Yong, Lee Young Ki, Kim Byung Hyuk, Cho Dae Hyun. Large scale production method of glycolipid biosurfactant by fermentation. 2006, KR Patent 100599248, Korea Res Inst of Bioscience.

Kim In Seon, Chae Dong Hyun, Kim Seul Ki. Rhamnolipids-producing *Pseudomonas* sp. EP-3, and its use for control of aphids. 2010, KR Patent 100940231, Univ Nat Chonnam Ind Found.

Kim Ji Yeon and Joo Myeong Hoon. Culture method of biosurfactant-producing microorganism, *Bacillus subtilis* JK-1. 2011, KR Patent 20110016159, Univ Inje Ind Acad Cooperation.

Kim Jong Shik, Park Nyun Ho, Kim Choong Gon. Novel biosurfactant produced by *Aureobasidium pullulans*. 2013, KR Patent 101225110, Gyeongbuk Inst for Marine Bioindustry.

Kim Jong Woo, Lee Sang Wook, Han Jae Jin, Lee Geun Hyung, Park Sang Jin, Park Eul Yong, Shin Joong Chul, Suh Joo Won, Yun Tea Mi, Kim Kyoung Rok. *Bacillus subtilis* BC1212 from traditional food and its production of surfactin. 2006c, KR Patent 20060045201, B N C Bio Pharm Co Ltd, Suh Joo Won.

Kim Sang-Jong, Lee Ki-Seung, Shin Jung-Eun. Novel *Pseudomonas* sp. which produce biosurfactant. 1998, KR Patent 141065, Kim Sang Jong.

Kim Yeong Sik, Park Gi Yeong, Park Jin Hui. Method and apparatus for purifying contaminated soil. 2004, KR Patent 20040063515, Samsung Everland Inc.

Kim Young Hee, Park Sun Young. Immunosuppressive agent containing surfactin as an effective component. 2011, KR Patent 20110013613, Pusan Nat Univ Ind Coop Found.

Kim Young Hee, Park Sun Young, Lee Sang Joon. Composition for the anti-metastasis containing surfactin. 2013, KR Patent 20130012855, Pusan Nat Univ Ind Coop Found.

Kim Young-Kwon, Cho Kyung-Duk, Lee Dong-Soo, Kim Dae-Woong, Kwon Sung-Jin. Isolation and purification method of sophorose lipid. 1999, KR Patent 0181952, Korea Chemical Co Ltd.

Kitagawa Masaru, Inamori Kazunori. Cosmetic. 2011, JP Patent 2011148731, Toyo Boseki.

Kitagawa Masaru Inamori Kazunori. Oil-in-water type emulsified composition. 2011, JP Patent 2011168527, Toyo Boseki.

Kitagawa Masaru, Inamori Kazunori. Oil-in-water type emulsified cosmetic. 2011, JP Patent 2011168548, Toyo Boseki.

Kitagawa Masaru, Inamori Kazunori. Oil-in-water type emulsified cosmetic. 2011, JP Patent 2011173843, Toyo Boseki.

Kitagawa Masaru, Inamori Kazunori. Skin care preparation. 2011, JP Patent 2011132134, Toyo Boseki.

Kitagawa Masaru, Inamori Kazunori. Skin cosmetic. 2011, JP Patent 2011126790, Toyo Boseki.

Kitagawa Masaru, Inamori Kazunori. Skin external preparation composition 2011, JP Patent 2011184322, Toyo Boseki.

Kitagawa Masaru, Kondo Motohiro. Hair wash using biosurfactant. 2011, JP Patent 2011026279, Toyo Boseki.

Kitagawa Masaru, Kondo Motohiro. Oil-based solid cosmetic. 2011, JP Patent 2011063558, Toyo Boseki.

Kitagawa Masaru, Kondo Motohiro. Oil-based solid cosmetic. 2011, JP Patent 2011063559, Toyo Boseki.

Kitagawa Masaru, Kondo Motohiro. Hair cosmetic using biosurfactant. 2011, JP Patent 2011026275, Toyo Boseki.

Kitagawa Masaru, Kondo Motohiro. Hair cosmetic using biosurfactant. 2011, JP Patent 2011026276, Toyo Boseki.

Kitagawa Masaru, Kondo Motohiro. Hair cosmetic using biosurfactant. 2011, JP Patent 2011026277, Toyo Boseki.

Kitagawa Masaru, Kondo Motohiro. Hair cosmetic using biosurfactant. 2011, JP Patent 2011026278, Toyo Boseki.

Kitagawa Masaru, Kondo Motohiro. Hair cosmetic using biosurfactant. 2011, JP Patent 2011026280, Toyo Boseki.

Kitagawa Masaru, Kondo Motohiro. Hair cosmetic using biosurfactant. 2011, JP Patent 2011026281, Toyo Boseki.

Kitagawa Masaru, Kondo Motohiro. Hair cosmetic using biosurfactant. 2011, JP Patent 2011026282, Toyo Boseki.

Kitagawa Masaru, Kondo Motohiro. Hair cosmetic using biosurfactant. 2011, JP Patent 2011046634, Toyo Boseki.

Kitagawa Masaru, Yamamoto Shuhei. Biosurfactant-containing oil-in-water type emulsion cosmetic composition. 2009, JP Patent 2009275017, Toyo Boseki.

Kitagawa Masaru, Kyo Motoki, Sogabe Atsushi. Cosmetic composition containing biosurfactant. 2009, JP Patent 2009149566, Toyo Boseki.

Kitagawa Masaru, Kyo Motoki, Sogabe Atsushi. Cosmetic composition containing biosurfactant. 2009, JP Patent 2009149567, Toyo Boseki.

Kitagawa Masaru, Kyo Motoki, Sogabe Atsushi. Cosmetic composition containing biosurfactant. 2009, JP Patent 2009167157, Toyo Boseki.

Kitagawa Masaru, Kyo Motoki, Sogabe Atsushi. Cosmetic composition containing biosurfactant. 2009, JP Patent 2009167158, Toyo Boseki.

Kitagawa Masaru, Kyo Motoki, Sogabe Atsushi. Cosmetic composition containing biosurfactant. 2009, JP Patent 2009167159, Toyo Boseki.

Kitagawa Masaru, Suzuki Michiko, Yamamoto Shuhei, Sogabe Atsushi, Kitamoto Dai, Imura Tomohiro, Fukuoka Tokuma, Morita Tomotake. Skin care cosmetic and skin and agent for preventing skin roughness containing biosurfactants. 2008, EP Patent 1964546, Toyo Boseki, Nat Inst of Advanced Ind Scien.

Kitagawa Masaru, Suzuki Michiko, Yamamoto Shuhei, Sogabe Atsushi, Kitamoto Masaru, Morita Tomotake, Fukuoka Tokuma. Method for producing glycolipid. 2008, JP Patent 2008043210, Toyo Boseki, Nat Inst of Adv Ind & Technol.

Kitagawa Masaru, Suzuki Michiko, Yamamoto Shuhei, Sogabe Atsushi. Ameliorating agent for skin roughening. 2008, JP Patent 2008044857, Toyo Boseki.

Kitagawa Masaru, Yamamoto Shuhei, Hirashima Naohide. Agent for inhibiting allergic reaction late phase, comprising biosurfactant as active ingredient. 2011, JP Patent 2011105607, Toyo Boseki, Nagoya City Univ.

Kitagawa Masaru, Yamamoto Shuhei, Kamikura Yoshiko. Cosmetic composition containing biosurfactant. 2010, JP Patent 2010018559, Toyo Boseki.

Kitagawa Masaru, Yamamoto Shuhei, Suzuki Michiko, Sogabe Atsushi, Kitamoto Masaru, Morita Tomotake, Imura Tomohiro, Fukuoka Tokuma Method for producing deacylated product of glycolipid type biosurfactant. 2008, JP Patent 2008187902, Toyo Boseki, Nat Inst of Adv Ind & Technol.

Kitamoto Masaru, Nakanishi Mamoru. Liposome, and carrier composed of the same for introducing material to cell. 2009, JP Patent 2009203240, Nat Inst of Adv Ind & Technol, Nakanishi Mamoru.

Kitamoto Masaru, Nakanishi Mamoru. Liposome, and carrier for introducing material to cell, comprising the same. 2003, JP Patent 2003040767, Nat Inst of Adv Ind & Technol, Nakanishi Mamoru.

Kitamoto Masaru, Nakane Takashi, Akitani Yoji, Nobuchika Kazuo, Nakada Tatsu, Tomiyama Sumiko, Nagai Jotaro. Composition for high-density cold storage transportation. 2001, JP Patent 2001131538, Toho Chem Ind Co Ltd, Nat Inst of Advanced Ind Scien, Kitamoto Masaru, Nakane Takashi, Akitani Yoji.

Kobayashi Toru, Koike Kenzo, Ito Susumu, Okamoto Kimihiko. Production of glycolipid 1988, JP Patent S6363389, Kao Corp.

Konishi Masao, Fukuoka Tokuma, Morita Tomotake, Imura Tomohiro, Kitamoto Masaru, Iwabuchi Hiroyuki. Method for producing acid-type sophorose lipid. 2008, JP Patent 2008247845, Nat Inst of Adv Ind & Technol, Lion Corp.

Koppe Juergen, Lausch Hartmut, Buesching Kai, Walter Regina. Method of removing recurring formation of bio-films, useful for disinfecting water, comprises exposing germs with hydrogen peroxide in presence of catalyst and conducting biosurfactant suspension in to the water-transporting system. 2007, DE Patent 102005062337, Mol Katalysatortechnik GmbH.

Kopp-Holtwiesche Bettina, Weiss Albrecht, Boehme Adelheid. Nutrient mixtures for the bioremediation of polluted soils and waters. 1997, US Patent 5635392.

Lal Banwari, Reddy Mula Ramajaneya Varapras, Agnihotri Anil, Kumar Ashok, Sarbhai Munish Prasad, Singh Nimmi, Khurana Raj Karan, Khazanchi Shinben Kishen, Misra Tilak Ram. A process for enhanced recovery of crude oil from oil wells using novel microbial consortium. 2005, WO Patent 2005005773, Energy Res Inst, Institue of Reservoir Studies, Lal Banwari, Reddy Mula Ramajaneya Varapras, Agnihotri Anil, Kumar Ashok, Sarbhai Munish Prasad, Singh Nimmi, Khurana Raj Karan, Khazanchi Shinben Kishen, Misra Tilak Ram.

Lang Siegmund, Brakemeier Andreas, Wullbrandt Dieter, Seiffert-Stoeriko Andreas. New sophoroselipids, method for their production and use. 1999, WO Patent 1999024448, Aventis Res & Tech Gmbh & Co, Lang Siegmund, Brakemeier Andreas, Wullbrandt Dieter, Seiffert Stoeriko Andreas.

Lang Siegmund, Brakemeier Andreas, Wullbrandt Dieter, Seiffert-Stoeriko Andreas. New sophoroselipids, method for their production and use. 1999, WO Patent 9924448, Aventis Res & Tech GMBH & Co, Lang Siegmund, Brakemeier Andreas, Wullbrandt Dieter, Seiffert-Stoeriko Andreas.

Lauer Ursula, Kopp-Holtwiesche Bettina. A process for the simplified biological rehabilitation of land polluted with long-term contamination based on mineral oil. 1999, CA Patent 2303413, Cognis Deutschland GmbH, Industrieanlagen Betr SGMBH.

Lee Jong Gi, Kim Byeong Jo, Choi Kyu Yong, Lee Hyeon Shin. A manufacturing method of phospholipids biosurfactant. 2013, KR Patent 20130000705, Chemtech Co Ltd Ak.

Lee Myung Jin, Park Byeong Deog, Lee Sung Taik, Kim Myung Kyum. Novel *Tetragenococcus koreensis* having rhamnolipid-producing ability and producing method for rhamnolipid using the same. 2007, KR Patent 20070027151, Neopharm Co Ltd, Korea Advanced Inst Sci & Tech.

Leighton Anton. Method for treating rhinitis and sinusitis by rhamnolipids. 2010, WO Patent 2010080406, Adrem Biotech Inc, Leighton Anton.

Leighton Anton. Method for treating rhinitis and sinusitis by rhamnolipids. 2011, US Patent 2011257115, Adrem Biotech Inc, Leighton Anton.

Leitermann Frank Paul, Hausmann Rudolf, Syldatk Christoph. Biosurfactants and production thereof. 2008, WO Patent 2008151615, Univ Karlsruhe, Leitermann Frank Paul, Hausmann Rudolf, Syldatk Christoph.

Lemal Jeannune, Marchal Remy, Davila Anne-Marie, Sulzer Caroline. Fermentative production of sophorolipid composition. 1994, DE Patent 4319540, Inst Francais Du Petrole.

Li Ren, Guoyuan Li, Jinyan Liu, Yingjun Wang. Polycaprolactone/calcium sulfate composite material and preparation method thereof. 2011, CN Patent 101974212, Univ South China Tech.

Li Zuyiand Shi Yiping. Production of sucrose ester by homogeneous solventless process. 1999, CN Patent 1232036, Shanghai Inst Organic Chem.

Liang Qu, Guangru Li, Ming Li, Yu Chen, Wenyu Lu, Lingqing Zhu, Xue Yang. Method for producing rhamnolipid by virtue of fermentation and separation of *Pseudomonas aeruginosa*. 2012, CN Patent 102796781, China Nat Offshore Oil Corp, Cnooc Energy Technology Co Ltd, China Bluechemical Environmental Prot Service Tianjin Co Ltd.

Liao Guangzhi, Liu Yi, Yue Jianjun. Method for increasing petroleum recovery ratio using three-component composite displacement emulsification. 2003, CN Patent 1420255, Daqing Oil Field Co Ltd.

Lili Bai, Yang Yu, Li Zhao, Zhiqiang Sui, Yanfang Jin, Shaojun Chen, Sheng Zhang. Rhamnolipid auxin. 2011, CN Patent 101948354, Daqing Vertex Chemical Co Ltd.

Lin Chi-Tsan. Method for removing pollutants in pollutants-contaminated soil. 2012, US Patent 2012021499, Chi Tsan Lin.

Lindoerfer Walter, Sewe Kai-Udo, Oberbremer Axel, Mueller-Hurtig Reinhard, Wagner Fritz. Microbial decontamination of soils contaminated with hydrocarbons, in particular mineral oils by microbial oxidation. 1992, US Patent 5128262, Wintershall AG.

Liu Zhipei, Li Xiwu, Liu Shuangjiang. *Erythro micrococcus* Em and usage for generating biologic emulsifier as well as degrading polycyclic aromatic hydrocarbon. 2004, CN Patent 1519312, Microorganism Inst Chinese Aca.

Liwen Liu, Sihai Liu, Yong Ning, Jiming Lu. Method for preparing high belite cement retarder. 2012, CN Patent 102584083, Liwen Liu.

Lixiang Zhou, Jie Ren. Bacterium S2 for efficiently generating biosurfactant and fermentation culture medium thereof. 2011, CN Patent 102250790, Univ Nanjing Agricultural.

Lixing Dai, Hui Feng, Wenli Li. Electrostatic spinning method for polyhydroxylated polymer. 2008, CN Patent 101275291, Univ Soochow.

Lu Jenn-Kan. Peptide having antimicrobial and/or biosurfactant property. 2008, TW Patent 200823231, Umo Internat Co Ltd.

Lu Jenn-Kan. Peptide having antimicrobial and/or biosurfactant property. 2009, WO Patent 2009044279, Umo Inc Co Ltd, Lu Jenn-Kan.

Marchal Remy, Lemal Jeannine, Sulzer Caroline, Davila Annne-Marie. Procédé de production de sophorolipides acétyles sous leur forme acide à partir d'un substrat consistant et une huile ou un ester. 1993, FR Patent 2692593, Inst Francais Du Petrole.

Mager Herbert, Rothlisberger Rudi, Wagner Fritz. Use of sophoroselipid-lactone for the treatment of dandruffs and body odour. 1987, EP Patent 0209783, Wella AG.

Maingault Martine. No title available. 1997, FR Patent 2735979, Inst Francais Du Petrole.

Maingault Martine. Utilization of sophorolipids as therapeutically active substances or cosmetic products, in particular for the treatment of the skin. 1997, WO Patent 9701343, Inst Francais Du Petrole, Maingault Martine.

Manresa Presas Angeles, Mercade Gil M Elena, Robert Sampietro Marta, Guinea Sanchez Jesus, Bosch Verderol M Pilar, Parra Juez Jose Luis, Espuny Tomas M Jose. Improvements made to the method for obtaining biosurfactants in a substrate of refined oils and fat. 1991, ES Patent 2018637, Manresa Presas Angeles, Mercade Gil M Elena, Robert Sampietro Marta, Guinea Sanchez Jesus, Bosch Verderol M Pilar, Parra Juez Jose Luis, Espuny Tomas M Jose.

Marchal Remy, Lemal Jeannine, Sulzer Caroline, Davila Anne-Marie. Fed batch prodn. of sophorolipid(s) from *Candida*—including pre-culture step, giving prod. in acetylated acid form, useful as emulsifier e.g. in sec. oil recovery. 1993, FR Patent 2691975, Inst Francais Du Petrole.

Marchal Remy, Lemal Jeannine, Sulzer Caroline. Process for the production os sophorosids by fermentation with continuous fatty acids ester or oil supply. 1992, CA Patent 2075177, Inst Francais Du Petrole.

Marchal Remy, Warzywoda Michel, Chaussepied Bernard. Process for the production of sophorolipids by cyclic fermentation with fatty acid esters or oil supply. 1998, EP Patent 0837140, Inst Francais Du Petrole.

Marchal Remy, Warzywoda Michel, Chaussepied Bernard. Process for the production of sophorolipids by cyclic fermentation with feed of fatty acid esters or oils. 1999, US Patent 5879913, Inst Francais Du Petrole.

Marchenko A.I, Vorob'ev A.V, Djadishchev N.R, Rybalkin S.P, Blokhin V.A, Marchenko S.A. Strain of bacterium *Pseudomonas alcaligenes* MEV used for treatment of soil, ground and surface water from petroleum and products of its processing. 2004, RU Patent 2002122613.

Marchenko A.I, Vorob'ev A.V, Borovik R.V, Aldobaev V.N, Zharikov G.A, Kapranov V.V. Strain of bacterium *Pseudomonas stutzeri* MEV-S1 used for treatment of soil, ground and surface water from petroleum and products of its processing. 2004, RU Patent 2002122612.

Masaru Kitagawa, Michiko Suzuki, Shuhei Yamamoto, Atsushi Sogabe, Dai Kitamoto, Tomohiro Imura, Tokuma Fukuoka, Tomotake Morita. Skin care cosmetic and skin and agent for preventing skin roughness containing biosurfactants. 2008, CN Patent 101316574, Toyo Boseki.

Masuoka Kazuyuki, Obata Toru, Fukuda Yoshinori, Taharu Noboru. Improvement in cellulosic fiber and composition. 1998, JP Patent H1096174, Nagase Seikagaku Kogyo KK.

Matsuda Hitoshi, Amashita Kazuo, Kitano Nozomi. Nematode attractant and method for nematode control. 2012, WO Patent 2012115225, JX Nippon Oil & Energy Corp.

Matsufuji Motoko, Nakada Kunio. Production of rhamnolipid by using ethanol. 1998, JP Patent H1075796, Tsushosangyosho Kiso Sangyo KY.

Matsuura Fumito, Ota Masaya, Tamai Masahiro, Tamura Kokichi. Glycolipid and method for producing the same. 2005, JP Patent 2005104837, Hiroshima Prefecture, Maruzen Pharma.

Mattei Georges, Bertrand Jean-Claude. Processes and devices for the continuous production of biosurfactants. 1986, FR Patent 2578552, Inst Fs Rech Expl Mer.

Mayama Shigeyuki, Tosa Yukio, Otsu Yasunari, Toyoda Hideyoshi, Matsuda Katsunori, Nonomura Teruo. Plant insect pest controlling agent and method for controlling plant insect pest utilizing insect pathogenic bacterium *Pseudomonas fluorescens* KPM-018P. 2005, JP Patent 2005102510, Univ Kobe.

Mcinerney Michael J, Jenneman Gary E, Knapp Roy M, Menzie Donald E. Biosurfactant and enhanced oil recovery. 1985, US Patent 4522261, The Board of Regents for the University of Oklahoma.

Melent'ev Aleksandr Ivanovich, Kuz Mina Ljudmila Jur Evna, Jakovleva Ol Ga Valerievna, Kurchenko Vladimir Petrovich. Strain of bacterium *Bacillus subtilis* as producer of surfactin. 2006, RU Patent 2270858.

Melo Santa Anna Lidia Maria, Guimaraes Freire Denise Maria, De Araujo Kronemberger Frederico, Piacsek Borges Cristiano, Machado De Castro Aline. System for obtaining biological products. 2012, WO Patent 2012079138, Petroleo Brasileiro SA, Melo Santa Anna Lidia Maria, Guimaraes Freire Denise Maria, De Araujo Kronemberger Frederico, Piacsek Borges Cristiano, Machado De Castro Aline.

Meng Qin, Jiang Lifang. Application of rhamnolipid serving as protective agent of low-temperature or ultralow-temperature cell preservation. 2013, CN Patent 103027032, Univ Zhejiang.

Meng Qin, Zhang Guoliang. Application of rhamnolipid as biological cleaning agent. 2012, CN Patent 102399644, Huzhou Gemking Biotech Co Ltd.

Meng Qin, Zhang Guoliang. Application of rhamnolipid as crude oil pipeline drag reducer and viscosity reducer. 2012, CN Patent 102399547, Huzhou Gemking Biotech Co Ltd.

Meng Qin, Zhang Guoliang. Application of rhamnolipid in serving as printing and dyeing auxiliary. 2013, CN Patent 103061164, Huzhou Gemking Biotech Co Ltd.

Meng Qin, Zhang Guoliang. Preparation method and application of rhamnolipid. 2010, CN Patent 101845468, Hangzhou Gemking Biotech Co Ltd, Qin Meng.

Meng Qin, Zhang Guoliang. Rhamnolipid crude extract prepared by fermenting food and rink waste oil and application thereof. 2007, CN Patent 1908180, Univ Zhejiang.

Meng Qin, Zhang Guoliang. Separation method of rhamnolipid by using ultrafiltration membrane. 2012, CN Patent 102432643, Huzhou Gemking Biolog Technology Co Ltd.

Meng Qin, Jiang Lifang, Zhang Guoliang. Application of rhamnolipid as oral medicine absorbent accelerant. 2013, CN Patent 103007287, Huzhou Gemking Biotech Co Ltd.

Meng Qin, Long Xuwei, Zhang Guoliang. Application of rhamnolipid as demulsifier. 2013, CN Patent 102851059, Univ Zhejiang, Huzhou Gemking Biotech Co Ltd.

Mercier Jean Luc, Lemarchand Pierre. Cleaning foliage of agricultural plants to improve photosynthesis and increase crop yields, by adding acid, alkylsulfonate and biosurfactant to irrigation water. 2000, FR Patent 2787439, Mercier Jean Luc.

Middelberg Anton Peter Jacob, Dimitrijev-Dwyer Mirjana, Brech Michael. Designed biosurfactants, their manufacture, purification and use. 2012, WO Patent 2012079125, Univ Queensland, Middelberg Anton Peter Jacob, Dimitrijev-Dwyer Mirjana, Brech Michael.

Mingxiu Xin, Ziyu Song, Qian Li. Method for preparing rhamnolipid from macroporous resin. 2009, CN Patent 101407831, Mingxiu Xin.

Mixich Johann, Rapp Knut M, Vogel Manfred. Method for the preparation of rhamnose monohydrate from rhamnolipids.1996, US Patent 5550227, Suedzucker AG.

Mixich Johann, Rapp Knut Martin,Vogel Manfred. Process for producing rhamnose from rhamnolipids. 1992, WO Patent 9205182, Suedzucker AG.

Miyazaki Yusuke, Senba Takashi. Succinoyl trehalose lipid composition, solution and emulsion composition thereof, and method for producing the same. 2009, JP Patent 2009013160, Nippon Catalytic Chem Ind.

Miyazaki Yusuke, Mukoyama Masaharu, Watanabe Noriyuki. Oily gel composition and method of preparation of emulsified composition. 2009, JP Patent 2009079030, Nippon Catalytic Chem Ind.

Mohamado Osuman, Ishigami Yutaka, Ishizuka Yasuko. Straight-chain surfactin. 1996, JP Patent H0892279, Agency Ind Science Techn.

Morishima Seiji, Monoi Noriyuki. Composition for oral cavity. 2005, JP Patent 2005298357, Lion Corp.

Morishima Seiji. Composition for oral cavity. 2003, JP Patent 2003246717, Lion Corp.

Morita Tomotake, Fukuoka Tokuma, Imura Kazuhiro, Kitamoto Masaru. Method for producing biosurfactant. 2007, JP Patent 2007209332, Nat Inst of Adv Ind & Technol.

Morita Tomotake, Fukuoka Tokuma, Imura Tomohiro, Kitamoto Masaru, Wada Koji, Matsuzaki Goro, Takara Kensaku, Hirose Naoto, Teruya Akira, Inafuku Keiichiro, Takahashi Makoto. Method for producing biosurfactant. 2010, JP Patent 2010200695, Nat Inst of Advanced Ind Scien, Univ Ryukyus, Okinawa Prefecture, Kanehide Bio KK.

Morita Tomotake, Fukuoka Tokuma, Imura Tomohiro, Kitamoto Masaru. Method for producing biosurfactant. 2007, JP Patent 2007209333, Nat Inst of Adv Ind & Technol.

Morita Tomotake, Fukuoka Tokuma, Imura Tomohiro, Kitamoto Masaru. Method for highly efficiently producing mannosylerythritol lipid. 2007c, JP Patent 2007252279, Nat Inst of Adv Ind & Technol.

Morita Tomotake, Fukuoka Tokuma, Imura Tomohiro, Kitamoto Masaru. New microorganism and method for producing sugar type biosurfactant using the same. 2011, JP Patent 2011182660, Nat Inst of Advanced Ind Scien.

Morita Tomotake, Fukuoka Tokuma, Imura Tomohiro, Kitamoto Masaru. New microorganism and method for producing sugar type biosurfactant therewith. 2011, JP Patent 2011182740, Nat Inst of Advanced Ind Scien.

Mulligan Catherine N, Chow Terry Y K. Enhanced production of biosurfactant through the use of a mutated B. subtilis strain. 1992, CA Patent 2025812, Her Majesty the Queen, in Right of Canada, as represented by the Minister of the National Research.

Mulligan Catherine N, Chow Terry Y, Chow Terry Y-K. Enhanced production of biosurfactant through the use of a mutated B subtilis strain. 1991, US Patent 5037758, United Kingdom Government.

Nakayama Kazuo, Masuda Takashi, Ishigami Yutaka, Iwahashi Hitoshi, Takizawa Yasutomi. Functional polybutylene succinate resin composition. 2003, JP Patent 2003253105, Nat Inst of Adv Ind & Technol.

Negishi Keisoku, Matsuo Noriki, Miyadera Keisuke, Yajima Masae, Esumi Yasuaki. EB-162 material and its production. 2000, JP Patent 2000273100, Basic Ind Bureau Miti.

Nero Michael D. Systems and methods for cleaning materials. 2006, US Patent 2006080785, Nero Michael D.

Nero Michael D. Systems and methods for spot cleaning materials. 2006, US Patent 2006084587, Nero Michael D.

Nero Micheal. Methods for cleaning materials. 2009, US Patent 7556654, Naturell.

No inventor available. Skin cosmetic. 2011, JP Patent 2011225453, Toyo Boseki.

No inventor available. Absorption-enhancing composition containing composite of basic physiologically active protein and sophorolipid. 2012, JP Patent 2012232963, Saraya KK.

No inventor available. Microorganism and its use for microbial breakdown of petroleum hydrocarbons and mineral oil in contaminated soils, water and aerosols. 1990, DE Patent 3909324, Mikro-Bak Biotechnik GmbH, 3338 Schoeningen, De.

No inventor available. Microorganism. 1985,GB Patent 8516702, Kao Corp Publication.

No inventor available. Rhamnolipids with improved cleaning. 2012, EP Patent 2410039, Unilever PLC.

No inventor available. Process for producing a hydroxyfatty acid ester 1979, GB Patent 2002756, Kao Corp.

No inventor available. Process for dehydrating purification of a fermentation product. 1979, GB Patent 2002369, Kao Corp.

No inventor available. Hydroxypropyletherified glycolipid esters. 1983, GB Patent 2033895, Kao Corp.

O'Connell Timothy, Siegert Petra, Evers Stefan, Bongaerts Johannes, Weber Thomas, Maurer Karl-Heinz, Bessler Cornelius. Method for improving the cleaning action of a detergent or cleaning agent. 2011, US Patent 2011201536, Henkel Ag & Co KGAA.

Okubo Takashi, Saida Yoshihiro, Naijo Shuichi. Antistatic agent, antistatic film and product coated with antistatic film. 2006, WO Patent 2006016670, Showa Denko KK, Ohkubo Takashi, Saida Yoshihiiro, Naijo Shuichi.

Okubo Takashi, Saida Yoshihiro, Uchijiyou Shiyuuichi. Antistatic agent, antistatic film, and article coated with the antistatic film. 2006, JP Patent 2006077236, Showa Denko KK.

Okada Gentarou, Nakakuki Teruo, Kainuma Seishiro, Unno Takehiro. Ameliorative substance for enteral flora. 1991, JP Patent H03262460, Japan Maize Prod.

Okura Ichiro, Kubo Miki, Hasumi Fumihiko, Yamamoto Etsuo. New microorganism assimilating petroleum and waste oil. 1995, JP Patent H078270, Gen Sekiyu KK.

Okura Ichiro, Kubo Miki, Hasumi Fumihiko, Yamamoto Etsuo. New microorganism assimilating petroleum and waste oil. 1995, JP Patent H078271, Gen Sekiyu KK.

Omori Toshiro, Maruoka Ikuyuki, Furuta Yoshifumi, Konishi Masao, Fukuoka Tokuma, Morita Tomotake, Imura Tomohiro, Kitamoto Masaru. Microorganism having ability of producing saccharide type biosurfactant and method for producing saccharide type biosurfactant using the same. 2010, JP Patent 2010158192, Sanwa Shurui Co Ltd, Nat Inst of Advanced Ind Scien.

Omori Toshiro, Takeshima Naoki, Hayashi Kei, Maruoka Ikuyuki, Konishi Masao, Fukuoka Tokuma, Morita Tomotake, Imura Tomohiro, Kitamoto Masaru. Method for producing biosurfactant using fermentation product of plant as medium. 2009, JP Patent 2009207493, Sanwa Shurui Co Ltd, Nat Inst of Adv Ind & Technol.

Ootani Yasuhito, Ukiana Toshinao, Yamane Kouichi. Purification of sophorolipid derivatives. 1980, JP Patent S554344, Kao Corp.

Osugi Takao. Bathing agent. 1988, JP Patent S63156714, Lion Corp.

Owen Donald, Fan Lili. Oligomeric biosurfactants in dermatocosmetic compositions. 2009, WO Patent 2009148947, Owen Donald, Fan Lili.

Owen Donald, Fan Lili. Polymeric biosurfactants. 2007, WO Patent 2007143006, Therapeutic Peptides Inc, Owen Donald, Fan Lili.

Park Ae-Ran, Kim Jung-Ill, Jang Chang-Ho. Novel microorganism *Acinetobacter* sp. KRC-K4 and process for preparing bioemulsifier using the same. 1999, KR Patent 170107, Korea Kumho Petrochem Co Ltd.

Park Byeong-Deog, Lee Myung-Jin, Kwon Mi-Jong, Gwak Hyung-Sub, Kim Myung-Kyum, Kim Yoon. Novel microorganisms having oil biodegradability and method for bioremediation of oil-contaminated soil. 2008, US Patent 2008020947, Park Byeong-Deog, Lee Myung-Jin, Kwon Mi-Jong, Gwak Hyung-Sub, Kim Myung-Kyum, Kim Yoon.

Park Hyun Seok. Conditioning shampoo composition containing biosurfactant. 2009, KR Patent 20090117081, LG Household & Health Care Ltd.

Park Je Hyeon, Park Jin Hui Bacterium with higher biosurfactant activity for biological soil remediation. 2002, KR Patent 20020003460, Samsung Everland Inc.

Park Jin Hui.Bioremediation method for oil-spoiled soil bioreactor. 2003, KR Patent 20030066072, Samsung Everland Inc.

Park Jin Hui. Biosurfactant-secreting bacteria and its utilization method. 2002, KR Patent 20020011251, Samsung Everland Inc.

Park Kyeong Ryang, Shim So Hee. Novel *Pseudomonas* sp. G314 producing biosurfactant. 2008, KR Patent 20080017148, Park Kyeong Ryang.

Parry Alyn James, Parry Neil James, Peilow Anne Cynthia, Stevenson Paul Simon. Detergent compositions comprising biosurfactant and enzyme. 2012, WO Patent 2012010405, Unilever PLC, Unilever NV, Unilever Hindustan, Parry Alyn James, Parry Neil James, Peilow Anne Cynthia, Stevenson Paul Simon.

Parry Alyn James, Parry Neil James, Peilow Anne Cynthia, Stevenson Paul Simon. Detergent compositions comprising biosurfactant and lipase, 2012, WO Patent 2012010407, Unilever PLC, Unilever NV, Unilever Hindustan, Parry Alyn James, Parry Neil James, Peilow Anne Cynthia, Stevenson Paul Simon.

Parry Alyn James, Parry Neil James, Pelow Anne Cynthia, Stevenson Paul Simon. Combinations of rhamnolipids and enzymes for improved cleaning. 2012, WO Patent 2012010406, Unilever PLC, Unilever NV, Unilever Hindustan, Parry Alyn James, Parry Neil James, Peilow Anne Cynthia, Stevenson Paul Simon.

Pastore Glaucia Maria, Moraes Diane Teo De. Method for producing a modified strain of *Bacillus subtilis*, modified strain, method for producing surfactin and use of said modified strain. 2012, WO Patent 2012151647, Unicamp, Pastore Glaucia Maria, Moraes Diane Teo De.

Patil J. R, Chopade B. A. Bioemulsifier production by *Acinetobacter* strains isolated from healthy human skin. 2004, US Patent 2004138429, Council Scient Ind Res.

Patil J. R, Chopade B. A. Bioemulsifier production by *acinetobacter* strains isolated from healthy human skin. 2005, US Patent 2005163739, Council Scient Ind Res.

Pekin Gulseren. Two stage sophorolipids production with *Candida bombicola* ATCC 22214. 2006, TR Patent 200601906, Pekin G.

Pellicier Francoise, Andre Patrice. Cosmetic use of sophorolipids as subcutaneous adipose cushion regulating agents and slimming application. 2004, WO Patent 2004108063, LVMH Rech, Pellicier Francoise, Andre Patrice.

Pellicier Francoise, Andre Patrice. No title available. 2004, FR Patent 2855752, LVMH Rech.

Peng Chen, Leilei Jiang, Xiang Jin, Ying Liang, Xiangyang Liu, Xiaodan Mei, Xinglai Shi, Qiang Sun, Wei Wang, Wei Ye. Method for preparing *Bacillus subtilis* lipopeptid biosurfactant. 2010, CN Patent 101838621, Dalian Biteomics Inc.

Peng Li. Method for synthesizing LiFePO4 by using biosurfactant. 2009, CN Patent 101555003, Shandong Inst Light Industry.

Pesce Luciano. A biotechnological method for the regeneration of hydrocarbons from dregs and muds, on the base of biosurfactants. 2002, WO Patent 02062495, Idrabel Italia S R L, Jeneil Biosurfactant Company L.

Pierce Deborah, Heilman Timothy J. Germicidal Composition. 1998, CA Patent 2267678, Alterna Inc.

Piljac Goran, Piljac Visnja. Immunological activity of rhamnolipids. 1995, US Patent 5466675.

Piljac Goran, Piljac Visnja. Immunological activity of rhamnolipids. 1996, US Patent 5514661, Piljac Goran, Piljac Visnja.

Piljac Goran, Piljac Visnja. Pharmaceutical preparation based on rhamnolipid. 1995, US Patent 5455232.

Piljac Goran, Piljac Visnja. Pharmaceutical preparation based on rhamnolipid against dermatological diseases, for example, papilloma virus infections. 1993, WO Patent 9314767, Innovi NV.

Piljac Goran. Rhamnolipid based surface-active compound. 1994, BE Patent 1005825, Piljac Goran.

Piljac Goran. The use of rhamnolipids as a drug of choice in the case of nuclear disasters in the treatment of the combination radiation injuries and illnesses in humans and animals. 2011, WO Patent 2011109200, Piljac Goran.

Piljac Tatjana, Piljac Bran. Use of rhamnolipids in wound healing, treating burn shock, atherosclerosis, organ transplants, depression, schizophrenia and cosmetics. 2007, US Patent 20070155678, Piljac Tatjana, Piljac Bran.

Piljac Tatjana, Piljac Goran. Use of rhamnolipids in wound healing, treating burn shock, atherosclerosis, organ transplants, depression, schizophrenia and cosmetics. 2008, EP Patent 1889623, Paradigm Biomedical Inc.

Piljac Tatjana, Piljac Goran. Use of rhamnolipids in wound healing, treating burn shock, atherosclerosis, organ transplants, depression, schizophrenia and cosmetics. 2007, US Patent 7262171, Paradigm Biomedical Inc.

Piljac Tatjana, Piljac Goran. Use of rhamnolipids in wound healing, treating burn shock, atherosclerosis, organ transplants, depression, schizophrenia and cosmetics. 1999, WO Patent 1999043334, Piljac Tatjana, Piljac Goran.

Potter Anne, Malle Gerard, Donovan Mark. Non-therapeutic cosmetic method, useful for improving barrier function of skin and/or for hydrating skin using sulfated polysaccharide with rhamnose pattern, comprises applying a composition comprising sulfated polysaccharide on skin. 2013, FR Patent 2982152, Oreal.

Powell Jr John E. System and process for in tank treatment of crude oil sludges to recover hydrocarbons and aid in materials separation. 2000, US Patent 6033901, APLC Inc.

Prosperi Giulio, Camilli Marcello, Crescenzi Francesco, Fascetti Eugenio, Porcelli Filippo, Sacceddu Pasquale. New lipopolysaccharide biosurfactant. 1999, EP Patent 0924221, Enitecnologie Spa.

Pyroh Tetiana Pavlivna, Voloshyna Iryna Mykolaivna, Ihnatenko Serhii Viktorovych. Strain of *Rhodococcus erythropolis* EK-1 bacteria—producer of the surface-active substances. 2006, UA Patent 77345, Nat Univ Food Technologies.

Qi Shuhua, Zhang Xiaoyong, Peng Jiang. Methylotrophic bacillus for producing surfactins and iturin A compounds and application of methylotrophic bacillus. 2013, CN Patent 102994418, South China Sea Inst Oceanolog.

Qing Hong, Qian Wang, Jiandong Jiang, Rong Li, Shunpeng Li. Bacterial strain for generating rhamnolipid biosurfactant and generated microbial inoculum thereof. 2011, CN Patent 101948793, Univ Nanjing Agricultural.

Qinglu Huang, Fangming Luo. Biological heavy metal adsorbing and flocculating agent and preparation method thereof. 2012, CN Patent 102764632, Hao Yuan Te Control Co Ltd.

Reid Gregor, Bruce Andrew W, Busscher Henk J, Van Der Mei Henny C. *Lactobacillus* therapies. 2000, US Patent 6051552, Urex Biotech Inc.

Ren Weidong, Yu Haiqing, Dong Zhiqiang, Yu Tao. Method for preparing microemulsion by mixing pesticide, liquid fertilizer and bioemulsifier. 2012, CN Patent 102826916, Heilongjiang Hetianfengze Agriculture Science and Technology Dev Co Ltd.

Reus Matthias, Syldatk Christoph, Binder Michael, Schuetz Martina. Production of sophorose glyco-lipid bio-surfactants. 1995, DE Patent 19518768, Reus Matthias, Syldatk Christoph, Binder Michael, Schuetz Martina.

Rocha Carlos Ali, Gonzalez Dosinda, Iturralde Maria Lourdes, Lacoa Ulises Leonardo, Morales Fernando Antonio. Production of oily emulsions mediated by a microbial tenso-active agent. 1999, US Patent 5866376, Univ Simon Bolivar.

Rosenberg Eugene and Ron Eliora Z. Bioemulsifiers. 1998, US Patent 5840547, Univ Ramot.

Rosenberg Eugene, Ron Eliora. Novel bioemulsifiers. 1996, WO Patent 9620611, Univ Ramot.

Rosito Filho Jose. Composition for microbial inactivation used in chirurgical apparatus. 2011, WO Patent 2011143726, Lebon Produtos Quimicos Farmaceuticos Ltda, Rosito Filho Jose.

Rothmel Randi K. Method of biodegrading hydrophobic organic compounds, particularly PCB S, in the presence of *Pseudomonas cepacia*. 1996, US Patent 5516688, Envirogen Inc.

Rui Yin, Jing Zhang, Xiangui Lin, Weiwei Liu. Combined restoring method of polycyclic aromatic hydrocarbon contaminated soil. 2011, CN Patent 101972772, Nanjing Inst Soil Sci Cas.

Ryu Mizuyuki and Hirata Yoshihiko. Adsorption suppressing composition containing sophorolipid. 2009, JP Patent 2009275145, Saraya KK.

Ryu Mizuyuki, Igarashi Keisuke, Hirata Yoshihiko, Kihara Koji, Furuta Taro. Extract of *Momordica grosvenori* swingla and skin care composition for external use containing sophorolipid. 2007, JP Patent 2007106733, Saraya KK.

Saeki Takeshi, Komura Hajime, Ohwada Takuji, Kinoshita Mikio. Composition for controlling common scab of agricultural crop containing surfactin. 2008, EP Patent 1929869, Suntory Ltd.

Sai Eikoku, Hidaka Hisao, Koike Takaki, Mitsuzuka Yoshihiro, Harada Hisashi, Ishigami Yutaka, Fukuoka Takao. Thick nanocolloidal gold liquid, fine gold particle, and their manufacturing methods. 2009, JP Patent 2009057627, Ishigami Yutaka, Choi Young Kook, Hidaka Hisao, Koike Takaki, Mitsuzuka Yoshihiro, Harada Hisashi, Fukuoka Takao.

Sakurai Shoji, Imanaka Tadayuki, Morikawa Masaaki. Method for concentrating biosurfactant. 1993, JP Patent H05211876, Nikko Bio Technica Co.

Sakurai Shoji, Manabe Yohei, Imanaka Tadayuki, Morikawa Masaaki. Method for screening microorganism capable of producing biosurfactant and measuring activity of biosurfactant. 1993, JP Patent H05211892, Nikko Bio Technica Co.

Schaffer Steffen, Wessel Mirja, Thiessenhusen Anja, Stein Nadine. Cells and methods for producing rhamnolipids. 2012, CA Patent 2806430, Evonik Goldschmidt Gmbh.

Schiedel Marc-Steffen, Giesen Brigitte, Plantikow Petra, Bellomi Luca, Scheffler Karl-Heinz. Detergent. 2012, DE Patent 102011004771, Henkel AG & Co KGAA.

Schofield Mark H, Thavasi Thavasi Renga, Gross Richard A. Modified sophorolipids for the inhibition of plant pathogens. 2013, US Patent 2013085067, Politechnic Inst Univ New York.

Shaojun Chen, Sheng Zhang, Zhiqiang Sui, Yang Yu. Industrial production method of rhamnolipid biosurfactant dry powder. 2012, CN Patent 102766172, Daqing Vertex Chemical Co Ltd.

Sheehy Alan. Recovery of oil from oil reservoirs. 1992, US Patent 5083610, BWN Live Oil.

Shi Jingang, Yuan Xingzhong, Zeng Guangming. Biosurfactant of rhamnolipid and its application in artificial manure turned from house hold garbage. 2003, CN Patent 1431036, Univ Hunan.

Shi Jingang, Zeng Guangming, Huang Guohe. Lipopeptide biosurfactant and its application in making domestic garbage turn to compost. 2003, CN Patent 1431314, Univ Hunan.

Shi Zhou Ma. Process of extracting rhanolipid as biosurfactant. 2007, CN Patent 1974589, Univ Hunan.

Shigeta Akira, Yamashita Akira. Method of modifying quality of wheat flour product. 1986, JP S61205449, Kao Corp.

Shilo Moshe, Fattom Ali. Cyanobacterium-produced bioemulsifier composition and solution thereof. 1987, US Patent 4693842, Solmat Syst, Yissum Res Dev Co.

Shimizu Yoshihisa, Yasukagawa Tsunetaka, Tashiro Eiichi. Method for assisting production of biosurfactant and method for cleaning contaminated soil. 2003, JP Patent 2003320367, Tashiro Eiichi.

Shoham Yuval, Rosenberg Eugene, Gutnick David L. Enzymatic degradation of lipopolysaccharide bioemulsifiers. 1989, US Patent 4818817, Petroleum Fermentations.

Shoham Yuval, Rosenberg Eugene, Gutnick David l. Enzymatic degradation of lipopolysaccharide bioemulsifiers. 1987, US Patent 4704360, Petroleum Fermentations.

Shulin Yang, Chong Sun, Fengxiang Ai, Xinxin Meng, Xiaosong Niu, Yongbin Wang, Guangrong Meng, Ying Wang. Method for sifting and producing generation agent of dual-rhamnolipid. 2008, CN Patent 101173210, Univ Nanjing Science & Tech.

Simaev Ju M, Bazekina L V, Tukhteev R M, Tujgunov M R, Kalinskij B A. Composition for increase of oil recovery. 1999, RU Patent 2143553, Otkrytoe Aktsionernoe Obshches, Eftjanaja Kompanija Bashneft.

Sloma Alan, Sternberg David, Adams Lee F, Brown Stephen. Methods for producing polypeptides in mutants of Bacillus cells. 1999, US Patent 5958728, Novo Nordisk Biotech Inc.

Sloma Alan, Sternberg David, Adams Lee F, Brown Stephen. Methods for producing polypeptides in surfactin mutants of Bacillus cells. 1998, WO Patent 9822598. Novo Nordisk Biotech Inc.

Soetaert Wim, Van Bogaert Inge. Sophorolipid transporter protein. 2011, WO 2011070113, Univ Gent, Soetaert Wim, Van Bogaert Inge.

Soetaert Wim, De Maeseneire Sofie, Saerens Karen, Roelants Sophie, Van Bogaert Inge. Yeast strains modified in their sophorolipid production and uses thereof. 2011, WO 2011154523, Univ Gent, Soetaert Wim, De Maeseneire Sofie, Saerens Karen, Roelants Sophie, Van Bogaert Inge.

Song Xin Qu. Application of sophorolipid in preparing antibiotic medicine. 2007, CN Patent 101019875, Univ Shandong.

Song Xin Qu. Preparation method of yeast *W. domercqiae* Y2A variation waufa glucolipid crude extract with antineoplastic activity. 2006, CN Patent 1839892, Univ Shandong.

Song Xin Qu. *Wickerhamiella domercqiae* Y2A for producing sophorose lipid and its uses. 2006, CN Patent 1807578, Univ Shandong.

Stanghellini Michael E, Miller Raina Margaret, Rasmussen Scott Lynn, Kim Do Hoon, Zhang Yimin. Microbially produced rhamnolipids (biosurfactants) for the control of plant pathogenic zoosporic fungi. 1997, WO Patent 9725866, Univ Arizona.

Stipcevic Tamara, Piljac Tihana, Piljac Jasenka, Dujmic Tatjana, Piljac Goran. Use of rhamnolipids in wound healing, treatment and prevention of gum disease and periodontal regeneration. 2001, WO Patent 2001010447, Stipcevic Tamara, Piljac Tihana, Piljac Jasenka, Dujmic Tatjana, Piljac Goran.

Sun Wen, Liu Jie, Wang Zejian, Liu Shanshan. Method for purifying sodium surfactin. 2013, CN Patent 103059108, Anhui Province Kingorigin Biotechnology Co Ltd.

Sun Wen, Wang Zejian, Liu Jie, Liu Shanshan. Method for purifying sodium surfactin. 2013, CN Patent 103059107, Anhui Province Kingorigin Biotechnology Co Ltd.

Suzuki Michiko, Kitagawa Masaru, Yamamoto Shuhei, Sogabe Atsushi, Kitamoto Dai, Morita Tomotake, Fukuoka Tokuma, Imura Tomohiro. Activator comprising biosurfactant as the active ingredient mannosyl erythritol lipid. 2009, EP Patent 2055314, Toyo Boseki, Nat Inst of Advanced Ind Scien.

Suzuki Michiko, Kitagawa Masaru, Yamamoto Shuhei, Sogabe Atsushi, Kitamoto Masaru, Morita Tomotake, Fukuoka Tokuma, Imura Tomohiro. Activator comprising biosurfactant as active ingredient. 2008, JP Patent 2008044855, Toyo Boseki, Nat Inst of Adv Ind & Technol.

Tabuchi Takeshi, Kayano Masako. Production of succinyl-trehalose lipid by microorganism. 1987, JP Patent S6283896, Nippon Oils & Fats Co Ltd.

Tagami Hidetoshi, Yoshihara Toru, Tada Kiyotake, Furukawa Hisashi. Composition for dyeing keratinous fiber. 1994, JP Patent H06247833, Kao Corp.

Tamai Masahiro, Tamura Kokichi. Method for producing mannosyl erythritol lipid. 2004, JP Patent 2004254595, Hiroshima Prefecture, Maruzen Pharma.

Tanaka Yoshimasa, Fukui Tomoko, Negi Tahee. Novel *Acinetobacter calcoaceticus* and novel biosurfactant. 1990, EP Patent 401700, Lion Corp.

Teng Zhang, Yanlin Zhou, Yanjie Sun. Wheat protein detergent. 2011, CN Patent 102051270, Beijing Reward Home Care Chemical Co Ltd.

Then Johann, Giani Carlo, Wohner Gerhard, Wink Joachim, Buchholz Rainer, Voelskow Hartmut, Schlingmann Merten, Rapp Knut, Vogel Manfred. Rhamnose-containing polysaccharide, process for its preparation and its use. 1989, EP Patent 0339445, Hoechst AG.

Theodore Spencer John Francis, Patrick Tulloch Alexander, James Gorin Philip Albert. Oil glycosides of sophorose and fatty acid esters thereof. 1967, US Patent 3312684, Canada Nat Res Council.

Tianbo Weng. Oil production method using biosurfactant. 2008, CN Patent 101131086, Shanghai Zhongyou Entpr Group.

Tsutsumi Hisao, Abe Yoshiaki, Inoue Shigeo, Ishida Atsuo. Skin-protecting cosmetic composition. 1982, US Patent 4309447, Kao Corp.

Tsutsumi Hisao, Kawano Junichi, Inoue Shigeo, Hayashi Shizuo. Cosmetic composition. 1981, US Patent 4305961, Kao Corp.

Uzawa Hirotaka, Usui Yasuichi. Sulfated oligosaccharide compound and intermediate therefor. 2000, JP Patent 2000143687, Agency Ind Science Techn.

Uzawa Hirotaka, Usui Yasuichi. Sulfated Oligosaccharide Compound. 2000, JP Patent 2000143686, Agency Ind Science Techn.

Van Haesendonck Ingrid Paula H & Vanzeveren Emmanuel Claude Alb. Rhamnolipids in bakery products. 2004, EP Patent 1415538, Puratos NV.

Vanzeveren Emmanuel Claude Alb. Rhamnolipids in bakery products. 2005, MX Patent PA05004797, Puratos Nv.

Voelskow Hartmut, Schlingmann Merten. Process for the preparation of rhamnose or fucose. 1984, ZA Patent 8305832, Hoechst AG.

Wadgaonkar Raj, Gross Richard A, Butnariu Daniel, Patel Vipul, Somnay Kaumudi. Lung injury treatment. 2008, WO Patent 2008137891, Univ New York State Res Found, Wadgaonkar Raj, Gross Richard A, Butnariu Daniel, Patel Vipul, Somnay Kaumudi.

Wagner Fritz Rapp Peter, Bock Hans, Lindoerfer Walter, Schulz Walther, Gebetsberger Wilhelm. Method and installation for flooding petroleum wells and oil-sands. 1982, CA Patent 1119794, Biotechnolog Forschung GmbH Wintershall AG.

Wagner Fritz, Syldatk Christoph, Matulowic Uwe, Hofmann Hans-Jurgen, Sewe Kai-Udo, Lindorfer Walter. Process for the biotechnological production of rhamnolipids and rhamnolipids with only one beta-hydroxydecanecarboxylic acid moiety in the molecule. 1985, EP Patent 0153634, Wintershall AG.

Wagner Fritz, Ristau Egbert, Li Zu-Yi, Lang Siegmund, Schulz Walther, Hofmann Hans-Juergen, Sewe Kai-Udo, Lindoerfer Walter. Trehalose-lipid-tetraesters. 1987, CA Patent 1226545, Wintershall AG.

Wagner Fritz, Ristau Egbert, Li Zu-Yi, Lang Siegmund, Schulz Walther, Hofmann Hans-Juergen, Sewe Kai-Udo, Lindoerfer Walter. Trehalose-lipid-tetraesters. 1988, US Patent 4720456, Wintershall AG.

Wang Lei, Liu Rulin, Feng Lu, Liang Fenglai. *Geobacillus thermodenitrificans* as well as the screening method and the uses thereof. 2009, US Patent 2009148881, Univ Nankai.

Wei Yu-Hong, Chang Jo-Shu, Xin Wen-Fu, Cheng Chieh-Lun. Method for producing rhamnolipid and medium of the same. 2010, TW Patent 201009080, Univ Nat Cheng Kung.

Weidong Wang, Caifeng Li, Ximing Li, Yongting Song, Yan Jiang, Zhiyong Song, Gongze Cao, Jing Wang, Dengting Xu, Guangjun Gao, Liaoyuan Guo. Method for improving crude oil recovery ratio by utilization of industrial sewage and waste gas. 2009, CN Patent 101503956, Oil Ext Tech Institute Sinopec.

Weimer Bart C. Compositions and methods for using syringopeptin 25A and rhamnolipids. 2008, US Patent 2008261891, Univ Utah State.

Weimin Cai, Qiuzhuo Zhang. Online production method of rhamnolipid biosurfactant in cellulose hydrolyzation. 2009, CN Patent 101538604, Univ Shanghai Jiaotong.

Weimin Cai, Qiuzhuo Zhang. Two-stage cohydrolysis method of cellulase. 2009, CN Patent 101358226, Univ Shanghai Jiaotong.

Wen Sun, Guohong Gong, Jifeng Fan, Mei Wang, Junjun Li, Jie Liu, Mei Liu, Shiqi Ding. Fermentation device. 2011, CN Patent 102061282, Anhui Province Kingorigin Biotechnology Co Ltd.

Wenjie Xia, Hanping Dong, Li Yu, Lixin Huang, 2011a. Method for selecting rhamnolipid-producing bacterium and used medium. CN Patent 101948787, Petrochina Co Ltd.

Wenjie Xia, Hanping Dong, Li Yu, Lixin Huang, Qingfeng Cui. *Pseudomonas aeruginosa* for producing rhamnolipid with high yield and application thereof. 2011b, CN Patent 101948786, Petrochina Co Ltd.

Witek-Krowiak Anna, Witek Joanna. Method for the rhamnolipid membrane separation. 2011, PL Patent 390696, Politechnika Wroclawska.

Wullbrandt Dieter, Giani Carlo, Brakemeier Andreas Dch, Lang Siegmund, Wagner Fritz. New lipids of glucose and sophorose, their preparation and their use. 1996, EP Patent 745608, Hoechst AG.

Wuxing Liu, Yongming Luo, Xiaobing Wang, Dianxi Wang. *Bacillus amyloliquefaciens* BZ6-1 strain for producing biosurfactant and application thereof in oil sludge elution. 2011, CN Patent 101935633, Nanjing Inst of Soil Chinese Academy of Sciences.

Xiang Xiao, Chuncan Si, Ying Lin, Daolin Du, Danming Cao, Mingna Wang. Remediation method of soil polycyclic aromatic hydrocarbon by combining surfactant producing bacteria and mycorrhiza. 2012, CN Patent 102652957, Univ Jiangsu.

Xiangui Lin, Rui Yin, Jing Zhang. Method for remediating polycyclic aromatic hydrocarbon-polluted soil by jointly enhancing plants through edible fungus residue and biosurfactant. 2010, CN Patent 101780465, Nanjing Inst Soil Sci Cas.

Xiaodan Mei, Bufei Fu, Ping Wang, Jing Zhao, Ping Yang, Chen Chen, Ying Liu, Guofeng Du, Wenjing Liu, Runhai Liu. *Rhodococcus equi* strain and use thereof in petroleum microorganism yield increase. 2011, CN Patent 101935630, Dalian Biteomics Inc.

Xiaodan Mei, Bufei Fu, Ying Liu, Jing Zhao, Ping Wang, Chen Chen, Ping Yang, Guofeng Du, Runhai Liu, Wenjing Liu. Composite bacterial agent and biological method for treating flow-back fracturing fluid to obtain oil displacement active water. 2011, CN Patent 101935615, Dalian Biteomics Inc.

Xiaolan Liu, Guangming Zeng, Hua Zhong, Haiyan Fu, Zhifeng Liu, Yaoning Chen, Haiyi Huang, Jing Wang. Method for enhancing yield of rhamnolipid produced by copper green pseudomonas. 2008, CN Patent 101182560, Univ Hunan.

Xin Song, Xiaojing Ma. Method for producing sophorolipid through fermentation of lignocellulose material. 2012, CN Patent 102492753, Univ Shandong.

Xingzhong Yuan, Lili Jiang, Guangming Zeng, Zhifeng Liu, Yaoning Chen. Heavy metal biological absorbent, preparation method thereof and application in treating cadmium-containing wastewater. 2011, CN Patent 102151551, Univ Hunan.

Xingzhong Yuan, Xin Peng, Huajun Huang, Lingzhi Guo, Guangming Zeng. Method for extracting and purifying lignin peroxidase by using reverse micelles. 2012, CN Patent 102533688, Univ Hunan.

Xinzheng Qin, Xinping Yang, Xian Tang, Huitao Zhang, Xinqiang Hou, Aihemaiti Guli, Jing Chen. Fermentation production technique for rhamnolipid biological surface activator. 2008, CN Patent 101265488, Urumchi United Biotechnology C.

Yalpani Manssur. Halogenated emulsans. 2004, US Patent 2004171128.

Yamamoto Shuhei, Kitagawa Masaru, Kamikura Yoshiko. Cosmetic composition containing biosurfactant. 2010, JP Patent 2010018558, Toyo Boseki.

Yamamoto Shuhei, Yanagiya Shusaku, Nishiya Yoshiaki. Composition comprising biosurfactant and polyhydric alcohol. 2011, JP Patent 2011001312, Toyo Boseki.

Yamamoto Shuhei, Yanagiya Shusaku, Nishiya Yoshiaki. Composition containing biosurfactant and surfactant. 2011, JP Patent 2011001313, Toyo Boseki.

Yamamoto Shuhei, Yanagiya Shusaku, Nishiya Yoshiaki. Composition containing biosurfactant and antiinflammatory agent. 2011c, JP Patent 2011001314, Toyo Boseki.

Yanagisawa Satohiro, Kawano Shigeru, Yasohara Yoshihiko. Method for producing sophorose lipid. 2011, EP Patent 2351847, Kaneka Corp.

Yanfang Jin, Yumei Li, Yanling Ma, Libin Han, Yang Sun, Shaojun Chen. Culture medium prescription for the industrial production of rhamnolipid fermentation liquor. 2008, CN Patent 101173238, Daqing Vertex Chemical Co Ltd.

Yang Lin, Li Qianqiu, Li Huabin. Rhamnolipid biological surfactant its preparation process and use in tertiary oil recovery. 2000, CN Patent 1275429, Daqing Petroleum Admin.

Yang Shulin Ning. Method for preparing rhamnolipid. 2007, CN Patent 1891831, Nanjing Technology Univ.

Yarong Fu, Huayun Ma, Huaiyu Hu, Dongqing Li, Hui Hu, Lixia Fu, Jin Cao et al. Pressure reducing and injection increasing agent containing lipopeptide biological enzyme. 2012, CN Patent 102690640, Petrochina Co Ltd.

Yin Xihou, Nie Maiqian, Shen Qirong. Rhamnolipid biosurfactant from *Pseudomonas aeruginosa* strain NY3 and methods of use. 2011, US 20110306569, Univ Oregon State, Oregon State.

Yoneda Tadashi, Miyota Yoshiaki, Furuya Kazuo, Tsuzuki Toshi. Production process of surfactin. 2002, WO Patent 2002026961, Showa Denko KK, Yoneda Tadashi, Myota Yoshiaki, Furuya Kazuo, Tsuzuki Toshi.

Yoneda Tadashi, Miyota Yoshiaki, Furuya Kazuo. Production process of subtilin surfactin. 2004, CN Patent 1466626, Showa Denko KK.

Yonghu Xu, Ruiying Li, Hua Guo, Yongle Wang, Yun Liu, Yajun Hu, Xiaoqiong Wang et al. Method for producing sophorose ester through fermenting waste molasses and waste glycerin. 2012, CN Patent 102329833, Tianjin Sf Bio Ind Bio Tech Co Ltd.

Yonghu Xu, Yun Liu, Ruiying Li, Hua Guo, Yajun Hu, Hui Han, Xiaoqiong Wang, Ting Fan. Method for producing cellulase by adding sophorose lipid for inducing fermentation. 2011, CN Patent 102250859, Tianjin Sf Bio Ind Bio Tech Co Ltd.

Yoshida Koji, Tamiya Eiichi, Tsubouchi Naoki, Ishio Ichihiro. Method for treating leaked oil. 2003 JP Patent 2003183635, Gate KK.

Yu Zebin, Huang Jun, Hu Xiao. Method for electrically repairing heavy metal As (Arsenic) contaminated soil by using rhamnolipid. 2013, CN Patent 102886374, Univ Guangxi.

Yu Zhang, Guopei Zhang, Shaohua Zhang, Haiyan Liu. Method for preparing rhamnolipid by utilizing microorganism fermentation. 2009, CN Patent 101613725, Hebei Xinhe Biochemical Co Ltd.

Yu Zhang, Guopei Zhang, Shaohua Zhang, Xuezhen Liu, Jingmin Jia. Post extraction method of rhamnolipid. 2009, CN Patent 101613381, Hebei Xinhe Biochemical Co Ltd.

Yu Zhang, Yanqin Deng, Min Yang, Shanmu Hei, Wenzhou LV. Yeast strain for producing biosurfactant and application thereof. 2012, CN Patent 102766580, Res Ct Eco Environmental Scien.

Yuan Huang, Chunhui Chen, Tingting Fan, Zhirong Zhang. Novel solid lipid nanoparticle medicament delivery system for protein-loaded medicaments. 2011, CN Patent 102106821, Univ Sichuan.

Yujiang Shen, Yanda Sun, Guojun Li, Zhu Wang, Liru Zhang, Baoyan Duan. Industrialized preparation method of a lipopeptide biosurfactant. 2012, CN Patent 102373258, Daqing Huali Energy Biolog Technology Co Ltd.

Yujiang Shen, Yanda Sun, Guojun Li, Zhu Wang, Liru Zhang, Baoyan Duan. Lipopetide biosurfactant oil extraction agent for improving extraction rate of crude oil. 2012, CN Patent 102504789, Daqing Huali Energy Biolog Technology Co Ltd.

Yum Kyu-Jin. Microbial materials for degradation of oils and toxic chemicals. 2008, US Patent 2008032383, Yum Kyu-Jin.

Yumei Li, Yanfang Jin, Yanling Ma, Chao Shen, Yujiang Shen, Tingting Kang, Fang Wang, Wei Li, Yuanyang Xu, Shaojun Chen. Industrial preparation method of rhamnolipid biological fermentation liquor. 2008, CN Patent 101177696, Daqing Vertex Chemical Co Ltd.

Yun Han Dae, Hong Su Young, Cho Kye Man. Rapid production method of kanjang and chungkookjang containing high level surfactin using *Bacillus pumilus* HY1 strain. 2008, KR Patent 20080070127, Ind Academic Coop.

Zajic James E, Gerson Donald F. Hydrocarbon extraction agents and microbiological processes for their production. 1981, CA Patent 1114759, Canadian Patents Dev.

Zajic James E, Gerson Donald F. Hydrocarbon extraction agents and microbiological processes for their production. 1987, US Patent 4640767, Canadian Patents Dev.

Zajic James E, Knettig Eva. Emulsifying agents of microbiological origin. 1976, CA Patent 990668, Canadian Patents Dev.

Zajic James E, Knettig Eva. Emulsifying agents of microbiological origin. 1976, US Patent 3997398, Canadian Patents Dev.

Zajic James E, Gerson Donald F, Gerson Richard K, Panchal Chandrakant J. Microbiological production of novel biosurfactants. 1982, US 4355109, Zajic James E, Gerson Donald F, Gerson Richard K, Panchal Chandrakant J.

Zerkowski Jonathan, Solaiman Daniel, Ashby Richard D, Foglia Thomas. Charged sophorolipids and sophorolipid containing compounds. 2007, US Patent 2007027106, Zerkowski Jonathan, Solaiman Daniel, Ashby Richard D, Foglia Thomas.

Zhang Gang Xi. Method for producing extracellular mucopoly-saccharide using rhamnose *Lactobacillus* and product. 2006, CN Patent 1781950, Tianjin City Ind Microbe I.

Zhang Min-Zheng. Apparatus and method for producing strain of rhamnolipid. 2009, TW Patent 200909586, Univ Nat Cheng Kung.

Zhaoxin Lu, Huigang Sun, Xiaomei Bie, Fengxia Lu, Yu Wang, Lin Cao. Promotor replacement method for improving volume of production of *Bacillus subtilis* surfactin. 2009, CN Patent 101402959, Univ Nanjing Agricultural.

Zhenshan Deng, Li Jun, Yongjie Su. Method for recovering petroleum-polluted water source by using composite bacterial solution. 2010, CN Patent 101851027, Yan An Inst of Microbiology.

Zhifeng Liu, Guangming Zeng, Xingzhong Yuan, Hua Zhong, Jianbing Li, Meifang Zhou, Xiaoling Ma, Lu Huang. Method for removing phenol in waste water by enzyme catalysis. 2011, CN Patent 102020347, Univ Hunan.

Zhiqiang Cai, Guanghua Yang, Xiaolin Zhu, Liang Li, Yucai He, Xiyue Zhao, Eryang Li. Method for preparing lipopeptid biosurfactant by utilizing high temperature *Bacillus* spp. 2012, CN Patent 102399847, Univ Changzhou.

Zhuowei Cheng, Jianmeng Chen, Yifeng Jiang, Runye Zhu, Xinna Gu. Method of generating biosurfactant by using *Pseudomonas*. 2012, CN Patent 102732573, Univ Zhejiang Technology.

12 Industrial Applications of Biosurfactants

Letizia Fracchia, Chiara Ceresa, Andrea Franzetti,
Massimo Cavallo, Isabella Gandolfi, Jonathan Van Hamme,
Panagiotis Gkorezis, Roger Marchant, and Ibrahim M. Banat

CONTENTS

12.1 Introduction 245
12.2 Biosurfactants: A Real Prospect for Industrial Use? 246
    12.2.1 Biomedical and Pharmaceutical Applications..247
    12.2.2 Uses in the Cosmetics Industry 249
    12.2.3 Uses in Detergent Formulations 251
12.3 Biosurfactant Potential in the Food Industry..252
12.4 Biosurfactants in the Textile Industry: Opportunities for Future Exploitation ...254
12.5 Application in Enhanced Oil Recovery 256
12.6 Biosurfactants in Agriculture 257
12.7 Use in Other Industries..259
12.8 Conclusions..259
References..260

12.1 INTRODUCTION

Chemically synthesized surfactants, derived from either petrochemical or oleochemical sources, are important constituents of many everyday products and are integral to numerous industrial, agricultural, and food-related processes (Desai and Banat 1997). The industrial demand for surfactants is high (more than 13 million tons per annum), but in view of their environmental sustainability, many companies are trying to replace some or all of their chemical surfactant components with similar compounds of natural origin principally produced by microorganisms from sustainable feedstock (Marchant and Banat 2012a). These molecules, commonly known as biosurfactants and bioemulsifiers, have advantages over chemical surfactants some of which are biodegradability, low toxicity, biocompatibility, and digestibility in addition to improved stability at relatively high temperature and in adverse environments with the possibility of being produced from cheap raw materials (Makkar et al. 2011). Another striking advantage is the ability to modify their chemical composition through genetic engineering or the use of biological and biochemical techniques to alter metabolic end products, thus tailoring them to meet specific functional requirements.

During the past few years, several books and reviews have been dedicated to the subject of *Biosurfactants* (Sen 2010; Soberón-Chávez 2011; Marchant and Banat 2012a,b) outlining and predicting the future role and importance that these natural molecules may have in the search for new bioactive molecules and/or sustainable alternatives to chemical surfactants.

As has been described in earlier chapters, biosurfactants are amphiphilic compounds produced mostly on microbial cell surfaces, or excreted extracellularly and characterized by having both hydrophobic and hydrophilic moieties that confer the ability to accumulate at or orient themselves on interfaces, thus reducing surface tension and interfacial tension of liquids. They have the ability to form molecular aggregates including micelles. The micellar aggregation of biosurfactants is initiated at the critical micelle concentration typically at concentrations from 1 to 200 mg/L and interestingly, about 10- to 40-fold lower than that for most chemical surfactants (Martinotti et al. 2013).

Biosurfactants are generally classified according to their chemical composition into two major groups. The low-molecular weight, which can lower surface tension and interfacial tension effectively; these include glycolipids such as rhamnolipids, mannosylerythritol lipids (MELs), and sophorolipids or lipopeptides such as the well-known surfactin and polymyxin. The high molecular weight polymers, which are more efficient emulsion-stabilizing agents, include lipoproteins, lipopolysaccharides, and amphipathic forms (Smyth et al. 2010a,b).

Biosurfactants are naturally synthesized by a multiplicity of microorganisms on different substrates such as sugars, oils, alkanes, and wastes. Lipopeptides are synthesized mainly by *Bacillus* spp. and other species such as *Brevibacterium aureum*, *Nocardiopsis alba*, glycolipids by species of *Pseudomonas*, *Burkholderia*, *Mycobacterium*, *Rhodococcus*, *Arthrobacter*, *Nocardia*, *Gordonia*, the yeasts *Starmerella*, *Yarrowia*, and *Pseudozyma*, and fungi such as *Ustilago scitaminea*, phospholipids by *Thiobacillus thiooxidans*, lipid–polysaccharide complexes by *Acinetobacter* species (Desai and Banat 1997; Franzetti et al. 2009; Banat et al. 2010; Morita et al. 2011).

The diverse functional properties of biosurfactants including emulsification, wetting, foaming, cleansing, phase separation, surface activity, and reduction in viscosity of heavy liquids, such as crude oil, make them appropriate for many industrial and domestic application purposes (Franzetti et al. 2010; Perfumo et al. 2010a; Satpute et al. 2010).

The versatility and diversity of these molecules has led to increased interest in industrial and environmental applications such as bioremediation, soil washing, enhanced oil recovery, and other general oil-processing and related industries (Perfumo et al. 2010b). Other potential commercial applications in additional industries including paint, cosmetics, textile, detergent, agrochemical, food, and pharmaceutical industries have also emerged (Banat et al. 2000, 2010).

In this chapter, the actual industrial applications of microbial biosurfactants will be surveyed with special emphasis on commercial products containing these molecules. Moreover, some patents concerning biosurfactant exploitation and production will be reviewed.

12.2   BIOSURFACTANTS: A REAL PROSPECT FOR INDUSTRIAL USE?

Increasing environmental concerns, the advancements in biotechnology, and the emergence of increasingly stringent laws have led to biosurfactants becoming a credible alternative to the chemical surfactants available on the market (Henkel et al. 2012). The number of such related publications and patents involving biosurfactants has recently increased considerably (Müller et al. 2012).

In 2006, Shete and coauthors mapped the patents on biosurfactants and bioemulsifiers and reported 255 patents covering various aspects issued worldwide. The highest number of patents was granted to petroleum-related industries (33%), followed by cosmetics (15%), pharmaceutical (12%), and bioremediation (11%), with sophorolipids, surfactin, and rhamnolipids representing the largest portion of the patents as confirmed by Reis et al. (2013).

The application of biosurfactants in industry is fast becoming a reality; however, several problems remain before further widespread use can be envisaged. These problems relate to yield and cost of production, including downstream processing and the time and effort needed for tailoring selected molecules to specific applications (Marchant and Banat 2012a,b).

The biosurfactant industry and market trends have recently been the object of intensive evaluation carried out by Transparency Market Research™. According to this study, the global biosurfactant market amounted to USD 1735.5 million in 2011 and is expected to reach USD 2210.5 million in 2018. Europe is expected to enjoy 53.3% of the global biosurfactant market revenue share in 2018 followed by North America.

A number of well-known surfactants vendors such as BASF-Cognis and Ecover have already ventured into the biosurfactant market, where BASF-Cognis is the leader with over a 20% share of the market in 2011. Other major producers include Ecover, Urumqi Unite, Saraya, and MG Intobio. The top three biosurfactant vendors accounted for more than 60% of the market share in 2011 (http://www.transparencymarketresearch.com/biosurfactants-market.html).

In the following paragraphs, the current industrial applications in the areas of healthcare, cosmetics, and detergents will be surveyed, and the presence of commercial products and patenting concerning these market sectors will be highlighted.

12.2.1 BIOMEDICAL AND PHARMACEUTICAL APPLICATIONS

The numerous and interesting physicochemical and biological features of biosurfactants have led to a wide range of potential pharmaceutical and biomedical applications. In particular, their ability to destabilize cell membranes by disturbing their integrity and permeability (Ortiz et al. 2009; Zaragoza et al. 2009; Sánchez et al. 2010), and their ability to affect microorganism adhesion by modifying surface characteristics are valuable properties for biomedical application. Furthermore, some experimental results in the literature suggest that they are nontoxic or less toxic than synthetic surfactants (Edwards et al. 2003; Muthusamy et al. 2008; Cochis et al. 2012).

Literature reports on potential biomedical applications of biosurfactants are numerous and have been described in many reviews (Rodrigues et al. 2006; Seydlová and Svobodová 2008; Banat et al. 2010; Rodrigues and Teixeira 2010; Gharaei-Fathabad 2011; Kalyani et al. 2011; Seydlová et al. 2011; Fracchia et al. 2012; Cortés-Sánchez et al. 2013). However, although many patents have been issued concerning biosurfactant usage for health improvement, real applications in the biomedical and pharmaceutical industry remain quite limited.

The search for novel antimicrobial agents to combat the emerging pathogens and to elude increased resistance shown by others against existing antimicrobial drugs has drawn attention to natural products with different modes of action, as suitable alternatives or adjuvants to some synthetic medicines (Banat et al. 2010).

Lipopeptides form the most widely reported class of biosurfactants with antimicrobial activity due to their ability to disrupt lipid membranes. Studies on lipopetide mechanisms of action have shown that pore formation in membranes occurs after lipopeptide oligomer binding, some of which are Ca^{2+}-dependent multimers (Scott et al. 2007). These pores may cause trans-membrane ion influxes, which result in membrane disruption and cell death (Ostroumova et al. 2010; Mangoni and Shai 2011). These properties may lead to applications in the pharmaceutical industry, where lipopeptides have been used when conventional antibiotics were no longer working against resistant bacteria or fungi (Mandal et al. 2013).

Polymyxin A, isolated from the soil bacterium *Bacillus polymyxa*, was the first lipopeptide discovered with an antimicrobial function (Jones 1949). Similarly, surfactin, fengycin, iturin, bacillomycins, and mycosubtilins produced by *Bacillus subtilis* (Vater et al. 2002); lichenysin, pumilacidin, and polymyxin B produced by *B. licheniformis, Bacillus pumilus,* and *B. polymyxa,* respectively (Naruse et al. 1990; Yakimov et al. 1995; Grangemard et al. 2001; Landman et al. 2008); daptomycin, a cyclic lipopeptide from *Streptomyces roseosporus* (Baltz et al. 2005) and viscosin, a cyclic lipopeptide from *Pseudomonas* (Saini et al. 2008) are all known antimicrobial lipopeptides.

Some lipopeptides have reached a commercial antibiotic status, like daptomycin (Robbel and Marahiel 2010), echinocandins caspofungin (Ngai et al. 2011), micafungin (Emiroglu 2011), and anidulafungin (George and Reboli 2012).

Daptomycin (Cubicin®, Cubist Pharmaceuticals) is a branched cyclic lipopeptide antibiotic of nonribosomal origin, isolated as a member of an antibiotic complex from cultures of *S. roseosporus* (Robbel and Marahiel 2010). Studies intended to elucidate its mechanisms of action hypothesized that the daptomycin/calcium complex interacts with the membrane phosphatidylglycerol head groups and undertakes a second conformational alteration that induces oligomerization of the membrane, leading to membrane penetration (Muraih et al. 2011).

Daptomycin was approved in 2003 for the nontopical treatment of skin structure infections caused by Gram-positive pathogens, including methicillin-resistant *Staphylococcus aureus* and in 2006 for the treatment of bactremia and endocarditis caused by *S. aureus* strains and methicillin-resistant *S. aureus*. Daptomycin shows potent antibacterial activity against other clinically relevant resistant pathogens, such as vancomycin-resistant *Enterococci*, glycopeptide-intermediate-susceptible *S. aureus* (GISA), coagulase-negative *Staphylococci* (CNS), and penicillin-resistant *Streptococcus pneumoniae* (Tally et al. 1999).

The echinocandins caspofungin, micafungin, and anidulafungin are synthetically modified lipopeptides, originally derived from the fermentation broths of various fungi, respectively, *Glarea lozoyensis*, Coleophomaempetri, and *Aspergillus nidulans* (Wagner et al. 2006). Echinocandins can inhibit fungal cell wall formation (Schneider and Sahl 2010a,b) by specific and noncompetitive inhibition of the enzyme β-(1,3)-D-glucan synthase. The lack of β-(1,3)-D-glucan, an essential carbohydrate component for the fungal cell wall, leads to cell wall deterioration and consequent cell death (Yao et al. 2012). Their major characteristics include low toxicity, rapid fungicidal activity against most isolates of *Candida* spp and predictable favorable kinetics allowing a once daily dose. In addition to *Candida* spp, their inhibitory spectrum includes *Aspergillus* spp and *Pneumocystis carinii*, but not *Cryptococcus neoformans* (Denning 2002).

Caspofungin was the first licensed echinocandin product. It has been approved since 2001 in over 80 countries worldwide for the treatment in adults of esophageal and invasive candidiasis, invasive aspergillosis in patients refractory to or intolerant of standard therapy and empirical therapy of suspected fungal infections in neutropenic patients. Since 2008, caspofungin has been approved for the same indications in pediatric patients (Ngai et al. 2011). Micafungin, as well, is used in immune compromised children to combat invasive fungal infections by *Candida* and *Aspergillus* species (Emiroglu 2011). Anidulafungin is the newest antifungal drug. It is well tolerated and its clinical efficacy has been demonstrated in the treatment of candidemia and other forms of candidiasis. Moreover, unlike micafungin and caspofungin, anidulafungin does not undergo any degree of hepatic metabolism and does not require dosage adjustment in renal or hepatic impairment (George and Reboli 2012).

Lipopeptides with antimicrobial activity suitable for the treatment and prevention of microbial infections were also described in several preparations (Deleu et al. 2004; Cameron et al. 2005; Neuhof et al. 2006; Fardis et al. 2007; Hill et al. 2008) showing potential for pharmaceutical applications. Beside antimicrobial action, lipopeptides were claimed to have activity toward T lymphocytes and were patented for treating or preventing skin diseases and diseases of the mucosa, in topical medicinal products (Groux et al. 2005). The lipopeptides viscosin and analogues have been patented as antibacterial, antiviral, antitrypanosomal therapeutic compounds that inhibit the growth of *Mycobacterium tuberculosis*, *Herpes simplex* virus 2, and/or *Trypanosoma cruzi* (Burke et al. 1999).

A relatively new area of application for biosurfactants has appeared with the suggestion that rhamnolipid biosurfactants may aid in wound healing (Stipcevic et al. 2006; Piljac et al. 2008). The replacement of chemical surfactants, incorporated into skin care products, with biosurfactants could have the added benefit of aiding the healing of minor skin lesions (Marchant and Banat 2012a).

Compositions containing rhamnolipids have been invented for inducing re-epithelization in adult skin tissue, to provide wound healing with reduction of fibrosis, and for the treatment of burn shock. Interestingly, these rhamnolipids compositions were also indicated in the treatment and prevention of atherosclerosis, rejection of transplanted organ, and in the treatment of depression and schizophrenia (Piljac and Piljac 1999). More recently, a patent has been released about antimicrobial biosurfactant peptides produced by probiotic strains such as *Streptococcus* spp, *Bifidobacterium* spp, and lactic acid bacteria able to selectively bind to collagen and inhibit infections around wounds at the site of implants and biofilms associated with infections in mammals (Howard et al. 2002).

Paradigm Biomedical Inc. (the USA), however, is committed to the development of pharmaceutical products derived from rhamnolipids. The company has reported developing an array of applications for skin disorders such as lichen planus, seborrheic dermatitis, and atopic eczema, as well as specialized rhamnolipid applications for treating wound healing and burns.

In addition to their direct action against pathogens, biosurfactants are able to interfere with biofilm formation, modulating microbial interaction with surfaces, and, consequently, their adhesion (Rodrigues et al. 2007; Quinn et al. 2013). Microbial biofilm formation on medical devices is an important and mostly hazardous occurrence, especially as bacteria within such biofilms usually become highly resistant to antibiotics and adverse environmental challenges. Current biofilm preventive strategies are essentially aimed at coating medical surfaces with antimicrobial agents, a process not always successful (Francolini and Donelli 2010). From this point of view, it could be useful to increase the efficacy of known antibiotics and biocides with alternative strategies aimed at decreasing bacterial adhesion and reducing the biofilm populations on medical device surfaces.

Rivardo et al. (2011) observed that a lipopeptide biosurfactant produced by the strain *B. subtilis* V9T14 in association with antibiotics led to a synergistic increase in the efficacy of antibiotics against pathogenic *Escherichia coli* CFT073 biofilm formation and, in some combinations, to the total biofilm eradication of this pathogenic strain. An international patent on this application has also been published (Ceri et al. 2010); the biosurfactant composition can be used in combination with biocides, as an adjuvant, to aid in preventing formation and/or eradicating bacterial growth planktonically or as a biofilm on biotic and abiotic surfaces.

Lactobacillus biosurfactants have also been patented as inhibitors of adherence and colonization of bacterial pathogens on medical devices, in particular for preventing urogenital infection in mammals (Bruce et al. 2000).

Sugar-based biosurfactants such as MELs and rhamnolipids have also been reported to lead to molecular self-assembly (SA), which is the spontaneous and reversible organization of molecular units into ordered structures by noncovalent interactions without the use of external force or stimulus (Worakitkanchanakul et al. 2008; Kitamoto et al. 2009; Rehman et al. 2010). MELs have been of recent interest in the area of nanotechnology for gene transfection and drug-delivery applications (Ueno et al. 2007a,b; Kitamoto 2008). MELs have been successfully employed in the enhancement of the gene transfection efficiency of cationic liposomes (Igarashi et al. 2006; Ueno et al. 2007a; Kitamoto et al. 2009; Nakanishi et al. 2009) and the mechanism of action was suggested to be the induction of highly efficient membrane fusion between liposomes and the plasma membrane of the target cells and subsequent DNA release (Ueno et al. 2007b; Inoh et al. 2010). In 1988, rhamnolipid liposomes were patented as drug-delivery systems, useful as microcapsules for drugs, proteins, nucleic acids, dyes, and other compounds, as biomimetic models for biological membranes and as sensors for detecting pH variations. These novel liposomes were described as safe and biologically decomposable, with a suitable affinity for biological organisms and with extended stability and shelf life (Gama et al. 1990). More recently, rhamnolipids and sophorolipids have been mixed with lecithins to prepare biocompatible micro-emulsions for cosmetic and drug-delivery applications (Nguyen et al. 2010). Another recent invention is directed to polymeric acylated biosurfactants that can self-assemble or auto-aggregate into polymeric micellar structures useful in topically applied dermatologic products containing pharmaceutical ingredients such as antimicrobial agents, anti-acne agents, and external analgesics (Owen et al. 2007).

12.2.2 Uses in the Cosmetics Industry

Chemically synthesized surfactants are commonly used in the cosmetic industry for their detergency, wetting, emulsifying, solubilizing, dispersing, and foaming properties. Long-term use of such chemicals may have adverse effects on the environment and humans. Using natural alternative compounds is of particular interest because of the current demand for environmental and animal friendly natural cosmetics (Corley 2007). During the last decades, biosurfactant-containing marketable products and patents have been reported for application in the healthcare and cosmetic industries (Lourith and Kanlayavattanakul 2009; Banat et al. 2010; Kanlayavattanakul and Lourith 2010; Morita et al. 2013).

Glycolipid biosurfactants such as sophorolipids, rhamnolipids, and MELs are very attractive in these fields. Sophorolipid derivatives containing propylene glycol are applied in cosmetics as moisturizers or softeners utilizing their hygroscopic properties (Faivre and Rosilio 2010). A product containing 1 mol of sophorolipid and 12 mol of propylene glycol, showing excellent skin compatibility, is used commercially as a skin moisturizer (Yamane 1987).

Sophorolipids are produced commercially by Kao Co. Ltd. as humectants for cosmetic makeup brands such as Sofina. The product has application in lipstick and as moisturizer for skin and hair products (Inoue et al. 1979a,b), eye shadow as well as in compressed powder cosmetics and in aqueous solutions (Kawano et al. 1981).

The French company Soliance (http://www.groupesoliance.com) develops and sells sophorolipid-based active ingredients for the cosmetic industry. Sophorolipids, produced from rapeseed oil fermentation, are contained in Sopholiance S, an antibacterial and sebo-regulator formulation for applications in deodorants, face cleansers, shower gels, make up removers, and for the treatment of acne-prone skin and in Sophogreen, a high-performance bio-solubilizer.

The Korean biotech company MG Intobio Co. (http://mgintobio.en.makepolo.com) commercializes Sopholine cosmetics, a brand name for functional soaps containing sophorolipids specific for acne treatments.

Patents concerning sophorolipids being used as moisturizing agents for the amelioration of skin physiology, skin restructuring and repair, as an activator of macrophages, and as an agent in fibrinolytic healing, for desquamating and depigmenting processes have also been reported (Maingault, 1999). A germicidal composition containing fruit acids, a surfactant, and a sophorolipid biosurfactant, able to totally eradicate *E. coli, Salmonella,* and *Shigella* within 30 s, has also been patented for cleaning fruits, vegetable, skin, and hair (Pierce and Heilman 2001).

Rhamnolipids have also been claimed to be suitable in healthcare products in a number of different formulations, such as insect repellents, antacids, acne pads, antidandruff products, contact lens solutions, deodorants, nail care products, and toothpastes (Maier and Soberon-Chavez 2000). They also are reported to have low skin irritation (Haba et al. 2003), and some rhamnolipid-containing formulations have been patented as antiwrinkle and antiaging products (Piljac and Piljac 1999). They have also been suggested for use as antimicrobial agents in commercial skin care cosmetics, in personal hygiene and care products (sprays, soaps, shampoo, and creams) and for animal cleaning (Desanto 2008). Other applications for use in making liposomes and emulsions, both important in the cosmetic industry, have also been patented (Ishigami and Suzuki 1997; Ramisse et al. 2000).

MELs, produced by *Pseudozyma* yeasts, have recently gained attention due to their favorable production conditions, structural diversity, self-assembling properties, and versatile biochemical functions. In particular, their potential use as cosmetic ingredients such as moisturization of dry skin, repair of damaged hair, activation of fibroblast and papilla cells and protective effects in skin cells, and as antioxidant have been demonstrated (Morita et al. 2013).

MELs have a similar amphiphilic structure to that of ceramide-3, an essential component of the intracellular lipids of the stratum corneum, thus suggesting possible ceramide-like skin care properties (Kitamoto et al. 2009). Yamamoto et al. (2012), using a three-dimensional cultured human skin model, recently demonstrated that MELs had recovery effects on sodium dodecylsulfate (SDS)-damaged cells that were comparable to those of ceramide-3. Moreover, MEL-B treatment of human forearm skin considerably increased the stratum corneum water content and suppressed the perspiration on the skin surface. These reports highlight potential uses in skin moisturizing. Patents for skin care formulations containing MELs as the surfactant in antiwrinkle cosmetics (Kato and Tsuzuki 2008) and as the active ingredient in skin care cosmetics for roughness (Masaru et al. 2007) and another as an ingredient in an antiaging agent have been issued in Japan (Suzuki et al. 2010). More recently, the Japanese company, Daito Kasei Kogyo Co., Ltd., started to provide a new foundation powder containing metal oxide particles coated with MELs as a product with excellent moisture retention properties (Morita et al. 2013).

The consumption of cosmetics-containing lipopeptides is also increasing as a result of their exceptional surface properties and diverse biological activities that include antiwrinkle and moisturizing activities (Kanlayavattanakul and Lourith 2010) and their broad spectrum antimicrobial activity (Sun et al. 2006). Surfactin is the most effective in terms of interfacial properties and is available in several Japanese cosmetic products including a skin preparation, invented by SHOWA DENKO, comprising tocopherol and ascorbic acid derivatives with lipopeptides (Kato et al. 2005).

Lipopeptides were used as emulsifiers and claimed to have low skin irritation suitable for external skin preparations such as transparent cosmetics with a sequestering function (Yoneda et al. 1999). Other major applications for lipopeptides have been in anti-wrinkle cosmetics (Guglielmo and Montanari 2003), in the treatment and prevention of skin stretch marks (Montanari and Guglielmo 2008) and in cleansing products exhibiting excellent washability, with extremely low skin irritating property (Yoneda 2006). Lipopeptides have been formulated by SHOWA DENKO in several dosage forms that include oil in water emulsified compositions with moisture retention and emollient properties (Yoneda et al. 2005) for skin care cosmetics.

Bioemulsifiers produced by *Acinetobacter calcoaceticus* have also been reported to have been used in shampoos and soaps against acne and eczema and in personal care products. The skin cleansing cream and lotion containing these bioemulsifiers have, among other properties, the ability to interfere with microbial adhesion on skin or hair (Hayes 1991).

12.2.3   Uses in Detergent Formulations

Almost half of all surfactants produced are for use in the washing and cleaning sectors (Daniel et al. 1998). A number of classes of chemical surfactant have been used for many years as detergent active materials, including anionic and nonionic materials (Hall et al. 1996). Detergent compositions traditionally contain one or more detergent active materials in addition to various other ingredients such as detergency builders, bleaches, fluorescers, and perfumes. The major applications of detergent compositions are to clean fabric, crockery, cooking utensils, and hard surfaces such as glass, glazed surfaces, plastics, metals, and enamels (Hall et al. 1996).

Increased attention to environmental protection, use of compounds with low toxicity, low carbon footprint, and high biodegradability is promoting interest in selecting and including "green" components in the detergent formulations (Furuta et al. 2004). The challenge for manufacturers, however, lies in the ability to increase the level of such ingredients in formulations without compromising cost or performance.

Biosurfactant versatility as cleaning agents and the additional properties of antimicrobial effects and reduction of solvent uses in manufacturing are making them more attractive. A leading detergent manufacturer, Henkel, started using sophorolipid surfactant in some of its regional branded glass-cleaning products, such as Sidolin, Instanet, Sonasol, Tenn and Breff, all of which were sold in Europe. Henkel is also looking in to include rhamnolipids and MELs in their products as reported by Karl-Heinz Maurer, global director of biotechnology at Henkel (http://www.icis.com/Articles/2010/10/04/9396996/surfactant-manufacturers-look-for-green-but-cheap-petro-alternatives.html).

The selection of a surfactant for a detergent composition depends on its purpose of end use. For example, while facial washing products necessitate a surfactant with high foaming power and which is mild to the skin, laundry detergents require a surfactant with high washing power and able to form foam that is easily removed (Furuta et al. 2004).

Sophorolipids are said to be well suited for low-foaming applications such as hard surface and auto-dish cleaning products. A laundry detergent composition containing a synergistic combination of glycolipids or, alternatively, a combination of sophorolipids and nonionic surfactant has been invented (Hall et al. 1996) and assigned to the Lever Brothers Co., New York. The detergent compositions show enhanced oily soil detergency in fabric washing. The invention also provides a method of washing that comprises contacting fabrics, or inanimate surface to be cleaned, such as dishes and household surfaces.

A biodegradable low-foaming dish-washing product containing sophorolipids, having a good washing power across a wide temperature range, has been patented (Furuta et al. 2004). This new detergent composition is particularly designed for dish washing machines with jet washing technology, which requires low-foaming and high-temperature-resistant surfactants.

Another interesting patent has been issued concerning a surfactant comprising a carbohydrate group that results in superior cleaning in a dry cleaning system that employs densified carbon dioxide. The surfactant has a hydrocarbon group that is more solvent-philic than a carbohydrate group, and can result in reverse micelle formation in densified carbon dioxide (Murphy and Binder 2002).

The US company SyntheZyme™ has developed yeast strains for conversion of natural biobased lipids to industrial monomers and sophorolipids to be used in oil dispersants, biopesticides, and antimicrobials. Moreover, the company is developing bio-cleaners using modified sophorolipids as washing and cleaning agents (in liquid and powder form) to replace currently used petrochemically derived detergents (http://www.synthezyme.com/Company.html).

Sophorolipids are also found in cleaning soap mixtures and all-purpose cleaners (such as interior cleaning spray, a window spray, a heavy duty power cleaner, a car wash, and wax cleaner) produced by Ecover™. The company claims that these "Eco-Surfactants" are produced by yeast from sugar and rapeseed using an entirely biochemical, low-energy process (www.ecover.com/).

Cleaning products containing biosurfactants are commercialized by several other companies. The American company Naturell® (http://www.naturellclean.com/) has developed and patented a product for carpet cleaning containing enzymes and biosurfactants derived from the fermentation of sea kelp.

The biotechnology company Z BioScience has developed environmentally friendly microorganism and biosurfactant-based cleaning products formulated specifically for use in commercial and office environments, clinical and healthcare settings, and household (http://www.z-bioscience.com/). In particular, two biosurfactant-based products—A1+ Biosurfactant Cleaner and B1+ Low Foaming Biosurfactant Cleaner—are claimed to provide a barrier to pathogens, eliminate the biofilm, and prevent re-colonization of surfaces.

The French company HTS BIO has created the product line Ecoway® for professional cleaning and maintenance that includes the cleaning product Biosurf 12. In India, Akshay Intensive Marketing (http://www.indiamart.com/akshay-intensive-marketing/) supplies Acticlean, a product containing nonspecified biosurfactants, whose functions are claimed to be the improvement of wetting and spreading, detergency activity, modification and control of foam, emulsification, solubilization, and dispersion.

Taylor Mclure produces Drain-Zyme a liquid product for the removal of starch, fats, proteins, and so on in septic tanks, which contains a blend of microorganisms specially selected for their ability to produce unspecified biosurfactants that can help to emulsify the grease, that is, giving some advantages to overall degradation (http://www.taylormclure.co.uk/pages/more_information__drain_zyme_ 159396.cfm).

12.3   BIOSURFACTANT POTENTIAL IN THE FOOD INDUSTRY

Biosurfactants have several promising applications in the food industry as food additives. Particularly, they may be used as emulsifiers for the processing of raw materials. Emulsification plays an important role in producing the right consistency and texture as well as in phase dispersion. Lecithin and its derivatives, fatty acid esters containing glycerol, sorbitan, or ethylene glycol, and ethyoxylated derivatives of monoglycerides including a recently synthesized oligopeptide are currently in use as emulsifiers in the food industries worldwide. The desire to increase the range of emulsifiers and decrease dependency on plant emulsifiers particularly with the ever-diminishing sources of nongenetically modified soybean, the main source for lecithin, and the wish to benefit from the favorable antioxidant, antiadhesive, antimicrobial, and biofilm disruption properties has led to an increase in interest in finding alternative natural sources of amphiphilic molecules suitable for use in some food industries (Campos et al. 2013)

High-molecular mass biosurfactants are better emulsifiers than low-molecular mass ones. Sophorolipids from *Torulopsis bombicola* (now *Starmerella bombicola*) have been shown to reduce surface and interfacial tension but not to be good emulsifiers. By contrast, liposan has been shown to reduce surface tension and to emulsify edible oils (Cirigliano and Carman 1985). Polymeric surfactants coat the oil droplets forming very stable emulsions that never coalesce (Rosenberg and Ron 1999). Biosurfactants control the agglomeration of fat globules, stabilize aerated systems, improve texture and shelf-life of starch-containing products, modify rheological properties of wheat dough and improve consistency and texture of fat-based products (Kachholz and Schlingmann 1987). Biosurfactants also retard staling, solubilize flavor oils, and improve organoleptic properties in bakery and ice cream formulations and act as fat stabilizers during cooking of fats (Kosaric 2001). An improvement of dough stability, texture, volume, and conservation of bakery products was obtained by the addition of rhamnolipid surfactants (Van Haesendonck and Vanzeveren 2004). The authors also suggested the use of rhamnolipids to improve properties of butter cream, croissants, and frozen confectionery products. Recently, a bioemulsifier isolated from a marine strain of *Enterobacter cloaceae* was used as a potential viscosity enhancement agent of interest in the food industry especially due to the good viscosity observed at lower pH allowing its use in food products containing citric acid or ascorbic acid (Iyer et al. 2006). Although the addition of rhamnolipids has been suggested to improve dough characteristics of bakery products, the use as food ingredients of compounds derived from an opportunistic pathogen such as *Pseudomonas aeruginosa* is not practically feasible or ethically acceptable. Instead, it has been suggested to use biosurfactants obtained from yeasts or *Lactobacilli*, which are generally recognized as safe and are already involved in several food-processing technologies (Nitschke and Costa 2007). A lipopeptide surfactant produced by *B. subtilis* forms stable emulsions with soybean oil and coconut fat (Nitschke and Pastore 2006) and with sunflower, linseed, olive, palm, babassu, and Brazilian nut oils (Costa et al. 2006). Furthermore, a mannoprotein obtained from *Kluyveromyces marxianus* can form emulsions with corn oil that are stable for 3 months (Lukondeh et al. 2003).

Biosurfactants have also gained interest in the food industry as potential antioxidant agents. A biosurfactant produced by *B. subtilis* RW-1 was able to scavenge free radicals (Yalcin and Cavusoglu 2010) and a polysaccharide emulsifier from *Klebsiella* was observed as a potent inhibitor of the auto oxidation of soybean oil by encapsulating it from the surrounding medium (Kawaguchi et al. 1996).

Furthermore, a wide range of biosurfactants have shown antimicrobial activity against bacteria, yeast, fungi, algae, and viruses (Nitschke and Costa 2007) and can be used to avoid food contamination directly, as additives, or indirectly, as detergent formulations to clean surfaces that enter into contact with the food.

Bacterial biofilms present on food industry surfaces are potential sources of contamination, which may lead to food spoilage and disease transmission. *Salmonella enteritidis*, *Listeria monocytogenes*, and *Enterobacter sakazakii* are examples of pathogenic bacteria implicated in outbreaks associated with the ingestion of contaminated food. Numerous studies have shown that these bacteria that are able to adhere and to form biofilms on food-contact surfaces are more resistant to sanitation than free-living cells (Stepanović et al. 2004; Kim et al. 2006). The preconditioning of surfaces using microbial surface-active compounds could be an interesting strategy to prevent adhesion of food-borne pathogens to solid surfaces. A surfactant released by *Streptococcus thermophilus* has been used for fouling control of heat-exchanger plates in pasteurizers as it retards the colonization of other thermophilic strains of *Streptococcus* responsible for fouling (Busscher et al. 1996).

Zeraik and Nitschke (2010) recently reported the antiadhesive activity of surfactin and rhamnolipids on polystyrene surfaces against *L. monocytogenes*, *S. aureus*, and *M. luteus*. An interesting work regarding the use of biosurfactants to inhibit the adhesion of the pathogen *L. monocytogenes* on two types of surfaces classically used in food industry has been conducted by Meylheuc et al. (2001). The preconditioning of stainless steel and PTFE surfaces with a biosurfactant obtained from *Pseudomonas fluorescens* inhibited the adhesion of *L. monocytogenes* L028 strain. A significant reduction (>90%)

was observed in microbial adhesion on stainless steel, whereas no significant effect was observed on PTFE (Meylheuc et al. 2001). Further work demonstrated that the prior adsorption of this surfactant on stainless steel also favored the bactericidal effect of disinfectants (Meylheuc et al. 2006b). In addition, the ability of adsorbed biosurfactants, obtained from *P. fluorescens* and *Lactobacillus helveticus* bacteria isolated from foodstuffs, to inhibit the adhesion of *L. monocytogenes* on stainless steel was investigated. Both biosurfactants were able to decrease the level of contamination of the surface, reducing either the total adhering flora or the viable/cultivable adherent *L. monocytogenes* on stainless steel surfaces (Meylheuc et al. 2006a). Furthermore, preliminary studies regarding the corrosion effect of *P. fluorescens* surfactant on stainless steel suggested that it also has a good potential for use as a corrosion inhibitor (Dagbert et al. 2006). Kuiper et al. (2004) have studied the ability of two lipopeptide biosurfactants produced by *P. putida* PCL1445 to inhibit biofilm formation of *Pseudomonas* species and strains on polyvinyl chloride (PVC) surfaces, showing that they are able not only to reduce the biofilm formation but also to disrupt preexisting biofilms. In another work, it was observed that a 24-h pretreatment with surfactin reduced *E. sakazakii* ATCC 29004 cell adhesion on stainless steel 304 surfaces by one logarithm (Pires et al. 2007). In addition, Ferreira et al. (2007) compared the surfactin effect on *L. monocytogenes* ATCC 19112 adhesion on stainless-steel surfaces and polypropylene, revealing that surfactin was able to reduce the number of adhered cell on stainless steel by two orders of magnitude (from ≈ 8 to 5 log CFU/cm^3), and one order of magnitude from ≈ 6 to 5 log CFU/cm^2 on polypropylene.

Despite the advantages of biosurfactants, few reports are available regarding their use in food products and food processing. Biosurfactants are not yet used in food processing on a large scale due to the numerous regulations set by governmental agencies for new food ingredients and the lengthy approval process (Nitschke and Costa 2007). Nevertheless, some patents are being issued on biosurfactants demonstrating the current interest in using these microbial-derived products in the food field.

Imura et al. (2007) have obtained a patent regarding the use of biosurfactant as emulsifier or solubilizer for foodstuffs. They discovered that a combination of nisin with rhamnolipids extended the shelf life and inhibited thermophilic spores in UHT soymilk. In addition, the use of compositions containing natamycin and rhamnolipids in salad dressing extended its shelf life and inhibited mold growth. They also discovered that compositions with natamycin, nisin, and rhamnolipids have extended shelf life of cottage cheese by the inhibition of mold and bacterial growth, especially Gram-positive and spore-forming bacteria. Rhamnolipids can be used to retain moisture and/or maintain bakery texture. A formulation composed of rhamnolipids with amylose and amylopectin has been reported to be suitable in the context of bread or bakery products to inhibit amylose and amylopectin retrogradation or bakery staling (Gandhi and Palmer-Skebba 2011).

Jeneil Biosurfactant Corporation has also developed a rhamnolipid biofungicine formulation to prevent crop attack by pathogenic fungi. The product has been approved by FDA for use on vegetables, legumes, and fruits crops (Nitschke and Costa 2007).

Pierce and Heilman (2001) identified a germicidal mixture composed of fruit acids and a surfactant for the cleaning of fruits and vegetables. The surfactant may be an anionic surfactant (such as sodium lauryl sulfate), a sophorolipid biosurfactant, or a combination of the two surfactants. The composition was able to kill 100% of *E. coli, Salmonella,* and *Shigella* in 30 s after application to the surface of the object.

Cheminova a supplier of products and solutions to cane sugar industry has developed Dextrin 100, a complex mixture of organic polymers and unspecified bio-surfactant, for the control of dextran and color in mixed juice and syrup.

12.4   BIOSURFACTANTS IN THE TEXTILE INDUSTRY: OPPORTUNITIES FOR FUTURE EXPLOITATION

The textile finishing industry is a high water-consuming industry. The pretreatment of textiles is a very important step for the production of high-quality end-products in the textile finishing processes. Various fiber admixtures and special formulations of lipophilic substances, used as lubricants to obtain optimal friction behavior during the production of a fabric, must be removed from the fiber surface to prepare textiles for the next production step (Kesting et al. 1996). Furthermore, conventional washing processes with detergent compounds have the usual undesired effects on the environment.

Taking the above into consideration, the use of biosurfactants was proposed due to their favorable characteristics of increasing the bioavailability of water insoluble substrates while providing a wider range of surfactant types and properties than synthetic surfactants. Biosurfactant use, therefore, has been reported in textile finishing for emulsification, solubilization, dispersing, wetting, and detergency (Kesting et al. 1996) with the added advantage of reducing the potential for environmental pollution (Mohan et al. 2006).

Kesting et al. (1996) investigated the potential application of microbial surfactants in textile washing processes to remove various lipophilic preparations from fiber surfaces. Although *Rhodococcus globerulus* was able to decompose various pure substrates such as triglycerides, fatty acid esters, and polyoxyethylene esters with some success, only spinning oil was degraded successfully from various commercial fiber preparations. When they evaluated the detergency activity in removing oils from fibers using *Rhodococcus erythropolis* biosurfactants (trehalosetetraester and trehalosedicorynomycolate), the amount of the oil removed from the fabric was higher compared with the control experiment where a surfactant-free medium was used as washing agent.

The application of biosurfactants in commercial detergents, also used for fabric cleaning, is already a reality; however, biosurfactant application has other applications in the textile dyeing industry.

One of the main problems in the textile industry is dye solubility. Poor water solubility of the dye causes nonhomogenous distribution of the dye throughout the fabric solid phase and the accumulation of the dye preferentially on the fiber surfaces (Quagliotto et al. 2006). Enhanced dye water solubility can usually be achieved by the addition of surfactants, which also enhance the dispersion of dyes to achieve a more uniform and better dye penetration into the fiber (Quagliotto et al. 2006; Montoneri et al. 2008). The number and amount of dyes used by the textile industry is vast and about 15% of the world dye production is lost in the environment during the dyeing process (Robinson et al. 2001). The environmental impact of dyes is not only aesthetic but may also involve their metabolism or reduction producing carcinogenic aromatic amines (Gottlieb 2003).

Recently a comparison was made of dyeing performance for nylon 6 microfiber using an unidentified biosurfactant (cHAL) and commercial SDS and sodium dodecylbenzenesulfonate and water soluble and insoluble dyes. The biosurfactant cHAL allowed the same quality of dyed product as the synthetic surfactants, but at lower dye concentrations. This advantage was enhanced when the additional gains in process cost and environmental impact were considered (Savarino et al. 2007). Similar observations using another unspecified biosurfactant (named cHAL2) were reported. However, the photosensitizing effect of cHAL2 in the dye exhaust bath added further potential value to its use in dyeing technology, and should not conflict with the stability of the dyed fabric, provided that the additive was washed out of the dyed fabric upon completion of the dyeing process (Montoneri et al. 2008).

Savarino et al. (2009) also investigated six compost-isolated humic acid-like biosurfactants (cHAL2 to 7), for their potential to act as auxiliaries for dyeing cellulose acetate fabric with water-insoluble dyes, monitoring solubility in water, as well as dyeing efficiency under a variety of experimental conditions representing a range of challenge levels. The cHAL biosurfactants were effective in enhancing the color homogeneity and intensity of the cellulose acetate fabric as well as in maintaining or improving the dye penetration and fixation to the fabric. There were no differences among the biosurfactants used, probably as they were all above their critical micelle concentration values, and there were no differences observed when compared with the commercial surfactants such as sodium dodecylbenzenesulfonate and Ethofor. They concluded that the cHAL biosurfactants can be considered as promising chemical auxiliaries compared with commercial surfactants (Savarino et al. 2009).

12.5 APPLICATION IN ENHANCED OIL RECOVERY

Productive oil wells are usually initiated using traditional primary and secondary recovery technologies and depending on the well can only yield an average of 20%–30% of the total oil in the well. Up to two-thirds of the oil present in reservoirs can remain in place once these technologies have been exhausted some years later. Tertiary oil recovery technologies collectively termed as enhanced oil recovery (EOR) are employed at that stage to allow recovery of an additional 10%–15% of the residual oil. These include both chemical and microbial-based methods; the latter are referred to as microbial EOR or (MEOR) (Banat 1995; Sen 2008; Marchant and Banat 2012a). MEOR takes advantage of several microbiological processes, such as partial break down of large oil molecules, production of gases, selective plugging, and production of biosurfactants. In the last case, the reduction of the oil/water interfacial tension and the formation of an oil-in-water emulsion can lead to an improvement in the mobility of the oil through rock fractures. The use of biosurfactants in MEOR can be achieved through different strategies: *ex situ* production in offsite fermenters and injection into the oil reservoir; *in situ* production by injected allochthonous microorganisms; and injection of nutrients to stimulate *in situ* production by indigenous bacteria (Perfumo et al. 2010b).

One major obstacle to the development of *in situ* production strategies is the difficulty of isolating microbial strains adapted to the extreme environment of the reservoirs, which features high pressure and salinity, temperatures up to 85°C and extreme pH values. Moreover, when injecting microorganisms into wells, some operators experienced plugging and corrosion problems. For these reasons, the addition of *ex situ* produced biosurfactants has been recently proposed as an option for such applications (Sen 2008; Brown 2010). Although rhamnolipids were the first candidates for such application, all the main types of microbial surface-active compounds have been proposed for a MEOR application. In particular, lipopeptides, such as surfactin, lichenysin, and emulsan, have also proved to be very effective in enhancing oil recovery (Sen 2008). Recently, the stability of different biosurfactants produced by *B. subtilis, P. aeruginosa,* and *Bacillus cereus* was determined at temperatures, pHs, and salinity conditions mimicking those found in oil reservoirs. Oil displacement experiments in glass micromodels were also carried out on the biosurfactant from *B. subtilis,* which was suggested to be the most promising. In this case, the amount of oil recovery was enhanced by up to 25% (Amani et al. 2010).

Most laboratory experiments on MEOR applications using biosurfactants were carried out in core flooding systems to simulate oil reservoir characteristics. The lipopeptidic biosurfactant produced by a *Bacillus mojavensis* strain was used in core flooding tests to evaluate oil recovery from carbonate reservoirs: the treatment was reported to achieve values up to 60% of initial oil in place in such cores (Ghojavand et al. 2012).

To date, we are not aware of a well-documented direct injection of biosurfactants into the oil reservoirs carried out in the field that has fully exploited this technology for MEOR. This is probably due to the high production cost of these compounds. A more cost-effective alternative would be an *in situ* production of biosurfactants, either by injected bacteria or by stimulated autochthonous microorganisms. Therefore, despite the technical difficulties, research has recently turned to the isolation of new surfactant-producing microbial strains using extreme conditions similar to those in oil reservoirs (Gudiña et al. 2012). The potential performance of the isolated strains can be subsequently evaluated in core flooding systems.

She et al. (2011) observed incremental oil recoveries of 4.89%–6.96% after the addition of different *Bacillus* cultures, while Xia et al. (2012) obtained an efficiency of 9.02% with the injection of *P. aeruginosa* cells. Some efforts have been directed to the isolation of unconventional biosurfactant-producing microorganisms from reservoirs, which can produce molecules particularly effective in oil mobilization. Castorena-Cortés et al. (2012) isolated a stable consortium, whose predominant genus was *Thermoanaerobacter.* When tested in small oil recovery assays, it was responsible for an increase of 12% in oil release from the granular porous medium. In some cases, the injection of surfactant-producing microorganisms and the application of biosurfactants were compared in the same laboratory system. Zheng et al. (2012) found that the biosurfactant of *Rhodococcus ruber* Z25 gave higher percentages of oil recovery, up to 25.78% when it was injected in a sand-pack system as a cell-free supernatant, while Darvishi et al. (2011) observed a higher efficiency when the consortium ERCPPI-2, formed by an *Enterobacter* cloacae strain and a *Pseudomonas* sp. strain, were directly injected to a core flooding system.

Currently, there is no clear and conclusive evidence to identify the best biosurfactant-based MEOR strategy. Brown (2010) warned against the difficulty of extrapolating laboratory data to the field and the inadequacy of most field experiments. He also pointed out that mechanisms involved in MEOR processes have been elucidated only in a very few cases, so that available information on this topic is still scarce. From this point of view, an insight was provided by Armstrong and Wildenschild (2012), who recently demonstrated that flooding in a micro-model system with both biomass and biosurfactant was the optimal solution for oil recovery due to the combined effects of bio-clogging of the pore space and interfacial tension reduction.

A further limitation to progress in MEOR techniques is the lack of knowledge on the structure and diversity of microbial communities present in the reservoirs. In particular, a deeper knowledge about the metabolic and physiologic potential of the autochthonous microorganisms might further promote the strategy to stimulate *in situ* biosurfactant production, without the need to add external bacteria. To evaluate the potential to produce biosurfactants such as surfactin or lichenisin, Simpson et al. (2011) detected the presence of the srfA3/licA3 genes in brines collected from nine wells using polymerase chain reaction and concluded that a biostimulation approach for biosurfactant-mediated oil recovery might be effective. In some cases, the analysis of microbial community composition through cloning of the 16S rRNA gene gave important information that helped to estimate the feasibility of biosurfactant-based MEOR processes (Zhang et al. 2012; Zhao et al. 2012).

Beside applications in MEOR, biosurfactants can also be exploited for other applications in the oil industry. For example, the use of several microbial surfactants has recently been optimized for oil extraction from oil sludge, demonstrating that it was possible to obtain an oil recovery of up to 74.55%, depending on the washing conditions (Zheng et al. 2012). Furthermore, the rhamnolipid of *P. aeruginosa* F-2 was proved to remove up to 91.5% of oil from oil sludge in pilot-scale field tests (Yan et al. 2012). Some formulations are already available in the market, both for specific oil dispersant activity and for cleaning residual oil (http://www.kambiotechnology.com/kam-products/80-cellozyme-1000-hc) and with a broad spectrum of applications (http://www.synthezyme.com/Biosurfactants.html).

12.6 BIOSURFACTANTS IN AGRICULTURE

The numerous roles of biosurfactants based on their distinct characteristics and their involvement in biological control by acting either as antifungal agents or elicitors of induced systemic resistance render them potential candidates for future applications in crop protection. In general, the mode of action of biosurfactants in biological control includes the formation of channels in the cell wall and disturbance of the cell surface of the pathogen (Raaijmakers et al. 2006). From the many classes of biosurfactants, protection of plants through their antifungal properties against phytopathogenic fungi is provided mainly by glycolipids, such as cellobiose lipids, rhamnolipids, and cyclic lipopeptides, for example, surfactin, iturin, and fengycin (Banat et al. 2010).

Teichmann et al. (2007) demonstrated that infection of tomato leaves by the pathogenic fungus *Botrytis cinerea* is counteracted by co-inoculation with wild-type *Ustilago maydis* sporidia. This phytopathogenic fungus produces the first reported cellobiose lipid named ustilagic acid. Cellobiose lipids are natural detergents, which at relatively low concentrations can induce cell death of yeast and mycelial fungi and that effect is produced by their membrane-damaging properties (Kulakovskaya et al. 2010).

Stanghellini and Miller (1997) mentioned the potential use of biosurfactants as biological control agents; they describe how rhamnolipids can disrupt zoospore membranes and cause lysis of zoospores of many oomycete plant pathogens. Since then, numerous reports have discussed the important role of rhamnolipids against various phytopathogenic fungi. Debode et al. (2007) reported that rhamnolipids produced by *Pseudomonas* spp. facilitate the suppression of *Verticillium microsclerotia* viability. Perneel et al. (2008) hypothesized break down of *Pythium* hyphae after incubation in liquid medium amended with both phenazines and rhamnolipids produced by *P. aeruginosa* PNA1; the results indicated that both metabolites are acting synergistically in controlling soil-borne diseases caused by *Pythium* spp.

Rhamnolipids biosurfactants were found to have direct antifungal properties by inhibiting spore germination and mycelial growth of *B. cinerea*. Furthermore, the response of the grapevine against this disease was associated with Ca^{2+} influx, mitogen-activated protein kinase activation and reactive oxygen species production as early events. Induction of plant defenses including expression of a wide range of defense genes, hypersensitive-like response explained parts of the mechanisms involved in plant resistance. Moreover, rhamnolipids potentiated defense responses induced by a chitosan elicitor and by the culture filtrate of *B. cinerea* suggest that the combination of rhamnolipids with other effectors could participate in grapevine protection against the gray mold disease (Varnier et al. 2009).

Recent work by Sha et al. (2012) demonstrated that crude rhamnolipids in the form of a cell-free culture medium showed high activity against colony growth and biomass accumulation of seven plant pathogens comprising two *Oomycetes*, three *Ascomycota,* and two *Mucor* spp. fungi. The substantial better antifungal efficiency of cell-free culture medium of rhamnolipids could be attributed to the major component of di-rhamnolipid in this cell-free medium, which is characterized by better lysis traits over mono-rhamnolipid to rupture the spore membranes especially for zoospore-producing plant pathogens. Vatsa et al. (2010) presented an excellent review about the antimicrobial properties of rhamnolipids and the involvement of these molecules in the stimulation of "immunity" in plants.

Cyclic lipopeptides represent another class of biosurfactants with antifungal activity toward phytopathogenic fungi equally important as rhamnolipids. In short, biosynthesis of cyclic lipopeptides is governed by nonribosomal peptide synthetases, encoded by very large gene clusters and composed of modules; one for each amino acid that needs to be incorporated in the oligopeptide. Each module consists of several conserved domains responsible for recognition, activation, transport, and binding of the amino acid to the peptide chain. A special thioesterase domain in the last module coordinates cyclization and release of the peptide product. Owing to this unconventional biosynthesis scheme, incorporation of unusual amino acids is possible (D'Aes et al. 2010).

Grover et al. (2010) reported the ability of *B. subtilis* strain RP24 isolated from the rhizoplane of field grown pigeon pea to act as a biocontrol agent against a wide range of phytopathogenic fungi. Iturin antibiotics increase the membrane permeability of the target microorganism due to the formation of ion channels on the cell membranes thereby increasing the permeability to K^+ that is associated with fungicidal activity; therefore, the occurrence of an iturin operon in its genome explains the antifungal nature of the isolate.

In addition to iturin, surfactin is another lipopeptide with well-documented antifungal activity. Tendulkar et al. (2007) reported that secretion of a lipopeptide by *Bacillus licheniformis* BC98, identified as surfactin, induced morphological changes in *Magnaporthe grisea*, inhibiting its further growth, and thus exhibiting fungicidal activity.

Isolation and characterization of Leu7-Surfactin biosurfactant from the endophytic bacterium, *B. mojavensis* RRC 101, indicated that this bacterium can be a useful antagonist to assist in the control of endophytic infection and resulting diseases caused by *Fusarium verticillioides* (Snook et al. 2009). Consistent with these results, Bacon et al. (2012) reported that strains of *B. mojavensis* produce a considerably higher number of isoforms of surfactin A compared with B and C isoforms.

Genome sequencing provided insights about structure, genes, biosynthesis, and regulation of known cyclic lipopeptides produced by different bacterial genera. The sequenced genome of the plant-associated *Bacillus amyloliquefaciens* GA1 identified four gene clusters to direct the synthesis of the cyclic lipopeptides surfactin, iturin A, and fengycin (Arguelles-Arias et al. 2009). Recent reports confirmed the existence of biosynthetic gene clusters encoding secondary metabolites, such as lipopeptides (surfactin, iturin, and fengycin), associated with biocontrol activity among the genus *Bacillus* (Blom et al. 2012; Dunlap et al. 2013).

Pseudomonas spp. comprise another important group of plant-associated bacteria showing biocontrol activities mediated by the production of cyclic lipopeptides. The interaction between glycolipid-type biosurfactants and fungi has been the main focus of some researchers and the zoo-sporicidal properties of cyclic lipopeptides have also been reported (Hultberg et al. 2010). Excellent reviews about the role of biosurfactants in plant–*Pseudomonas* interactions, emphasizing their potential use as biocontrol agents are presented in Höfte and Altier (2010) and D'Aes et al. (2010). The above-mentioned Modified Sophorolipids produced by SyntheZyme™ are claimed to be able to treat leaf infection in different plants (http://www.synthezyme.com/Biosurfactants.html).

12.7 USE IN OTHER INDUSTRIES

Opportunities for industrial exploitation of biosurfactants are also envisaged in several other manu-facturing industries such as pulp and paper processing, paint, leather, plastics, and metals.

In the paper processing and related industries, biosurfactants can be used for the deresinification and washing of pulp, as defoaming, color leveling, and dispersing agents and, in the paper industry for calendaring, as wetting and leveling, coating and coloring agents (Kosaric 2001). For example, a biodispersan produced by *A. calcoaceticus* A2 was shown to be effective in grinding limestone to fine particles and successfully used as a filler in laboratory-made paper (Rosenberg et al. 1989). In another work, the extracellular polymeric substances from waste sludge of pulp and paper mills showed potential for utilization as wood adhesive (Pervaiz and Sain 2010). A composition including a cellulase enzyme and a culture solution of *Pseudomonas rubescens* containing a biosurfactant was patented as a process to enable the reduction of added cellulase and shorten the cellulosic fiber-treating time (Masuoka et al. 1998).

In the paint and protective coatings industry, biosurfactants can be employed in dispersing and wetting of pigment during grinding and for emulsification, dispersion of pigment, stabilization of latex, retard sedimentation, and pigment separation in latex paints (Kosaric 2001). An international patent has been recently granted describing the use of a biosurfactant, produced by *Cobetia marina*, a Gram-negative strain used in aquaculture to be used as an additive in paint formulations for sub-mersible surfaces (Dinamarca-Tapia et al. 2012). Biosurfactants have also generated interest in the field of biodyes. Hybrids of pyrene and biosurfactants have been prepared to monitor the micro-environmental conditions of various kinds of colloidal surfaces and bio-interfaces with respect to their polarity and fluidity (Ishigami and Suzuki 1997).

Biosurfactant potential applications in the leather industry include their use as skin detergents, and emulsifiers in degreasing, wetting and penetration, and promoters in tanning and dyeing. Kilic (2013) investigated a saponin biosurfactant as an alternative low-cost natural option to chemical surfactants for the degreasing of sheep skins and concluded that it was a viable option with promis-ing ecological gains. In the plastic industry, they can be used as antistatic agents, wetting agents, solubilizers, and emulsifiers (Kosaric 2001).

12.8 CONCLUSIONS

As we have seen in this chapter, a myriad of proposals have been put forward for commercial appli-cations of biosurfactants; many of these have not and will not reach any point of significant com-mercial application now or in the future. The reasons for this are complex, but involve the following specific points. Most publications on biosurfactants use the phrase "environmentally friendly" to describe these compounds, based on the assumption that a sustainable technology must have less environmental impact than a conventional nongreen technology. However, a critical Life Cycle

Assessment that considers the industrial process from the acquisition of raw materials, through the manufacturing of product to consumer use and final disposal, will not necessarily show that a biosurfactant will have less environmental impact, in terms of greenhouse gas emission, than a petrochemically derived surfactant. In addition, the key consideration for the replacement of chemical surfactants with biosurfactants lies in the functionality of the molecules in the specific formulations required. Biosurfactants will not be substituted for other surfactants unless they can perform at least as well in the selected applications. Following on from functionality is the important consideration of production and downstream processing costs. Low initial fermentation yields and the production of a mixture of congeners, which may need to be separated, are severe problems for some of the biosurfactant-producing systems. At present, a few biosurfactants have found use in commercial products, but these are either niche products in the market or are products from smaller companies in the global market. Biosurfactants will only come of age commercially when the large global companies such as Procter & Gamble and Unilever find ways to incorporate them in their bulk fast-moving consumer goods such as laundry detergents, surface cleaners, and personal care products.

REFERENCES

Amani, H., Sarrafzadeh, M. H., Haghighi, M., and M. R. Mehrnia. 2010. Comparative study of biosurfactant producing bacteria in MEOR applications. *Journal of Petroleum Science and Engineering* 75:209–214.

Arguelles-Arias, A., Ongena, M., Halimi, B. et al. 2009. *Bacillus amyloliquefaciens* GA1 as a source of potent antibiotics and other secondary metabolites for biocontrol of plant pathogens. *Microbial Cell Factories* 8:63.

Armstrong, R. T., and D. Wildenschild. 2012. Investigating the pore-scale mechanisms of microbial enhanced oil recovery. *Journal of Petroleum Science and Engineering* 94:155–163.

Bacon, C. W., Hinton, D. M., Mitchell, T. R., Snook, M. E., and B. Olubajo. 2012. Characterization of endophytic strains of *Bacillus mojavensis* and their production of surfactin isomers. *Biological Control* 62:1–9.

Baltz, R. H., Miao, V., and S. W. Wrigley. 2005. Natural products to drugs: Daptomycin and related lipopeptide antibiotics. *Natural Product Reports* 22:717–741.

Banat, I. M. 1995. Biosurfactants production and use in microbial enhanced oil recovery and pollution remediation: A review. *Bioresource Technology* 51:1–12.

Banat, I. M., Franzetti, A., Gandolfi, I. et al. 2010. Microbial biosurfactants production, applications and future potential. *Applied Microbiology and Biotechnology* 87:427–444.

Banat, I. M., Makkar, R. S., and S. S. Cameotra. 2000. Potential commercial applications of microbial surfactants. *Applied Microbiology and Biotechnology* 53:495–508.

Blom, J., Rueckert, C., Niu, B., Wang, Q., and R. Borriss. 2012. The complete genome of *Bacillus amyloliquefaciens* subsp. *plantarum* CAU B946 contains a gene cluster for nonribosomal synthesis of iturin A. *Journal of Bacteriology* 194:1845–1846.

Brown, L. R. 2010. Microbial enhanced oil recovery (MEOR). *Current Opinion in Microbiology* 13:316–320.

Bruce, A. W., Busscher, H. J., Reid G., and H. C. Van der Mei. 2000. Lactobacillus therapies. U.S. Patent US6051552A, April 18.

Burke, T., Chandrasekhar, B., and M. Knight. 1999. Analogs of viscosin and uses thereof. United States Patent US5965524, October 12.

Busscher, H. J., van der Kuijl-Booij, M., and H. C. van der Mei. 1996. Biosurfactants from thermophilic dairy streptococci and their potential role in the fouling control of heat exchanger plates. *Journal of Industrial Microbiology & Biotechnology* 16:15–21.

Cameron, D. R., Boyd, V. A., Leese, R. A. et al. 2005. Compositions of lipopeptides antibiotic derivatives and methods of use thereof. World Patent WO2005000878 A3, June 16.

Campos, J. M., Montenegro Stamford, T. L., Sarubbo, L. A., de Luna, J. M., Rufino, R. D., and I. M. Banat. 2013. Microbial biosurfactants as additives for food industries. *Biotechnology Progress* 29:1097–1108.

Castorena-Cortés, G., Zapata-Peñasco, I., Roldán-Carrillo, T. et al. 2012. Evaluation of indigenous anaerobic microorganisms from Mexican carbonate reservoirs with potential MEOR application. *Journal of Petroleum Science and Engineering* 81:86–93.

Ceri, H., Turner, R., Martinotti, M. G., Rivardo, F., and G. Allegrone. 2010. Biosurfactant composition produced by a new *Bacillus licheniformis* strain, uses and products thereof. World Patent WO2010067345A1, June 17.

Cirigliano, M. C., and G. M. Carman. 1985. Purification and characterization of liposan, a bioemulsifier from *Candida lipolytica*. *Applied and Environmental Microbiology* 50:846–850.

Cochis, A., Fracchia, L., Martinotti, M. G., and L. Rimondini. 2012. Biosurfactants prevent *in vitro Candida albicans* biofilm formation on resins and silicon materials for prosthetic devices. *Oral Surgery, Oral Medicine, Oral Pathology, Oral Radiology* 113:755–761.

Corley, J. W. 2007. All that is good—Naturals and their place in personal care. In *Naturals and Organics in Cosmetics: From R & D to the Market Place*, ed. A. C.Kozlowski, pp. 7–12. Allured Businnes Media, Carol Stream, IL.

Cortés-Sánchez, A. D. J., and T. Mosqueda-Olivares. 2013. Una mirada a los organismos fúngicos: Fábricas versátiles de diversos metabolitos secundarios de interés biotecnológico. *Revista QuímicaViva* 2:64–90.

Costa, S. G. V. A. O., Nitschke, M., Haddad, R., Eberlin, M. N., and J. Contiero. 2006. Production of *Pseudomonas aeruginosa* LBI rhamnolipids following growth on Brazilian native oils. *Process Biochemistry* 41:483–488.

D'Aes, J., De Maeyer, K., Pauwelyn, E., and M. Höfte. 2010. Biosurfactants in plant-*Pseudomonas* interactions and their importance to biocontrol. *Environmental Microbiology Reports* 2:359–372.

Dagbert, C., Meylheuc T., and M. N. Bellon-Fontaine. 2006. Corrosion behavior of AISI 304 stainless steel in presence of a biosurfactant produced by *Pseudomonas fluorescens*. *Electrochimica Acta* 51:5221–5227.

Daniel, H. J., Reuss, M., and C. Syldatk. 1998. Production of sophorolipids in high concentration from deproteinized whey and rapeseed oil in a two stage fed batch process using *Candida bombicola* ATCC 22214 and *Cryptococcus curvatus* ATCC 20509. *Biotechnology Letters* 20:1153–1156.

Darvishi, P., Ayatollahi, S., Mowla, D., and A. Niazi. 2011. Biosurfactant production under extreme environmental conditions by an efficient microbial consortium, ERCPPI-2. *Colloids and Surfaces B: Biointerfaces* 84:292–300.

Debode, J., De Maeyer, K., Perneel, M., Pannecoucque, J., De Backer, G., and M. Höfte. 2007. Biosurfactants are involved in the biological control of *Verticillium microsclerotia* by *Pseudomonas* spp. *Journal of Applied Microbiology* 103:1184–1196.

Deleu, M., Brasseur, R., Paquot, M. et al. 2004. Novel use of lipopeptides preparations. World Patent WO2004002510 A1, January 8.

Denning, D. W. 2002. Echinocandins: A new class of antifungal. *Journal of Antimicrobial Chemotherapy* 49:889–891.

Desai, J. D., and I. M. Banat. 1997. Microbial production of surfactants and their commercial potential. *Microbiology and Molecular Biology Reviews* 61:47–64.

Desanto, K. 2008. Rhamnolipid-based formulations. World Patent WO2008013899A2, January 31.

Dinamarca-Tapia, M. A., Ojeda-Herrera, J. R., and C. L. Ibacache-Quiroga. 2012. Strain of *Cobetia marina* and biosurfactant extract obtained from same. World Patent WO 2012164508 A1, December 6.

Dunlap, C. A., Bowman, M. J., and Schisler, D. A. 2013. Genomic analysis and secondary metabolite production in *Bacillus amyloliquefaciens* AS 43.3: A biocontrol antagonist of Fusarium head blight. *Biological Control* 64:166–175.

Edwards, K. R., Lepo, J. E., and M. A. Lewis. 2003. Toxicity comparison of biosurfactants and synthetic surfactants used in oil spill remediation to two estuarine species. *Marine Pollution Bulletin* 46:1309–1316.

Emiroglu, M. 2011. Micafungin use in children. *Expert Review of Anti-Infective Therapy* 9:821–834.

Faivre, V., and V. Rosilio. 2010. Interest of glycolipids in drug delivery: From physicochemical properties to drug targeting. *Expert Opinion on Drug Delivery* 7:1031–1048.

Fardis, M., Cameron, D. R., and V. A. Boyd. 2007. DAB9 Derivatives of lipopeptides antibiotics and method of making and using the same. US Patent US20070167357, July 19.

Ferreira, F. S., Pires, R. C., Araùjo, L. V., Siqueira R. S. De, and M. Nitschke. 2007. Surfactin inhibits the adhesion of *Listeria monocytogenes* ATCC 19112 to solid surfaces. Paper presented at II Simpósio Latino Americano de Ciências de Aliments, Campinas, São Paolo.

Fracchia, L., Cavallo, M., Martinotti, M. G., and I. M. Banat. 2012. Biosurfactants and bioemulsifiers: Biomedical and related applications-present status and future potentials. In *Biomedical Science, Engineering and Technology*, ed. D. N. Ghista, pp. 325–370. Rijeka: InTech.

Francolini, I., and G. Donelli. 2010. Prevention and control of biofilm-based medical-device-related infections. *FEMS Immunology and Medical Microbiology* 59:227–238.

Franzetti, A., Caredda, P., Ruggeri, C. et al. 2009. Potentials applications of surface active compounds by *Gordonia* sp. strain BS29 in soil-remediation technologies. *Chemosphere* 75:801–807.

Franzetti, A., Tamburini, E., and I. M. Banat. 2010. Applications of biological surface active compounds in remediation technologies. In *Biosurfactants. Advances in Experimental Medicine and Biology*, ed. R. Sen, pp. 121–134. Austin: Landes Bioscience.

Furuta, T. Hirata, Y., and K. Igarashi. 2004. Low-foaming detergent compositions. U.S. Patent US20040171512 A1, September 2.

Gama, Y., Hongu, T., Ishigami, Y., Nagahora, H., and M. Yamaguchi. 1990. Rhamnolipid liposomes. U.S. Patent US4902512A, February 20.

Gandhi, N. R., and V. L. Palmer Skebba. 2011. Rhamnolipid compositions and related methods of use. U.S. Patent US7968499 B2, June 28.

George, J., and A. C. Reboli. 2012. Anidulafungin: When and how? The clinician's view. *Mycoses* 55:36–44.

Gharaei-Fathabad, E. 2011. Biosurfactants in pharmaceutical industry: A mini-review. *American Journal of Drug Discovery and Development* 1:58–69.

Ghojavand, H., Vahabzadeh, F., and A. K. Shahraki. 2012. Enhanced oil recovery from low permeability dolomite cores using biosurfactant produced by a *Bacillus mojavensis* (PTCC 1696) isolated from Masjed-I Soleyman field. *Journal of Petroleum Science and Engineering* 81:24–30.

Gottlieb, A. Shaw C., Smith A., Wheatley A., and S. Forsythe. 2003. The toxicity of textile reactive azo dyes after hydrolysis and decolourisation. *Journal of Biotechnologies* 101:49–56.

Grangemard, I., Wallach, J., Maget-Dana, R., and F. Peypoux. 2001. Lichenysin: A more efficient cation chelator than surfactin. *Applied Biochemistry and Biotechnology* 90:199–210.

Groux, H., Brun, V., and A. Foussat. 2005. Use of lipopeptides for activating T lymphocytes through the skin. World Patent WO2005046729 A2, May 26.

Grover, M., Nain, L., Singh, S. B., and A. K. Saxena. 2010. Molecular and biochemical approaches for characterization of antifungal trait of a potent biocontrol agent *Bacillus subtilis* RP24. *Current Microbiology* 60:99–106.

Gudiña, E. J., Pereira, J. F., Rodrigues, L. R., Coutinho, J. A., and J. A. Teixeira. 2012. Isolation and study of microorganisms from oil samples for application in Microbial Enhanced Oil Recovery. *International Biodeterioration & Biodegradation* 68:56–64.

Guglielmo, M., and D. Montanari. 2003. Cosmetic preparation with anti-wrinkle action. World Patent WO2003000222 A2, January 3.

Haba, E., Pinazo, A., Jauregui, O., Espuny, M. J., Infante, M. R., and A. Manresa. 2003. Physicochemical characterization and antimicrobial properties of the rhamnolipids products by *Pseudomonas aeruginosa* 47T2 NCIMB 40044. *Journal of Surfactants and Detergents* 6:155–161.

Hall, P. J., Haverkamp, J., Michael, S., and C. G. Van Kralingen, 1996. Laundry detergent composition containing synergistic combination of sophorose lipid and nonionic surfactant. U.S. Patent US005520839A, May 28.

Hayes, M. E. 1991. Personal care products containing bioemulsifiers. U.S. Patent US4999195A, March 12.

Henkel, M., Müller, M. M., Kügler, J. H. et al. 2012. Rhamnolipids as biosurfactants from renewable resources: Concepts for next-generation rhamnolipid production. *Process Biochemistry* 47:1207–1219.

Hill, J., Parr, I., Morytko, M. et al. 2008. Lipopeptides as antibacterial agents. US Patent US7335725 B2, February 26.

Höfte, M., and N. Altier. 2010. Fluorescent pseudomonads as biocontrol agents for sustainable agricultural systems. *Research in Microbiology* 161:464–471.

Howard, J., Reid, G., and B. S. Gan. 2002. Treatment of microbial infections with bacterial proteins and peptides. U S Patent US20020120101 A1, August 29.

Hultberg, M., Alsberg, T., Khalil, S., and B. Alsanius 2010. Suppression of disease in tomato infected by *Pythium ultimum* with a biosurfactant produced by *Pseudomonas koreensis*. *Biocontrol* 55:435–444.

Igarashi, S., Hattori, Y., and Y. Maitani. 2006. Biosurfactant MEL-A enhances cellular association and gene transfection by cationic liposome. *Journal of Controlled Release* 112:362–368.

Imura, T., Kitamoto, M., Fukuoka T., and T. Morita. 2007. Emulsifier or solubilizing agents. Japan Patent JP2007181789-A, July 19.

Inoh, Y. Furuno, T., Hirashima, N., Kitamoto, D., and M. Nakanishi. 2010. The ratio of unsaturated fatty acids in biosurfactants affects the efficiency of gene transfection. *International Journal of Pharmaceutics* 398:225–230.

Inoue, S., Kimura, Y., and M. Kinta. 1979a. Process for producing a glycolipid methyl ester. German Patent DE2905252A, August 23.

Inoue, S., Kimura, Y., and M. Kinta. 1979b. Process for producing a glycolipid ester. German Patent DE2905295A1, August 30.

Ishigami, Y., and S. Suzuki. 1997. Development of biochemicals—Functionalization of biosurfactants and natural dyes. *Progress in Organic Coatings* 31:51–61.

Iyer, A., Mody, K., and J. Bhavanath. 2006. Emulsifying properties of a marine bacterial exopolysaccharide. *Enzyme and Microbial Technology* 38:220–222.

Jones, T. S. 1949. Chemical evidence for the multiplicity of the antibiotics produced by *Bacillus polymyxa*. *Annals of the New York Academy of Sciences* 51:909–916.

Kachholz, T. and M. Schlingmann. 1987. Possible food and agricultural applications of microbial surfactants: An assessment. In *Biosurfactants and Biotechnology*, ed. N. Kosaric, W. L. Cairns, and N. C. C. Gary, pp. 183–210. New York: Marcel Dekker.

Kalyani, R., Bishwambhar, M., and V. Seneetha. 2011. Recent potential usage of biosurfactant from microbial origin in pharmaceutical and biomedical arena: A perspective. *International Research Journal of Pharmacy* 2:11–15.

Kanlayavattanakul, M., and N. Lourith. 2010. Lipopeptides in cosmetics. *International Journal of Cosmetic Science* 32:1–8.

Kato, E., and T. Tsuzuki. 2008. Dermatological anti-wrinkle agent. World Patent WO2008001921A2, January 3.

Kato, E., Tsuzuki, T., and E. Ogata. 2005. Tocopherol derivative, ascorbic acid derivative and skin preparation for external use comprising surfactant having lipopeptides structure, Japan Patent 2005336171, December 8.

Kawaguchi, K., Satomi, K., Yokoyama, M., and Y. Ishida. 1996. Antioxidative properties of an extracellular polysaccharide produced by a bacterium *Klebsiella* sp. Isolated from river water. *Nippon Suisan Gakkaishi* 42:243–251.

Kawano, J., Utsugi, T., Inoue, S., and S. Hayashi. 1981. Powered compressed cosmetic material. U.S. Patent US4305931 A, December 15.

Kesting, W., Tummuscheit, M., Schacht, H., and E. Schollmeyer. 1996. Ecological washing of textiles with microbial surfactants. *Progress in Colloid and Polymer Science* 101:125–130.

Kilic, E. 2013. Evaluation of degreasing process with plant derived biosurfactant for leather making: An ecological approach. *Tekstil Ve Konfeksiyon* 23:181–187.

Kim, H., Ryu, J. H., and L. R. Beuchat. 2006. Attachment of and biofilm formation by *Enterobacter sakazakii* on stainless steel and enteral feeding tubes. *Applied and Environmental Microbiology* 72:5846–5856.

Kitamoto, D. 2008. Naturally engineered glycolipid biosurfactants leading to distinctive self-assembling properties. *Yakugaku Zasshi* 128:695–706.

Kitamoto, D., Morita, T., Fukuoka, T., Konishi, M., and T. Imura. 2009. Self-assembling properties of glycolipid biosurfactants and their potential applications. *Current Opinion in Colloid and Interface Science* 14:315–328.

Kosaric, N. 2001. Biosurfactants and their application for soil bioremediation. *Food Technololgy and Biotechnology* 39:295–304.

Kuiper, I., Lagendijk, E. L., Pickford, R. et al. 2004. Characterization of two *Pseudomonas putida* lipopeptide biosurfactants, putisolvin I and II, which inhibit biofilm formation and break down existing biofilms. *Molecular Microbiology* 51:97–113.

Kulakovskaya, T. V., Golubev, W. I., Tomashevskaya, M. A. et al. 2010 Production of antifungal cellobiose lipids by *Trichosporon porosum*. *Mycopathologia* 169:117–123.

Landman, D., Georgescu, C., Martin, D. A., and J. Quale. 2008. Polymyxins revisited. *Clinical Microbiology Reviews* 21:449–465.

Lourith, M., and M. Kanlayavattanakul. 2009. Natural surfactants used in cosmetics: Glycolipids. *International Journal of Cosmetic Science* 31:255–261.

Lukondeh, T., Ashbolt, N. J., and P. L. Rogers. 2003. Evaluation of *Kluyveromyces marxianus* FII 510700 grown on a lactose-based medium as a source of natural bioemulsifiers. *Journal of Industrial Microbiology and Biotechnology* 30:715–720.

Maier, R. M., and G. Soberon-Chavez. 2000. *Pseudomonas aeruginosa* rhamnolipids: Biosynthesis and potential applications. *Applied Microbiology and Biotechnology* 54:625–633.

Maingault, M. 1999. Utilization of sophorolipids as therapeutically active substances or cosmetic product, in particular for the treatment of the skin. United States Patent US5981497A, November 9.

Makkar, R. S., Cameotra, S. S., and I. M. Banat. 2011. Advances in utilization of renewable substrates for biosurfactant production. *Applied Microbiology and Biotechnology Express* 1:5.

Mandal, S. M., Barbosa, A. E. A. D., and O. L. Franco. 2013. Lipopeptides in microbial infection control: Scope and reality for industry. *Biotechnology Advances* 31:338–345.

Mangoni, M. L., and Y. Shai. 2011. Short native antimicrobial peptides and engineered ultrashort lipopeptides: Similarities and differences in cell specificities and modes of action. *Cell and Molecular Life Science* 68:2267–2280.

Marchant, R., and I. M. Banat. 2012a. Microbial biosurfactants: Challenges and opportunities for future exploitation. *Trends in Biotechnology* 30:558–565.

Marchant, R., and I. M. Banat. 2012b. Biosurfactants: A sustainable replacement for chemical surfactants? *Biotechnology Letters* 34:1597–1605.

Martinotti, M. G., Allegrone, G., Cavallo, M., and L. Fracchia. 2013. Biosurfactants. In *Sustainable Development in Chemical Engineering—Innovative Technologies*, ed. V. Piemonte, M. De Falco, and A. Basile, pp. 199–240. Chichester: John Wiley & Sons.

Masaru, K., Michiko, S., and Shuhei, Y. 2007. Skin care cosmetic and skin and agent for preventing skin roughness containing biosurfactants. World Patent WO2007060956 A1, May 31.

Masuoka, K., Obata, T., Fukuda, Y., and N. Taharu. 1998. Improvement in cellulosic fiber and composition. Japan Patent JP 10–96174 A, April 14.

Meylheuc, T., Methivier, C., Renault, M., Herry, J. M., Pradier, C. M., and M. N. Bellon-Fontaine. 2006a. Adsorption on stainless steel surfaces of biosurfactants produced by gram-negative and gram-positive bacteria: Consequence on the bioadhesive behavior of *Listeria monocytogenes*. *Colloids and Surface B Biointerfaces* 52:128–137.

Meylheuc, T., Renault, M., and M. N. Bellon-Fontaine. 2006b. Adsorption of a biosurfactant on surfaces to enhance the disinfection of surfaces contaminated with *Listeria monocytogenes*. *International Journal of Food Microbiology* 109:71–78.

Meylheuc, T., van Oss, C. J., and M. N. Bellon-Fontaine. 2001. Absorption of biosurfactant on solid surfaces and consequences regarding the bioadhesion of *Listeria monocytogenes* LO28. *Journal of Applied Microbiology* 91:822–832.

Mohan, P. K., Nakhla, G., and E. K. Yanful. 2006. Biokinetics of biodegradability of surfactants under aerobic, anoxic and anaerobic conditions. *Water Research* 40:533–540.

Montanari, D., and M. Guglielmo. 2008. Cosmetic composition for the treatment and/or prevention of skin stretch marks. World patent WO2008080443 A2, July 10.

Montoneri, E., Savarino, P., Bottigliengo, S. et al. 2008. Humic acid-like matter isolated from green urban wastes. Part II: Performance in chemical and environmental technologies. *Bioresources* 3:217–233.

Morita, T., Fukuoka, T., Imura, T., and D. Kitamoto. 2013. Production of mannosylerythritol lipids and their application in cosmetics. *Applied Microbiology and Biotechnology* 97:4691–4700.

Morita, T., Ishibashi, Y., Hirose, N. et al. 2011. Production and characterization of a glycolipid biosurfactant, mannosylerythritol lipid B, from sugarcane juice by *Ustilago scitaminea* NBRC 32730. *Bioscience, Biotechnology, and Biochemistry* 75:1371–1376.

Müller, M. M., Kügler, J. H., Henkel, M. et al. 2012. Rhamnolipids-Next generation surfactants? *Journal of Biotechnology* 162:366–380.

Muraih, J. K., Pearson, A., Silverman, J., and M. Palmer. 2011. Oligomerization of daptomycin on membranes. *Biochimica et Biophysica Acta* 1808:1154–1160.

Murphy, D. S., and D. A. Binder. 2002. Dry cleaning system comprising carbon dioxide solvent and carbohydrate containing cleaning surfactant. United States Patent US6369014 B1, April 9.

Muthusamy, K., Gopalakrishnan, S., Ravi, T. K., and P. Sivachidambaram. 2008. Biosurfactants: Properties, commercial production and application. *Current Science* 94:736–747.

Nakanishi, M., Inoh, Y., Kitamoto, D., and T. Furuno. 2009. Nano vectors with a biosurfactant for gene transfection and drug delivery. *Journal of Drug Delivery Science and Technology* 19:165–169.

Naruse, N., Tenmyo, O., Kobaru, S. et al. 1990. Pumilacidin, a complex of new antiviral antibiotics: Production, isolation, chemical properties, structure and biological activity. *Journal of Antibiotics* 43:267–280.

Neuhof, T., Dieckmann, R., von Doehren, H. et al. 2006. Lipopeptides having pharmaceutical activity. World patent 2006/092313, September 8.

Ngai, A. L., Bourque, M. R., Lupinacci, R. J., Strohmaier, K. M., and N. A. Kartsonis. 2011. Overview of safety experience with caspofungin in clinical trials conducted over the first 15 years: A brief report. *International Journal of Antimicrobial Agents* 38:540–544.

Nguyen, T. T. L., Edelen, A., Neighbors, B., and D. A. Sabatini. 2010. Biocompatible lecithin-based microemulsions with rhamnolipid and sophorolipid biosurfactants: Formulation and potential applications. *Journal of Colloid and Interface Science* 348:498–504.

Nitschke, M., and S. G. V. A. Costa. 2007. Biosurfactants in food industry. *Trends Food Science and Technology* 18:252–259.

Nitschke, M., and G. M. Pastore. 2006. Production and properties of a surfactant obtained from *Bacillus subtilis* grown on cassava wastewater. *Bioresource Technology* 97:336–341.

Ortiz, A., Teruel, J. A., Espuny, M. J., Marqués, A., Manresa, A., and F. J. Aranda. 2009. Interactions of a bacterial biosurfactant trehalose lipid with phosphatidylserine membranes. *Chemistry and Physics Lipids* 158:46–53.

Ostroumova, O. S., Malev, V. V., Ilin, M. G., and L. V. Schagina. 2010. Surfactin activity depends on the membrane dipole potential. *Langmuir* 26:15092–15097.

Owen, D., L. Fan., and L. C. Paul. 2007. Polymeric biosurfactants. World Patent WO2007143006 A3, February 28.

Perfumo, A., Rancich, I., and I. M. Banat. 2010b. Possibilities and challenges for biosurfactants uses in petroleum industry. In *Biosurfactants. Advances in Experimental Medicine and Biology*, ed. R. Sen, pp. 135–145. Austin: Landes Bioscience.

Perfumo, A., Smyth, T. J. P., Marchant, R., and I. M. Banat. 2010a. Production and roles of biosurfactants and bioemulsifiers in accessing hydrophobic substrates. In *Handbook of Hydrocarbon and Lipid Microbiology*, ed. K. N. Timmis, pp. 1501–1512. Berlin Heidelberg: Springer-Verlag.

Pervaiz, M., and M. Sain. 2010. Extraction and characterization of extra cellular polymeric substances (EPS) from waste sludge of pulp and papermill. *International Review of Chemical Engineering* 2:550–554.

Pierce, D., and T. J. Heilman. 2001. Germicidal composition. United States Patent 6262038, July 17.

Piljac, A., Stipcević, T., Piljac-Zegarac, J., and G. Piljac. 2008. Successful treatment of chronic decubitus ulcer with 0.1% dirhamnolipid ointment. *Journal of Cutaneous Medicine and Surgery* 12:142–146.

Piljac, T., and G. Piljac. 1999. Use of rhamnolipids in wound healing, treating burn shock, atherosclerosis, organ transplants, depression, schizophrenia and cosmetics. European Patent EP1889623 A3, February 20.

Pires, R. C., Araújo, L. V., Ferreira, F. S., de Siqueira, R. S., and M. Nitschke. 2007. Potential application of surfactin to inhibit the adhesion oh pathogens on stainless steel surfaces. *Revista Higiene Alimentar* 21:530–531.

Quagliotto, P. L., Montoneri, E., Tambone, F., Adani, F., Gobetto, R., and G. Viscardi. 2006. Chemicals from wastes: Compost-derived humic acid-like matter as surfactant. *Environmental Science and Technology* 40:1686–1692.

Quinn, G. A., Maloy, A. P., Banat, M. M., and I. M. Banat. 2013. A comparison of effects of broad-spectrum antibiotics and biosurfactants on established bacterial biofilms. *Current Microbiology* 67:614–623.

Raaijmakers, J. M., De Bruijn, I., and M. J. D. De Kock. 2006. Cyclic lipopeptide production by plant-associated *Pseudomonas* spp.: Diversity, activity, biosynthesis, and regulation. *Molecular Plant-Microbe Interactions* 19:699–710.

Ramisse, F., Delden, C., and S. Gidenne. 2000. Decreased virulence of a strain of *Pseudomonas aeruginosa* O12 overexpressing a chromosomal type 1 β-lactamase could be due to reduced expression of cell-to-cell signalling dependent virulence factors. *FEMS Immunology and Medical Microbiology* 28:241–245.

Rehman, A., Raza, Z. A., Rehman, S. U. et al. 2010. Synthesis and use of self-assembled rhamnolipid microtubules as templates for gold nanoparticles assembly to form gold microstructures. *Journal of Colloid and Interface Science* 347:332–335.

Reis, R. S., Pacheco, G. J., Pereira, A. G., and D. M. G. Freire. 2013. Biosurfactants: Production and applications. In *Biodegradation—Life of Science*, ed. R. Chamy, pp. 31–61. InTech. Available from: http://www.intechopen.com/books/biodegradation-life-of-science/biosurfactants-production-and-applications.

Rivardo, F., Martinotti, M. G., Turner, R. J., and H. Ceri 2011. Synergistic effect of lipopeptide biosurfactant with antibiotics against *Escherichia coli* CFT073 biofilm. *International Journal of Antimicrobial Agents* 37:324–331.

Robbel, L., and M. A. Marahiel. 2010. Daptomycin, a bacterial lipopeptide synthesized by a nonribosomal machinery. *Journal of Biological Chemistry* 285:27501–27508.

Robinson, T., McMullan, G., Marchant, R., and P. Nigam. 2001. Remediation of dyes in textile effluent: A critical review on current treatment technologies with a proposed alternative. *Bioresource Technology* 77:247–255.

Rodrigues, L., Banat, I. M., Teixeira, J., and R. Oliveira. 2006. Biosurfactants: Potential applications in medicine. *Journal of Antimicrobial Chemotherapy* 57:609–618.

Rodrigues, L. R., Banat, I. M., Teixeira, J., and R. Oliveira. 2007. Strategies for the prevention of microbial biofilm formation on silicone rubber voice prostheses. *Journal of Biomedical Materials Research Part B: Applied Biomaterials* 81B:358–370.

Rodrigues, L. R., and J. A. Teixeira. 2010. Biomedical and therapeutic applications of biosurfactants. *Advance in Experimental Medicine and Biology* 672:75–87.

Rosenberg, E., and E. Z. Ron. 1999. High- and low-molecular-mass microbial surfactants. *Applied Microbiology and Biotechnology* 52:154–162.

Rosenberg, E., Schwartz, Z., Tenenbaum, A., Rubinovitz, C., Legmann R., and E. Z. Ron. 1989. A microbial polymer that changes the surface properties of limestone: Effect of biodispersan in grinding limestone and making paper. *Journal of Dispersion Science and Technology* 10:241–250.

Saini, H. S., Barragán-Huerta, B. E., Lebrón-Paler, A. et al. 2008. Efficient purification of the biosurfactant viscosin from *Pseudomonas libanensis* strain M9-3 and its physicochemical and biological properties. *Journal of Natural Products* 71:1011–1015.

Sánchez, M., Aranda, F. J., Teruel, J. A. et al. 2010. Permeabilization of biological and artificial membranes by a bacterial dirhamnolipid produced by *Pseudomonas aeruginosa*. *Journal of Colloid and Interface Science* 341:240–247.

Satpute, S. K., Banpurkar, A. G., Dhakephalkar, P. K., Banat, I. M., and B. A. Chopade. 2010. Methods for investigating biosurfactants and bioemulsifiers: A review. *Critical Reviews in Biotechnology* 30:127–144.

Savarino, P., Montoneri, E., Biasizzo, M., Quagliotto, P., Viscardi, G., and V. Boffa. 2007. Upgrading biomass wastes in chemical technology. Humic acid-like matter isolated from compost as chemical auxiliary for textile dyeing. *Journal of Chemical Technology and Biotechnology* 82:939–948.

Savarino, P., Montoneri, E., Bottigliengo, S. et al. 2009. Biosurfactants from urban wastes as auxiliaries for textile dyeing. *Industrial and Engineering Chemistry Research* 48:3738–3748.

Schneider, T., and H. G. Sahl. 2010a. Lipid II and other bactoprenol-bound cell wall precursors as drug targets. *Current Opinion in Investigational Drugs* 11:157–164.

Schneider, T., and H. G. Sahl. 2010b. An oldie but a goodie—Cell wall biosynthesis as antibiotic target pathway. *International Journal of Medical Microbiology* 300:161–169.

Scott, W. R., Baek, S. B., Jung, D., Hancock, R. E., and S. K. Straus. 2007. NMR structural studies of the antibiotic lipopeptide daptomycin in DHPC micelles. *Biochimica et Biophysica Acta* 1768:3116–3126.

Sen, R. 2008. Biotechnology in petroleum recovery: The microbial EOR. *Progress in Energy and Combustion Science* 34:714–724.

Sen, R. 2010. *Biosurfactant. Advances in Experimental Medicine and Biology.* New York: Landes Bioscience and Springer Science.

Seydlová, G., Čabala, R., and J. Svobodová. 2011. Surfactin—Novel solutions for global issues. In *Biomedical Engineering, Trends, Research and Technologies*, ed. M. A. Komorowska, and S. Olsztynska-Janus, pp. 305–330. Rijeka: InTech.

Seydlová, G., and J. Svobodová. 2008. Review of surfactin chemical properties and the potential biomedical applications. *Central European Journal of Medicine* 3:123–133.

Sha, R. Y., Jiang, L. F., Meng, Q., Zhang, G. L., and Z. R. Song. 2012. Producing cell-free culture broth of rhamnolipids as a cost-effective fungicide against plant pathogens. *Journal of Basic Microbiology* 52: 458–466.

She, Y. H., Zhang, F., Xia, J. J. et al. 2011. Investigation of biosurfactant-producing indigenous microorganisms that enhance residue oil recovery in an oil reservoir after polymer flooding. *Applied Biochemistry and Biotechnology* 163:223–234.

Shete, A. M., Wadhawa, G., Banat, I. M., and B. A. Chopade. 2006. Mapping of patents on bioemulsifier and biosurfactant: A review. *Journal of Scientific and Industrial Research* 65:91–115.

Simpson, D. R. Natraj, N. R., McInerney, M. J., and K. E. Duncan. 2011. Biosurfactant-producing *Bacillus* are present in produced brines from Oklahoma oil reservoirs with a wide range of salinities. *Applied Microbiology and Biotechnology* 91:1083–1093.

Smyth, T. J. P., Perfumo, A., McClean, S., Marchant, R., and I. M. Banat. 2010a. Isolation and analysis of lipopeptides and high molecular weight biosurfactants. In *Handbook of Hydrocarbon and Lipid Microbiology*, ed. K. N. Timmis, pp. 3689–3704. Berlin Heidelberg: Springer-Verlag.

Smyth, T. J. P., Perfumo, A., McClean, S., Marchant, R., and I. M. Banat. 2010b. Isolation and analysis of low molecular weight microbial glycolipids. In *Handbook of Hydrocarbon and Lipid Microbiology*, ed. K. N. Timmis, pp. 3705–3723. Berlin Heidelberg: Springer-Verlag.

Snook, M. E., Mitchell, T., Hinton, D. M., and C. W. Bacon. 2009. Isolation and characterization of Leu7-surfactin from the endophytic bacterium *Bacillus mojavensis* RRC 101, a biocontrol agent for *Fusarium verticillioides*. *Journal of Agricultural and Food Chemistry* 57:4287–4292

Soberón-Chávez, G. 2011. *Biosurfactant: From Genes to Applications.* Berlin Heidelberg: Springer-Verlag.

Stanghellini, M. E., and R. M. Miller. 1997. Their identity and potential efficacy in the biological control of zoosporic plant pathogens. *Plant Disease* 81:4–12.

Stepanović, S. Cirković, I., Ranin, L., and M. Svabić-Vlahović. 2004. Biofilm formation by *Salmonella* spp. and *Listeria monocytogenes* on plastic surface. *Letters in Applied Microbiology* 38:428–432.

Stipcevic, T., Piljac, A., and G. Piljac. 2006. Enhanced healing of full-thickness burn wounds using di-rhamnolipid. *Burns* 32:24–34.

Sun, L., Lu, Z., Bie, X., Lu, F., and S. Yang. 2006. Isolation and characterization of a co-producer of fengycins and surfactins, endophytic *Bacillus amyloliquefaciens* ES-2, from *Scutellaria baicalensis* Georgi. *World Journal of Microbiology and Biotechnology* 22:1259–1266.

Suzuki, M., Kitagawa, M., Yamamoto, S. et al. 2010. Activator including biosurfactant as active ingredient, mannosyl erythritol lipid, and production method publication. United States Patent US20100168405A1, July 1.

Tally, F. P., Zeckel, M., Wasilewski, M. M. et al. 1999. Daptomycin: A novel agent for Gram-positive infections. *Expert Opinion on Investigation Drugs* 8:1223–1238.

Teichmann, B., Linne, U., Hewald, S., Marahiel, M. A., and M. Bölker. 2007. A biosynthetic gene cluster for a secreted cellobiose lipid with antifungal activity from *Ustilago maydis*. *Molecular Microbiology* 66:525–533.

Tendulkar, S. R., Saikumari, Y. K., Patel, V. et al. 2007. Isolation, purification and characterization of an anti-fungal molecule produced by *Bacillus licheniformis* BC98, and its effect on phytopathogen *Magnaporthe grisea*. *Journal of Applied Microbiology* 103:2331–2339.

Ueno, Y., Hirashima, N., Inoh, Y., Furuno, T., and M. Nakanishi. 2007a. Characterization of biosurfactant-containing liposomes and their efficiency for gene transfection. *Biological and Pharmaceutical Bulletin* 30:169–172.

Ueno, Y., Inoh, Y., Furuno, T., Hirashima, N., Kitamoto, D., and M. Nakanishi. 2007b. NBD-conjugated bio-surfactant (MEL-A) shows a new pathway for transfection. *Journal of Controlled Release* 123:247–253.

Van Haesendonck, I. P. H., and E. C. A. Vanzeveren. 2004. Rhamnolipids in bakery products. World Patent WO2004040984, May 21.

Varnier, A. L., Sanchez, L., Vatsa, P. et al. 2009. Bacterial rhamnolipids are novel MAMPs conferring resistance to *Botrytis cinerea* in grapevine. *Plant, Cell and Environment* 32:178–193.

Vater, J., Kablitz, B., Wilde, C., Franke, P., Mehta, N., and S. S. Cameotra. 2002. Matrix-assisted laser desorp-tion ionization—Time of flight mass spectrometry of lipopeptide biosurfactants in whole cells and culture filtrates of *Bacillus subtilis* C-1 isolated from petroleum sludge. *Applied and Environmental Microbiology* 68:6210–6219.

Vatsa, P., Sanchez, L., Clement, C., Baillieul, F., and S. Dorey. 2010. Rhamnolipid biosurfactants as new players in animal and plant defense against microbes. *International Journal of Molecular Sciences* 11: 5095–5108.

Wagner, C., Graninger, W., Presterl, E., and C. Joukhadar. 2006. The echinocandins: Comparison of their phar-macokinetics, pharmacodynamics and clinical applications. *Pharmacology* 78:161–77.

Worakitkanchanakul, W., Imura, T., Fukuoka, T. et al. 2008. Aqueous-phase behavior and vesicle formation of natural glycolipid biosurfactant, mannosylerythritol lipid-B. *Colloids and Surfaces B Biointerfaces* 65:106–112.

Xia, W. J., Luo, Z. B., Dong, H. P., Yu, L., Cui, Q. F., and Y. Q. Bi. 2012. Synthesis, characterization, and oil recovery application of biosurfactant produced by indigenous *Pseudomonas aeruginosa* WJ-1 using waste vegetable oils. *Applied Biochemistry and Biotechnology* 166:1148–1166.

Yakimov, M. M., Timmis, K. N., Wray, V., and H. L. Fredrickson. 1995. Characterization of a new lipopep-tide surfactant produced by thermotolerant and halotolerant subsurface *Bacillus licheniformis* BAS 50. *Applied and Environmental Microbiology* 61:1706–1713.

Yalcin, E., and K. Cavusoglu. 2010. Structural analysis and antioxidant activity of a biosurfactant obtained from *Bacillus subtilis* RW-I. *Turkish Journal of Biochemistry* 35:243–247.

Yamamoto, S., Morita, T., Fukuoka, T. et al. 2012. The moisturizing effects of glycolipid biosurfactants, man-nosylerythritol lipids, on human skin. *Journal of Oleo Science* 61:407–412.

Yamane, T. 1987. Enzyme technology for the lipid industry. An engineering overview. *Journal of the American Oil Chemists' Society* 64:1657–1662.

Yan, P., Lu, M., Yang, Q., Zhang, H. L., Zhang, Z. Z., and R. Chen. 2012. Oil recovery from refinery oily sludge using a rhamnolipid biosurfactant-producing *Pseudomonas*. *Bioresource Technology* 116:24–28.

Yao, J., Liu, H., Zhou, T. et al. 2012. Total synthesis and structure-activity relationships of new echinocandin-like antifungal cyclolipohexapeptides. *European Journal of Medicinal Chemistry* 50:196–208.

Yoneda, T. 2006. Cosmetic composition comprising a and a lipopeptides. United States Patent US0222616A1, October 5.

Yoneda, T., Ito, S., Masatsuji, E. et al. 1999. Surfactant for use in external preparations for skin and external preparation for skin containing the same. World patent WO1999062482 A1, December 9.

Yoneda, T., Ito, N., and T. Yoneda. 2005. Oil-in-water emulsified composition and external preparation for skin and cosmetics using the composition. World patent WO2005089708 A1, September 29.

Zaragoza, A., Aranda, F. J., Espuny, M. J. et al. 2009. Mechanism of membrane permeabilization by a bacterial trehalose lipid biosurfactant produced by *Rhodococcus* sp. *Langmuir* 25:7892–7898.

Zeraik, A. E., and M. Nitschke. 2010. Biosurfactants as agents to reduce adhesion of pathogenic bacteria to polystyrene surfaces: Effect of temperature and hydrophobicity. *Current Microbiology* 61:554–559.

Zhang, F., She, Y. H., Li, H. M. et al. 2012. Impact of an indigenous microbial enhanced oil recovery field trial on microbial community structure in a high pour-point oil reservoir. *Applied Microbiology and Biotechnology* 95:811–821.

Zhao, L., Ma, T., Gao, M. et al. 2012 Characterization of microbial diversity and community in water flooding oil reservoirs in China. *World Journal of Microbiology and Biotechnology* 28:3039–3052.

Zheng, C., Yu L., Huang, L., Xiu, J., and Z. Huang. 2012. Investigation of a hydrocarbon-degrading strain, *Rhodococcus ruber* Z25, for the potential of microbial enhanced oil recovery. *Journal of Petroleum Science and Engineering* 81:49–56.

13 Biological Applications of Biosurfactants and Strategies to Potentiate Commercial Production

Mohd Sajjad Ahmad Khan, Brijdeep Singh, and Swaranjit Singh Cameotra

CONTENTS

13.1 Introduction 270
13.2 Classifications of Biosurfactants..271
13.3 Applications of Biosurfactants 271
    13.3.1 Role of Biosurfactants in Biodegradation Processes 271
    13.3.2 Biosurfactants and Metals Remediation...274
    13.3.3 Biosurfactants in the Cosmetic Industry 275
    13.3.4 Oil Recovery and Processing..275
        13.3.4.1 Microbial Enhanced Oil Recovery 275
    13.3.5 Biosurfactants in the Food Industry 276
        13.3.5.1 Food Emulsifier...276
        13.3.5.2 Food Stabilizer..276
    13.3.6 Application in Medicine 276
        13.3.6.1 Genetic Manipulation 277
        13.3.6.2 Immune Modulatory Action 277
        13.3.6.3 Toxic Activity against Microorganisms...277
        13.3.6.4 Antiadhesive Agents 277
        13.3.6.5 Antimicrobial Activity of Biosurfactants 278
    13.3.7 Application in Agriculture...278
        13.3.7.1 Improvement of Soil Quality 279
        13.3.7.2 Plant–Pathogen Elimination 279
13.4 Strategies to Enhance Production of Biosurfactants for Its Industrial Applications...........279
    13.4.1 Cheap Substrates: Economical and Promising Alternatives 280
        13.4.1.1 Vegetable Oils and Oil Wastes...280
        13.4.1.2 Lactic Whey and Distillery Wastes 283
        13.4.1.3 Starchy Substrates..285
        13.4.1.4 Biosurfactant Production from Lignocellulosic Waste...................285
    13.4.2 Bioprocess Development: Optimum Production and Recovery 286
    13.4.3 Process Optimization: The Best Combination of Essential Factors.....................286
    13.4.4 Downstream Processing: Fast, Efficient, and Cheap Product Recovery 287
    13.4.5 Mutant and Recombinant Strains: The Hyper-Producers................................287
13.5 Conclusion 288
References...288

13.1 INTRODUCTION

Biosurfactants are amphiphilic compounds produced by microorganisms such as bacteria, yeasts, and molds with pronounced surface and emulsifying activities (Singh et al. 2007). The presence of hydrophobic and hydrophilic groups confers these molecules the ability to accumulate between interfaces of dissimilar polarities like liquid–air, liquid–liquid, or liquid–solid interface, and thereby reducing surface and interfacial tension at the surface and interface regions, respectively (Karanth et al. 1999). Microbially produced surfactants are soluble in both organic (nonpolar) and aqueous (polar) solvents and categorized by their chemical composition and microbial origin. They include glycolipids, lipopeptides, polysaccharide–protein complexes, protein-like substances, lipopolysaccharides, phospholipids, fatty acids, and neutral lipids (Van Hamme et al. 2006).

Variation in the chemical nature of biosurfactants is responsible for its diverse properties and physiological functions such as increasing the surface area and bioavailability of hydrophobic water-insoluble substrates, heavy metal binding, bacterial pathogenesis, quorum sensing, and biofilm formation (Singh and Cameotra 2004). These compounds can be synthesized by the microorganisms growing on water-immiscible hydrocarbons as well as on water-soluble compounds (Mukherjee et al. 2006). But optimization of environmental conditions and nutrient sources or use of genetically altered microorganisms may lead to increased production of biosurfactants. Enhancing the yield of these biomolecules is highly esteemed for their exploitation in industries.

Primarily, biosurfactants attracted attention as hydrocarbon dissolution agents in the 1960s, and their applications have been greatly extended in the past five decades as an improved alternative to chemical surfactants (carboxylates, sulfonates, and sulfate acid esters), especially in agriculture, cosmetics, food, pharmaceutical, and oil industry (Banat et al. 2000). As far as biological applications of biosurfactants are considered, these molecules have an edge over their chemically synthesized counterparts such as

1. *Biodegradability:* Because of the low toxicity and simple chemical structure, these compounds do not persist in the environment and are degraded easily. This property prevents the problem of accumulation of biosurfactants such as shown by chemical surfactants. Therefore, these molecules are greatly environmental friendly and termed as green chemicals (Abdel-Mawgoud et al. 2010).
2. *Biocompatability and digestibility:* Biological origin of these molecules imparts them an inherent characteristic of compatibility with organisms. This property allows their unabated usage in cosmetics, pharmaceuticals, and as functional food additives.
3. *Availability of raw materials and economic production:* Biosurfactants can be produced from relatively cheap raw materials and renewable feedstocks by microorganisms. These materials such as industrial wastes and by-products are available in abundance and provide carbon source ranging from hydrocarbons, carbohydrates to lipids and may be used separately or in combination with each other for microbial production.
4. *Environmental control:* The amphiphillic nature of biosurfactant is exploited in the processes for stabilization of industrial emulsions, control of oil-spills, biodegradation, and detoxification of industrial effluents. Also, bioremediation of contaminated soil can be favored with the use of biosurfactants.
5. *Specificity:* The presence of specific functional groups imparts specificity in the action by the biosurfactant molecules. This property can be of paramount importance in detoxification of specific pollutants, de-emulsification of industrial emulsions, development of specific cosmetic, specialized pharmaceutical, and food applications.
6. *Stability at extreme conditions:* The most significant properties of these microbial products are effectiveness at the extremes of temperature, pH, and salinity. Along with these parameters and the unique structure markedly differing from the chemical surfactants can promote alternate usage for which the classical surfactants fail.

During the past few years, biosurfactant production by various microorganisms has been studied extensively and also the various aspects of biosurfactants such as their biomedical and therapeutic properties, natural roles, production using cheap alternate substrates, and commercial potential, which has been reviewed recently. Biosurfactants are diverse in their applications and are becoming beneficial molecules for industrial uses. When these are used in the food-processing, cosmetic, and pharmaceutical industries, biosurfactants have been included in formulations to serve as emulsifiers and solubilizers, as well as foaming, wetting, antiadhesive, antimicrobial, and antiviral agents (Rodrigues et al. 2006; Shete et al. 2006; Muthusamy et al. 2008). The principle aim of this chapter is to focus special emphasis on the exploitation of biosurfactants by various industries utilizing cheap substrates as carbon source. The reason for use of biosurfactants in vast applications is that biosurfactants possess the characteristic property of reducing the surface and interfacial tension using the same mechanisms as that for chemical surfactants. Unlike chemical surfactants, which are mostly derived from petroleum feedstock, these molecules can be produced by microbial fermentation processes using cheaper agro-based substrates and waste materials. The screening of overproducer microbial strains and optimization of cultural parameters are other strategies to increase biosurfactants yield and applications.

13.2 CLASSIFICATIONS OF BIOSURFACTANTS

Unlike chemically synthesized surfactants, which are usually classified according to the nature of their polar groups, biosurfactants are generally categorized mainly by their chemical composition and microbial origin. Biosurfactants are generally classified into two categories: (i) low-molecular-mass molecules, which efficiently lower surface and interfacial tension; and (ii) high-molecular-mass polymers, more effective as emulsion-stabilizing agents. The major classes of low-mass surfactants include glycolipids, lipopeptides, and phospholipids, whereas high-mass surfactants include polymeric and particulate surfactants. Most biosurfactants derived from diverse microbial sources are either anionic or neutral and the hydrophobic moiety is based on long-chain fatty acids or fatty acid derivatives, whereas the hydrophilic portion can be a carbohydrate, amino acid, phosphate, or cyclic peptide (Nitschke and Coast 2007) (Table 13.1).

13.3 APPLICATIONS OF BIOSURFACTANTS

Biosurfactants produced by microorganisms are becoming important biotechnology products for industrial and medical applications due to their specific modes of action, low toxicity, relative ease of preparation, and widespread applicability. They can be used as emulsifiers, de-emulsifiers, wetting and foaming agents, functional food ingredients, and as detergents in petroleum, petrochemicals, environmental management, agrochemicals, foods, and beverages, cosmetics and pharmaceuticals, and in mining and metallurgical industries. The addition of a surfactant of chemical or biological origin accelerates or sometimes inhibits the bioremediation of pollutants. Surfactants also play an important role in enhanced oil recovery by increasing the apparent solubility of petroleum components and effectively reducing the interfacial tensions of oil and water *in situ*. For medical applications, biosurfactants are useful as antimicrobial agents and immunomodulatory molecules. The properties of biosurfactants that make them useful in various applications are presented in Figure 13.1 and Table 13.2. Beneficial applications of biosurfactants in various industries are discussed in this chapter.

13.3.1 ROLE OF BIOSURFACTANTS IN BIODEGRADATION PROCESSES

Bioremediation typically involves augmentation of soil or other media, contaminated with pollutants with nutrients and sometimes microorganisms, to improve processes for biodegradation of the contaminants. Biodegradation rate of a contaminant in soil depends on its bioavailability

TABLE 13.1

Classification and Types of Biosurfactant Produced by Various Microorganisms

Biosurfactant		Microorganisms
Group	Class	
Glycolipids	Rhamnolipids	*Pseudomonas aeruginosa*
	Trehalolipids	*Mycobacterium tuberculosis, Rhodococcus erythropolis*
	Sophorolipids	*Torulopsis bombicola, Torulopsis petrophilum*
Fatty acids, phospholipids, and neutral lipids	Corynomycolic acid	*Corynebacterium lepus*
	Spiculisporic acid	*Penicillium spiculisporum*
	Phosphatidylethanolamine	*Acinetobacter* sp., *Rhodococcus erythropolis*
Lipopeptides	Surfactin	*Bacillus subtilis*
	Lichenysin	*Bacillus licheniformis*
Polymeric biosurfactants	Emulsan	*Acinetobacter calcoaceticus* RAG-1
	Alasan	*Acinetobacter radioresistens* KA-53
	Biodispersan	*Acinetobacter calcoaceticus* A2
	Liposan	*Candida lipolytica*
	Mannoprotein	*Saccharomyces cerevisiae*

to the metabolizing organisms, which is influenced by factors such as desorption, diffusion, and dissolution. A promising method that can improve bioremediation effectiveness of hydrocarbon-contaminated environments is the use of biosurfactants. Many of the most persistent contaminants exhibit low water solubility and hence, bioavailability of contaminants can often be improved by addition of emulsifiers. By reducing surface and interfacial tension among liquids, solids, and gases,

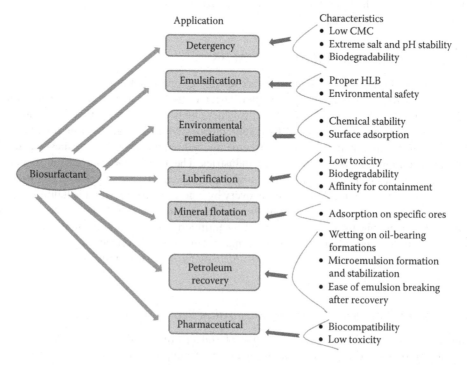

FIGURE 13.1 Characteristics of biosurfactants useful in various applications.

TABLE 13.2
Industrial Applications of Chemical Surfactants and Biosurfactants

Industry	Application	Role of Surfactants
Petroleum	Enhanced oil recovery	Improving oil drainage into well bore; stimulating release of oil entrapped by capillaries; wetting of solid surfaces; reduction of oil viscosity and oil pour point; lowering of interfacial tension; dissolving of oil
	De-emulsification	De-emulsification of oil emulsions; oil solubilization; viscosity reduction, wetting agent
Environmental	Bioremediation	Emulsification of hydrocarbons; lowering of interfacial tension; metal sequestration
	Soil remediation and flushing	Emulsification through adherence to hydrocarbons; dispersion; foaming agent; detergent; soil flushing
Food	Emulsification and de-emulsification	Emulsifier; solubilizer; demulsifier; suspension, wetting, foaming, defoaming, thickener, lubricating agent
	Functional ingredient	Interaction with lipids, proteins, and carbohydrates, protecting agent
Biological	Microbiological	Physiological behavior such as cell mobility, cell communication, nutrient accession, cell–cell competition, plant and animal pathogenesis
	Pharmaceuticals and therapeutics	Antibacterial, antifungal, antiviral agents; adhesive agents; immunomodulatory molecules; vaccines; gene therapy
Agricultural	Biocontrol	Facilitation of biocontrol mechanisms of microbes such as parasitism, antibiosis, competition, induced systemic resistance, and hypovirulence
Bioprocessing	Downstream processing	Biocatalysis in aqueous two-phase systems and microemulsions; biotransformations; recovery of intracellular products; enhanced production of extracellular enzymes and fermentation products
Cosmetic	Health and beauty products	Emulsifiers, foaming agents, solubilizers, wetting agents, cleansers, antimicrobial agent, mediators of enzyme action

allowing them to disperse readily as emulsions, chemical or biological surfactants may have variable effects on contaminant biodegradation (Banat et al. 2000).

They can enhance hydrocarbon bioremediation by two mechanisms. The first includes the increase of substrate bioavailability for microorganisms, while the other involves interaction with the cell surface that increases the hydrophobicity of the surface allowing hydrophobic substrates to associate more easily with bacterial cells. Addition of biosurfactants can be expected to enhance hydrocarbon biodegradation by mobilization, solubilization, or emulsification (Urum and Pekdemir 2004; Nguyen et al. 2008; Nievas et al. 2008).

The capability of biosurfactants and biosurfactant-producing bacterial strains to enhance organic contaminants' availability and biodegradation rates was reported by many authors (Rahman et al. 2003; Inakollu et al. 2004). Martínez-Checa et al. (2007) investigated the usefulness of the V2-7 bioemulsifier-producing strain F2-7 of *Halomonas eurihalina* in oil bioremediation process. First, they studied the capacity of strain F2-7 to grow and produce bioemulsifier in the presence of different hydrocarbon compounds. They observed that all analyzed hydrocarbons supported the growth of F2-7 strain and the production of V2-7 bioemulsifier. The ability of the analyzed strain to remove polycyclic aromatic hydrocarbons was investigated during the growth of this strain for 96 h in liquid medium supplemented with naphthalene, phenanthrene, fluoranthene, and pyrene. After the experiment, the obtained residual concentrations of fluoranthene (56.6%) and pyrene (44.5%) were higher than naphthalene (13.6%) and phenanthrene (15.6%). Efficiency of strain F2-7 in removing poly aromatic hydrocarbons (PAHs) confirmed its potential applicability in oil bioremediation technology.

Obayori et al. (2009) investigated the biodegradative properties of biosurfactant produced by *Pseudomonas* sp. LP1 strain on crude oil and diesel. The results obtained confirmed the ability of strain LP1 to metabolize the hydrocarbon components of crude and diesel oil. They reported 92.34% degradation of crude oil and 95.29% removal of diesel oil. Biodegradative properties of biosurfactant-producing *Brevibacterium* sp. PDM-3 strain were tested by Reddy et al. (2010). They reported that this strain could degrade 93.92% of the phenanthrene and also had the ability to degrade other polyaromatic hydrocarbons such as anthracene and fluorene.

Kang et al. (2010) used sophorolipid in studies on the biodegradation of aliphatic and aromatic hydrocarbons and Iranian light, crude oil under laboratory conditions. The addition of this biosurfactant to soil also increased biodegradation of tested hydrocarbons with the rate of degradation ranging from 85% to 97% of the total amount of hydrocarbons. Their results indicated that sophorolipid may have the potential for facilitating the bioremediation of sites contaminated with hydrocarbons having limited water solubility and increasing the bioavailability of microbial consortia for biodegradation.

13.3.2 BIOSURFACTANTS AND METALS REMEDIATION

The contamination of soil environments with heavy metals is very hazardous for human and other living organisms in the ecosystem. Owing to their extremely toxic nature, the presence of even low concentrations of heavy metals in the soil has been found to have serious consequences. Application of microorganisms was discovered many years ago to help in the reduction of metal contamination. Heavy metals are not biodegradable; they can only be transferred from one chemical state to another, which changes their mobility and toxicity. Microorganisms can influence metals in several ways. Some forms of metals can be transformed either by redox processes or by alkylation.

Metal wastes are produced by a variety of sources including mines, tanneries, and electroplating facilities and through the manufacture of paints, metal pipes, batteries, and ammunition. Metal contamination has been linked to birth defects, cancer, skin lesions, mental and physical retardation, learning disabilities, liver and kidney damage, and a host of other maladies. Microorganisms and their products, namely, biosurfactants are extensively used for the enhanced metal and metal co-contaminated site bioremediation (Singh and Cameotra 2004).

Biosurfactants can be applied to a small part of the contaminated soil: the soil is placed in a huge cement mixer, the biosurfactant–metal complex is flushed out, soil gets deposited back, and the biosurfactant–metal complex is treated to precipitate the biosurfactant, leaving behind the metal. The bond formed between the positively charged metal and the negatively charged surfactant is so strong that flushing water through soil removes the surfactant–metal complex from the soil matrix. This method can also be carried out for deeper subsurface contamination only with more pumping activities (Pacwa-Płociniczak et al. 2011).

According to Miller (1995), biosurfactants enhance the desorption of heavy metals from soils in two ways: (i) complexation of the free form of the metal residing in solution. This decreases the solution phase activity of the metal and, therefore, promotes desorption according to Le Chatelier's principle. (ii) Direct contact of biosurfactant to sorbed metal at solid solution interface under conditions of reduced interfacial tension, which allows biosurfactants to accumulate at solid solution interface.

Considerable work has been done on rhamnolipid biosurfactant produced by various *Pseudomonas aeruginosa* strains capable of selectively complexing cationic metal species such as Cd, Pb, and Zn. Studies have shown that this surfactant complexes preferentially with toxic metals such as cadmium and lead than with normal soil metal cations such as calcium and magnesium, for which it has a much lower affinity (Herman et al. 1995; Torrens et al. 1998). The feasibility of using surfactin, rhamnolipid, and sophorolipid for the removal of heavy metals, Cu and Zn, from sediments has been evaluated by Mulligan et al. (1999, 2001). In one study by Jeong-Jin et al. (1998), biosurfactant-based ultrafiltration was used to remove the divalent metal ions, $Cu\flat2$, $Zn\flat2$, $Cd\flat2$, and $Ni\flat2$, from aqueous solution containing either a single metal species or a mixture of metal ions. Todd et al. (2000) studied the effectiveness of rhamnolipid biosurfactants in the remediation of cadmium and naphthalene co-contaminated site. They observed reduced cadmium toxicity by *P. aeruginosa* rhamnolipid leading to an enhanced naphthalene biodegradation by a *Burkholderia* species.

Parthasarathi and Sivakumaar (2011) investigated the effects of biosurfactant produced by a mangrove isolate on a heavy metal-spiked soil remediation using two different methods of biosurfactant addition (pretreatment and direct application) at different concentrations (0.5–5%) for 10 days employing column and batch method of washings. The pre-addition of biosurfactant at 0.5% concentrations and further incubation for a month resulted in better chromium removal than the direct biosurfactant washing method. A maximum recovery of lead (99.77%), nickel (98.23%), copper (99.62%), and cadmium (99.71%) was achieved with a column washing method at 1% biosurfactant concentration. Release of 26% soluble fractions of nickel (pre-addition with biosurfactant) and 40% copper (direct application) was achieved by the column washing method at 1.0% concentration of biosurfactant. A total of 0.034 mg/10 g of lead, 0.157 mg/10 g of nickel, 0.022 mg/10 g of copper, 0.025 mg/10 g of cadmium, and 0.538 mg/10 g of chromium were found to remain in the spiked soil after column washing with 1.0% biosurfactant solution. However, pre-addition of 0.5% biosurfactant treatment helps in maximum removal of chromium metal leaving a residual concentration of 0.426 mg/10 g of soil, suggesting effective removal at very low concentration. The average extraction concentration of metals in batch washings was between 93% and 100%, irrespective of the concentration of biosurfactant studied.

13.3.3 BIOSURFACTANTS IN THE COSMETIC INDUSTRY

Multifunctional biosurfactants have several cosmetic applications because of their exceptional surface properties such as detergency, wetting, emulsifying, solubilizing, dispersing, and foaming effects. The most widely used biosurfactant glycolipids in cosmetics are sophorolipids, rhamnolipids, and mannosylerythritol lipids. Sophorolipids have good skin compatibility and excellent moisturizing properties, rhamnolipids are natural surfactants and emulsifiers that can replace petrochemical-based surfactants used in most of the cosmetic products. They also have been used in acne pads antidandruff, antiwrinkle, and antiaging products, deodorants, nail care products, and toothpastes in several different formulations, because of its high surface and emulsifying activities (Piljac and Piljac, 1999). Mannosylerythritol lipids are generally used in skin care formulations as the active ingredient to prevent skin roughness (Masaru et al. 2007).

13.3.4 OIL RECOVERY AND PROCESSING

Chemical surfactants and biosurfactants can increase the pseudo solubility of petroleum components in water (Pekdemir et al. 2005). Surfactants are effective in reducing the interfacial tensions of oil and water *in situ* and they can also reduce the viscosity of the oil and remove water from the oil prior to processing (Liu et al. 2004). Biosurfactants can be as effective as the synthetic chemical surfactants and for certain applications they have advantages such as high specificity. Most of the biosurfactants and many chemical surfactants employed for bioremediation purposes are biodegradable. Pollution of sea by crude oil, mostly caused by stranding of tankers, is one of the urgent and serious environmental issues across the world (Olivera et al. 2003). Chemical surfactants are used over few decades for the degradation of these hydrocarbons due to its high toxicity and low degradability.

13.3.4.1 Microbial Enhanced Oil Recovery

Poor oil recovery in oil-producing wells may be due to either low permeability of some reservoirs or high viscosity of the crude oil resulting in poor mobility. The concept of microbial-enhanced oil recovery (MEOR) was first proposed nearly 80 years ago but received only limited attention until the early 1980s (Stosur 1991). The ability of indigenous or injected microorganisms to synthesize useful fermentation products to improve oil recovery from the oil reservoirs is exploited in MEOR processes. Three strategies are recognized for biosurfactant application:

1. Biosurfactant production in batch or continuous culture and addition to the reservoir using the conventional way of MEOR
2. Production of biosurfactant by injected microbes at the cell–oil interface within the reservoir
3. Injection of selected nutrients into the reservoir to stimulate growth of indigenous biosur-factant-producing bacteria

The potential application of biosurfactants produced by the thermo- and halo-tolerant species of *Bacillus licheniformis* JF-2 and *Bacillus subtilis* has been explored for enhanced oil recoveries in labo-ratory columns and reservoirs with oil recoveries from 9.3% to 62% (Lin et al. 1994; Yakimov et al. 1997; Makkar and Cameotra 1998). Increases in oil recovery by about 30% have been reported from underground sandstone by using trehalolipids from *Nocardia rhodochrus* (Rapp et al. 1979). Oilfield emulsions, both oil in water and water in oil, are formed at various stages of exploration, production, and oil recovery and processing, represent a major problem for the petroleum industry (Manning and Thompson 1995). A process of de-emulsification is required to recover oil from these emulsions.

13.3.5 BIOSURFACTANTS IN THE FOOD INDUSTRY

Biosurfactants as biocompatible, biodegradable, and/or nontoxic compounds have the combination of particular characteristics that exhibit a variety of useful properties for the food industry, espe-cially as emulsifiers, foaming, wetting, solubilizers (Banat et al. 2000), antiadhesive, and antimicro-bial agents (Singh and Cameotra 2004).

13.3.5.1 Food Emulsifier

Biosurfactants show several properties such as emulsion-based formulations that have great poten-tial applications in many fields of food industry. An emulsion is a heterogeneous system, consisting of at least one immiscible liquid intimately dispersed in another in the form of droplets, having dispersed and continuous phase. The addition of emulsifiers is of special value for low-fat products (Rosenberg and Ron 1999) as it improves the texture and creaminess of dairy products. However, polymeric surfactants coat the droplets of oil and form very stable emulsions that never coalesce. This property is especially useful for making oil/water emulsions for cosmetics and food.

13.3.5.2 Food Stabilizer

Biosurfactants act as controlling consistency in bakery and ice cream formulations. They are also utilized as fat stabilizer and antispattering agent during cooking of oil and fats (Kosaric 2001). In food processing, the addition of rhamnolipid surfactants improves the texture and shelf-life of starch-containing products, modifies rheological properties and stability of wheat dough (Van Haesendonck and Venzeveren 2004). Surfactants can also be used to control the agglomeration of fat globules, stabilize aerated systems, and texture of fat-based products (Kachholz and Schlingmann 1987). L-Rhamnose, which already has an industrial application as precursor of high-quality flavor components like Furaneol (trademark of Firmenich SA, Geneva), is obtained by hydrolyzing rham-nolipid surfactants produced by *P. aeruginosa* (Linhardt et al. 1989).

13.3.6 APPLICATION IN MEDICINE

One of the earliest noted antimicrobial activities of biosurfactants was that of iturin A, a potent antifungal lipopeptide produced by strains of *B. subtilis*. Iturin A has been proposed as an effective antifungal agent for profound mycosis. Other members of the iturin group, including bacillomycin D and bacillomycin Lc, were also found to have antimicrobial activities (Eshita et al. 1995; Moyne et al. 2001; Singh and Cameotra 2004). Apart from antifungal and moderate antibacterial properties, surfactin (i) inhibits fibrin clot formation, (ii) induces formation of ion channels in lipid bilayer membranes, (iii) inhibits cyclic adenosine monophosphate (cAMP), (iv) inhibits platelet and spleen cytosolic phospholipase A2 (PLA2), and (v) exhibits antiviral and antitumor activities. Surfactin also has antimycoplasma properties and has been used in a fast and simple method for complete and permanent inactivation of mycoplasmas in mammalian monolayer and suspension cell cultures (Vollenbroich et al. 1997). Surfactin is active against several viruses, including semliki forest virus, herpes simplex virus (HSV-1 and HSV-2), suid herpes virus, vesicular stomatitis virus, simian immunodeficiency virus, feline calicivirus, and murine encephalomyocarditis virus (Singh and Cameotra 2004).

13.3.6.1 Genetic Manipulation

It is stated by Gharaei-Fathabad (2011) that the establishment of an efficient and safe method for introducing exogenous nucleotides into mammalian cells is critical for basic sciences and clinical applications such as gene therapy. Lipofection using cationic liposomes is considered to be a safe way to deliver foreign gene to the target cells without side effects among various known methods of gene transfection (Fujita et al. 2009; Zhang et al. 2010). The use of liposomes made from biosurfactants has been used as an important strategy for gene transfection. In a study, Inoh et al. (2001) found that MEL-A dramatically increased the efficiency of gene transfection mediated by cationic liposomes with a cationic cholesterol derivative. Studies like these could lead to the development of effective and safe nonviral vector-mediated gene transfection and gene therapy procedures. Kitamoto et al. (2002) demonstrated that liposome based on biosurfactants shows increasing efficiency of gene transfection in comparison with commercially available cationic liposome. In the last decade, for the liposome-based gene transfection some techniques and methodologies have been developed. Members of the *Candida antarctica* strain produce two kinds of mannosylerythritol lipids (MEL-A and MEL-B) that exhibit antimicrobial activity, particularly against Gram-positive bacteria (Kitamoto et al. 1993). Ueno et al. in 2007 examined MEL-A-containing liposome for gene transfection by introducing biosurfactants in this field.

13.3.6.2 Immune Modulatory Action

Bacterial lipopeptides constitute potent nontoxic and nonpyrogenic immunological adjuvant when mixed with conventional antigens (Gharaei-Fathabad 2011). In rabbits and chickens (Rodrigues et al. 2006), a marked enhancement of the humoral immune response was obtained by coupling of poly-L-lysine (MLR-PLL) with the low-molecular mass antigens iturin-AL, herbicolin-A, and microcystin (MLR).

13.3.6.3 Toxic Activity against Microorganisms

Several biosurfactants have strong antimicrobial, antifungal, and antiviral activity; this versatile performance is conferred due to the diverse structures of biosurfactants (Zhao et al. 2010). The structure of biosurfactant is supposed to exert its toxicity on the cell membrane permeability as a detergent-like effect. Fernandes et al. (2007) investigated the antimicrobial activity of biosurfactants from *Bacillus subtilus* demonstrating that lipopeptides have a broad spectrum of antimicrobial activity against microorganisms with multidrug-resistant profile (Fernandes et al. 2007). Rodrigues et al. (2006) mentioned many biosurfactants produced by *Candida antartica*, *P. aeruginosa*, *B. subtilis,* and *B. licheniformis* that have shown toxic activities against microorganisms.

13.3.6.4 Antiadhesive Agents

Adhesion of biosurfactants to solid surfaces might constitute a new and effective means of combating colonization by pathogenic microorganisms (Rivardo et al. 2009). Biosurfactants have been found to inhibit the adhesion of pathogenic organisms to the site of infection (Das et al. 2009). These surfactants can play the significant role of antiadhesive agent making them useful for treating many diseases as well as in use as therapeutic and probiotic agent. In addition, biosurfactant in pharmaceutical fields can be used as agents for stimulating stem fibroblast metabolism, while in premature infants the deficiency of pulmonary surfactant, a phospholipids protein complex also cause the failure of respiration. Moreover, isolation of genes for protein molecules of these surfactants and cloning in bacteria has made possible its fermentation production for medical application.

Biosurfactants play an essential natural role in the swarming motility of microorganisms and participate in cellular physiological processes of signaling and differentiation as well as in biofilm formation. Swarming motility and biofilm formation are the key actions in the colonization of a surface by bacteria, and increase the likelihood of nosocomial infections. Biosurfactants have been found to inhibit the adhesion of pathogenic organisms to solid surfaces or to infection sites. Surfactin decreases the amount of biofilm formed by *Salmonella typhimurium*, *Salmonella enterica*, *Escherichia coli*, and *Proteus mirabilis* in polyvinyl chloride wells, as well as in vinyl urethral catheters (Mireles et al. 2001). Precoating the catheters by running the surfactin solution through them before inoculation with media was just as effective as including surfactin in the growth medium. Given the importance of opportunistic infections with *Salmonella* species, including urinary tract infections of AIDS patients, these results have potential for practical applications. A biosurfactant of *Pseudomonas fluorescens* was found to inhibit the adhesion of *Listeria monocytogenes* LO28 to polytetrafluoroethylene and stainless-steel surfaces (Meylheuc et al. 2001). There are reports of inhibition of biofilm formed by uropathogens and yeast on silicone rubber by biosurfactants produced by *Lactobacillus acidophilus* (Velraeds et al. 1998).

13.3.6.5 Antimicrobial Activity of Biosurfactants

The antimicrobial activity of several biosurfactants has been reported in the literature for many different applications (Cameotra and Makkar 2004). For instance, the antimicrobial activity of two biosurfactants obtained from probiotic bacteria, *Lactococcus lactis* (Naruse et al. 1990) and *Streptococcus thermophilus* A, against a variety of bacterial and yeast strains isolated from explanted voice prostheses was evaluated (Rodrigues et al. 2004). Antibacterial and antiphytoviral effects of various rhamnolipids have been described in the literature (Bai et al. 1997; Benincasa et al. 2004). Seven different rhamnolipids were identified in cultures of *P. aeruginosa* AT10 from soybean oil refinery wastes and these showed excellent antifungal properties against various fungi (Abalos et al. 2001). Golubev et al. (2001) reported the production of an extracellular, low-molecular weight, protease-resistant thermostable glycolipid fungicide from the yeast *Pseudozyma fusiformata* (Ustilaginales). This fungicide was active against >80% of the 280 yeast and yeast-like species tested under acidic conditions (pH 4.0) at 20–30°C (Kulakovskaya et al. 2003). The purified glycolipids enhanced nonspecific permeability of the cytoplasmic membrane in sensitive cells, which resulted in ATP leakage.

Gomma (2013) investigated the antimicrobial effect of the lipopeptide biosurfactants produced by *B. licheniformis* strain M104 grown on whey. The biosurfactant was investigated for potential antimicrobial activity against different Gram-positive bacteria (*B. subtilis*, *B. thuringiensis*, *B. cereus*, *Staphylococcus aureus*, and *L. monocytogenes*), Gram-negative bacteria (*P. aeruginosa*, *E. coli*, *S. typhimurium*, *Proteous vulgaris*, and *Klebsiella pneumoniae*), and a yeast (*Candida albicans*).

13.3.7 APPLICATION IN AGRICULTURE

The dual hydrophobic/hydrophilic nature of biosurfactant can be widely exploited in areas related to agriculture for the enhancement of biodegradation of pollutants to improve the quality of agriculture soil, for indirect plant growth promotion as these biosurfactants have antimicrobial activity and to increase the plant microbe interaction beneficial for plant (Sachdev and Cameotra 2013). These biosurfactants can replace the harsh surfactant presently used in pesticide industries as these natural surfactants are found to be utilized as carbon source by soil-inhabiting microbes (Lima et al. 2011) and this accounts for the biological removal of biosurfactants from the agricultural soil.

13.3.7.1 Improvement of Soil Quality

The productivity of agricultural land is affected by the presence of organic and inorganic pollutants that impart abiotic stress on the cultivated crop plant. To increase the quality of such soil contaminated by hydrocarbon and heavy metals, process of bioremediation is required. Microorganism-producing biosurfactants and/or biosurfactants can be effectively used for the removal of hydrocarbons as well as heavy metals (Sun et al. 2006). A very important phenomenon of desorption of hydrophobic pollutants tightly bound to soil particles is accelerated by biosurfactants. This is very crucial for bioremediation process. Biosurfactants can also enhance the degradation of certain chemical insecticides that are accumulated in the agricultural soil (Zhang et al. 2011). Biosurfactant from *Lactobacillus pentosus* has demonstrated reduction by 58.6–62.8% of octane hydrocarbon from soil (Moldes et al. 2011), thus exhibiting the biodegradation accelerator property of biosurfactant. It has been observed that a biosurfactant-producing species of *Burkholderia* isolated from oil-contaminated soil may be a potential candidate for bioremediation of variety of pesticide contamination (Wattanaphon et al. 2008).

13.3.7.2 Plant–Pathogen Elimination

Several biosurfactants from microbes have antimicrobial activity against plant pathogens and therefore they are considered to be a promising biocontrol molecule for achieving sustainable agriculture. An agricultural application of biosurfactants also facilitates biocontrol mechanism of plant growth-promoting microbes such as parasitism, antibiosis, competition, induced systemic resistance, and hypovirulence (Singh et al. 2007). Biological control involves the exploitation of selected microorganisms (termed antagonistic), using naturally occurring mechanisms, to suppress harmful organisms. The modes of action are parasitism, antibiosis, competition, induced systemic resistance, and hypovirulence. In many instances, surfactants enhance the effects of the microbial biocontrol agent.

Brevibacillus brevis strain HOB1 produces surfactin isoform and this lipopeptide biosurfactant has demonstrated strong antibacterial and antifungal property which can be exploited for control of phytopathogens (Haddad 2008). Plant growth-promoting *Pseudomonas putida* produces biosurfactants that can cause lysis of zoospores of the oomycete pathogen *Phytophthora capsici*; causative agent of damping-off of cucumber (Kruijt et al. 2009). The lipopeptide biosurfactant produced by strains of *Bacillus* exhibits growth inhibition of phytopathogenic fungi like *Fusarium* spp., *Aspergillus* spp., and *Biopolaris sorokiniana*. Such biosurfactant can be used as biocontrol agent (Velho et al. 2011). Kim et al. (2011) have isolated biosurfactant from a strain of *Pseudomonas*, which has demonstrated insecticidal activity against green peach aphid (*Myzus persicae*). A possible plant pathogen *P. aeruginosa* was reported to be inhibited by biosurfactant produced by *Staphylococcus* sp., isolated from crude oil-contaminated soil (Eddouaouda et al. 2012). Biosurfactant-producing rhizospheric isolates of *Pseudomonas* and *Bacillus* have exhibited biocontrol of soft rot causing *Pectobacterium* and *Dickeya* spp. (Krzyzanowska et al. 2012).

13.4 STRATEGIES TO ENHANCE PRODUCTION OF BIOSURFACTANTS FOR ITS INDUSTRIAL APPLICATIONS

Improvement in production procedures and technologies has helped to some extent and can lead to further improvements. Researchers have emphasized the key parameters affecting the efficiency of biosurfactant production in terms of higher yields and lower production costs (Mukherjee et al. 2006). According to Syldatk and Hausmann (2010) the reasons for limited use of microbial surfactants in industry are the use of expensive substrates, limited product concentrations, low yields, and formation of product mixtures rather than pure compounds. All these factors and other growth and upscale problems like the use of antifoaming agents add on to the high costs of the downstream processing. The main strategies to achieve this are through (i) the assessment of the substrate and product output with focus on appropriate organism, nutritional balance, and the use of cheap or waste substrates to lower the initial raw material costs involved in the process; (ii) the development of efficient bioprocesses, including optimization of the culture conditions and cost-effective separation processes to maximize recovery; and (iii) the development and use of overproducing mutant or recombinant strains for enhanced yields (Mukherjee et al. 2006; Makkar et al. 2011).

The use of the alternative substrates such as agro-based industrial wastes is one of the attractive strategies for economical biosurfactant production. Kosaric (1992) suggested the use of industrial and/or municipal wastewaters, rich in organic pollutants, to achieve a double benefit of reducing the pollutants while producing useful products. Another approach involves using raw substrates with negligible or no value. An approach for reducing the production costs is developing processes that use renewable low-cost raw materials or high-pollutant wastes. A wide variety of alternative raw materials are currently available as nutrients for industrial fermentations, namely, various agricultural and industrial by-products and waste materials (Makkar and Cameotra 2002; Savarino et al. 2007; Ferreira 2008; da Silva et al. 2009; Montoneri et al. 2009).

However, the production cost of synthetic surfactants is not affordable for their use in larger ecosystems. The literature evidenced that the marine microbes are scarcely explored for the production of biosurfactants. The sponge-associated marine bacteria are current focus of bioactive leads from the marine environment (Selvin et al., 2009).

13.4.1 Cheap Substrates: Economical and Promising Alternatives

Production economy is the major bottleneck in biosurfactant production, as is the case with most biotechnological processes. Often, the amount and type of a raw material can contribute considerably to the production cost; it is estimated that raw materials account for 10–30% of the total production costs in most biotechnological processes. Thus, to reduce this cost it is desirable to use low-cost raw materials (Makkar and Cameotra 2002). One possibility explored extensively is the use of cheap and agro-based raw materials as substrates for biosurfactant production. A variety of cheap raw materials, including plant-derived oils, oil wastes, starchy substances, lactic whey, and distillery wastes (DWs), have been reported to support biosurfactant production.

The availability of raw materials for scaled-up production processes and acceptable production economics has widened the scope of biosurfactants. Most of the biosurfactants are produced from agricultural residues and from industrial waste products. The main problem related to the use of alternative substrates as culture medium is to find a waste with the right balance of nutrients that permits cell growth and product accumulation (Makkar and Cameotra 1999). Thus they have additional advantages from the viewpoint of resource replacement and recycling.

Modern society produces high quantity of waste materials through activity related to industries, forestry, agriculture, and municipalities (Martins et al. 2006). These inexpensive agro-industrial waste substrates include olive oil mill effluent, plant oil extracts and waste, distillery and whey wastes (WWs), potato process effluent, and cassava wastewater. These waste materials are some examples of food industry by-products or wastes that can be used as feedstock for biosurfactant production. The use of such waste materials serves a dual role of generating a usable product and reducing waste disposal (Makkar et al. 2011).

13.4.1.1 Vegetable Oils and Oil Wastes

Several studies with plant-derived oils have shown that they can act as effective and cheap raw materials for biosurfactant production, for example, rapeseed oil, babassu oil, and corn oil. Similarly, vegetable oils such as sunflower and soybean oils were used for the production of rhamnolipid, sophorolipid, and mannosylerythritol lipid biosurfactants by various microorganisms. Apart from various vegetable oils, oil wastes from vegetable oil refineries and the food industry were also reported as good substrates for biosurfactant production (Trummler et al. 2003; Pekin et al. 2005; Kim et al. 2006).

Several plant-derived oils, for example, jatropha oil, mesua oil, castor oils, ramtil oil, and jojoba oil, are not suitable for human consumption due to their unfavorable odor, color, and composition and are, therefore, available at much cheaper rates. Incorporation of these cheaper oils and oil wastes in the industrial production media might potentially reduce the overall costs of biosurfactant production, making them challenging targets for future R&D activities (Mukherjee et al. 2006).

13.4.1.1.1 Biosurfactant Production Using By-Products of Vegetable Industries

Vegetable oils are a lipidic carbon source and mostly comprise saturated or unsaturated fatty acids with 16–18 carbon atoms chain. Researchers have used variety of vegetable oils from canola, corn, sunflower, safflower, olive, rapeseed, grape seed, palm, coconut, fish, and soybean. The crude or unrefined oils extracted from oilseeds are generally rich in free fatty acids, mono-, di-, and triacylglycerides, phosphatides, pigments, sterols, tocopherols, glycerol, hydrocarbons, vitamins, protein fragments, trace metals, glycolipids, pesticides, resinous, and mucilaginous materials (Dumont and Narine 2007).

13.4.1.1.2 Biosurfactant Production Using a Single Substrate of Vegetable-Processing Industries

Mercade et al. (1993) were the first group to show the production of rhamnolipids by *P. aeruginosa* 47T2 when grown on olive oil mill effluent as the sole carbon source (a major waste problem in Spain). This study was important in demonstrating the possibility of using other lipophillic wastes for wider application. Kitamoto et al. (1993) studied the interfacial and antimicrobial properties of two kinds of mannosylerythritol lipids (MEL-A and B), biosurfactants, produced by *C. antarctica* T-34, when grown on soybean oil as substrate. As the biosurfactant produced in this study exhibited antimicrobial activity particularly against Gram-positive bacteria, the process could be more economical because of high-value application in pharmaceutical industry. Sim et al. (1997) have tested mixture of vegetable oils (canola oil, soybean, and glucose) for rhamnolipid production by *P. aeruginosa* UW-1 and reported 10–12-fold increase in rhamnolipid production on vegetable oils in comparison to glucose. Camargo-de-Morais et al. (2003) studied the production of a glycolipid with emulsifier properties during cultivation of *Penicillium citrinum* on mineral medium with 1% olive oil as carbon source. The growth-associated emulsifier production reached maximal activity at 60 h of cultivation with the production yield (Yp/s) of 0.54. An emulsifier that was stable in a wide range of pH and temperature was stimulated by high salt concentration implying a possible application in industrial waste or marine remediation. Chang et al. (2005) reported the production of biosurfactant by *Pseudoxanthomonas kaohsiungensis* sp. nov. strain J36T during cultivation on olive oil as the sole carbon and energy source.

Rufino et al. (2007) studied the cultivation of *Candida lipolytica* grown on groundnut oil for the production of a new biosurfactant. The preliminary investigation of chemical composition suggested that it was a lipopeptide in nature. The biosurfactant had a yield of 4.5 g/L and exhibited good surface activity, emulsification ability, and could withstand high salt concentration but was not thermo stable. They later also applied sequential factorial design to optimize biosurfactant production by *C. lipolytica* using soybean oil refinery residue as substrate (Rufino et al. 2008). In this study, they evaluated the impact of three cultivation factors, amounts of refinery residue, glutamic acid, and yeast extract. The biosurfactant product showed high surface activity and emulsifying ability and was very stable at wide range of pH (2–12), temperatures (0–120°C), and salinity (2–10% NaCl). They concluded that combination of an industrial waste and a cheap substrate is a promising approach to reduce production cost.

Coimbra et al. (2009) also showed the biosurfactant production by six *Candida* strains cultivated in insoluble (*n*-hexadecane) and soluble substrates (soybean oil, groundnut oil refinery residue, corn steep liquor, and glucose). These biosurfactants were able to remove 90% of the hydrophobic contaminants from sand. Oliveira et al. (2009) used palm oil, a low-cost agricultural by-product, which is used in as raw material for soap and food industries, for biosurfactant production using *Pseudomonas alcaligenes* (a strain isolated from crude oil-contaminated soil). They achieved a biosurfactant concentration of 2.3 g/L and E24 more than 70% with the hexane, jet fuel, and crude oil.

Plaza et al. (2011) aimed at the development of economical methods for biosurfactant production by the use of unconventional substrates. The research investigated the potential of utilizing agro-industrial wastes to replace synthetic media for cultivation of *Bacillus* strains and biosurfactant production. In total, 21 of the waste products from dairy, sugar, fatty, fruit and vegetable-processing industries, breweries, and distillery were examined. Three bacterial strains were identified by 16S rRNA gene sequencing: *B. subtilis* (I′-1a), *Bacillus* sp. (T-1), and *Bacillus* sp. (T′-1). The best unconventional substrates for bacteria growing and biosurfactant production at 30°C under aerobic conditions were molasses, brewery effluents, and fruit and vegetable decoction from the processing factory.

13.4.1.1.3 Biosurfactant Production Using Mixed Substrates of Vegetable Industries

To make processes more economical some researchers followed an approach of mixed substrates as carried out by Casas and Garcia-Ochoa (1999), who utilized the capability of *Candida bombicola* to produce sophorolipid biosurfactant properties when grown in medium composed of two different carbon sources and a nitrogen source. One of the carbon sources was a readily available sugar to maximize biomass production and the second was sunflower oil and they were able to achieve 120 g/L sophorolipid in 8 days under the best operational conditions. Bednarski et al. (2004) reported the synthesis of biosurfactants by *C. antarctica* or *Candida apicola* in the cultivation medium supplemented with oil refinery waste (either with soap stock or post-refinery fatty acids). Enrichment of the medium with the oil refinery waste resulted in a 7.5–8.5-fold greater concentration of glycolipids in comparison to the medium without addition of oil refinery waste. Costa et al. (2006) evaluated the possible use of oil from Buriti (*Mauritia flexuosa*), Cupuaçu (*Theobroma grandiflora*), Passion Fruit (*Passiflora alata*), Andiroba (*Carapa guianensis*), Brazilian nut (*Bertholletia excelsa*), and Babassu (*Orbignya* spp.) for rhamnolipid production by *P. aeruginosa* LBI. They observed extensive surface tension reduction and good emulsification. The highest rhamnolipid concentrations were obtained from Brazilian nut (9.9 g/L) and passion fruit (9.2 g/L) oils. Another Brazilian group led by Prieto (Prieto et al. 2008) isolated *P. aeruginosa* LBM10 from a southern coastal zone in Brazil, which could produce a rhamnolipid-type biosurfactant growing on different cheap carbon sources, such as soybean oil, soybean oil soapstock, fish oil, and glycerol. A combination of sugarcane molasses and three different oils (soybean, sunflower, or olive oil) was used a low-cost media by Daverey and Pakshirajan (2009), for the production of sophorolipids from the yeast *C. bombicola*. They achieved a yield approx. 24 g/L in this mixed media in comparison to media with single constituents. This yield was comparable to the costly conventional synthetic medium containing yeast extract, urea, soybean oil, and glucose.

Fontes et al. (2012) studied the production of a biosurfactant synthesized by *Yarrowia lipolytica* using different renewable resources as carbon source was investigated. Crude glycerol, a biodiesel co-product, and clarified cashew apple juice (CCAJ), an agro-industrial residue, were applied as feedstocks for the microbial surfactant synthesis. The microorganism was able to grow and produce biosurfactant on CCAJ and crude glycerol, achieving maximum emulsification indexes of 68.0% and 70.2% and maximum variations in surface tension of 18.0 mN/m and 22.0 mN/m, respectively. Different organic solvents (acetone, ethyl acetate, and chloroform–methanol) were tested for biosurfactant extraction. Maximum biosurfactant recovery was obtained with chloroform–methanol (1:1), reaching 6.9 g/L for experiments using CCAJ and 7.9 g/L for media containing crude glycerol as carbon source. The results herein obtained indicate that CCAJ and the co-product of biodiesel production are appropriate raw materials for biosurfactant production by *Y. lipolytica*.

Luna et al. (2012) studied the use of two industrial wastes, corn steep liquor and groundnut oil refinery residue, as low-cost nutrients for the production of a biosurfactant by *Candida sphaerica* (UCP 0995). They used an optimized medium with distilled water supplemented with 9% groundnut oil refinery residue and 9% corn steep liquor as substrates to produce biosurfactants by *C. sphaerica*, at 28°C during 144 h under 200 rpm. The isolated biosurfactant was formed with a yield of 9 g/L. The biosurfactant showed high surface tension-reducing activity the 25 mN/m, a small CMC value (0.025%), thermal (5–120°C), and pH (2–12) stability with respect to surface tension reducing activity and to emulsification activity and tolerance under high salt concentrations (2–10%). The biosurfactant was characterized as glycolipid and recovered 95% of the motor oil adsorbed in a sand sample, showing great potential to be used in bioremediation processes, especially in the petroleum industry.

13.4.1.1.4 Biosurfactant Production from Vegetable Oil Industries' Wastes

Benincasa et al. (2002) reported isolating a rhamnolipids producing *P. aeruginosa* strain LBI using soap stock as the sole carbon source. Soap stock is the waste from the sunflower oil process, the main co-product from the seed-oil-refining industry. Rhamnolipid concentration in the range of 15.9 g/L was achieved. Nitschke et al. (2005) evaluated the oil wastes as alternative low-cost substrates for the production of rhamnolipids by *P. aeruginosa* LBI strain. They used wastes obtained from soybean, cottonseed, babassu, palm, and corn oil refinery. The soybean soap stock waste was the best substrate, generating 11.7 g/L of rhamnolipids and a production yield of 75%. Another soybean-associated waste, which has been utilized for biosurfactant production, is soy molasses, a by-product of soybean oil processing. It contains high fermentable carbohydrate (30% w/v) and is about 60% of solids carbohydrate that makes it well suited for economical production of biosurfactants. Increased interest in consumption of healthy soy protein-based foods and drinks has established a sustained growing soy-based industry and as a result an abundance of waste by-products (Deak and Johnson 2006). It has been found that soy molasses act as carbon and nitrogen source for the fermentative production of sophorolipids by *C. bombicola* with yields of 55 g/L (Solaiman et al. 2007). In this study, they achieved a further cost reduction by substitution of expensive yeast extract and urea from the growth medium. The study opened a new frontier for applicability of low-cost carbon reach substrates as a combined source of carbon and nitrogen for other industrial bioprocesses. In an effort to economize biosurfactant production, Thavasi et al. (2008) used a mixture of peanut oil cake and waste motor lubricant oil as a substrate for the biosurfactant production. Peanut oil cake, a rich source of carbohydrate, protein, and lipids, is a by-product during the peanut oil manufacturing process. The cost of peanut cake is negligible compared with other pure carbon sources and waste motor oil is a waste product generated by the geared motor vehicles' after long use. They confirmed that *Bacillus megaterium*, *Azotobacter chroococcum*, and *Corynebacterium kutscheri* had the capability of using these substrates for biosurfactant production with better yields achieved with peanut oil cake. Recently, the authors have reported the biosurfactant production by *Lactobacillus delbrueckii* using peanut oil cake as the carbon source. The biosurfactant produced (5.35 mg/mL) was capable of promoting biodegradation to a large extent (Thavasi et al. 2011). These studies showed the suitability of peanut oil cake as a substrate for glycolipid biosynthesis (Makkar et al. 2011).

Govindammal and Parthasarathi (2013) studied the production of biosurfactant by *Pseudomonas fluorescence* MFS03 isolated from mangrove forest soil, Pitchavaram, Tamilnadu, India, using renewable substrates. The maximum biomass (11.73 mg/mL) and biosurfactant production (9.23 mg/mL) was observed with coconut oil cake at 120 and 132 h, respectively. Characterization of the biosurfactant revealed that it is a glycolipid with chemical composition of carbohydrate (48.5 µg 0.1/mL) and lipid (50.2 µg 0.1/mL). The biosurfactant shows higher emulsification activity (89%) with crude oil and coconut oil (84%) among the different hydrocarbon tested. Emulsification activity of the biosurfactant against different hydrocarbons showed its possible application in insecticide cleaning in vegetables. Monocrotophos with initial concentration of 100 ppm was washed out with 10 ppm concentration of the biosurfactant.

13.4.1.2 Lactic Whey and Distillery Wastes

Lactic whey from dairy industries has also been reported to be a cheap and viable substrate for biosurfactant production. The effluent from the dairy industry, known as dairy wastewater, supports good microbial growth and is used as a cheap raw material for biosurfactant production (Dubey and Juwarkar 2004). Furthermore, the potential use of dairy wastewaters provides a stratagem for the economical production of biosurfactants and efficient dairy wastewater management.

In a study by Dubey et al. (2012), combinations of DW with other industrial wastes, namely, curd WW, fruit-processing waste (FPW), and sugar industry effluent (SIE) were evaluated to replace the use of water that was reported earlier for biosurfactant production from 1:3 diluted DW by using four new bacterial cultures BS-A, BS-J, BS-K, and BS-P, isolated from soil collected from a distillery unit. These isolates have the potential to produce biosurfactant from these individual wastes and in their combinations. Highest biomass and biosurfactant yields with higher reduction in the chemical oxygen demand (COD), total sugars, nitrogen, and phosphate levels were obtained in 1:1:1 proportion of DW + WW + FPW followed by DW + WW + SIE and individual wastes. The combinations of wastes improved the yields of biosurfactants by 18–41% and reduced COD of the combined wastes by 76–84.2%. Total sugars, nitrogen, and phosphate levels reduced in the range of 79–86%, 58–71%, and 45–59%, respectively. Among the four microbial isolates tested, BS-J and BS-P were the efficient biosurfactant producers and were identified as *Kocuria turfanesis* and *P. aeruginosa* based on the 16S rDNA sequence and phylogenetic analyses. Benefits derived by using combined DW with other wastes are improved production of biosurfactant as resource and saving precious water and the costly nutrients with concomitant reduction in pollution load of the wastes.

13.4.1.2.1 *Biosurfactant Production from Dairy and Sugar Industry Wastes*

The dairy industry has a considerable amount of by-products such as buttermilk, whey, and their derivatives. Whey is a liquid by-product of cheese production, rich in lactose (75% of dry matter), and containing other organic water-soluble components (12–14% protein). Daniel et al. (1998) reported the high yields of sophorolipids production with whey concentrate and rapeseed oil as substrate. However, in this study the organisms did not utilize lactose. Daverey and Pakshirajan (2010) also reported the production of sophorolipids by the yeast *C. bombicola* on medium containing mixed hydrophilic substrate (deproteinized whey and glucose), yeast extract, and oleic acid. They could achieve a yield up to 34 g/L under experimental conditions in optimized medium formulation.

Molasses is a co-product of sugar production, both from the sugar cane and sugar beet industry in India as runoff syrup from the final step of sugar crystallization after which further sugar crystallization becomes uneconomical. The main reasons for the widespread use of molasses as substrate are its low price compared with other sources of sugar and the presence of several other compounds and vitamins (Makkar et al. 2011). Molasses are mainly composed of sugars (sucrose 48–56%), non-sugar organic matter (9–12%), proteins, inorganic components, and vitamins. The total fermentable sugars are in the range of 50–55% by weight. Traditionally, molasses was used as an animal feed, production of pullulan, xanthan gum, citric acid, and in ethanol industries (Maneerat 2005). Molasses with its high sugar content is a good substrate for biosurfactant production as evidenced by many studies covering two decades. The possibility of using soy molasses as a relatively inexpensive and easily available resource to produce rhamnolipids was investigated by Rashedi et al. (2005). They reported that biosurfactant production by the bacterial strain on soy molasses was growth related. The specific production rate of rhamnolipid when using 2%, 4%, 6%, 8%, and 10% of molasses were 0.00065, 4.556, 8.94, 8.85, and 9.09, with rhamnolipids/biomass yield of 0.003, 0.009, 0.053, 0.041, and 0.213, respectively. Others such as Raza et al. (2007) reported the production of a microbial surfactant by growing *P. aeruginosa* EBN-8 mutant on clarified blackstrap molasses as a sole carbon and energy source. Maximum rhamnolipid (1.45 g/L) yields were observed, at 96 h of incubation on 2% total sugar-based molasses amended with sodium nitrate. In an effort to reduce the cost of surfactin production by *B. subtilis* BS5, Abdel-Mawgoud et al. (2008) optimized the environmental and nutritional production conditions for economizing of the production process. Optimized medium containing 16% molasses, 5 g/L NaNO$_3$, and the trace elements solution of ZnSO$_4 \cdot$7H$_2$O (0.16 g/L), FeCl$_3 \cdot$6H$_2$O (0.27 g/L), and MnSO$_4 \cdot$H$_2$O (0.017 g/L) gave surfactin yield of 1.12 g/L. In conclusion, both molasses and whey have been successfully utilized as substrate for biosurfactant production. In addition, rhamnolipid biosurfactant production using 18 strains of *Pseudomonas* sp. were investigated by Onbasli and Aslim (2009). The two strains with the highest yield of rhamnolipids production (*Pseudomonas luteola* B17 and *P. putida* B12) were further examined for rhamnolipid production on different sugar beet molasses concentrations. Maximum rhamnolipid production was achieved with 5% (w/v) of molasses and occurred after 12 h incubation. More studies, however, are required to overcome the problems associated with batch variability and ways to standardize the pretreatment requirement of these substrates for more productive output.

13.4.1.3 Starchy Substrates

Waste starchy materials are also potential alternative raw materials for the production of biosurfactants. Potato process effluents (wastes from potato-processing industries) were used to produce biosurfactant by *B. subtilis*. Cassava wastewater, another carbohydrate-rich residue, which is generated in large amounts during the preparation of cassava flour, is also an attractive substrate in biotechnological processes and has been used for surfactin production by *B. subtilis* (Noah et al. 2005; Nitschke and Pastore, 2006). These wastes are obtained at low cost from the respective processing industries and are as potent as low-cost substrates for industrial level biosurfactant production. Several other starchy waste substrates, such as rice water (effluent from rice-processing industry and domestic cooking), cornsteep liquor, and wastewater from the processing of cereals, pulses, and molasses, have tremendous potential to support microbial growth and biosurfactant production. Extensive research is needed to establish the suitability of these carbohydrate-rich substrates in industrial-level biosurfactant production process (Mukherjee et al. 2006).

13.4.1.3.1 Biosurfactant from Starch-Rich Substrates

Starch is a major agricultural product of corn, tapioca, wheat, and potatoes, which are major crops. Other sources include sugar plants such as sugar beet, sugar cane, or sugar sorghum. Sugar and starch-processing industries also produce large amount of solid residues of starch-containing wastewater. The high fiber content of the solid residue makes them a good source for paper and packaging industries, whereas the carbohydrate-rich wastewater is a suitable substrate for production of microbial products. Biological wastes rich in starchy materials are suitable for biosurfactant production.

One such substrate is potato which is one of the important staple foods and a lucrative cash crop in many countries. Processing of potatoes results in starch-rich wastewater, potato peels, and unconsumable potatoes, which are rich substrates for the microbes. Fox and Bala (2000) attempted to produce biosurfactants utilizing potato-associated waste. They evaluated potato substrate as a carbon source for biosurfactant production using *B. subtilis* ATCC 21332. They compared growth, surface activity, and carbohydrate utilization of *B. subtilis* ATCC 21332 on an established potato medium, simulated liquid, and solid potato waste media and a commercially prepared potato starch in a mineral salts medium. The results obtained indicated the utilization of potato substrate and production of surfactant as indicated by high surface tension reduction. The efficiency of two *B. subtilis* strains for the production of biosurfactants in two fermentation systems using powdered potato peels as substrate was investigated (Das and Mukherjee 2007). Potato peels were immersed in very hot water followed by oven drying. The dried peels were grinded to a paste and stored at 4°C before further use. Both the fermentation process resulted in biosurfactant (lipopeptides) with good surface activity and yield. Wang et al. (2008) applied a *B. subtilis* strain B6-1, for the production of biosurfactant using soybean and sweet potato residues in solid-state fermentation.

13.4.1.4 Biosurfactant Production from Lignocellulosic Waste

Lignocellulosic materials are among the most abundant organic carbon available on earth (Kukhar 2009), and they are the major components of different waste streams from various industries, forestry, agriculture, and municipalities. Such waste materials are mostly burned releasing CO_2 that contributes to the greenhouse effect. Lignocellulose consists of mainly three types of polymers—cellulose, hemicellulose, and lignin—that are strongly intermeshed and chemically bonded by both noncovalent forces and covalent cross-linkages. From an economical point of view, ligoncellulosic-rich agricultural residues can be employed for producing useful biomolecules such as biosurfactants. There have been reports of some forms of lignocellulosic wastes for the production of biosurfactant. Portilla-Rivera et al. (2007) were the first to look into the capability of *Lactobacillus* sp. to use hemicellulosic hydrolyzates from various agricultural residues for simultaneous production of biosurfactants and lactic acid. Such dual production strategy makes biosurfactant more economically viable in market and reduce the effects of waste burning on environment. In their efforts they achieved reduced surface tension and biosurfactant yield of 0.71 g/g of biomass, when hemicellulosic hydrolyzates from trimming wine shoots were used. This study is important considering the large amount of pruning wastes of vine stocks generated worldwide and the resulting constitutive monomeric sugar solutions, which are potential renewable sources for the other biomolecules like lactic acid. They concluded that hemicellulosic sugars from the agricultural residues are interesting substrates for the competitive cost production of biosurfactants.

13.4.2 Bioprocess Development: Optimum Production and Recovery

An efficient and economical bioprocess is the foundation for every profit-making biotechnology industry; hence, bioprocess development is the primary step toward commercialization of all biotechnological products, including biosurfactants. Any attempt to increase the yield of a biosurfactant demands optimal addition of media components and selection of the optimal culture conditions that will induce the maximum or the optimum productivity. Similarly, efficient downstream-processing techniques and methods are needed for maximum product recovery.

13.4.3 Process Optimization: The Best Combination of Essential Factors

Several elements, media components, and precursors are reported to affect the process of biosurfactant production and the final quantity and quality. Different elements such as nitrogen, iron, and manganese are reported to affect the yield of biosurfactants, for example, the limitation of nitrogen is reported to enhance biosurfactant production in *P. aeruginosa* strain BS-2 (Dubey Juwarkar, 2004) and *Ustilago maydis* (Hewald et al. 2005). Similarly, the addition of iron and manganese to the culture medium was reported to increase the production of biosurfactant by *B. subtilis* (Wei et al. 2003). The ratios of different elements such as C:N, C:P, C:Fe, or C:Mg affected biosurfactant production and their optimization enhanced it (Amézcua-Vega et al. 2007).

Makkar and Cameotra (2002) studied the effects of various factors on growth and biosurfactant production by *B. subtilis* MTCC 2423. They found that sucrose (2%) and potassium nitrate (0.3%) were the best carbon and nitrogen sources. The addition of various metal supplements (magnesium, calcium, iron, and trace elements) greatly affected growth and biosurfactant production. The effect of the metal cations, used together, is greater than when they are used individually. The biosurfactant production increased considerably (almost double) by addition of metal supplements. Very high concentrations of metal supplements, however, inhibited biosurfactant production. Amino acids such as aspartic acid, asparagine, glutamic acid, valine, and lysine increased the final yield of biosurfactant by about 60%. The organism could produce biosurfactant at 45°C and within the pH range of 4.5–10.5. The biosurfactant was thermostable and pH stable (from 4.0 to 12.0). The capability of the organism to produce biosurfactant under thermophilic, alkaliphilic, and halophilic conditions makes it a suitable candidate for field applications. Infrared, nuclear magnetic resonance, and mass spectroscopy studies showed the surfactant to be identical to surfactin.

Nutritional requirements for maximal production of biosurfactant by an oil field bacterium *P. putida* were determined by Pruthi and Cameotra (2003). The optimal concentrations of nitrogen, phosphate, sulfur, magnesium, iron, potassium, sodium, calcium, and trace elements for maximal production of biosurfactants were ascertained, and they formulated a new "Pruthi and Cameotra" salt medium. Data from their study showed that maximal biomass (2.4 g/L) and biosurfactant production (6.28 g/L) takes place after 72 h of growth on 2% hexadecane. The biosurfactant was produced optimally over pH and temperature ranges of 6.4–7.2°C and 30–40°C, respectively.

The biosurfactant production of a marine actinobacterium *Brevibacterium aureum* MSA13 was optimized by Kiran et al. (2010) using industrial and agro-industrial solid waste residues as substrates in solid-state culture (SSC). On the basis of the optimization experiments of their study, they reported that the biosurfactant production by MSA13 was increased to threefold over the original isolate under SSC conditions with pretreated molasses as substrate and olive oil, acrylamide, FeCl3, and inoculums size as critical control factors. The strain *B. aureum* MSA13 produced a new lipopeptide biosurfactant with a hydrophobic moiety of octadecanoic acid methyl ester and a peptide part predicted as a short sequence of four amino acids including pro-leu-gly-gly. The biosurfactant produced by the marine actinobacterium MSA13 can be used for the microbially enhanced oil recovery processes in the marine environments.

13.4.4 Downstream Processing: Fast, Efficient, and Cheap Product Recovery

Even if optimum production is obtained using optimal media and culture conditions, the production process is still incomplete without an efficient and economical means for the recovery of the products. Thus, one important factor determining the feasibility of a production process on a commercial scale is the availability of suitable and economic recovery and downstream procedures. For many biotechnological products, the downstream processing costs account for ~60% of the total production costs. Several conventional methods for the recovery of biosurfactants are: acid precipitation, solvent extraction, crystallization, and ammonium sulfate precipitation and centrifugation. A few unconventional and interesting recovery methods have also been reported in recent years. These procedures take advantage of some of the other properties of biosurfactants—such as their surface activity or their ability to form micelles and/or vesicles—and are particularly applicable for large-scale continuous recovery of extracellular biosurfactants from culture broth. A few examples of such biosurfactant recovery strategies include foam fractionation (Davis et al. 2001; Noah et al. 2005), ultrafiltration (Sen and Swaminathan, 2005), adsorption–desorption on polystyrene resins and ion exchange chromatography, and adsorption–desorption on wood-based activated carbon (Dubey et al. 2005).

13.4.5 Mutant and Recombinant Strains: The Hyper-Producers

The genetics of the producer organism is an important factor affecting the yield of all biotechnological products because the capacity to produce a metabolite is bestowed by the genes of the organism. The bioindustrial production process is often dependent on the use of hyper-producing microbial strains: even with cheap raw materials, optimized medium and culture conditions, and efficient recovery processes, a production process cannot be made commercially viable and profitable until the yield of the final product by the producer organisms is naturally high. Moreover, the industrial production process is dependent on the availability of recombinant and mutant hyper-producers if good yields are lacking from the natural producer strains. Even if high-yielding natural strains are available, the recombinant hyper-producers are always required, to economize further the production process and to obtain products with better commercially important properties.

Besides the natural biosurfactant producer strains, a few mutant and recombinant varieties with enhanced biosurfactant production characteristics are reported in the literature.

Sekhon et al. (2011) utilized olive oil as a carbon source, which has been explored by many researchers. However, studying the concomitant production of biosurfactant and esterase enzyme in the presence of olive oil in the *Bacillus* species and its recombinants is a relatively novel approach. In their study, *Bacillus* species isolated from endosulfan-sprayed cashew plantation soil was cultivated on a number of hydrophobic substrates. Olive oil was found to be the best inducer of biosurfactant activity. The protein associated with the release of the biosurfactant was found to be an esterase. There was a twofold increase in the biosurfactant and esterase activities after the successful cloning of the biosurfactant genes from *B. subtilis* SK320 into *E. coli*. Multiple sequence alignment showed regions of similarity and conserved sequences between biosurfactant and esterase genes, further confirming the symbiotic correlation between the two. Biosurfactants produced by *B. subtilis* SK320 and recombinant strains BioS a, BioS b, BioS c were found to be effective emulsifiers, reducing the surface tension of water from 72 dynes/cm to as low as 30.7 dynes/cm. The attributes of enhanced biosurfactant and esterase production by hyper-producing recombinant strains have many utilities from industrial viewpoint. This study for the first time has shown a possible association between biosurfactant production and esterase activity in any *Bacillus* species. Biosurfactant–esterase complex has been found to have powerful emulsification properties, which shows promising bioremediation, hydrocarbon biodegradation, and pharmaceutical applications.

13.5   CONCLUSION

Surfactants, both chemical and biological, are amphiphilic compounds, which can reduce surface and interfacial tensions by accumulating at the interface of immiscible fluids and increase the solubility, mobility, bioavailability, and subsequent biodegradation of hydrophobic or insoluble organic compounds. Investigations on their impacts on microbial activity have generally been limited in scope to the most common and best characterized surfactants. Recently, a number of new biosurfactants have been described and accelerated advances in molecular and cellular biology are expected to expand our insights into the diversity of structures and applications of biosurfactants. Biosurfactants also exhibit natural physiological roles in increasing bioavailability of hydrophobic molecules and can complex with heavy metals, and some also possess antimicrobial activity. They have been exploited in this way, for example, as antimicrobial agents in disease control and to improve degradation of chemical contaminants. Considering the growing awareness on the climate change issues, the greener processes for the production of biosurfactants from industrial waste and bioremediation of petroleum hydrocarbons using biosurfactants will greatly reduce the uses of chemicals and xenobiotics in the environment.

The application of economical technologies and process based on utilization of waste conversion to products is also gaining ground. The commercial realization of the biosurfactants that are restricted by high production costs can be equipoise by optimized production conditions provided by utilization of the cheaper renewable substrates and application of novel and efficient multistep downstream processing methods. Recombinant and mutant hyper-producer microbial strains, able to grow on a wide range of cheap substrates, may produce biosurfactants in high yield and potentially bring the required breakthrough for their economic production. The true significance of these processes will be justified only when these studies will be scaled up to commercially viable processes.

REFERENCES

Abalos A, Pinazo A, Infante MR, Casals M, Garcia F, Manresa A. 2001. Physicochemical and antimicrobial properties of new rhamnolipids produced by *Pseudomonas aeruginosa* AT10 from soybean oil refinery wastes. *Langmuir* 17:1367–1371.

Abdel-Mawgoud A, Aboulwafa M, Hassouna N. 2008. Optimization of surfactin production by *Bacillus subtilis* isolate BS5. *Appl Biochem Biotechnol* 150:305–325.

Abdel-Mawgoud AM, Lepine F, Deziel E. 2010. Rhamnolipids: Diversity of structutre, microbial origins and roles. *Appl Microbiol Biotechnol* 86:1323–1336.

Amézcua-Vega C, Poggi-Varaldo HM, Esparza-García F, Ríos-Leal E, Rodríguez-Vázquez R. 2007. Effect of culture conditions on fatty acids composition of a biosurfactant produced by *Candida ingens* and changes of surface tension of culture media. *Bioresource Technol* 1:237–240.

Bai G, Brusseau ML, Miller RM. 1997. Influence of a rhamnolipid biosurfactant on the transport of bacteria through a sandy soil. *Appl Environ Microbiol* 63:1866–1873.

Banat IM, Makkar RS, Cameotra SS. 2000. Potential commercial applications of microbial surfactants. *Appl Microbiol Biotechnol* 53:495–508.

Bednarski W, Adamczak M, Tomasik J, Plaszczyk M. 2004. Application of oil refinery waste in the biosynthesis of glycolipids by yeast. *Bioresour Technol* 95:15–18.

Benincasa M, Abalos A, Oliveira I, Manresa A. 2004. Chemical structure, surface properties and biological activities of the biosurfactant produced by *Pseudomonas aeruginosa* LB1 from soapstock. *Antonie Van Leeuwenhoek* 85:1–8.

Benincasa M, Contiero J, Manresa MA, Moraes IO. 2002. Rhamnolipid production by *Pseudomonas aeruginosa* LBI growing on soapstock as the sole carbon source. *J Food Eng* 54:283–288.

Camargo-de-Morais M, Ramos SAF, Pimentel M, de Morais M Jr, Lima Filho J. 2003. Production of an extracellular polysaccharide with emulsifier properties by *Penicillium citrinum*. *World J Microbiol Biotechnol* 19:191–194.

Cameotra S, Makkar R. 2004. Recent applications of biosurfactants as biological and immunological molecules. *Curr Opin Microbiol* 7:262–266.

Casas J, Garcia-Ochoa F. 1999. Sophorolipid production by *Candida bombicola*: Medium composition and culture methods. *J Biosci Bioeng* 88:488–494.

Chang JS, Chou CL, Lin GH, Sheu S-Y, Chen WM. 2005. *Pseudoxanthomonas kaohsiungensis*, sp. nov., a novel bacterium isolated from oil-polluted site produces extracellular surface activity. *Syst Appl Microbiol* 28:137–144.

Coimbra CD, Rufino RD, Luna JM, Sarubbo LA. 2009. Studies of the cell surface properties of *Candida* species and relation to the production of biosurfactants for environmental applications. *Curr Microbiol* 58:245–251.

Costa SGVAO, Nitschke M, Haddad R, Eberlin MN, Contiero J. 2006. Production of *Pseudmonas aeruginosa* LBI rhamnolipids following growth on Brazilian native oils. *Process Biochem* 41:483–488.

da Silva GP, Mack M, Contiero J. 2009. Glycerol: A promising and abundant carbon source for industrial microbiology. *Biotechnol Adv* 27:30–39.

Daniel HJ, Otto RT, Reuss M, Syldatk C. 1998. Sophorolipid production with high yields on whey concentrate and rapeseed oil without consumption of lactose. *Biotechnol Lett* 20:805–807.

Das K, Mukherjee AK. 2007. Comparison of lipopeptide biosurfactants production by *Bacillus subtilis* strains in submerged and solid state fermentation systems using a cheap carbon source: Some industrial applications of biosurfactants. *Pro Biochem* 42:1191–1199.

Das P, Mukherjee S, Sen R. 2009. Antiadhesive action of a marine microbial surfactant. *Colloids Surf B: Biointerfaces* 71:183–186.

Daverey A, Pakshirajan K. 2009. Production, characterization, and properties of sophorolipids from the yeast *Candida bombicola* using a low-cost fermentative medium. *Appl Biochem Biotechnol* 158:663–674.

Daverey A, Pakshirajan K. 2010. Sophorolipids from *Candida bombicola* using mixed hydrophilic substrates: Production, purification and characterization. *Coll Surf B, Biointerfaces* 79:246–253.

Davis DA, Lynch HC, Varley J. 2001. The application of foaming for recovery of surfactin from *B. subtilis* ATCC 21332. *Enzyme Microb Technol* 28:346–354.

Deak N, Johnson L. 2006. Functional properties of protein ingredients prepared from high-sucrose/low-stachyose soybeans. *JOCOS* 83:811–818.

Dubey K, Juwarkar A. 2004. Determination of genetic basis for biosurfactant production in distillery and curd whey wastes utilizing *Pseudomonas aeruginosa* strain BS2. *Ind J Biotechnol* 3:74–81.

Dubey KV, Charde PN, Meshram SU, Yadav SK, Singh S, Juwarkar AA. 2012. Potential of new microbial isolates for biosurfactant production using combinations of distillery waste with other industrial wastes. *J Petrol Environ Biotechnol* S1:002. doi:10.4172/2157-7463.S1-002.

Dubey KV, Juwarkar AA, Singh SK. 2005. Adsorption–desorption process using wood based activated carbon for recovery of biosurfactant from fermented distillery wastewater. *Biotechnol Prog* 21:860–867.

Dumont MJ, Narine SS. 2007. Soapstock and deodorizer distillates from North American vegetable oils: Review on their characterization, extraction and utilization. *Food Res Int* 40:957–974.

Eddouaouda K, Mnif S, Badis A, Younes SB, Cherif S, Ferhat S, Mhiri N, Chamkha M, Sayadi S. 2012. Characterization of a novel biosurfactant produced by *Staphylococcus* sp. strain 1E with potential application on hydrocarbon bioremediation. *J Basic Microbiol* 52:408–418.

Eshita SM, Roberto NH, Beale JM, Mamiya BM, Workman RF. 1995. Bacillomycin Lc, a new antibiotic of the iturin group: Isolations, structures, and antifungal activities of the congeners. *J Antibiot (Tokyo)* 48:1240–1247.

Fernandes PAV, de Arruda IR, de Santos AFAB, de Araújo AA, Maior AMS, Ximenes EA. 2007. Antimicrobial activity of surfactants produced by Bacillus subtilis R14 against multidrug-resistant bacteria. *Braz J Microbiol* 38:704–709.

Ferreira NL. 2008. Industrial exploitation of renewable resources: From ethanol production to bio products development. *J Soc Biol* 202:191–199.

Fontes GC, Ramos NM, Amaral PFF, Nele M, Coelho MAZ. 2012. Renewable resources for biosurfactant production by *Yarrowia lipolytica*. *Braz J Cheml Eng* 29:483–493.

Fox SL, Bala GA. 2000. Production of surfactant from *Bacillus subtilis* ATCC 21332 using potato substrates. *Bioresour Technol* 75:235–240.

Fujita T, Furuhata M, Hattori Y, Kawakami H, Toma K, Maitani Y. 2009. Calcium enhanced delivery of tetraarginine-PEG-liquid coated DNA/Protamine complexes. *Int J Pharm* 368:186–192.

Gharaei-Fathabad E. 2011. Biosurfactants in pharmaceutical industry: A mini review. *Am J Drug Disc Develop* 1:58–69.

Golubev WI, Kulakovskaya TV, Golubeva W. 2001. The yeast *Pseudozyma fusiformata* VKM Y-2821 producing an antifungal glycolipid. *Microbiology* 70:553–556.

Gomma EZ. 2013. Antimicrobial activity of a biosurfactant produced by *Bacillus licheniformis* strain M104 grown on whey. *Braz Arch Biol Technol* 56:259–268.

Govindammal M, Parthasarathi R. 2013. Production and characterization of biosurfactant using renewable substrates by *Pseudomonas fluorescence* isolated from mangrove ecosystem. *J Applic Chem* 2:55–62.

Haddad NI. 2008. Isolation and characterization of a biosurfactant producing strain, *Brevibacilis brevis* HOB1. *J Ind Microbiol Biotechnol* 35:1597–1604.

Herman DC, Artiola JF, Miller RM. 1995. Removal of cadmium, lead and zinc from soil by a rhamnolipid biosurfactant. *Environ Sci Technol* 29:2280–2285.

Hewald S, Josephs K, Bolker M. 2005. Genetic analysis of biosurfactant production in *Ustilago maydis*. *Appl Environ Microbiol* 71:3033–3040.

Inakollu S, Hung H, Shreve GS. 2004. Biosurfactant enhancement of microbial degradation of various structural classes of hydrocarbon in mixed waste systems. *Environ Eng Sci* 21:463–469.

Inoh Y, Kitamoto D, Hirashima N, Nakanishi M. 2001. Biosurfactants of MEL-A increase gene transfection mediated by cationic liposomes. *Biochem Biophys Res Commun* 289:57–61.

Jeong-Jin H, Yang S, Lee C, Choi Y, Kajuichi T. 1998. Ultrafiltration of divalent metal cations from aqueous solution using polycarboxylic acid type biosurfactant. *J Coll Inter Sci* 202:63–73.

Kachholz T, Schlingmann M. 1987. Possible food and agricultural applications of microbial surfactants: An assessment. In N. Kosaric, W. L. Carns, N. C. C. Gray (Eds.), *Biosurfactants and Biotechnology*. New York: Marcel Dekker (pp. 183–210).

Kang SW, Kim YB, Shin JD, Kim EK. 2010. Enhanced biodegradation of hydrocarbons in soil by microbial biosurfactant, sophorolipid. *Appl Biochem Biotechnol* 160:780–790.

Karanth NGK, Deo PG, Veenanadig NK. 1999. Microbial production of biosurfactants and their importance. *Curr Sci* 77:116–123.

Kim HS, Jeon JW, Kim BH, Ahn CY, Oh HM, Yoon BD. 2006. Extracellular production of a glycolipid biosurfactant, mannosylerythritol lipid, by *Candida* sp. SY16 using fed batch fermentation. *Appl Microbiol Biotechnol* 70:391–396.

Kim SK, Kim YC, Lee S, Kim JC, Yun MY, Kim IS. 2011. Insecticidal activity of rhamnolipid isolated from *Pseudomonas* sp. EP-3 against green peach aphid (*Myzus persicae*). *J Agric Food Chem* 59:934–938.

Kiran GS, Thomas TA, Soelvin J, Sabarathnam B, Lipton AP. 2010. Optimization and characterization of a new lipopeptide biosurfactant produced by marine *Brevibacterium aureum* MSA13 in solid state culture. *Biores Technol* 101:2389–2396.

Kitamoto D, Isoda H, Nakahara T. 2002. Functions and potential applications of glycolipid biosurfactants from energy-saving materials to gene delivery carriers. *J Bio Sci Bio Eng* 94:187–201.

Kitamoto D, Yanagishita H, Shinbo T, Nakane T, Kamisawa C, Nakahara T. 1993. Surface active properties and antimicrobial activities of mannosylerythritol lipids as biosurfactants produced by *Candida antarctica*. *J Biotechnol* 29:91–96.

Kosaric N. 1992. Biosurfactants in industry. *Pure Appl Chem* 64:1731–1737.

Kosaric N. 2001. Biosurfactants and their application for soil bioremediation. *Food Technol Biotechnol* 39:295–304.

Kruijt M, Tran H, Raaijmakers JM. 2009. Functional, genetic and chemical characterization of biosurfactants produced by plant growth-promoting *Pseudomonas putida* 267. *J Appl Microbiol* 107:546–556.

Krzyzanowska DM, Potrykus M, Golanowska M, Polonis K, Gwizdek-Wisniewska A, Lojkowska E, Jafra S. 2012. Rhizosphere bacteria as potential biocontrol agents against soft rot caused by various *Pectobacterium* and *Dickeya* spp. strains. *J Plant Pathol* 94(2):367–378. doi: 10.4454/JPP.FA.2012.042.

Kukhar V. 2009. Biomass—feedstock for organic chemicals. *Kem Ind* 58:57–71.

Kulakovskaya T, Kulakovskaya E, Golubev W. 2003. ATP leakage from yeast cells treated by extracellular glycolipids of *Pseudozyma fusiformata*. *FEMS Yeast Res* 3:401–404.

Lima TM, Procópio LC, Brandão FD, Leão BA, Tótola MR, Borges AC. 2011. Evaluation of bacterial surfactant toxicity towards petroleum degrading microorganisms. *Bioresour Technol* 102:2957–2964.

Lin SC, Carswell KS, Sharma MM, Georgiou G. 1994. Continuous production of the lipopeptide biosurfactant of *Bacillus licheniformis* JF-2. *Appl Microbiol Biotechnol* 41:281–285.

Linhardt RJ, Bakhit R, Daniels L, Mayerl F, Pickenhagen W. 1989. Microbially produced rhamnolipid as a source of rhamnose. *Biotechnol Bioeng* 33:365–368.

Liu Q, Dong M, Zhoua W, Ayub M, Zhang YP, Huang S. 2004. Improved oil recovery by adsorption–desorption in chemical flooding. *J Petrol Sci Eng* 43:75–86.

Luna JM, Rufino RD, Campos-Takaki GM, Sarubbo LA. 2012. Properties of the biosurfactant produced by *Candida sphaerica* cultivated in low-cost substrates. *Chem Eng Transac* 27.

Makkar RS, Cameotra SS. 1998. Production of biosurfactant at mesophilic and thermophilic conditions by a strain of *Bacillus subtilis*. *J Ind Microbiol Biotechnol* 20:48–52.

Makkar RS, Cameotra SS. 1999. Biosurfactant production by microorganisms on unconventional carbon sources—A review. *J Surfactants Deterg* 2:237–241.

Makkar RS, Cameotra SS. 2002a. An update on use of unconventional substrates for biosurfactants production and their new applications. *Appl Microbiol Biotechnol* 58:428–434.

Makkar RS, Cameotra SS. 2002b. Effects of various nutritional supplements on biosurfactant production by a strain of *Bacillus subtilis* at 45°C. *J Surf Deterg* 5:11–17.

Makkar RS, Cameotra SS, Banat IM. 2011. Advances in utilization of renewable substrates for biosurfactant production. *AMB Express* 1:5.

Maneerat S. 2005. Production of biosurfactants using substrates from renewable resources. *Songklanakarin J Sci Techno* 27:675–683.

Manning FC, Thompson RE. 1995. *Oilfield Processing. Crude Oil*, vol. 2. Tulsa, Oklahoma: PennWell.

Martínez-Checa F, Toledo FL, Mabrouki KE, Quesada E, Calvo C. 2007. Characteristics of bioemulsifier V2-7 synthesized in culture media added of hydrocarbons: Chemical composition, emulsifying activity and rheological properties. *Bioresour Technol* 98:3130–3135.

Martins VG, Kalil SJ, Bertolin TE, Costa JA. 2006. Solid state biosurfactant production in a fixed-bed column bioreactor. *Z Naturforsch [C]* 61:721–726.

Masaru K, Michiko S, Shuhei Y, Atsushi S, Dai K, Tomohiro I, Tokuma F, Tomatake M. 2007. Skin care cosmetic and skin and agent for preventing skin roughness containing biosurfactants (World Patent 2007/060956). Toyo Boseki Kabu Shiki Kaisha and National Industrial Science and Technology, Osaka, Japan.

Mercade ME, Manresa MA, Robert M, Espuny MJ, de Andres C, Guinea J. 1993. Olive oil mill effluent (OOME). New substrate for biosurfactant production. *Bioresour Technol* 43:1–6.

Meylheuc T, vanOss CJ, Bellon-Fontaine MN. 2001. Adsorption of biosurfactant on solid surfaces and consequences regarding the bioadhesion of *Listeria monocytogenes* LO28. *J Appl Microbiol* 91:822–832.

Miller RM. 1995. Biosurfactant facilitated remediation of contaminated soil. *Environ Health Perspect* 103 (Suppl.1):59–62.

Mireles JR 2nd, Toguchi A, Harshey RM. 2001. *Salmonella enterica* serovar typhimurium swarming mutants with altered biofilm forming abilities: Surfactin inhibits biofilm formation. *J Bacteriol* 183:5848–5854.

Moldes AB, Paradelo R, Rubinos D, Devesa-Rey R, Cruz JM, Barral MT. 2011. Ex situ treatment of hydrocarbon-contaminated soil using biosurfactants from *Lactobacillus pentosus*. *J Agric Food Chem* 59:9443–9447.

Montoneri E, Savarino P, Bottigliengo S, Boffa V, Prevot AB, Fabbri D, Pramauro E. 2009. Biomass wastes as renewable source of energy and chemicals for the industry with friendly environmental impact. *Fresenius Environ Bull* 18:219–223.

Moyne AL, Shelby R, Cleveland TE, Tuzun S. 2001. Bacillomycin D: An iturin with antifungal activity against *Aspergillus flavus*. *J Appl Microbiol* 90:622–629.

Mukherjee S, Das P, Sen R. 2006. Towards commercial production of microbial surfactants. *Trends Biotechnol* 24:509–515.

Mulligan CN, Yong CN, Gibbs BF. 1999. Removal of heavy metals from contaminated soil and sediments using the biosurfactant surfactin. *J Soil Contam* 8:231–254.

Mulligan CN, Yong CN, Gibbs BF. 2001. Heavy metal removal from sediments by biosurfactants. *J Hazard Mat* 85:111–125.

Muthusamy K, Gopalakrishnan S, Ravi TK, Sivachidambaram P. 2008. Biosurfactants: Properties, commercial production and application. *Curr Sci* 94:736–747.

Naruse N, Tenmyo O, Kobaru S, Kamei H, Miyaki T, Konishi M, Oki T. 1990. Pumilacidin, a complex of new antiviral antibiotics: Production, isolation, chemical properties, structure and biological activity. *J Antibiot (Tokyo)* 43:267–280.

Nguyen TT, Youssef NH, McInerney MJ, Sabatini DA. 2008. Rhamnolipid biosurfactant mixtures for environmental remediation. *Water Res* 42:1735–1743.

Nievas ML, Commendatore MG, Estevas JL, Bucala V. 2008. Biodegradation pattern of hydrocarbons from a fuel oil-type complex residue by an emulsifier-producing microbial consortium. *J Hazard Mater* 154:96–104.

Nitschke M, Costa SG. 2007. Biosurfactants in food industry. *Trends Food Sci Technol* 18:252–259.

Nitschke M, Costa SG, Haddad R, Goncalves LA, Eberlin MN, Contiero J. 2005. Oil wastes as unconventional substrates for rhamnolipid biosurfactant production by *Pseudomonas aeruginosa* LBI. *Biotechnol Prog* 21:1562–1566.

Nitschke M, Pastore G. 2006. Production and properties of a surfactant obtained from *Bacillus subtilis* grown on cassava wastewater. *Bioresource Technol* 97:336–341.

Noah KS, Bruhn DF, Bala GA. 2005. Surfactin production from potato process effluent by *Bacillus subtilis* in a chemostat. *Appl Biochem Biotechnol* 122:465–474.

Obayori OS, Ilori MO, Adebusoye SA, Oyetibo GO, Omotayo AE, Amund O. 2009. Degradation of hydrocarbons and biosurfactant production by *Pseudomonas* sp. strain LP1. *World J Microbiol Biotechnol* 25:1615–1623.

Oliveira FJS, Vazquez L, de Campos NP, de França FP. 2009. Production of rhamnolipids by a *Pseudomonas alcaligenes* strain. *Process Biochem* 44:383–389.

Olivera NL, Commendatore MG, Delgado O, Esteves JL. 2003. Microbial characterization and hydrocarbon biodegradation potential of natural bilge waste microflora. *J Ind Microbiol Biotechnol* 30:542–548.

Onbasli D, Aslim B. 2009. Determination of rhamnolipid biosurfactant production in molasses by some *Pseudomonas* spp. *New Biotechnol* 25:S255–S255.

Pacwa-Płociniczak M, Płaza GA, Piotrowska-Seget Z, Cameotra SS. 2011. Environmental applications of biosurfactants: Recent advances. *Int J Mol Sci* 12:633–654.

Parthasarathi R, Sivakumaar PK. 2011. Biosurfactant mediated remediation process evaluation on a mixture of heavy metal spiked topsoil using soil column and batch washing methods. *Soil Sedi Contam, Int J* 20:892–907.

Pekdemir T, Copur M, Urum K. 2005. Emulsification of crude oil–water systems using biosurfactants. *Process Saf Environ Prot* 83:38–46.

Pekin G, Vardar-Sukan F, Kosari N. 2005. Production of sophorolipids from *Candida bombicola* ATCC 22214 using Turkish corn oil and honey. *Eng Life Sci* 5:357–362.

Piljac T, Piljac G. 1999. Use of rhamnolipids in wound healing, treating burn shock, atherosclerosis, organ transplants, depression, schizophrenia and cosmetics (European Patent 1 889 623). Paradigm Biomedical Inc New York.

Plaza G, Pacwa-Plociniczak M, Piotrowska-Seget Z, Jangid K, Wilk KA. 2011. Agroindustrial wastes as unconventional substrates for growing of *Bacillus* strains and production of biosurfactant. *Environ Protec Eng* 37: 63–71.

Portilla-Rivera OM, Moldes Menduiña AB, Torrado Agrasar AM, Domínguez González JM. 2007. Biosurfactants from grape marc: Stability study. *J Biotechnol* 131:1010–1020.

Prieto LM, Michelon M, Burkert JF, Kalil SJ, Burkert CA. 2008. The production of rhamnolipid by a Pseudomonas aeruginosa strain isolated from a southern coastal zone in Brazil. *Chemosphere* 71:1781–1785. DOI: 10.1016/j.chemosphere.2008.01.003.

Pruthi V, Cameotra SS. 2003. Effect of nutrients on optimal production of biosurfactants by *Pseudomonas putida*—A Gujarat oil field isolate. *J Surf Deterg* 6:65–68.

Rahman KSM, Rahman TJ, Lakshmanaperumalsamy P, Marchant R, Banat IM. 2003. The potential of bacterial isolates for emulsification with range of hydrocarbons. *Acta Biotechnol* 4:335–345.

Rapp P, Bock H, Wray V, Wagner F. 1979. Formation, isolation and characterization of trehalose dimycolates from *Rhodococcus erythropolis* grown on n-alkanes. *J Gen Microbiol* 115:491–503.

Rashedi H, Assadi MM, Bonakdarpour B, Jamshidi E. 2005. Environmental importance of rhamnolipid production from molasses as a carbon source. *Int J Environ Sci Technol* 2:59–62.

Raza ZA, Rehman A, Saleem Khan M, and Khalid ZM. 2007. Improved production of biosurfactant by a Pseudomonas aeruginosa mutant using vegetable oil refinery wastes. *Biodegradation*, 18:115–121. DOI: 10.1007/s10532-006-9047-9.

Reddy MS, Naresh B, Leela T, Prashanthi M, Madhusudhan NC, Dhanasri G, Devi P. 2010. Biodegradation of phenanthrene with biosurfactant production by a new strain of *Brevibacillus* sp. *Bioresource Technol* 101:7980–7983.

Rivardo F, Turner RJ, Allegrone G, Ceri H, Martinotti MG. 2009. Anti-adhesion activity of two biosurfactants produced by *Bacillus* spp. prevents biofilm formation of human bacterial pathogens. *Appl Microbiol Biotechnol* 86:511–553.

Rodrigues L, Banat IM, Teixeira J, Oliveira R. 2006. Biosurfactants: Potential applications in medicine. *J Antimicrob Chemother* 57:609–618.

Rodrigues L, vanderMei HC, Teixeira J, Oliveira R. 2004. Influence of biosurfactants from probiotic bacteria on formation of biofilms on voice prostheses. *Appl Environ Microbiol* 70:4408–4410.

Rosenberg E, Ron EZ. 1999. High- and low-molecular-mass microbial surfactants. *Appl Microbiol Biotechnol* 52:154–162.

Rufino R, Sarubbo L, Campos-Takaki G. 2007. Enhancement of stability of biosurfactant produced by *Candida lipolytica* using industrial residue as substrate. *World J Microbiol Biotechnol* 23:729–734.

Rufino RD, Sarubbo LA, Neto BB, Campos-Takaki GM. 2008. Experimental design for the production of tensio-active agent by *Candida lipolytica*. *J Ind Microbiol Biotechnol* 35:907–914.

Sachdev DP, Cameotra SS. 2013. Biosurfactants in agriculture. *Appl Microbiol Biotechnol* 97:1005–1016.

Savarino P, Montoneri E, Biasizzo M, Quagliotto P, Viscardi G, Boffa V. 2007. Upgrading biomass wastes in chemical technology. Humic acid-like matter isolated from compost as chemical auxiliary for textile dyeing. *J Chem Technol Biotechnol* 82:939–948.

Sekhon KK, Khanna S, Cameotra SS. 2011. Enhanced biosurfactant production through cloning of three genes and role of esterase in biosurfactant release. *Microb Cell Fact* 10:49.

Selvin J, Shanmughapriya S, Gandhimathi R, Seghal Kiran G, Rajeetha Ravji T, Natarajaseenivasan K, Hema TA. 2009. Optimization and production of novel antimicrobial agents from sponge associated marine actinomycetes *Nocardiopsis dassonvillei* MAD08. *Appl Microbiol Biotechnol* 83:435–445.

Sen R, Swaminathan T. 2005. Characterization of concentration and purification parameters and operating conditions for the small-scale recovery of surfactin. *Process Biochem* 40:2953–2958.

Shete A, Wadhawa G, Banat I, Chopade B. 2006. Mapping of patents on bioemulsifier and biosurfactant: A review. *J Sci Ind Res* 65:91–115.

Sim L, Ward OP, Li ZY. 1997. Production and characterisation of a biosurfactant isolated from *Pseudomonas aeruginosa* UW-1. *J Ind Microbiol Biotechnol* 19:232–238.

Singh A, Van Hamme JD, Ward OP. 2007. Surfactants in microbiology and biotechnology: Part 2: Application aspects. *Biotechnol Adv* 25:99–121.

Singh P, Cameotra SS. 2004a. Enhancement of metal bioremediation by use of microbial surfactants. *Biochem Biophys Res Commun* 319:291–297.

Singh P, Cameotra S. 2004b. Potential applications of microbial surfactants in biomedical sciences. *Trends Biotechnol* 22:142–146.

Solaiman D, Ashby R, Zerkowski J, Foglia T. 2007. Simplified soy molasses-based medium for reduced-cost production of sophorolipids by *Candida bombicola*. *Biotechnol Lett* 29:1341–1347.

Stosur GJ. 1991. Unconventional EOR concepts. *Crit Rev Appl Chem* 33:341–373.

Sun X, Wu L, Luo Y. 2006. Application of organic agents in remediation of heavy metals-contaminated soil. *Ying Yong Sheng Tai Xue Bao* 17:1123–1128.

Syldatk C, Hausmann R. 2010. Microbial biosurfactants. *Eur J Lipid Sci Technol* 112:615–616.

Thavasi R, Jayalakshmi S, Balasubramanian T, Banat I. 2008. Production and characterization of a glycolipid biosurfactant from *Bacillus megaterium* using conomically cheaper sources. *World J Microbiol Biotechnol* 24:917–925.

Thavasi R, Jayalakshmi S, Banat IM. 2011. Application of biosurfactant produced from peanut oil cake by *Lactobacillus delbrueckii* in biodegradation of crude oil. *Bioresour Technol* 102:3366–3372.

Todd RS, Andrea MC, Maier RM. 2000. A rhamnolipid biosurfactant reduces cadmium toxicity during naphthalene biodegradation. *Appl Environ Microbiol* 66:4585–4588.

Torrens JL, Herman DC, Miler RM. 1998. Biosurfactants (rhamnolipid) sorption and the impact on rhamnolipid-facilitated removal of cadmium from various soils under saturated flow conditions. *Environ Sci Technol* 32:776–781.

Trummler K, Effenberger F, Syldatk C. 2003. An integrated microbial/enzymatic process for production of rhamnolipids and l-(+)-rhamnose from rapeseed oil with *Pseudomonas* sp. DSM 2874. *Eur J Lipid Sci Tech* 105:563–571.

Ueno Y, Hirashima N, Inoh Y, Furuno T, Nakanishi M. 2007. Characterization of biosurfactant-containing liposomes and their efficiency for gene transfection. *Biol Pharm Bull* 30:169–172.

Urum K, Pekdemir T. 2004. Evaluation of biosurfactants for crude oil contaminated soil washing. *Chemosphere* 57:1139–1150.

Van Haesendonck IPH, Venzeveren ECA. 2004. Rhamnolipids in bakery products. W.O. 2004/040984, International application patent (PCT).

Van Hamme JD, Singh A, Ward OP. 2006. Physiological aspects: Part 1 in a series of papers devoted to surfactants in microbiology and biotechnology. *Biotechnol Adv* 24:604–620.

Velho RV, Medina LF, Segalin J, Brandelli A. 2011. Production of lipopeptides among *Bacillus* strains showing growth inhibition of phytopathogenic fungi. *Folia Microbiol (Praha)* 56:297–303.

Velraeds MM, vandeBelt-Gritter B, vanderMei HC, Reid G, Busscher HJ. 1998. Interference in initial adhesion of uropathogenic bacteria and yeasts to silicone rubber by a *Lactobacillus acidophilus* biosurfactant. *J Med Microbiol* 47:1081–1085.

Vollenbroich D, Pauli G, Ozel M, Vater J. 1997. Antimycoplasma properties and applications in cell culture of surfactin, a lipopeptide antibiotic from *Bacillus subtilis*. *Appl Environ Microbiol* 63:44–49.

Wang Q, Chen S, Zhang J, Sun M, Liu Z, Yu Z. 2008. Co-producing lipopeptides and poly-[gamma]-glutamic acid by solid-state fermentation of *Bacillus subtilis* using soybean and sweet potato residues and its biocontrol and fertilizer synergistic effects. *Bioresour Technol* 99:3318–3323.

Wattanaphon HT, Kerdsin A, Thammacharoen C, Sangvanich P, Vangnai AS. 2008. A biosurfactant from *Burkholderia cenocepacia* BSP3 and its enhancement of pesticide solubilization. *J Appl Microbiol* 105:416–423.

Wei YH, Wang LF, Changy JS, Kung SS. 2003. Identification of induced acidification in iron enriched cultures of *Bacillus subtilis* during biosurfactant fermentation. *J Biosci Bioeng* 96:174–178.

Yakimov MM, Amor MM, Bock M, Bodekaer K, Fredrickson HL, Timmis KN. 1997. The potential of *Bacillus licheniformis* for *in situ* enhanced oil recovery. *J Petrol Sci Eng* 18:147–160.

Zhang C, Wang S, Yan Y. 2011. Isomerization and biodegradation of beta-cypermethrin by *Pseudomonas aeruginosa* CH7 with biosurfactant production. *Bioresour Technol* 102:7139–7146.

Zhang Y, Li H, Sun J, Gao J, Liu W, Li B, Guo Y, Chen J. 2010. Dc-chol/Dope cationic liposomes: A comparative study of the influence factors on plasmid pDNA and Si RNA gene delivery. *Int J Pharm* 390:198–207.

Zhao Z, Wang Q, Wang K, Brian K, Liu C, Gu Y. 2010. Study of the antifungal activity of *Bacillus vallismortis* ZZ 185 *in vitro* and identification of its antifungal components. *Bioresour Technol* 101:292–329.

14 Perspectives on Using Biosurfactants in Food Industry

Lívia Vieira de Araujo, Denise Maria Guimarães Freire, and Márcia Nitschke

CONTENTS

14.1 Introduction 295
14.2 Applications of Biosurfactants in the Food Industry...296
    14.2.1 Biofilm Control 296
        14.2.1.1 Biofilms on Food-Processing Surfaces 297
        14.2.1.2 Biosurfactants' Antibiofilm Properties 298
    14.2.2 Biosurfactants' Antimicrobial Properties..300
    14.2.3 Biosurfactants as Food Emulsifiers 301
    14.2.4 Production of Biosurfactants from Food Wastes...303
        14.2.4.1 Sugar and Fruit-Processing Industry 304
        14.2.4.2 Dairy Industry...304
        14.2.4.3 Edible Oils Industry..305
        14.2.4.4 Starch Products Industry...306
14.3 Future Trends...307
References...307

14.1 INTRODUCTION

Many fields of application could be fulfilled by biosurfactants (BSs) properties, such as agriculture, construction, bioremediation processes, lubricants, leather treatment, paper and metal industries, as well as textile, cosmetics, pharmaceutical, petrochemicals, industrial cleansing, and above all, within the scope of this chapter, the food industry. The properties demonstrated by microbial BSs such as high biodegradability, low toxicity, low critical micelle concentration (CMC), high surface activity, stability to extreme pH, temperature, salt concentrations, and biological activity are very useful for food industry. Thus, BSs can replace conventional synthetic surfactants with great advantages and additionally, their bio-based origin confers to these molecules the status of natural additives satisfying the actual market needs.

BSs antiadhesive property is desirable to reduce/remove biofilms at the surfaces that come in contact with food, therefore enhancing food shelf life due to less contamination during processing. Antimicrobial activity can be exploited by including them into food formulations to enhance product shelf life and avoid food contamination. As emulsifiers, BSs can be applied during food raw-material processing of bakery, meat, and dairy products, where the addition of emulsifiers enhances texture and creaminess. BSs can be potential candidates in the search for new products with a wide range of functionalities, inasmuch as they achieve the requirements of food additives.

Considering the potential of BSs as new promising alternatives to food industry, this chapter discusses their application as agents to form emulsions, to control microbial pathogens growth, adhesion, and biofilm formation, as well as their production using wastes or by-products from food-processing chain.

14.2 APPLICATIONS OF BIOSURFACTANTS IN THE FOOD INDUSTRY

14.2.1 BIOFILM CONTROL

Microbial communities that grow between phases, such as solid–liquid or air–liquid, are called biofilms (Jenkinson and Lappin-Scott, 2001). In food industries, any failure during cleaning processes enables some food residues to remain adhered to equipment and surfaces. Therefore, under certain conditions, microorganisms adhere, interact with the surface, and initiate cellular multiplication until the formation of a mass capable of aggregate nutrients, residues, and other microorganisms (Bagge-Ravn et al., 2003; Kim et al., 2006).

Biofilms contain proteins, lipids, carbohydrates, nucleic acids, mineral salts, and vitamins, forming layers where microorganisms develop. They are represented by the interaction and organization from a tridimensional structure of microorganisms involved by exopolysaccharides (EPS), with water channels and multiple cell layers (Jenkinson and Lappin-Scott, 2001).

The EPS matrix is responsible for cell-to-cell linkage, as well as adherence of individual cells to surfaces (Lindsay and Holy, 1997); it is hydrated and can be used as nutrient source (Lindsay and Holy, 1997). Under turbulent flow, biofilms spread through surfaces, due to the EPS property of being a viscoelastic fluid (Jenkinson and Lappin-Scott, 2001). Some factors that can interfere on microbial adhesion to surfaces are the presence of cellular appendix (Parizzi et al., 2004; Chae et al., 2006), hydrophobicity and charge of microorganisms and surfaces involved (Meylheuc et al., 2006a) as well as the capacity or not to produce EPS (Parizzi et al., 2004). Strains with a higher capacity to produce EPS are less hydrophobic and have a higher ability to form biofilms. It was verified that the higher the levels of extracellular carbohydrates produced by *Listeria monocytogenes*, the higher will be their ability to form biofilms (Borucki et al., 2003; Chae et al., 2006).

There are many theories about biofilm development; the most-studied process is observed in terms of distance from bacterial cell to the surface. When the distance is higher than 50 nm, only long-range forces act and the attachment is reversible. As the distance of separation is closer to 20 nm, long-range forces (van der Waals and electrostatic) and short-range interactions (chemical bonds and hydrophobic interactions) are operating. This phase can be reversible, but with time it becomes irreversible. At distances lower than 15 nm, additional forces begins to act (production of adhesive polymers) leading to irreversible fixation. The final phase depends on the ability of the microorganism to metabolize and produce the EPS matrix (Hood and Zottola, 1995).

Other general principles can be referred to bacterial adhesion to surfaces. The adhesion will occur if it results in a decrease of the system's free energy (Busscher et al., 1984; Carpentier and Cerf, 1993). Hydrophobic interactions are suggested as responsible for a wide range of adherence phenomena. Some evidence exists to suggest that hydrophobicity is involved in the initial phase of microbial adhesion (Hood and Zottola, 1995; Djordjevic et al., 2002). Cellular hydrophobicity is related to bacterial adhesion at hydrophobic surfaces as much as at hydrophilic surfaces (Sommer et al., 1999). At hydrophobic surfaces, hydrophobic interactions are the acting main forces, whereas at hydrophilic surfaces, electrostatic interactions are the major forces.

Microbial development at a surface is associated with massive agitation, regulated by bacterial cells' growth kinetics and amount of nutrients available. Bacterial cells send signals to each other (Kolter and Greenberg, 2006), which is necessary for the establishment of a multicellular behavior, reflected by coordinated activities of interaction and communication of the microorganisms (Jenkinson and Lappin-Scott, 2001; Takhistov and George, 2004).

The cells' dispersion from biofilms is also affected by many factors, such as time, temperature, and mechanical and chemical forces (Maukonen et al., 2003). After a period of time, biofilms release microbial cells enabling them to colonize other environments.

Biofilm microorganisms are more resistant than planktonic cells to chemical and physical agents commonly used in hygiene processes (Djordjevic et al., 2002). The biofilm cells' resistance to environmental influences such as antibiotics, host defense, antiseptics, and shear forces is a constant worry for industry and medicine (Jenkinson and Lappin-Scott, 2001). The mechanisms involved in the higher resistance of biofilm cells involve the following: EPS protection due to the ability from EPS to avoid the contact of the agent to the microorganisms; differentiated metabolism—microorganisms in biofilms present a differentiated metabolism (slow metabolism); and agent neutralization—mainly with multispecies biofilms, some strains are able to modify and/or neutralize the agent avoiding their expected effect. Therefore, EPS does not necessarily provide a diffusion barrier against inhibitory compounds (antimicrobial), but when in biofilm bacteria, it may be inherently more resistant. At multispecies biofilms, species can benefit each other through substrate exchange and/or mutual metabolite removal (Jenkinson and Lappin-Scott, 2001; Maukonen et al., 2003).

14.2.1.1 Biofilms on Food-Processing Surfaces

At food-processing industries, some microorganisms represent damage to food products, influencing the storage quality due to food deterioration, whereas some other microorganisms represent risk to food safety due to foodborne diseases if the food contamination occurs after a pasteurization/sterilization step (Hood and Zottola, 1995).

Many microorganisms are found forming biofilms/adhering to food and food contact surfaces, some of them are *Pseudomonas aeruginosa*, *Pseudomonas fragi*, *Pseudomonas fluorescens*, *Micrococcus* sp., *Enterococcus faecium*, *L. monocytogenes*, *Yersinia enterocolitica*, *Salmonella typhimurium*, *Escherichia coli* O157:H7, *Staphylococcus aureus*, and *Bacillus cereus* (Hood and Zottola, 1995; Maukonen et al., 2003).

L. monocytogenes is an example of a pathogenic bacteria often found adhering at food industries possessing a high ability to develop biofilms in a wide range of surfaces found in food processing; this microorganism also has the ability to grow under many environmental conditions such as low temperature, acidic pH, and high salt concentrations (Chae and Schraft, 2000; Kalmokoff et al., 2001; Meylheuc et al., 2001). Usual food involved in listeriosis (Listeria infection) is comprised of dairy (milk, ice creams, and cheese); meat products (bovine, ovine, swine, and poultry origin); fish and also vegetable products (Jay, 2005).

P. fluorescens is an example of a nonpathogenic bacterium commonly found in food industries. This microorganism also presents a high ability to form biofilms at different surfaces. Furthermore, it belongs to one of the most genera found at industrial water circuits (Machado, 2005). Water treatment and juice filtration systems that use membranes generally develop bioincrustation ("*biofouling*"), leading to performance losses in those processes. Biofouling is characterized by particles and microorganisms adsorbed to surfaces, contributing directly to the beginning of biofilm development. In this case, the growth of microorganisms such as *P. fluorescens* causes overloading of the system, increasing working pressure, leading to obstruction, membrane rupture, and unnecessary waste of energy.

The main goal to be achieved at food-processing industries is to furnish safe food, with nutritional and sensorial quality to consumers. Microorganisms' control is essential to achieve this goal (Hood and Zottola, 1995; Bagge-Ravn et al., 2003). Some pathogens like *Salmonella* spp. and *L. monocytogenes* are not tolerated at food industries consequently, a unique cell of those bacteria may be as important as an established biofilm (Hood and Zottola, 1995). Industrial processes involving food supply nutrients and conditions for bacterial growth (Maukonen et al., 2003) and so, surfaces sanitization is essential to prevent contamination. To decrease or eliminate microorganisms from surfaces, food-processing industries usually apply techniques including physical methods, as manual washing and high-pressure sprays, as well as chemical methods (hypochlorite, quaternary ammonia compounds). Both techniques remove and inactivate microorganisms at the equipment surfaces (Hood and Zottola, 1995). Bacterial fixation and biofilm development at food contact surfaces can occur within short periods of time (20 min), depending on the microorganism and environmental conditions (Lindsay and Holy, 1997). Unfortunately, evidences suggest that the usual sanitization procedures are less effective to adhered microorganisms than to planktonic cells (Lindsay and Holy, 1997; Bagge-Ravn et al., 2003). At food-processing plants, sanitization is preceded by cleaning with detergents, because it is well known that the efficacy of sanitizing agent is decreased when organic matter is present. Cleansing and sanitization recommendations are still the best way to fight contamination. Furthermore, detergents can be considered as the major important compound to control bacterial number, due to its ability to remove organic matter that would serve as substrate to microbial adhesion (Hood and Zottola, 1995; Somers and Wong, 2004). As stated before, even using recognized cleaning systems, bacteria can remain at equipment surfaces and survive for long periods, depending on the amount and nature of residual substrate, temperature, and relative humidity (Maukonen et al., 2003).

There is a wide range of surface types that comes in contact with food during processing enabling microbial adherence, some examples are stainless steel, glass, cast iron, rubber, polypropylene, polystyrene, low-density polyethylene, polycarbonate, and others. That is the reason why the choosing of a surface is an important factor to combat biofilm development. The most resistant surface to mechanical cleaning and chemical agents must be chosen to each processing step. Biofilms are commonly formed at dead spaces, joints, valves, seals, and corroded surfaces (Stopforth et al., 2002; Maukonen et al., 2003; Parizzi et al., 2004).

Owing to their great importance to food industries, new biofilm controlling strategies are constantly being investigated. The use of surfactant agents with natural, green, and nontoxic properties is a subject of increasing interest.

14.2.1.2 Biosurfactants' Antibiofilm Properties

Studies have demonstrated that surfaces pre-conditioned with BSs are able to decrease strongly the microbial contamination of materials and inhibit or reduce the subsequent biofilm development (Meylheuc et al., 2006b). When a surfactant adsorbs to hydrophobic surfaces, it usually turns the hydrophobic group to the surface and exposes the polar group to water phase. Therefore, the surface becomes hydrophilic and, as a result, the interfacial tension between water and surface is reduced (Machado, 2005).

Rhamnolipid surfactants plays a role at *P. aeruginosa* biofilm cells desorption (Boles et al., 2005). Lipopeptides are able to modify bacterial surface hydrophobicity and their adhesion to solid surfaces; such an effect depends on initial bacterial hydrophobicity, as well as the type of lipopeptide and its concentration, leading to an increase or decrease on bacterial surface hydrophobicity (Ahimou et al., 2000).

The preconditioning of urethral catheters with surfactin totally inhibited *Salmonella enterica*, *E. coli*, and *P. mirabilis* biofilm formation (Mirelles et al., 2001). A BS produced from *P. fluorescens* (noncharacterized) was used for conditioning of PTFE and Stainless Steel 304 AISI surfaces showing a significant decrease on *L. monocytogenes* LO28 adhesion to stainless steel at different temperatures (up to 92%); however, at the PTFE surface, adhesion decrease was not observed (Meylheuc et al., 2001).

Silicone rubber surfaces conditioned with rhamnolipids from *P. aeruginosa* (4 g/L) inhibited 60.9%, 53.1%, 58.2%, and 33.8% the adhesion of *Rothia dentocariosa*, *Staphylococcus epidermidis*, *Streptococcus salivarius*, and *Staphylococcus aureus*, respectively (Rodrigues et al., 2006a).

Stainless steel conditioned with surfactin (0.1%) was able to significantly reduce *L. monocytogenes* ATCC 19112 adhesion at approximately 10^2 CFU/cm^2 and *Enterobacter sakazakii* ATCC 29004 at approximately 10^1 CFU/cm^2. The polypropylene conditioned with the same surfactant showed a significant adhesion reduction of approximately 10^1 CFU/cm^2 from *L. monocytogenes*, *E. sakasakii*, and *Salmonella enteritidis* (Nitschke et al., 2009).

Polystyrene conditioning with crude rhamnolipids, purified rhamnolipids, and surfactin-decreased adhesion from different strains of *L. monocytogenes* (ATCC 19112, 19117, 15313, and 7644) was more effective than the surface conditioning with sodium-dodecyl sulfate (Araujo et al., 2011).

Surfactin and rhamnolipids were also able to decrease adhesion from *S. aureus*, *L. monocytogenes*, and *Micrococcus luteus* at a wide range of temperatures when used as polystyrene surface conditioners (Zeraik and Nitschke, 2010). Preconditioning of polystyrene with a lipopeptide produced by *Bacillus circulans* inhibited the adhesion of various pathogenic microorganisms from 15% to 89% (Das et al., 2009). An extract containing surfactin, used as conditioning agent on polystyrene surfaces, inhibited approximately 42% the biofilm formation of *L. monocytogenes* ATCC 19112 (Gomes and Nitschke, 2012).

Shakerifard et al. (2009) evaluated the ability of lipopeptide surfactants obtained from *Bacillus subtilis* to modify hydrophobicity of Teflon and stainless-steel surfaces and its correlation with the adhesion of *B. cereus* spores. They concluded that there is a good correlation between surface hydrophobicity modifications promoted by BSs and attachment of spores. The best results were shown to iturin A, which reduce the adhesion of spores to teflon by 6.5-fold at 100 mg/L.

The antiadhesive or inhibitory effect from BSs adsorbed to the surfaces may be linked to unique changes in the physicochemical properties, by altering the attractive interactions between surface and microorganism, combined or not with a desorption effect of the biocompound (Araujo et al., 2007; Nitschke et al., 2009; Araujo et al., 2011).

The BSs and other compounds that have been produced by microorganisms may act as a *quorum-sensing* factors (Yarwood et al., 2004), positively or negatively interfering in the process of adhesion and biofilm formation. It was reported in the literature that *P. aeruginosa* can produce other molecule that possesses anti-biofilm effect, an unsaturated fatty acid *cis*-2-decenoic; however, its action mechanism was not yet elucidated (Shank and Kolter, 2011). So, rhamnolipids' effect on biofilm inhibition can be optimized by the presence of these compounds.

Another important approach to be explored is the use of BSs to remove (disrupt) preformed biofilms. In order to be effective in removing biofilms, BSs have to penetrate into the interface between solid substrate and the biofilm so that they could adsorb and reduce interfacial tension. The attractive interactions between bacterial and solid surfaces may be decreased leading to the removal of biofilm (McLandsborough et al., 2006).

Kuiper et al. (2004) characterized two lipopeptide BSs produced by *Pseudomonas putida* PCL1445 and verified that both BSs were able to inhibit biofilm formation from different *Pseudomonas* species and strains on PVC surfaces. Furthermore, these BSs were also able to disrupt preexisting biofilms. Irie et al. (2005) reported the disruption of *Bordetella bronchiseptica* biofilms using rhamnolipids, and Dusane et al. (2010) have demonstrated that *Bacillus pumilus* biofilms were removed up to 93% using 100 mM rhamnolipids. A glycolipid BS produced by a marine *Brevibacterium casei* was effective in disruption of 24-hour-old biofilms of *Vibrio parahaemolyticus*, *P. aeruginosa*, and *E. coli* formed on glass coverslips (Kiran et al., 2010).

Gomes and Nitschke (2012) evaluated the disruption of 48-hour-old biofilms of individual and mixed cultures of food pathogens established on polystyrene surfaces using rhamnolipids and surfactin. After 2-h contact with 0.1% surfactin, the preformed biofilms of *S. aureus* were reduced by 63.7%, *L. monocytogenes* by 95.9%, *S. enteritidis* by 35.5%, and the mixed culture by 58.5%. The rhamnolipids at 0.25% concentration removed 58.5% the biofilm of *S. aureus*, 26.5% of *L. monocytogenes*, 23.0% of *S. enteritidis*, and 24.0% the mixed culture. Authors found that the increase in concentration of BS and time of contact decreased biofilm removal percentage. The different susceptibilities of each bacterial biofilm to the BSs can be related with the amount and chemical composition of the polymeric material produced by the strains.

Tahmourespour et al. (2011) observed that a BS produced by *L. fermentum* was able to inhibit the production of extracellular glucans by *Streptococcus mutans* a well-known biofilm-producing bacteria involved in dental caries. It was hypothesized that two enzymes produced by *S. mutans* for polymer synthesis, were inhibited by the BS. The BS showed substantial antibiofouling activity reducing the process of attachment and biofilm production. This work demonstrates that investigation of surface-active molecules and the study of their effect on biofilm population is an unexplored area of research that can lead to the discovery of new molecules applied to specific bacterial control.

14.2.2 BIOSURFACTANTS' ANTIMICROBIAL PROPERTIES

A wide range of BSs demonstrated antimicrobial activity against bacteria, fungi, yeasts, algae, and viruses (Nitschke and Costa, 2007). Among BSs, the lipopeptides are the most reported in terms of antimicrobial action. The genus *Bacillus* is responsible for the production of well-known lipopeptides surfactants (Fernandes et al., 2007; Das et al., 2008). *B. subtilis* is a surfactin producer, whereas *Bacillus licheniformis* is a lichenisin producer and *B. pumilus* a pumilacidin producer (Das et al., 2008).

Rivardo et al. (2010) reported that a lipopeptide from *B. licheniformis* V9T14 increases the bactericidal activity of silver ions against *E. coli* biofilms reducing the biofilm population. This synergistic interaction could be useful in disinfecting surfaces to reduce bacterial colonization and spreading of disease.

Another report demonstrated antimicrobial activity of two lipopeptides from *B. subtilis* against 29 bacteria (Fernandes et al., 2007). Lipopeptides produced by *B. subtilis* and *B. licheniformis* were tested against two pathogenic microorganisms, being the product of *B. subtilis* effective against Gram-positive and not effective against Gram-negative bacteria. The inverse occurred for the product obtained from *B. licheniformis* (Rivardo et al., 2009). Lipopeptides like surfactin intrude in membranes containing phospholipids that have shorter chains and/or are in fluid organization, thus interfering with their biological functions by inserting in lipid bilayers, modifying the permeability of the membrane by forming ionic channels, or by carrying mono- or di-valent cations and solubilizing the membrane due to detergent mechanism (Bouffioux et al., 2007). This antimicrobial effect is probably more efficient in Gram-positive than Gram-negative microorganisms because of differences in cell walls structures, causing a greater inhibition of cell growth, since usually the Gram-positive microorganisms have higher sensitity to anionic detergents than Gram-negative microorganisms (Tortora et al., 2005).

Other BSs classes that are reported with antimicrobial activity are the rhamnolipids of *P. aeruginosa* and sophorolipids of *Candida bombicola*. The mannosileritritol lipids (MEL and MEL-A-B) produced by a strain of *Candida antarctica* showed antimicrobial activity against Gram-positive bacteria (Das et al., 2008). A glycolipid from *B. casei* showed bactericidal effect against *E. coli* and *Vibrio alginolyticus* showing an MBC of 38 and 41 µg/mL, respectively (Kiran et al., 2010).

The antimicrobial activity of a mixture of seven rhamnolipids homologues produced by *P. aeruginosa* AT10 over a wide range of Gram-positive and Gram-negative bacteria, yeasts, and molds was evaluated and given to the physicochemical properties and high antimicrobial activity, this product could be seen as a useful tool in bioremediation processes, cosmetics, and applications in food industries (Abalos et al., 2001).

Araujo et al. (2011) demonstrated that the rhamnolipid produced by *P. aeruginosa* PA1 inhibits the growth of *L. monocytogenes* ATCC 19112 and ATCC 7644 suggesting that it could be exploited as an agent to control this important food pathogen. In a recent report, the antimicrobial activity of rhamnolipids was evaluated against several *L. monocytogenes* isolates. Among the 32 tested cultures, 90.6% were susceptible to rhamnolipids and the MIC values varied from 78.1 to 2500 µg/mL. Rhamnolipid activity was primarily bacteriostatic and its interaction with nisin was also investigated. The results showed a particular synergistic effect improving the efficacy of both antimicrobials. The mechanism of action for the combination of nisin and RL was not elucidated; however, as both antimicrobials act on the same target, authors hypothesized that the interaction could occur on the cytoplasm membrane (Magalhães and Nitschke, 2013).

Although many studies describe antimicrobial activity of BSs, there is a lack of experiments showing this activity directly in a food matrix. Huang et al. (2011) demonstrated that a *S. enteritidis* strain was sensitive to surfactin and polylysine, with MICs of 6.25 and 31.25 μg/mL, respectively. *S. enteritidis* was reduced to six log-cycles in milk when the temperature was 4.45°C, the action time 6.91 h, and the concentration (surfactin/polylysine ratio 1:1) was 10.03 μg/mL.

The combination of nisin with rhamnolipids was able to increase the shelf life and inhibit thermophilic spores in soy milk UHT. The use of natamycin associated with rhamnolipids in salad dressings was able to increase the shelf life and inhibit growth of yeasts. The combination of natamycin, nisin, and rhamnolipids also prolonged shelf life and inhibited the growth of bacteria and yeast in cottage cheese (Gandhi and Skebba, 2007).

Apart from antimicrobial action, another interesting application of the BSs is to inhibit metal corrosion. A BS produced by *P. fluorescens* was able to postergate corrosion of stainless steel 304 when applied as a surface conditioner (Dagbert et al., 2006). The metallic corrosion process results in the formation of corrosion products and energy liberation, in other words, the more protected against corrosion will be the surfaces that present the lowest surface total free energy values. BSs conditioning can be related to the formation of a pellicle of those molecules upon the surface guiding the hydrophobic tail to the surface and the hydrophilic head to the environment, keeping the surface protected against H^+ and O_2 interactions and consequently avoiding corrosion (Dagbert et al., 2006).

14.2.3 Biosurfactants as Food Emulsifiers

An emulsion is formed when two immiscible liquids (generally water and oil) are put together where one is dispersed in small drops into the other. According to the distribution of the oil and aqueous phases emulsions can be classified as oil-in-water (O/W), where droplets of oil are dispersed in an aqueous continuous phase and water-in-oil (W/O), where water droplets are dispersed in an oil phase. Many food products, even natural or processed, are emulsions including milk, butter, cream, margarine, mayonnaise, cream liqueur, whippable toppings, ice creams, desserts, sauces, and coffee whiteners (Kralova and Sjoblom, 2009). Emulsifiers are surface-active additives that allow normally immiscible liquids such as oil and water to form stable emulsion, preventing phase separation. Most emulsifiers are amphiphilic, having a polar and nonpolar region at the same molecule being also referred to as surfactants (Hasenhuettl, 2008). In addition to their major function of producing and stabilizing emulsions, food emulsifiers (or surfactants) play several roles as antistalling agents in bakery products; foaming stabilization in cakes and whipped toppings; crystal inhibition in salad oils; antisticking in candies; viscosity modification in chocolate; controlled fat agglomeration in ice cream; freeze–thaw stabilization in whipped toppings and coffee whiteners; gloss enhancement in confectionery coatings, canned and moist pet foods; antispattering in margarines; and solubilizing agents for color and flavor systems among others (Hasenhuettl, 2008). Therefore, they are indispensable additives in food formulations. The most common emulsifiers used in the food comprise phospholipids, amphiphilic proteins, and synthetic surfactants (Freire et al., 2010).

The identification of surfactant compounds with low toxicity and good surface activity properties is of great interest to food industry not only to accomplish regulatory rules but also to fulfill the increasing consumers' request for natural over synthetic ingredients. Microbial-derived surfactants can be explored as emulsifiers in food formulations with advantages when comparing to synthetic surfactants mainly due to their "natural or green" origin, environmental friendly nature (Mohan et al., 2006), and low toxicity (Flasz et al., 1998).

BSs can act controlling consistency, retarding staling, and solubilizing flavor oils in bakery and ice cream formulations (Kosaric, 2001). The addition of rhamnolipids surfactants improved stability, texture, volume, and conservation of bakery products (Van Haesendonck and Vanzeveren, 2004).

The addition of 0.10% rhamnolipid to muffins and croissants enhanced the moisture, improved the texture, and maintained freshness for a longer period of time (Gandhi and Skebba, 2007).

Lecithin is one of the most important food emulsifiers and can be obtained from soybean or egg, so it has a natural status that can satisfy organic and vegetarian requirements for foodstuffs. Egg lecithin is avoided by vegans, and natural emulsifiers may have a place in products directed to such communities. Moreover, Jewish and Islamic consumers demand Kosher-certified raw materials in their food, precluding the use of almost all animal fats (Hasenhuettl, 2008). The use of microbial-based emulsifiers represents an opening market to replace lecithin in such products and also to food requiring non-genetically modified (GM) ingredients.

Regarding their emulsifier properties, BSs can differ according to their nature; some present high emulsifier activity while others are more efficient detergents (reduce surface tension); however, they are not good emulsifiers (Dastgheib et al., 2008). High-molecular-mass BSs are in general better emulsifiers than low-molecular-mass BSs (Ron and Rosenberg, 2001).

Emulsifier activity of BSs is usually evaluated with hydrocarbons; however, some attempts have been made to use BSs as emulsifying agent for food materials. The high-molecular-mass BSs such as emulsan and liposan also referred as polymeric BS does not reduce the surface tension but form stable emulsions with edible oils with prolonged lifespan (Cirigliano and Carman, 1985). Alasan bioemulsifier obtained from *Acinetobacter radioresistens* KA53 was effective in stabilizing emulsions of food grade vegetable and coconut oils. Alasan was claimed to stabilize O/W mixtures, emulsions, and dispersions, to reduce fat content and to increase shelf life of foodstuffs (Rosenberg and Ron, 1998).

Candida glabrata UCP1002 produced a bioemulsifier that forms stable and compact emulsions with cotton seed oil showing emulsifying activity of 75% (Sarubbo et al., 2006). A mannoprotein emulsifier present in the cell wall of *Saccharomyces cerevisiae* was reported to stabilize O/W emulsions with corn oil (60%). O/W emulsions with 8 g/L of bioemulsifier and 5–50 g/L sodium chlorine were stable for 3 months at 4°C at pH 3–11. Such bioemulsifier was suggested as an alternative to mayonnaise production and also other food products such as meat, cakes, crackers, and ice creams (Torabizadeh et al., 1996). An extracellular emulsifier obtained from *Candida utilis* was employed in salad-dressing formulations (Shepherd et al., 1995) and a mannoprotein from *Kluyveromyces marxianus* was able to form emulsions with corn oil that are stable for 3 months (Lukondeh et al., 2003). The rhamnolipids from *P. aeruginosa* LBI were able to form stable emulsions with sunflower, linseed, olive, palm, babassu, and Brazilian nut oils (Costa et al., 2006) and a BS from *B. subtilis* MTCC 441 showed high emulsification index with mustard, coconut, gingelly, and sunflower oils suggesting their potential as emulsifying agent in food systems (Suresh Chander et al., 2012).

Marine bacterial species were reported to produce polymeric carbohydrate–protein molecules showing an emulsification index of 85–95% with soybean, sunflower, olive, peanut, sesame, and rice bran oils. The emulsions formed with the vegetable oils were stable for more than 90 days (Radhakrishnan et al., 2011).

The formation of micro and nanoemulsions has attracted attention in various fields of application. In food systems, they can be applied to delivering functional components (omega-3 fatty acids, vitamins, minerals, and probiotics) flavors, and antioxidants (Ré et al., 2010). Microemulsion systems produce high solubilization capacity and ultra-low interfacial tensions of oil and water, making them desirable in many practical applications (Kogan and Garti, 2006). Formation of microemulsions usually requires the use of surfactant mixtures with salt or alcohol, and only a few surfactants such as soybean lecithin are known to effectively form W/O microemulsion without addition the any co-surfactants (Kitamoto et al., 2009). The MEL-A surfactant obtained from *Pseudozyma* spp. was reported to form stable W/O microemulsion without the need of a co-surfactant showing great potential for future applications in this field (Worakitkanchanakul et al., 2008). Microemulsions of lecithin/rhamnolipid/sophorolipid BSs were successfully prepared using oils as isopropyl myristate and limonene. Ultra-low interfacial tension (<0.1 mN/m) was produced for microemulsions of oils (Nguyen et al., 2010), which is desirable in a wide variety of applications such as cosmetics, pharmaceuticals, and food.

Nanoemulsions consist of oil droplets in the nano-ranged size, between 10 and 100 nm dispersed within an aqueous continuous phase, with each oil droplet surrounded by surfactant molecules (Acosta, 2009; McClements et al., 2009). With the relatively small particle size, nanoemulsions are highly stable to gravitational phase separation and droplet aggregation.

Nanoemulsion production for encapsulation and delivery of functional compounds is one of the emerging fields of nanotechnology applied to food industry. Nanoemulsion systems can improve solubility, bioavailability, and functionality of hydrophobic compounds once they can act as carriers or delivery systems for lipophilic molecules, such as nutraceuticals, flavors, antioxidants, and antimicrobial agents (Silva et al., 2012).

BSs can be explored to formulate nanoemulsions not only to act as carrier but also to obtain antimicrobial nanoemulsions for decontamination of food equipment, packaging, or food itself. Oils such as sunflower, castor, coconut, groundnut, and sesame oils were screened for the development of a surfactin-based nanoemulsion formulation. Surfactin–sunflower-based nanoemulsion showed the highest antibacterial activity against *Salmonella typhi*, followed by *L. monocytogenes* and *S. aureus*. The nanoemulsion also demonstrated fungicidal activity against *Rhizopus nigricans*, *Aspergillus niger*, and *Penicillium* sp., and sporicidal activity against *B. cereus* and *B. circulans*. An *in situ* evaluation of antimicrobial activity of surfactin nanoemulsion on raw chicken, apple juice, milk, and mixed vegetables showed a significant reduction in the native cultivable bacterial and fungal populations of these products (Manoharan et al., 2012). Rhamnolipids from *P. aeruginosa* BS-161R were used to synthesize silver nanoparticles, which exhibited good activity against Gram-positive and Gram-negative pathogens and *Candida albicans* (Kumar et al., 2010).

BSs are versatile molecules that can be exploited by the food industry as cleaning agents or to avoid adhesion, corrosion, and remove biofilms, or also in food formulations as antimicrobial, antiadhesive, and emulsifiers demonstrating their potential as multipurpose ingredients or additives (Table 14.1).

One of the main goals of the food industry is to warrant the supply of safe products to their consumers; in this context, the toxicity of BSs is already a matter of concern. Some data are available about security levels of BSs; however, they are scarce and the microbial origin of these molecules such as the rhamnolipids obtained from the opportunistic pathogen *P. aeruginosa* can represent a limitation to their use. Rhamnolipids are considered nontoxic, noncarcinogenic, and were approved by the FDA to be used in fruit, vegetable, and legume crops as biofungicide (www.rhamnolipid. com). Non-pathogenic species of *Pseudomonas* such as *P. chlororaphis*, *P. luteola*, and *P. fluorescens* have been proposed to produce rhamnolipids (Nitschke et al., 2011) and other GRAS microbes as yeasts. *C. glabrata* (Sarubbo et al., 2006) have been considered for BS production avoiding the use of undesirable strains. A 2500 mg/kg oral administration of surfactin in rats did not show any toxicological effects (Hwang et al., 2009). Literature reviews regarding the use of BS in medical, cosmetics, and pharmaceutical products reinforce the idea of their low toxicity and safety (Kitamoto et al., 2009; Fracchia et al., 2012).

14.2.4 PRODUCTION OF BIOSURFACTANTS FROM FOOD WASTES

Despite their advantages, BSs prices are not yet competitive when compared with commonly used food additives. BS production costs are the main factors to be surpassed to permit their extensive use; however, food-processing industries have a great opportunity to have BSs be more competitive by exploring the production of those molecules from residues or by-products originated from their processes. The use of agro-industrial wastes can reduce surfactant production costs along with the waste treatment expense, rendering a new alternative for food and food-related industries, not only for valorizing their wastes but also to becoming microbial surfactant producers (Nitschke and Costa, 2007).

There are many examples in literature regarding the use of food-derived wastes as feedstock for BS production, particularly carbohydrate and lipid-rich industrial wastes. A brief discussion on the main food residues proposed is presented here. BS production from low-cost substrates is described in more detail in another chapter of this book.

TABLE 14.1

Food-Related Applications of Biosurfactants

Application	Biosurfactant	Reference
Anti-Adhesive		
Listeria monocytogenes	Surfactin and rhamnolipids	Araujo et al. (2011)
Staphylococcus aureus	Surfactin and rhamnolipids	Zeraik and Nitschke (2010)
Bacillus cereus	Iturin A	Shakerifard et al. (2009)
Biofilm Disruption		
Vibrio parahaemolyticus, Pseudomonas aeruginosa, and *Escherichia coli*	Glycolipid	Kiran et al. (2010)
S. aureus, Salmonella enteritidis, and *L.monocytogenes*	Rhamnolipids and surfactin	Gomes and Nitschke (2012)
Antimicrobial		
L. monocytogenes	Rhamnolipids	Magalhães and Nitschke (2013)
E. coli	Glycolipid	Kiran et al. (2010)
Salmonella enteritidis	Surfactin	Huang et al. (2011)
Metal corrosion inhibition	*P. fluorescens* BS	Dagbert et al. (2006)
Emulsifiers		
Mustard, coconut, gingelly, and sunflower oil emulsion	*B. subtilis* MTCC 441 BS	Suresh Chander et al. (2012)
Corn oil emulsion	Mannoprotein bioemulsifier	Lukondeh et al. (2003)
Microemulsions	MEL-A	Worakitkanchanakul et al. (2008)
Microemulsions	Sophorolipid and rhamnolipid	Nguyen et al. (2010)
Nanoemulsions	Surfactin	Manoharan et al. (2012)

14.2.4.1 Sugar and Fruit-Processing Industry

Molasses is a by-product of the sugar industry (even from sugar cane or sugar beet) that is low in price compared to other conventional sugar sources like sucrose or glucose and is rich in other nutrients such as minerals and vitamins (Makkar et al., 2011). Two *B. subtilis* strains were able to produce lipopeptide surfactants using minimal medium supplemented with molasses as carbon source (Makkar and Cameotra, 1997). A *B. subtilis* BS5 was also described to produce surfactin using an optimized medium containing 16% of molasses (Abdel-Mawgoud et al., 2008). Molasses was also described as a carbon source for the production of rhamnolipid BS by *P. aeruginosa* GS3; the interfacial tension of culture medium against crude oil was reduced from 21 to 0.47 mN/m (Patel and Desai, 1997). *P. luteola* and *P. putida* were reported to produce rhamnolipids from sugar beet molasses. The rhamnolipid production increased with the increase in the concentration of molasses and maximum production occurred when 5% (w/v) of molasses were used (Onbasli and Aslim, 2009).

The Brazilian cashew nut industry discards a significant amount of cashew apple juice, that is an inexpensive and attractive substrate rich in reducing sugar, vitamins, minerals, and salts. Strains of *Acinetobacter calcoaceticus*, *B. subtilis*, and *P. aeruginosa* were reported to grow at cashew apple juice-based media and produce BSs (Rocha et al., 2006, 2007, 2009).

14.2.4.2 Dairy Industry

The dairy industry involves processing raw milk into products including milk, butter, cheese, and yogurt, and typical by-products include buttermilk, whey, and their derivatives. Huge amounts of water are used during dairy process generating effluents rich in sugars, proteins, and fat characterized by their high biochemical oxygen demand (BOD) value, so their incorrect disposal represents a significant environmental problem. Only 50% of the cheese whey produced annually is recycled into useful products such as food ingredients and animal feed and the rest is regarded as waste (Makkar et al., 2011).

Cheese whey was used as substrate to BS production in a two-step batch cultivation process by *C. bombicola* and *Cryptococcus curvatus*. In the first step, *C. curvatus* was grown on deproteinized whey concentrates; the cultivation broth was disrupted with a glass bead mill and it served as a medium for growth and sophorolipid production by *C. bombicola* (Daniel et al., 1999). Also, whey showed as a potential substrate to BS production by *Lactobacillus pentosus* (Rodrigues et al., 2006b). *Yarrowia lipolytica, M. luteus*, and *Burkholderia cepacia* also demonstrated the ability to produce BS from whey wastewater. The BSs produced were biochemically characterized and the properties were analyzed showing good emulsification index, and surface activities (Yilmaz et al., 2009).

14.2.4.3 Edible Oils Industry

The world production of oils and fats is around 2.5–3.0 million tons, 75% of which are derived from plants and oil seeds (Dumont and Narine, 2007). Great quantities of waste are generated by the oil industries: post-refinery residual oils, soapstock, and frying oils. Vegetable oils and residues from vegetable oil refineries are among the most used low-cost substrates for BS production (Haba et al., 2000; Nitschke et al., 2005a; Makkar et al., 2011).

Olive oil mill effluent was the first oil residue that was proposed as substrate for BS production. *P. aeruginosa* 47T2 was able to grow and produce rhamnolipids using the substrate as the sole carbon source (Mercade et al., 1993). This work pointed out the great potential on using oil-derived wastes to produce BSs.

Soapstock is a residue from the oil neutralization process and is generated in large quantities by the vegetable oil-processing industry (amounts to 2–3% of the total oil production).

An example of BS production from soapstock and post-refinery fatty acids is the production of glycolipids by *C. antarctica* and *C. apicola* in a cultivation medium supplemented with these wastes. The efficiency of glycolipids synthesis was increased from 7.3 to 13.4 g/L and from 6.6 to 10.5 g/L in the medium supplemented with soapstock and post-refinery fatty acids, respectively (Bednarski et al., 2004).

Soapstock from soybean, cottonseed, babassu, palm, and corn oil refinery were tested to rhamnolipid production and the soybean soapstock waste was the best substrate, generating 11.7 g/L of rhamnolipids and a production yield of 75% (Nitschke et al., 2005b).

A fermentation medium composed of acidic wastewater (containing 0.2% of fatty acids) and soapstock (3% v/v) from a sunflower oil process was utilized for rhamnolipid production by *P. aeruginosa* LBI achieving the best final concentrations of 7.3 g/L (Benincasa and Accorsini, 2008).

Frying oil is produced in large quantities for use both in the food industry and at the domestic scale and their uses have great potential for microbial growth and transformation. Haba et al. (2000) studied a screening process for the selection of microorganism strains with capacity to grow on frying oils (sunflower and olive oils) and accumulate surface-active compounds in the culture media. Nine *Pseudomonas* spp. strains were selected based on a decrease of surface tension of the medium to 34–36 mN/m. *P. aeruginosa* 47T2 showed a final production of rhamnolipid of 2.7 g/L as rhamnose and a production yield of 0.34 g/g. A yield of 34 g/L of sophorolipids was obtained when *C. bombicola* was grown on restaurant oil waste (Shah et al., 2007). *Rhodococcus erythropolis* was tested to produce glycolipids using sunflower frying oil as a cheap renewable substrate and the final BS showed high surface activity and emulsification capability with potential application to clean-up of hydrocarbons contaminated sites (Sadouk et al., 2008). Rhamnolipid production by *P. aeruginosa* PACL using different waste frying soybean oils was investigated. A rhamnose concentration of 3.3 g/L, an emulsification index of 100%, and a surface tension of 26.0 mN/m were obtained. When comparing to fresh soybean, the waste oils resulted in 75–90% of rhamnolipid production (de Lima et al., 2009). These studies demonstrated the feasibility to reusing waste frying oil for both sophorolipids and rhamnolipids production on industrial scale.

14.2.4.4   Starch Products Industry

Biological wastes rich in starchy materials are suitable for BS production. The processing of carbohydrate-rich agriculture products (cassava, corn, wheat, and potato) generates large amount of wastes. Fox and Bala (2000) demonstrated that potato-processing effluent was a suitable alternative carbon source to generate surfactant from *B. subtilis*. Das and Mukherjee (2007) evaluated the BS production using powdered potato peels as the substrate and the lipopeptides BS showed good surface activity and yield.

Nitschke and Pastore (2003, 2004) described the surfactin production by *B. subtilis* from "manipueira" a liquid residue generated by the pressing of cassava roots during the production of cassava flour. The major nutrients present on cassava waste are sugars, nitrogen, and mineral salts, which are quite attractive as a substrate for biotechnological processes once nutritional supplementation is not necessary. After 48 h of process, 3.0 g/L of surfactin was obtained from cassava wastewater and the recovery product showed a surface tension of 26.6 mN/m and a CMC of 33 mg/L (Nitschke and Pastore, 2006). *P. aeruginosa* also was related to production of BS from cassava wastewater, and this waste showed potential as an alternative substrate for rhamnolipid production (Costa et al., 2009). Table 14.2 summarizes some examples of the main classes of food wastes that have been proposed as alternative low-cost substrates for BS production.

Wastes from food processing are cheap renewable substrates that showed excellent potential for biosurfatant production; however, their application also is limited because of the difficulty to achieve optimal composition for growth and production of BSs, due to largely different metabolic requirements of the different microorganisms. Additionally, agro-industrial wastes are subject to highly variable components, which account for major variability in different batches of these potential feedstocks; however, this problem can be overcome by strict controlling and standardization of waste media composition combined with process control (Henkel et al., 2012).

TABLE 14.2
Examples of Food Wastes Utilized as Substrates for Biosurfactant Production

Food Waste Substrate	Biosurfactant Type	Reference
Edible Oil Industry		
Olive oil mill effluent	Rhamnolipid	Mercade et al. (2003)
Post-refinery fatty acids and soapstock	Glycolipids	Bednarsky et al. (2004)
Soapstock from soybean, corn, palm, cottonseed, babassu	Rhamnolipid	Nitschke et al. (2005b)
Soapstock from sunflower oil and acidic wastewater	Rhamnolipid	Benincasa and Accorsini (2008)
Starch-Based Industry		
Potato-processing effluent	Surfactin	Fox and Bala (2000)
Powdered potato peels	Lipopeptides	Das and Mukherjee (2007)
Cassava wastewater	Surfactin, rhamnolipid	Nitschke and Pastore (2006), Costa et al. (2009)
Dairy Industry		
Deproteinized whey concentrates	Sophorolipids	Daniel et al. (1999)
Cheese whey	Not determined	Rodrigues et al. (2006b)
Whey	Glycolipid, lipopeptide	Yilmaz et al. (2009)
Sugar and Fruit-Processing Industry		
Molasses	Lipopeptide	Makkar and Cameotra (1997)
Molasses	Surfactin	Abdel-Mawgoud et al. (2008)
Sugar beet molasses	Rhamnolipids	Onbasli and Aslim (2009)
Cashew apple juice	Surfactin, rhamnolipid	Rocha et al. (2007, 2009)

14.3 FUTURE TRENDS

Consumer demands for foods with fewer synthetic additives and mild preservation techniques without compromising safety led the food manufacturers and researchers to search for alternative agents. Many efforts are being made to study BSs applications as antiadhesion, antibiofilms, and antimicrobial agents, and the results suggest that in the near future these products could be used and produced at large scale, as alternative to synthetic products considering their status of bio-based or green chemicals. BS production can also be explored as prospective high-value products obtained from food waste substrates. The main challenges to be overwhelmed to improve BS utilization by food industry comprise the study on their toxicity including *in vivo* tests; moreover, research has to be done applying BS in food matrix in order to evaluate their interaction with other food components and their sensorial impact.

REFERENCES

Abalos, A., Pinazo, A., Infante, M. R., Casals, M., Garcia, F., and Manresa, A. 2001. Physicochemical and antimicrobial properties of new rhamnolipids produced by *Pseudomonas aeruginosa* AT10 from soybean oil refinery wastes. *Langmuir*, 17: 1367–1371.

Abdel-Mawgoud, A., Aboulwafa, M., and Hassouna, N. 2008. Optimization of surfactin production by *Bacillus subtilis* isolate BS5. *Applied Biochemistry and Biotechnology*, 150: 305–325.

Acosta, E. 2009. Bioavailability of nanoparticles in nutrient and nutraceutical delivery. *Current Opinion in Colloid & Interface Science*, 14: 3–15.

Ahimou, F., Jacques, P., and Deleu, M. 2000. Surfactin and iturin A effects on *Bacillus subtilis* surface hydrophobicity. *Enzyme and Microbial Technology*, 27: 749–754.

Araujo, L. V., Abreu, F., Lins, U., Santa Anna, L. M. M., Nitschke, M., and Freire, D. M. G. 2011. Rhamnolipid and surfactin inhibit *Listeria monocytogenes* adhesion. *Food Research International*, 44: 481–488.

Araujo, L. V., Pires, R. C., Siqueira, R. S. De Freire, D. M. G., and Nitschke, M. 2007. Potential use of pre-conditioning plastic surfaces with biosurfactants to inhibit *Listeria monocytogenes* 19112 biofilms. In: *II International Conference on Environmental, Industrial and Applied Microbiology (BioMicroWorld 2007)*, Sevilla, Spain.

Bagge-Ravn, D., Yin, N., Hjelm, M., Christiansen, J. N., Johansen, C., and Gram, L. 2003. The microbial ecology of processing equipment in different fish industries: Analysis of the microflora during processing and following cleaning and disinfection. *International Journal of Food Microbiology*, 87: 239–250.

Bednarski, W., Adamczak, M., Tomasik, J., and Plaszczyk, M. 2004. Application of oil refinery waste in the biosynthesis of glycolipids by yeast. *Bioresource Technology*, 95: 15–18.

Benincasa, M. and Accorsini, F. R. 2008. *Pseudomonas aeruginosa* LBI production as an integrated process using the wastes from sunflower-oil refining as a substrate. *Bioresource Technology*, 99: 3843–3849.

Boles, B. R., Thoendel, M., and Singh, P. K. 2005. Rhamnolipids mediate detachment of *Pseudomonas aeruginosa* from biofilms. *Molecular Microbiology*, 57: 1210–1223.

Borucki, M. K., Peppin, J. D., White, D., Loge, F., and Call, D. R. 2003. Variation in biofilm formation among strains of *Listeria monocytogenes*. *Applied and Environmental Microbiology*, 69: 7336–7342.

Bouffioux, O., Berquand, A., Eeman, M. et al. 2007. Molecular organization of surfactin-phospholipid monolayers: Effect of phospholipid chain length and polar head. *Biochimica et Biophysica Acta*, 1768: 1758–1768.

Busscher, H. J., Weeerkamp, A. H., Van Der Mei, H. C., Van Pelt, A. W. J., De Jong, H. P., and Arends, J. 1984. Measurements of the surface free energy of bacterial cell surfaces and its relevance for adhesion. *Applied and Environmental Microbiology*, 48: 980–983.

Carpentier, B. and Cerf, O. 1993. Biofilms and their consequences, with particular reference to hygiene and food industry. *Journal of Applied Bacteriology*, 75: 499–511.

Chae, M. S. and Schraft, H. 2000. Comparative evaluation of adhesion and biofilm formation of different *Listeria monocytogenes* strains. *International Journal of Food Microbiology*, 62: 103–111.

Chae, M. S., Schraft, H., Hansen, L. T., and Mackereth, R. 2006. Effects of physicochemical surface characteristics of *Listeria monocytogenes* strains on attachment to glass. *Food Microbiology*, 23: 250–259.

Cirigliano, M. C. and Carman, G. M. 1985. Purification and characterization of liposan, a bioemulsifier from *Candida lipolytica*. *Applied and Environmental Microbiology*, 50: 846–850.

Costa, S. G. V. A. O., Lepine, F., Milot, S. Deziel, E., Nitschke, M., and Contiero, J. 2009. Cassava wastewater as substrate for the simultaneous production of rhamnolipids and polyhydroxyalkanoates by *Pseudomonas aeruginosa*. *Journal of Industrial Microbiology & Biotechnology*, 36: 1063–1072.

Costa, S. G. V. A. O., Nitschke, M., Haddad, R., Eberlin, M. N., and Contiero, J. 2006. Production of *Pseudomonas aeruginosa* LBI rhamnolipids following growth on Brazilian native oils. *Process Biochemistry*, 41: 483–488.

Dagbert, C., Meylheuc, T., and Bellon-Fontaine, M. N. 2006. Corrosion behavior of AISI 304 stainless steel in presence of a biosurfactant produced by *Pseudomonas fluorescens*. *Electrochimica Acta*, 51: 5221–5227.

Daniel, H. J., Otto, R. T., Binder, M., Reuss, M., and Syldatk, C. 1999. Production of sophorolipids of whey: Development of a two-stage process with *Cryptococcus curvatus* ATCC 20509 and *Candida bombicola* ATCC 22214 using deproteinized whey concentrates as substrates. *Applied Microbiology and Biotechnology*, 51: 40–45.

Das, K. and Mukherjee, A. K. 2007. Comparison of lipopeptide biosurfactants production by *Bacillus subtilis* strains in submerged and solid state fermentation systems using a cheap carbon source: Some industrial applications of biosurfactants. *Process Biochemistry*, 42: 1191–1199.

Das, P., Mukherjee, S., and Sen, R. 2008. Antimicrobial potential of a lipopeptide biosurfactant derived from a marine *Bacillus circulans*. *Journal of Applied Microbiology*, 104: 1675–1684.

Das, P., Mukherjee, S., and Sen, R. 2009. Antiadhesive action of a marine microbial surfactant. *Colloids and Surfaces B: Biointerfaces*, 71: 183–186.

Dastgheib, S. M. M., Amoozegar, M. A., Elahi, E., Asad, S., and Banat, I. M. 2008. Bioemulsifier production by a halothermophilic *Bacillus* strain with potential applications in microbially enhanced oil recovery. *Biotechnology Letters*, 30: 263–270.

de Lima, C., Ribeiro, E., Sérvulo, E., Resende, M., and Cardoso, V. 2009. Biosurfactant production by *Pseudomonas aeruginosa* grown in residual soybean oil. *Applied Biochemistry and Biotechnology*, 152: 156–168.

Djordjevic, D., Wiedmann, M., and Mclandsborough, L. A. 2002. Microtiter plate assay for assessment of *Listeria monocytogenes* biofilm formation. *Applied and Environmental Microbiology*, 68: 2950–2958.

Dumont, M. J. and Narine, S. S. 2007. Soapstock and deodorizer distillates from North American vegetable oils: Review on their characterization, extraction and utilization. *Food Research International*, 40: 957–974.

Dusane, D. H., Nancharaiah, V., Zinjarde, S. S., and Venugopalan, V. P. 2010. Rhamnolipid mediated disruption of marine *Bacillus pumilus* biofilms. *Colloids and Surfaces B: Biointerfaces*, 81: 242–248.

Fernandes, P. A. V., De Arruda, I. R., Dos Santos, A. F. A. B., Araújo, A. A., De Maior, A. N. S., and Ximenes, E. A. 2007. Antimicrobial activity of surfactants produced by *Bacillus subtilis* R14 against multidrug-resistant bacteria. *Brazilian Journal of Microbiology*, 38: 704–709.

Flasz, A., Rocha, C. A., Mosquera, B., and Sajo, C. 1998. A comparative study of the toxicity of a synthetic surfactant and one produced by *Pseudomonas aeruginosa* ATCC 55925. *Medical Science Research*, 26: 181–185.

Fox, S. L. and Bala, G. A. 2000. Production of surfactant from *Bacillus subtilis* ATCC 21332 using potato substrates. *Bioresource Technology*, 75: 235–240.

Fracchia, L., Cavallo, M., Martinotti, M. G., and Banat, I. M. 2012. Biosurfactants and bioemulsifiers biomedical and related applications—Present status and future potentials. In *Biomedical Science, Engineering and Technology*, ed. D. N. Ghista, 325–370. Croatia: InTech.

Freire, D. M. G., Araújo, L. V., Kronemberger, F. A., and Nitschke, M. 2010. Biosurfactants as emerging additives in food processing. In *Food Engineering: New Techniques and Products*, ed. C. P. Ribeiro and M. L. Passos, 685–705. Boca Raton, FL: CRC Press.

Gandhi, N. R. and Skebba, V. L. P. 2007. Rhamnolipid compositions and related methods of use. W. O. International Application Patent (PCT) 2007/095258 A3, filed Feb.12, 2007, and issued Aug. 23, 2007.

Gomes, M. Z. V. and Nitschke, M. 2012. Evaluation of rhamnolipid and surfactin to reduce the adhesion and remove biofilms of individual and mixed cultures of food pathogenic bacteria. *Food Control*, 25: 441–447.

Haba, E., Espuny, M. J., Busquets, M., and Manresa, A. 2000. Screening and production of rhamnolipids by *Pseudomonas aeruginosa* 47T2 NCBI 40044 from waste frying oils. *Journal of Applied Microbiology*, 88: 379–387.

Hasenhuettl, G. L. 2008. Synthesis and commercial preparation of food emulsifiers. In *Food Emulsifiers and their Applications*, ed. G. L. Hasenhuettl and R. W. Hartel, 1–9. New York: Springer.

Henkel, M., Muller, M. M., Kugler, J. H. et al. 2012. Rhamnolipids as biosurfactants from renewable resources: Concepts for next-generation rhamnolipid production. *Process Biochemistry*, 47: 1207–1219.

Hood, S. K. and Zottola, E. A. 1995. Biofilms in food processing. *Food Control*, 6: 9–18.

Huang, X., Suo, J., and Cui, Y. 2011. Optimization of antimicrobial activity of surfactin and polylysine against *Salmonella enteritidis* in milk evaluated by a response surface methodology. *Foodborne Pathogens and Disease*, 8: 439–443.

Hwang, Y. H., Kim, M. S., Song, I. B. et al. 2009. Subacute (28 day) toxicity of surfactin C, a lipopeptide produced by *Bacillus subtilis*, in rats. *Journal of Health Science*, 55: 351–355.

Irie, Y., O`Toole, G. A., and Yuk, M. H. 2005. *Pseudomonas aeruginosa* rhamnolipids disperse *Bordetella bronchiseptica* biofilms. *FEMS Microbiology Letters*, 250: 237–243.

Jay, J. M. *Microbiologia de Alimentos*. Artmed: Porto Alegre, 2005.

Jenkinson, H. F. and Lappin-Scott, H. M. 2001. Biofilms adhere to stay. *Trends in Microbiology*, 9: 9–10.

Kalmokoff, M. L., Austin, J. W., Wan, X. D., Sanders, G., Banerjee, S., and Farber, J. M. 2001. Adsorption, attachment and biofilm formation among isolates of *Listeria monocytogenes* using model conditions. *Journal of Applied Microbiology*, 91: 725–734.

Kim, H., Ryu, J. H., and Beuchat, L. R. 2006. Attachment of and biofilm formation by *Enterobacter sakazakii* on stainless steel and enteral feeding tubes. *Applied and Environmental Microbiology*, 72: 5846–5856.

Kiran, G. S., Sabarathnam, B., and Selvin, J. 2010. Biofilm disruption potential of a glycolipid biosurfactant from marine *Brevibacterium casei*. *FEMS Immunology & Medical Microbiology*, 59: 432–438.

Kitamoto, D., Morita, T., Fukuoka, T., Konishi, M., and Imura, T. 2009. Self-assembling properties of glycolipid biosurfactants and their potential applications. *Current Opinion in Colloid & Interface Science*, 14: 315–328.

Kogan, A. and Garti, N. 2006. Microemulsions as transdermal drug delivery vehicles. *Advances in Colloid and Interface Science*, 123–126: 369–385.

Kolter, R. and Greenberg, E. P. 2006. The superficial life of microbes. *Nature*, 441: 300–302.

Kosaric, N. 2001. Biosurfactants and their application for soil bioremediation. *Food Technology and Biotechnology*, 39: 295–304.

Kralova, I. and Sjoblom, J. 2009. Surfactants used in food industry: A review. *Journal of Dispersion Science and Technology*, 30: 1363–1383.

Kuiper, I., Lagendijk, E. L., Pickford, R., Derrick, J. P., Lamers, G. E. M., Thomas-Oates, J. E., Lugtenberg, B. J. J., and Bloemberg, G. V. 2004. Characterization of two *Pseudomonas putida* lipopeptide biosurfactants, putisolvin I and II, which inhibit biofilm formation and break down existing biofilms. *Molecular Microbiology*, 51: 97–113.

Kumar, C. G., Mamidyala, S. K., Das, B., Sridhar, B., Devi, G. S., and Karuna, M. S. 2010. Synthesis of biosurfactant-based silver nanoparticles with purified rhamnolipids isolated from *Pseudomonas aeruginosa* BS-161R. *Journal of Microbiology and Biotechnology*, 20: 1061–1068.

Lindsay, D. and Holy, A. 1997. Evaluation of dislodging methods for laboratory-grown bacterial biofilms. *Food Microbiology*, 14: 383–390.

Lukondeh, T., Ashbolh, N. J., and Rogers, P. L. 2003. Evaluation of *Kluyveromyces marxianus* FII 510700 grown on a lactose-based medium as a source of natural bioemulsifier. *Journal of Industrial Microbiology & Biotechnology*, 30: 715–720.

Machado, S. M. de O. 2005. Avaliação do efeito antimicrobiano do surfatante cloreto de benzalcônio no controlo da formação de biofilmes indesejáveis. Master Dissertation—*Universidade do Minho*, Portugal, 2005.

Magalhães, L. and Nitschke, M. 2013. Antimicrobial activity of rhamnolipids against *Listeria monocytogenes* and their synergistic interaction with nisin. *Food Control*, 29: 138–142.

Makkar, R. S. and Cameotra, S. S. 1997. Utilization of molasses for biosurfactant production by two *Bacillus* strains at thermophilic conditions. *Journal of the American Oil Chemists' Society*, 74: 887–889.

Makkar, R. S., Cameotra, S. S., and Banat, I. M. 2011. Advances in utilization of renewable substrates for biosurfactant production. *AMB Express*, 1: 5.

Manoharan, M. J., Bradeeba, K., Parthasarathi, R. et al. 2012. Development of surfactin based nanoemulsion formulation from selected cooking oils: Evaluation for antimicrobial activity against selected food associated microorganisms. *Journal of the Taiwan Institute of Chemical Engineers*, 43: 172–180.

Maukonen, J., Mättö, J., Wirtanen, G., Raaska, T., Mattila-Sandholm, T., and Saarela, M. 2003. Methodologies for the characterization of microbes in industrial environments: A review. *Journal of Industrial Microbiology Biotechnology*, 30: 327–356.

McClements, D. J., Decker, E. A., Park, Y., and Weiss, J. 2009. Structural design principles for delivery of bioactive components in nutraceuticals and functional foods. *Critical Reviews in Food Science and Nutrition*, 49: 577–606.

McLandsborough, L., Rodriguez, A., Pe´Rez-Conesa, D., and Weiss, J. 2006. Biofilms: At the interface between biophysics and microbiology. *Food Biophysics*, 1: 94–114.

Mercade, M. E., Manresa, M. A., Robert, M., Espuny, M. J., de Andres, C., and Guinea, J. 1993. Olive oil mill effluent (OOME). New substrate for biosurfactant production. *Bioresource Technology*, 43: 1–6.

Meylheuc, T., Methivier, C., Renault, M., Herry, J. M., Pradier, C. M., and Bellon-Fontaine, M. N. 2006a. Adsorption on stainless steel surfaces of biosurfactants produced by gram-negative and gram-positive bacteria: Consequence on the bioadhesive behavior of *Listeria monocytogenes*. *Colloids and Surfaces B: Biointerfaces*, 52: 128–137.

Meylheuc, T., Renault, M., and Bellon-Fontaine, M. N. 2006b. Adsorption of a biosurfactant on surfaces to enhance the disinfection of surfaces contaminated with *Listeria monocytogenes*. *International Journal of Food Microbiology*, 109: 71–78.

Meylheuc, T., Van Oss, C. J., and Bellon-Fontaine, M. N. 2001. Adsorption of biosurfactant on solid surfaces and consequences regarding the bioadhesion of *Listeria monocytogenes* LO28. *Journal of Applied Microbiology*, 91: 822–832.

Mirelles II, J. R., Toguchi, A., and Harshey, R. M. 2001. *Salmonella enterica* serovar typhimurium swarming mutants with altered biofilm-forming abilities: Surfactin inhibits biofilm formation. *Journal of bacteriology*, 183: 5848–5854.

Mohan, P. K., Nakhla, G., and Yanful, E. K. 2006. Biokinetics of biodegradability of surfactants under aerobic, anoxic and anaerobic conditions. *Water Research*, 40: 533–540.

Nguyen, T. T. L., Edelen, A., Neighbors, B., and Sabatini, D. A. 2010. Biocompatible lecithin-based microemulsions with rhamnolipid and sophorolipid biosurfactants: formulation and potential applications. *Journal of Colloid and Interface Science*, 348: 498–504.

Nitschke, M., Araujo, L. V., Costa, S. G. V. A. O. et al. 2009. Surfactin reduces the adhesion of food-borne pathogenic bacteria to solid surfaces. *Letters in Applied Microbiology*, 49: 241–247.

Nitschke, M. and Costa, S. G. V. A. O. 2007. Biosurfactants in food industry. *Trends in Food Science & Technology*, 18: 252–259.

Nitschke, M., Costa, S. G. V. A. O., and Contiero, J. 2005a. Rhamnolipid surfactants: An update on the general aspects of these remarkable biomolecules. *Biotechnology Progress*, 21: 1593–1600.

Nitschke, M., Costa, S. G. V. A. O., and Contiero, J. 2011. Rhamnolipids and PHAs: Recent reports on Pseudomonas-derived molecules of increasing industrial interest. *Process Biochemistry*, 46: 621–630.

Nitschke, M., Costa, S. G. V. A. O., Haddad, R., Gonçalves, L. A. G., Eberlin, M. N., and Contiero, J. 2005b. Oil wastes as unconventional substrates for rhamnolipid biosurfactant production by *Pseudomonas aeruginosa* LBI. *Biotechnology Progress*, 21: 1562–1566.

Nitschke, M. and Pastore, G. M. 2003. Cassava flour wastewater as a substrate for biosurfactant production. *Applied Biochemistry and Biotechnology*, 105: 295–301.

Nitschke, M. and Pastore, G. M. 2004. Biosurfactant production by *Bacillus subtilis* using cassava processing effluent. *Applied Biochemistry and Biotechnology*, 112: 163–172.

Nitschke, M. and Pastore, G. M. 2006. Production and properties of a surfactant obtained from *Bacillus subtilis* grown on cassava wastewater. *Bioresource Technology*, 97: 336–341.

Onbasli, D. and Aslim, B. 2009. Determination of rhamnolipid biosurfactant production in molasses by some *Pseudomonas* spp. *New Biotechnology*, 25: 255.

Parizzi, S. Q. F., Andrade, N. J. De Silva, C. A. De S., Soares, N. F. F., and Silva, E. A. M. da 2004. Bacterial adherence to different inert surfaces evaluated by epifluorescence microscopy and plate count method. *Brazilian Archives of Biology and Technology*, 47: 77–83.

Patel, R. M. and Desai, A. J. 1997. Biosurfactant production by *Pseudomonas aeruginosa* GS3 from molasses. *Letters in Applied Microbiology*, 25: 91–94.

Radhakrishnan, N., Kavitha, V., Madhavacharyulu, E., Gnanamani, A., and Mandal, A. B. 2011. Isolation, production and characterization of bioemulsifiers of marine bacteria of coastal Tamil Nadu. *Indian Journal of Geo-Marine Sciences*, 40: 76–82.

Ré, M. I., Santana, M. H. A., and d´Avila, M. A. 2010. Encapsulation technologies for modifying food performance. In *Food Engineering: New Techniques and Products*, ed. C. P. Ribeiro and M. L. Passos, 223–275. Boca Raton: CRC Press.

Rivardo, F., Martinotti, M. G., Turner, R. J., and Ceri, H. 2010. The activity of silver against *Escherichia coli* biofilm is increased by a lipopeptide biosurfactant. *Canadian Journal of Microbiology*, 56: 272–278.

Rivardo, F., Turner, R. J., Allegrone, G., Ceri, H., and Martinotti, M. G. 2009. Anti-adhesion activity of two biosurfactants produced by *Bacillus* spp. prevents biofilm formation of human bacterial pathogens. *Applied Microbiology and Biotechnology*, 83: 541–553.

Rocha, M. V. P., Barreto, R. G., Melo, V., and Gonçalves, L. R. B. 2009. Evaluation of cashew apple juice for surfactin production by *Bacillus subtilis* LAMI008. *Applied Biochemistry and Biotechnology*, 155: 63–75.

Rocha, M. V. P., Oliveira, A. H. S., Souza, M. C. M., and Gonçalves, L. R. B. 2006. Natural cashew apple juice as fermentation medium for biosurfactant production by *Acinetobacter calcoaceticus*. *World Journal of Microbiology & Biotechnology*, 22: 1295–1299.

Rocha, M. V. P., Souza, M. C. M., Benedicto, S. C. L. et al. 2007. Production of biosurfactant by *Pseudomonas aeruginosa* grown on cashew apple juice. *Applied Biochemistry and Biotechnology*, 136/140: 185–194.

Rodrigues, L. R., Banat, I. M., Van Der Mei, H. C., Teixeira, J. A., and Oliveira, R. 2006a. Interference in adhesion of bacteria and yeasts isolated from explanted voice prostheses to silicone rubber by rhamnolipid biosurfactants. *Journal of Applied Microbiology*, 100: 470–480.

Rodrigues, L., Moldes, A., Teixeira, J., and Oliveira, R. 2006b. Kinetic study of fermentative biosurfactant production by *Lactobacillus* strains. *Biochemical Engineering Journal*, 28: 109–116.

Ron, E. Z. and Rosenberg, E. 2001. Natural roles of biosurfactants. *Environmental Microbiology*, 3: 229–236.

Rosenberg, E. and Ron, E. Z. 1998. Bioemulsifiers. US Patent 5,840,547, filed Sep. 30, 1996, and issued Nov. 24, 1998.

Sadouk, Z., Hacene, H., and Tazerouti, A. 2008. Biosurfactants production from low cost substrate and degradation of diesel oil by a *Rhodococcus* strain. *Oil & Gas Science and Technology*, 63: 747–753.

Sarubbo, L. A., Luna, J. M., and Campos-Takaki, G. M. 2006. Production and stability studies of the bioemulsifier obtained from a new strain of *Candida glabrata* UCP 1002. *Eletronic Journal of Biotechnology*, 9: 400–406.

Shah, V., Jurjevic, M., and Badia, D. 2007. Utilization of restaurant waste oil as a precursor for sophorolipid production. *Biotechnology Progress*, 23: 512–515.

Shakerifard, P., Gancel, F., Jacques, P., and Faille, C. 2009. Effect of different *Bacillus subtilis* lipopeptides on surface hydrophobicity and adhesion of *Bacillus cereus* 98/4 spores to stainless steel and teflon. *Biofouling*, 25: 533–554.

Shank, E. A. and Kolter, R. 2011. Extracellular signaling and multicellularity in *Bacillus subtilis*. *Current Opinion in Microbiology*, 14: 741–747.

Shepherd, R., Rockey, J., Sutherland, I. W., and Roller, S. 1995. Novel bioemulsifiers from microorganisms for use in foods. *Journal of Biotechnology*, 40: 207–217.

Silva, H. D., Cerqueira, M. A., and Vicente, A. A. 2012. Nanoemulsions for food applications: Development and characterization. *Food and Bioprocess Technology*, 5: 854–867.

Somers, E. B. and Wong, A. C. L. 2004. Efficacy of two cleaning and sanitizing combinations of *Listeria monocytogenes* biofilms formed at low temperature on a variety of materials in the presence of ready-to-eat meat residue. *Journal of Food Protection*, 67: 2218–2229.

Sommer, P., Martin-Rouas, C., and Mettler, E. 1999. Influence of the adherent population level on biofilm population, structure and resistance to chlorination. *Journal of Food Microbiology*, 16: 503–515.

Stopforth, J. D., Samelis, J., Sofos, J. N., Kendall, P. A., and Smith, G. C. 2002. Biofilm formation by acid-adapted and nonadapted *Listeria monocytogenes* in fresh beef decontamination washings and its subsequent inactivation with sanitizers. *Journal of Food Protection*, 65: 1717–1727.

Suresh Chander, C. R., Lohitnath, T., Mukesh Kumar, D. J., and Kalaichelvan, P. T. 2012. Production and characterization of biosurfactant from *Bacillus subtilis* MTCC441 and its evaluation to use as bioemulsifier for food bio-preservative. *Advances in Applied Science Research*, 3: 1827–1831.

Tahmourespour, A., Salehib, R., Kermanshahic, R. K., and Eslamid, G. 2011. The anti-biofouling effect of *Lactobacillus fermentum*-derived biosurfactant against *Streptococcus mutans*. *Biofouling*, 27: 385–392.

Takhistov, P. and George, B. 2004. Linearized kinetic model of *Listeria monocytogenes* biofilm growth. *Bioprocess and Biosystems Engineering*, 26: 259–270.

Torabizadeh, H., Shojaosadatib, S. A., and Tehrani, H. A. 1996. Preparation and characterization of bioemulsifier from *Saccharomyces cerevisiae* and its application in food products. *LWT—Food Science and Technology*, 29: 734–737.

Tortora, G. J., Funke, B. R., and Case, C. L. *Microbiologia*. Artmed: Porto Alegre, 2005.

Van Haesendonck, I. P. H. and Vanzeveren, E. C. A. 2004. Rhamnolipids in bakery products. W. O. International Application Patent (PCT) 2004/040984, filed Nov. 4, 2003, and issued May, 21, 2004.

Worakitkanchanakul, W., Imura, T., Morita, T. et al. 2008. Formation of W/O microemulsion based on natural glycolipid biosurfactant, mannosylerythritol Lipid-A. *Journal of Oleo Science*, 57: 55–59.

Yarwood, J. M., Bartels, D. J., Volper, E. M., and Greenberg, E. P. 2004. *Quorum sensing* in *Staphylococcus aureus* biofilms. *Journal of Bacteriology*, 186: 1838–1850.

Yilmaz, F., Ergene, A., Yalçin, E., and Tan, S. 2009. Production and characterization of biosurfactants produced by microorganisms isolated from milk factory wastewaters. *Environmental Technology*, 13: 1397–1404.

Zeraik, A. N. and Nitschke, M. 2010. Biosurfactants as agents to reduce adhesion of pathogenic bacteria to polystyrene surfaces: Effect of temperature and hydrophobicity. *Current Microbiology*, 61: 554–559.

15 Biosurfactant Applications in Agriculture

Rengathavasi Thavasi, Roger Marchant, and Ibrahim M. Banat

CONTENTS

15.1 Introduction 313
15.2 Biosurfactants in Agriculture 314
    15.2.1 Biosurfactants in Plant Disease Control 315
    15.2.2 Antimicrobial Mechanism of Biosurfactants 317
    15.2.3 Biosurfactant-Based Biopesticides in the Market... 317
    15.2.4 Biosurfactants in Plant Disease Resistance/Immunity Development.................. 318
    15.2.5 Biosurfactants in Pest/Insect Control 318
    15.2.6 Biosurfactants in Agricultural Soil Remediation 319
    15.2.7 Role of Biosurfactants in Plant–Microbe Interactions 320
    15.2.8 Biosurfactants in Agrochemical Formulation 320
15.3 Concluding Remarks 321
References... 321

15.1 INTRODUCTION

Biosurfactants are amphiphilic surface-active compounds produced by bacteria, fungi, actinomycetes, and plants. They belong to various classes of compounds including glycolipids, glycolipoproteins, glycopeptides, lipopeptides, lipoproteins, fatty acids, phospholipids, neutral lipids, lipopolysaccharides (Banat et al., 2010), and glycoglycerolipids (Wicke et al., 2000). The properties and applications of biosurfactants include detergency, emulsification, foaming, dispersion, wetting, penetrating, thickening, microbial growth enhancement, antimicrobial agents, metal sequestering, and enhanced oil recovery. These important properties may allow biosurfactants to replace some of the most versatile chemical surfactants that are currently in use. In addition, biosurfactants offer several advantages over chemically synthesized surfactants, such as *in situ* production from renewable substrates, lower toxicity, biodegradability, and ecological compatibility (Marchant and Banat, 2012a,b). Several proposed physiological roles of biosurfactants have been put forward including the following: (1) increasing the surface area and bioavailability of hydrophobic water insoluble substrates (e.g., oil-degrading microbes) (Ron and Rosenberg, 1999; Kosaric, 2001), (2) bacterial pathogenesis and quorum sensing and biofilm formation and maintenance (e.g., *Pseudomonas aeruginosa*) (Davey et al., 2003), (3) antimicrobial activity for self-defense (e.g., antimicrobial activity of rhamnolipids) (Stanghellini and Miller, 1997), and (4) cell proliferation in the producing bacteria (e.g., viscosinamide production by *Pseudomonas fluorescens*) (Nielsen et al., 1999). Several microbial communities and members of genera such as *Acinetobacter, Arthrobacter, Alcanivorax, Pseudomonas, Halomonas, Bacillus, Rhodococcus, Enterobacter, Azotobacter, Corynebacteium, Lactobacillus,* and yeasts have been reported to produce biosurfactants (Schulz et al., 1991; Passeri et al., 1992; Banat, 1993; Abraham et al., 1998; Rosenberg and Ron, 1998; Thavasi et al., 2007, 2009, 2011a,b; Das et al., 2008a,b; Perfumo et al., 2010).

Agriculture relies heavily on chemical pesticides for eradication of plant pathogens, chemical fertilizers to improve soil fertility, and chemical surfactants in agrochemical formulations. In spite of the valuable role these synthetic chemical-based products have for many agricultural applications, their use poses significant problems. The main difficulties for chemical pesticide use are effects on nontarget organisms including humans, domestic animals, beneficial insects, and wildlife. Their residues are often long lasting as most can remain on crops and accumulate in soil, water, and air for extended periods and through entry to the food trophic pyramid. For these reasons, development of safe naturally derived alternatives to chemical pesticides is an important goal for scientific investigation and commercial development. The global agrochemical market is expected to grow from $134 billion in 2010 to $223 billion in 2015, with a high compound annual growth rate (CAGR) of 10.6% from 2010 to 2015 (http://www.marketsandmarkets.com/Market-Reports/global-agro-chemicals-market-report-132.html).

In the agrochemical market, the pesticide market was valued at approximately $40 billion in 2008. This figure increased to nearly $43 billion in 2009 and is expected to grow at a CAGR of 3.6% to reach $51 billion in 2014. Biopesticides represent a strong growth area in the global pesticide market. This market is expected to grow at a 15.6% CAGR from $1.6 billion in 2009 to $3.3 billion in 2014 (http://www.bccresearch.com/market-research/chemicals/biopesticides-market-chm029c.html). The beneficial properties of biopesticides have created a keen interest among scientists and industries to develop safe, reliable, and easy-to-produce biopesticides. Among various antimicrobial compounds and biopesticides, biosurfactants are known for their antimicrobial activity against human and plant pathogens. There are few biosurfactant-based commercial biopesticides available in the market although some investigations are currently in progress in academic and industrial institutions. This chapter discusses the literature on agricultural applications of microbial biosurfactants that include disease and pest control, plant immunity development, agrochemical formulation, soil remediation, and plant–microbe interactions.

15.2 BIOSURFACTANTS IN AGRICULTURE

Among the many interesting properties of biosurfactants, some are of particular interest and application in agriculture such as emulsification, foaming, dispersion, wetting, penetrating, thickening, microbial growth enhancement, antimicrobial, and metal sequestering properties (Figure 15.1). The following sections in this chapter will discuss the applications of biosurfactants in agriculture.

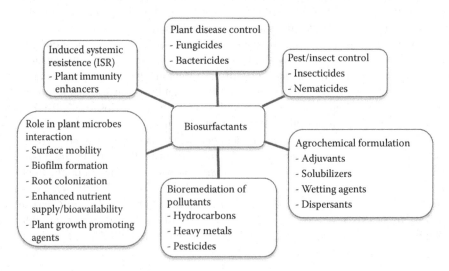

FIGURE 15.1 Areas of potential applications of biosurfactants in agriculture.

15.2.1 Biosurfactants in Plant Disease Control

The antimicrobial properties of some biosurfactants lead to interest in applications to combat plant diseases. Application of crude biosurfactant extracts and biosurfactant-producing cells of three bacterial species, *Pseudomonas fluorescences* (6519E01), *P. fluorescences biovar* I (6133D02), and *Serratia plymuthica* (6109D01), to control corn seed rots and seed blight was reported over two decades ago by Haefele et al. (1991), who concluded that corn seeds treated with culture broth for 30 min significantly reduced the disease impact on seeds. This approach was suggested to save loss of crops from seed diseases and prevent further spread of the disease to germinating plants. Use of biosurfactants for biological control of plant fungal diseases was later pioneered by Stanghellini and Miller (1997), who demonstrated that rhamnolipids (RL) can disrupt zoospore membranes and cause lysis of zoospores of many oomycete plant-pathogenic fungi. Zoospores are one of the main infecting agents produced by fungal pathogens, which can be devastating especially in hydroponic systems. In this investigation, purified mono- and dirhamnolipids were tested for zoosporicidal activity against the zoosporic plant pathogens: *Pythium aphanidermatum*, *Phytophthora capsici*, and *Plasmopara lactucae-radicis*. Treatments of zoospores of all three species with each biosurfactant at concentrations ranging from 5 to 30 µg/mL caused loss of spore motility and lysis of the entire zoospore population in less than 1 min. They also reported that application of biosurfactant-producing *Pseudomonas* isolates on leaf surfaces at 1×10^6 CFU/mL completely lysed the zoospores and inhibited their motility. It was also shown that the biosurfactant-producing bacteria survived on the leaf surface for 21 days and produced sufficient amounts of biosurfactant to maintain the inhibition, which indicated that zoospore-related plant diseases can be treated through biosurfactant application or in combination with biosurfactant-producing microbes.

Yoo et al. (2005) reported similar zoospore lysis and loss of mobility as disease-controlling activities by rhamnolipids and sophorolipids (SL). They reported 80% mycelial growth inhibition for *Phytophthora* sp. and *Pythium* sp. at 200 mg/L of RL or 500 mg/L of SL, while similar levels of zoospore motility inhibition in *Phytophthora* sp. occurred at 50 mg/L RL and 100 mg/L SL. The effective concentrations for zoospore lysis were generally two times higher than those for zoospore motility inhibition. Most effective zoospore lysis was observed with *Phytophthora capsici* with 80% lysis at 100 mg/L dirhamnolipid or lactonic SL indicating structure dependence for such activity. In plant pot tests the damping-off fungal pathogen occurrence decreased by exposure to 2000 mg/L RL and SL biosurfactants to 42% and 33%, respectively, of values obtained in the untreated control. These results showed the potential of RL and SL biosurfactants as effective antifungal agents against damping-off plant pathogens.

Two recent reports (Schofield et al., 2013; Gross and Thavasi, 2013) investigating the application SLs and SL derivatives on plant diseases added more evidence to the plant disease controlling properties of microbial biosurfactants. Investigators reported that SL derivatives and combinations of SL derivatives exhibited significant antifungal activity against 18 plant fungal pathogens (*Alternaria tomatophilia*, *A. solani*, *A. alternata*, *Aspergillus niger*, *Aureobasidium pullulans*, *Botrytis cinerea*, *Chaetomium globosum*, *Fusarium asiaticum*, *F. austroamericana*, *F. cerealis*, *F. graminearum*, *F. oxysporum*, *Penicillium chrysogenum*, *P. digitatum*, *P. funiculosum*, *Phytophthora infestans*, *P. capsici*, and *Ustilago maydis*) and 7 plant bacterial pathogens (*Acidovorax carotovorum*, *Erwinia amylovora*, *Pseudomonas cichorii*, *P. syringae*, *Pectobacterium carotovorum*, *Ralstonia solanacearum*, and *Xanthomonas campestris*). The minimum inhibitory concentrations ranged from 2.5 to 10 mg/mL for individual SL derivatives and 0.009–10 mg/mL for SL combinations. The same derivatives were also effective against grapevine downy mildew pathogen *Plasmopara viticola* zoospores, which showed loss of viability and lysis occurring at 50–500 µg/mL SL derivative concentration.

Another interesting area where biosurfactants have a role is the rhizosphere environment where some strains of biosurfactant-producing *Pseudomonas* and *Bacillus* isolated from the rhizosphere exhibited biocontrol of soft rot causing *Pectobacterium* and *Dickeya* spp. (Krzyzanowska et al., 2012).

Biocontrol activity of uncharacterized biosurfactants produced by two strains of *Pseudomonas* called CMR5c and CMR12a isolated from the rhizosphere of healthy cocoyam plants exhibited control of cocoyam root rot caused by *Pythium myriotylum* (Perneel et al., 2007). Antagonistic activity of cyclic lipopeptides (CLPs) produced by *Pseudomonas* spp., isolated from sugar beet fields was tested against the pathogenic fungi *Pythium ultimum* and *Rhizoctonia solani* and it was found that the CLPs exhibited significant antifungal activity (Nielsen et al., 2002). Details of biosurfactant-producing microbes isolated from the rhizosphere and their applications are listed in Table 15.1. In addition to their applications as biocontrol agent, biosurfactants can be applied as plant disease-preventing agents. De Jonghe et al. (2005) reported the prevention of brown root rot disease spread on Witloof chicory (*Cichorium intybus* var. *foliosum*) caused by *Phytophthora cryptogea*. A treatment with RLs (25% RLs formulated in oil) reduced the disease incidence significantly by controlling the zoospores, but the biosurfactant formulation was not effective on mycelial suspensions tested in an *in vitro* system, showing that the RL formulation is effective only for zoospore-related disease prevention.

Other lipopeptide-type biosurfactants have also been reported to have significant biological activities. Velho et al. (2011) reported that biosurfactant produced by a *Bacillus* strain caused growth inhibition of phytopathogenic fungi such as *Fusarium* spp., *Aspergillus* spp., and

TABLE 15.1
Biosurfactant-Producing Microbes Isolated from Rhizosphere and Agriculture Soil and Their Applications

Microbe	Source	Biosurfactant	Applications	Reference
Pseudomonas spp.	Pepper plant	RLs	Biocontrol of *P. capsici* and zoospores	Stanghellini and Miller (1997)
P. aeruginosa B5	Pepper plant	Di-RLs	Biocontrol of *P. capsici* and zoospores	Kim et al. (2000)
Pseudomonas sp.	Sugar beet	Viscosinamide	Biocontrol of *P. ultimum*	Thrane et al. (2000)
Pseudomonas fluorescens	Sugar beet rhizosphere	Cyclic lipopeptides (CLP)	Biocontrol of *Pythium ultimum* and *Rhizoctonia solani*	Nielsen et al. (2002)
P. aeruginosa PNA1	Tomato plant	RLs	Biocontrol of *P. nicotianae* and zoospores	De Jonghe et al. (2005)
Unidentified strain	Pepper plant	RLs	Biocontrol of *P. capsici* and zoospores	Nielsen et al. (2006)
P. fluorescens SS101	Tomato	Massetolide A	Biocontrol of *P. infestans* and ISR	Tran et al. (2007)
Pseudomonas strains CMR5c and CMR12a	Cocoyam rhizosphere	Uncharacterized biosurfactants	Biocontrol of *Pythium myriotylum*	Perneel et al. (2007)
P. aeruginosa	Bean plant	RLs + phenazines	Biocontrol of *Pythium splendens*	Perneel et al. (2008)
Unidentified strain	Grapevine	RLs	Biocontrol of *Botrytis cinerea*	Varnier et al. (2009)
Pseudomonas putida strain 267	Black pepper rhizosphere	CLPs	Contol *Phytophthora capsici* damping-off disease and zoospore lysis in cucumber	Kruijt et al. (2009)
Pseudomonas sp.	Agriculture soil	RLs	Biodegradation of Chlorpyrifos pesticide	Singh et al. (2009)
Pseudomonas fluorescens	Fique plants	RLs	Emulsification of kerosene	Sastoque-Cala et al. (2010)
Pseudomonas aeruginosa MA01	Spoiled apples	RLs	Emulsification of hydrocarbons	Abbasi et al. (2012)
Bacillus mojavensis	Maize	Cyclic lipopeptide	Biocontrol of fungal diseases	Snook et al. (2009)

Biopolaris sorokiniana. *Brevibacillus brevis* strain HOB1 produced surfactin isoform and this lipopeptide biosurfactant showed strong antibacterial and antifungal properties which can be used to control phytopathogens (Haddad, 2008). Romero et al. (2007) reported that lipopeptides produced by the two strains of *B. subtilis* were able to reduce cucurbit powdery mildew caused by *Podosphaera fusca* by inhibiting conidial germination. Leu7-surfactin, characterized as a cyclic heptapeptide linked to a β-hydroxy fatty acid, from the endophytic bacterium *Bacillus mojavensis* RRC 101 was reported as a biocontrol agent for *Fusarium verticillioides* by Snook et al. (2009).

15.2.2 ANTIMICROBIAL MECHANISM OF BIOSURFACTANTS

The antimicrobial activity mechanism of biosurfactants remains unclear, but there are few hypotheses, supported by some evidence, showing that loss of membrane integrity is the main process. The proposed hypotheses with evidence on antimicrobial activity of lipopeptide biosurfactants are as follows:

1. Antimicrobial effect of the biosurfactants is a result of the adhesion property of the biosurfactants to the cell surfaces causing deterioration of cell membrane integrity and eventual breakdown in the nutrition cycle (Hingley et al., 1986).
2. Insertion of fatty acid components of the biosurfactants into the cell membrane causing an increase in the size of the membrane and significant ultrastructural changes in the cells such as ability of the cell to interiorize plasma membrane (Gomaa, 2013).
3. Insertion of the shorter acyl tails of the biosurfactant into the cell membrane causing disruptions of the cytoskeletal elements and the plasma membrane, allowing the membrane to lift away from the cytoplasmic contents (Desai and Banat, 1997).
4. Disruption of cellular plasma membranes through the accumulation of intra-membranous particles in the cells increasing the electrical conductance of the membrane (Thimon et al., 1995).
5. Increasing membrane permeability through the interaction/disruption of the cell membrane phospholipids (Carrillo et al., 2003).

Despite all the above hypotheses, fully understanding the antimicrobial activity of biosurfactants remains a quest that needs further investigation.

15.2.3 BIOSURFACTANT-BASED BIOPESTICIDES IN THE MARKET

There are two biosurfactant-based biopesticides successfully launched in the biopesticides market. One is SERENADE from AgraQuest Inc, USA (recently acquired by Bayer Crop Science), a fungicide that contains *Bacillus subtilis* (strain QST 713) which produces over 30 different lipopeptides that work synergistically to destroy plant pathogens. The list of target plants includes vegetables, fruit, nut, and vine crops and the targeted pathogens are those causing diseases such as fire blight, botrytis, sour rot, rust, sclerotinia, powdery mildew, bacterial spot, and white mold. The second biosurfactant-based biopesticide in the market is Zonix, a rhamnolipid-based fungicide from Jeneil Biotech Inc, USA. Target protected plants include root, bulb, tuber, and cane crops, fruiting vegetables, leafy vegetables, fruit and nut trees, citrus fruits, tropical crops (e.g., banana), berry crops, grain, forage, fiber, and oil crops, vine crops, and herbs. The list of pathogens controlled by Zonix are any zoosporic plant-pathogenic microorganism including members of the following genera: *Achlya, Albugo, Aphanomyces, Basidiophora, Olpidium, Pachymetra, Peronophthora, Peronosclerospora, Physoderma, Phytophthora, Plasmodiophora, Plasmopara, Polymyxa, Pseudoperonospora, Pythium, Rhizophydium, Sclerophthora, Sclerospora, Spongospora, Synchytrium,* and *Trachysphaera*.

15.2.4 Biosurfactants in Plant Disease Resistance/Immunity Development

In addition to their antimicrobial properties, biosurfactants can act as inducers of plant disease resistance called induced systemic resistance (ISR) caused by pattern recognition receptors (PRRs) of molecular signatures that identify whole classes of microbes but are absent in hosts allowing non-self recognition (Boller and Felix, 2009; Boller and He, 2009). Upon recognition, these molecular signatures, conventionally named microbe-associated molecular patterns (MAMPs) (Mackey and McFall, 2006), trigger complex signaling pathways leading to transcriptional activation of defense-related genes and accumulation of antimicrobial metabolites in plant cells (Boller and Felix, 2009).

The ISR-based plant-mediated defense response against a broad spectrum of pathogens, caused by root colonization with nonpathogenic rhizobacteria, was reported by De Vleesschauwer and Hofte (2009). Varnier et al. (2009) reported that rhamnolipids Rha-C_{10}-C_{10} and Rha-Rha-C_{10}-C_{10} produced by *Pseudomonas aeruginosa* and Rha-Rha-C_{14}-C_{14} from *Burkholderia plantarii* trigger strong defense responses in grapevine including the early events of cell signaling like Ca^{2+} influx, reactive oxygen species (ROS) production, and mitogen-activated protein (MAP) kinases activation. Rhamnolipids also induce a large series of defense genes including pathogenesis-related protein genes and genes involved in oxylipins and phytoalexin biosynthesis pathways. It was also demonstrated that rhamnolipids potentiate defense responses induced by other elicitors (such as chitosan and culture filtrate of the fungus *Botrytis cinerea*). Another unique role of rhamnolipids consists of protecting grapevine against the necrotropic pathogen *B. cinerea*. However, the identification of the specific receptors in the plant plasma membrane for rhamnolipids and other biosurfactants remains unresolved.

Cyclic lipopeptide (CLP)-type biosurfactants produced by *Pseudomonas* and bacilli, made up of a cyclized oligopeptide lactone ring coupled to a fatty acid tail, also have ISR properties. Fengycins and surfactins produced by *Bacillus subtilis* S499 exhibited an ISR-mediated protective effect on tomato plants against *B. cinerea* (Ongena et al., 2007). However, the structure of the surfactant molecules is important for elicitation of ISR, since fengycins, but not surfactins, induced a defense response on potato tuber cells, while both surfactants induced defense-associated changes in tobacco cells (Ongena and Jacques, 2008). It seems that the ISR property of a biosurfactant can vary depending on plant and pathogen. The mechanism of ISR reported for surfactin was a disturbance of the plant plasma membrane, thereby activating a defense response in the plants (Jourdan et al., 2009). The ISR properties of biosurfactants could be used to trigger plant defense mechanisms against specific diseases, which could be a permanent solution for specific diseases and it may reduce the frequent application of pesticides/biopesticides to combat plant diseases.

15.2.5 Biosurfactants in Pest/Insect Control

The application of biosurfactants in plant pest/insect control is an emerging and developing field. There are very few reports available on insect control activity by biosurfactants. Two US patents were filed by AgSciTech (US 2005/0266036 A1 and US 2012/0058895 A1) (Awada et al., 2005, 2012) on insecticidal activity of rhamnolipid biosurfactants and biosurfactant-producing microbes. These patents suggest that two application protocols can be employed for insect control with biosurfactants:

1. Application of a biosurfactant-producing microbe together with purified/unpurified biosurfactant on plant pests
2. Application of purified/unpurified biosurfactant on plant pests

In the first method, it is assumed that the biosurfactant-producing microbes will produce biosurfactant on the surface where they are applied and the biosurfactant applied together with the microbe will help the microbe in the initial stage to act on the pests. *In situ* biosurfactant production on the surface was achieved by the addition of carbon substrates, such as oil or fatty acids, to target areas to support the growth and biosurfactant production of the organisms to achieve the desired pest control. In the second method, the pure/unpurified biosurfactant can be applied directly to, and will act on, the pests. Pests treated with the biosurfactants have included ants, aphids, thrips, whiteflies, scales, lice, cockroaches, termites, houseflies, mosquitoes, mites, spiders, ticks, nematodes, molluscs, amoebae, parasites, and algae. Experimental results indicated that significant pest-controlling activity was observed against all the pests tested and the effective concentrations of rhamnolipids ranged from 0.0075% to 5%.

Kim et al. (2011) reported the insecticidal activity of rhamnolipids isolated from *Pseudomonas* EP-3 against green peach aphid (*Myzus persicae*). Cell-free supernatant of EP-3 grown on glucose mineral salts medium showed >80% aphids mortality within 24 h of treatment. The dirhamnolipids used in this treatment showed a dose-dependent activity against aphids with 50% mortality at 40 µg/mL and 100% mortality at 100 µg/mL concentrations. Microscopy analyses of aphids treated with dirhamnolipid revealed the insecticidal mechanism as cuticle membrane damage. Another interesting finding on repellent activity was reported for nonmicrobial biosurfactant produced by beet armyworm *Spodoptera exigua* (Rostas and Blassmann, 2009). In this report, significant repellent activity was observed for the oral secretion (OS) of the *S. exigua* larvae against European fire ants. The claimed biosurfactant present in this oral secretion was not chemically identified and there was no information on its surface tension reducing activity. The only surfactant property information reported was contact angle measurement of these secretions on leaf surfaces and hydrophobic glass surface. The repellent property reported for OS shows that microbial biosurfactants could be explored for their insect-repellent properties.

15.2.6 BIOSURFACTANTS IN AGRICULTURAL SOIL REMEDIATION

Surface activity, detergency, emulsification, dispersion, and enhanced solubility properties of biosurfactants make them an ideal choice for remediation of agricultural soil polluted with hydrocarbons, heavy metals, and pesticides. In addition, application of biosurfactant-producing microbes together with biosurfactants could promote the degradation of the pollutants to a safer level, thus improving the soil quality and bringing the soil biology back to a pre-pollution condition. There are many reports on biosurfactant use in hydrocarbon degradation and bioremediation of soils (Rahman et al., 2002, 2003, 2006). Of particular interest in this area was the isolation of a nitrogen-fixing *Azotobacter chrococcum* strain with biosurfactant-producing ability and crude oil degrading potential (Thavasi et al., 2009). Strains like this degrade the hydrocarbons and also fix nitrogen, which can promote the fertility of soils.

The application of biosurfactants for the remediation of inorganic compounds such as heavy metals has also been described to remediate polluted soils. In this method, biosurfactants can be used as chelating and removal agents through mobilization of ions during washing steps facilitated by the chemical interactions between the amphiphiles and the metal ions. Biosurfactants enable better removal of heavy metals by affecting factors such as rate-limited mass transfer, sorption, and resistance to aqueous phase transport (Miller, 1995). Another interesting property of biosurfactants is their ability to encapsulate heavy metals or hydrocarbons by micelle formation, thus promoting the efficient removal of the pollutants. Further, encapsulation and mobilization can reduce the adverse effect of a pollutant on its surroundings (Chrzanowski et al., 2009).

Bioremediation through phytoremediation is another cost-effective and environment-friendly remediation method which involves the use of plants such as willows, poplar, and different types of grasses for remediation of metal-contaminated soils, and biosurfactants have been reported to enhance the on-site phytoremediation, which is otherwise a slow process (Abioye, 2011). Biosurfactants aid the removal of metals in soil in two ways: first, by forming a complex with free metal present in solution, reducing its phase activity, thus, enhancing desorption of metal. Second, the direct contact of biosurfactants and the sorbed metal solid solution interface under reduced interfacial tension enhances the plants' uptake ability of these metals (Das et al., 2009). Apart from mobilization, biosurfactants can also improve bacterial tolerance and resistance toward high concentrations of heavy metals by entrapping the heavy metals in micelles which allows the bacteria to continue their biodegradation process.

15.2.7 ROLE OF BIOSURFACTANTS IN PLANT–MICROBE INTERACTIONS

Biosurfactant production by plant-associated microbes is essential for producing microbes to create symbiotic/pathogenic relationship with plants such as motility, signaling, and biofilm formation (Sachdev and Cameotra, 2013). A detailed review on biosurfactants in plant–*Pseudomonas* interaction was published by D'aes et al. (2010) in which information on the role of biosurfactants in plant–microbe interactions is presented. In particular, cyclic lipopeptides and rhamnolipids are the most-studied biosurfactants for their role in plant–microbe interactions which include the following:

1. Facilitation of microbial mobility (swarming) on plant root surface (Deziel et al., 2003).
2. Adhesion and dispersion of biofilms to create microcolonies on plant surfaces (Davey et al., 2003).
3. Enhanced bioavailability of nutrients or hydrophobic molecules to plants, for example, biosurfactants produced by soil microbes increase wettability of soil and support better distribution of chemical fertilizers in soil thus assisting the nutrient uptake of plants (Sachdev and Cameotra, 2013).
4. Protection from toxic compounds (hydrocarbons and heavy metals) (Chrzanowski et al., 2009).

Plant-associated bacteria play a key role in the ability of the host plant to adapt to a polluted environment. Traditional interest in endophytic microorganisms has been focused on their commercial potential as biocontrol agents of plant pathogens and for increasing plant yield and the nutritive value of agricultural crops. For example, Kruijt et al. (2009) reported a plant growth-promoting *Pseudomonas putida* strain 267, originally isolated from the rhizosphere of black pepper which produces biosurfactants that cause lysis of zoospores of the oomycete pathogen *Phytophthora capsici*. A more novel application is the use of plant–endophyte associations to enhance *in situ* phytoremediation. Some endophytes have been found to be resistant to heavy metals and/or capable of degrading organic contaminants and endophyte-assisted phytoremediation has been documented as a promising technology for *in situ* remediation of contaminated soils. The heavy-metal-resistant endophytes can enhance plant growth, decrease metal phytotoxicity, and affect metal translocation and accumulation in plants. These microorganisms can alter the metabolism of plant cells, so that upon exposure to heavy metal stress, the plants are able to tolerate high concentrations of metals (Ma et al., 2011). Inoculating plants with plant growth-promoting bacteria has been shown to improve plant growth and their establishment in soils contaminated with both heavy metals and/or organic pollutants (Becerra-Castro et al., 2011). In general, these bacteria interact directly or indirectly with a host plant through a variety of mechanisms, including indole acetic acid, cytokinins, 1-aminocyclopropane-1-carboxylic acid deaminase, and biosurfactant production. Such bacterial strains could be used in agriculture to promote plant growth and disease control and lead to cost-effective agriculture practices free from chemical pesticides.

15.2.8 BIOSURFACTANTS IN AGROCHEMICAL FORMULATION

Emulsification, foaming, dispersion, wetting, penetrating, and thickening properties of biosurfactants are currently used for agrochemical formulations. Example 1: A glycolipid biosurfactant produced by *Burkholderia cenocepacia* BSP3 isolated from oil-contaminated soil has been reported to have a high emulsification index of $90 \pm 2\%$ and to enhance solubilization of three hydrophobic pesticides, namely, methyl parathion, ethyl parathion, and trifluralin (Wattanaphon et al., 2008). Example 2: Microbial biosurfactants can be used as penetrants and carriers for active agents in pesticide, insecticide, and herbicide formulations (Awada et al., 2011). In this study, the addition of rhamnolipids to a water-soluble insecticide Acephate formulation controlled its solubility; thereby providing the insecticide at a predetermined rate. Combining the rhamnolipids with the water-insoluble insecticide imidacloprid enhanced the solubility as well as the translocation of the insecticide into the treated plant or tree.

Another report on use of sophorolipids and derivatives as adjuvants for pesticide formulation has added more evidence for the use of biosurfactants in agro-formulation (Giessler-Blank et al., 2012). Adjuvants are substances other than water which are not in themselves active as an ingredient but which enhance or support the effectiveness of an active ingredient. In the above-mentioned study, SLs and SL derivatives were used as adjuvants in a formulation containing fungicide (Opus) and herbicide (Cato) to evaluate their ability to boost the activity of the fungicides and herbicides. It was reported that there was a 53% increase in fungicidal performance and 20% increase in herbicidal activity, when applied with sophorolipids as adjuvants. These results clearly indicate the potential application of biosurfactants in pesticide and herbicide formulations. Further, biosurfactants used in pesticide formulations could increase the wettability of the leaf surface thus increasing the residence time of the pesticide on the leaf surface to facilitate prolonged protection for the plants. In addition to the application in pesticide formulations, biosurfactants can be applied in agriculture as nutrient solubilizers in fertilizer formulations, which will help the plant to assimilate the nutrients more effectively.

Another interesting application for biosurfactants is their potential use as insecticide/pesticide residue cleaning agents on vegetables and fruits. Cheowtirakul and Linh (2010) reported the application of rhamnolipids as cleaning agents for insecticide residues from lettuce leaf. Results from this study indicated that at 25 ppm concentration RLs were able to reduce the pesticide cypermethrin concentration from 100 to 2 ppm in 5 min. This is another potential area for biosurfactants to be applied as cleaning agents to remove pesticide residues from food vegetables in a safer way.

15.3 CONCLUDING REMARKS

The scientific literature clearly indicates that there is substantial evidence to support the use of microbial biosurfactants in agriculture for disease and pest/insect control, induced systemic resistance, soil remediation, and agroformulations. So far only two biosurfactants, RLs and SLs, have been explored and successfully developed to commercial scale for biopesticide applications. There are many other microbial biosurfactants that remain unexplored for their application in agriculture and a detailed investigation of these unexplored biosurfactants could bring more value to microbial biosurfactants. Finally, the increased environmental regulations on chemical pesticides and agrochemicals provide impetus to the need for safe green chemicals in the agroindustry, where microbial biosurfactants could be a suitable choice.

REFERENCES

Abbasi, H., Hamedi, M.M., Lotfabad, T.B., Zahiri, H.S., Sharafi, H., Masoomi, F., Moosavi-Movahedi, A.A., Ortiz, A., Amanlou, M., and Noghabi, K.A. 2012. Biosurfactant-producing bacterium, *Pseudomonas aeruginosa* MA01 isolated from spoiled apples: Physicochemical and structural characteristics of isolated biosurfactant. *Journal of Bioscience and Bioengineering*, 113:211–219.

Abioye, O.P. 2011. Biological remediation of hydrocarbon and heavy metals contaminated soil. In: *Soil Contamination*. S. Pascucci, Editor, Croatia: InTech, 201, 127–142. DOI: 10.5772/24938, ISBN: 978-953-307-647-8.

Abraham, W.R., Meyer, H., and Yakimov, M. 1998. Novel glycine containing glucolipids from the alkane using bacterium *Alcanivorax borkumensis*. *Biochimica et Biophysica Acta*, 1393:57–62.

Awada, S.M., Awada, M., and Spendlove, R.S. 2011. Method of controlling pests with biosurfactant penetrants as carriers for active agents. US Patent No. 2011/0319341 A1.

Awada, S.M., Awada, M., and Spendlove, R.S. 2012. Compositions and methods for controlling pests with glycolipids. US Patent No. 2012/0058895 A1.

Awada, S.M., Spendlove, R.S., and Awada, M. 2005. Microbial biosurfactants as agents for controlling pests. US Patent No. 2005/0266036 A1.

Banat, I.M. 1993. The isolation of a thermophilic biosurfactant producing *Bacillus* sp. *Biotechnology Letters*, 15:591–594.

Banat, I.M., Franzetti, A., Gandolfi, I., Bestetti, G., Martinotti, M.G., Fracchia, L., Smyth, T.J., and Marchant, R. 2010. Microbial biosurfactants production, applications and future potential. *Applied Microbiology and Biotechnology*, 87:427–444.

Becerra-Castro, C., Kidd, P.S., Prieto-Fernández, Á., Weyens, N., Acea, M., and Vangronsveld, J. 2011. Endophytic and rhizoplane bacteria associated with *Cytisus striatus* growing on hexachlorocyclohexane contaminated soil: Isolation and characterization. *Plant Soil*, 340:413–433.

Boller, T. and Felix, G. 2009. A renaissance of elicitors: Perception of microbe-associated molecular patterns and danger signals by pattern-recognition receptors. *Annual Review of Plant Physiology*, 60:379–406.

Boller, T. and He, S.Y. 2009. Innate immunity in plants: An arms race between pattern recognition receptors in plants and effectors in microbial pathogens. *Science*, 324:742–744.

Carrillo, C., Teruel, J.A., Aranda, F.J., and Ortiz, A. 2003. Molecular mechanism of membrane permeabilization by the peptide antibiotic surfactin. *Biochimica et Biophysica Acta*, 611:91–97.

Cheowtirakul, C. and Linh, N.D. 2010. The study of biosurfactant as a cleaning agent for insecticide residue in leafy vegetables. *Assumption University Journal of Technology*, 14:75–87.

Chrzanowski, L., Wick, L.Y., Meulenkamp, R., Kaestner, M., and Heipieper, H.J. 2009. Rhamnolipid biosurfactants decrease the toxicity of chlorinated phenols to *Pseudomonas putida* DOT-T1E. *Letters in Applied Microbiology*, 48:756–762.

D'aes, J., De Maeyer, K., Pauwelyn, E., and Hofte, M. 2010. Biosurfactants in plant–*Pseudomonas* interactions and their importance to biocontrol. *Environmental Microbiology*, 2:359–372.

Das, P., Mukherjee, S., and Sen, R. 2008a. Improved bioavailability and biodegradation of a model polyaromatic hydrocarbon by a biosurfactant producing bacterium of marine origin. *Chemosphere*, 72:1229–1234.

Das, P., Mukherjee, S., and Sen, R. 2008b. Antimicrobial potentials of a lipopeptide biosurfactant derived from a marine *Bacillus circulans*. *Journal of Applied Microbiology*, 104:1675–1684.

Das, P., Mukherjee, S., and Sen, R. 2009. Biosurfactant of marine origin exhibiting heavy metal remediation properties. *Bioresource Technology*, 100:4887–4890.

Davey, M.E., Caiazza, N.C., and Tootle, G.A.O. 2003. Rhamnolipid surfactant production affects biofilms architecture in *Pseudomonas aeruginosa* PA-01. *Journal of Bacteriology*, 185:1027–1036.

De Jonghe, K., De Dobbelaere, I., Sarrazyn, R., and Höfte, M. 2005. Control of *Phytophthora cryptogea* in the hydroponic forcing of Witloof chicory with the rhamnolipid-based biosurfactant formulation PRO1. *Plant Pathology*, 54:219–226.

Desai, J.D. and Banat, I.M. 1997. Microbial production of surfactants and their commercial potential. *Microbiology and Molecular Biology Reviews*, 61:47–56.

De Vleesschauwer, D. and Hofte, M. 2009. Rhizobacteria induced systemic resistance. *Advances in Botanical Research*, 51:223–281.

Deziel, E., Lepine, F., Milot, S., and Villemur, R. 2003. rhlA is required for the production of a novel biosurfactant promoting swarming motility in *Pseudomonas aeruginosa*: 3-(3-hydroxyalkanoyloxy)alkanoic acids (HAAs), the precursors of rhamnolipids. *Microbiology*, 149:2005–2013.

Giessler-Blank, S., Schilling, M., Sieverding, E., and Thum, O. 2012. Use of sophorolipids and derivatives thereof in combination with pesticides as adjuvant/additive for plant protection and the industrial non-crop field. US Patent No. 2012/0220464 A1.

Gomaa, E.Z. 2013. Antimicrobial activity of a biosurfactant produced by *Bacillus licheniformis* strain M104 grown on whey. *Brazilian Archives of Biology and Technology*, 56:259–268.

Gross, R.A. and Thavasi, R. 2013. Modified sophorolipids combinations as antimicrobial agents. US Patent No. 2013/0142855 A1.

Haddad, N.I. 2008. Isolation and characterization of a biosurfactant producing strain, *Brevibacillis brevis* HOB1. *Journal of Industrial Microbiology and Biotechnology*, 35:1597–1604.

Haefele, D., Lamptey, J.C., and Marlow, J.L. 1991. Biological control of corn seed rot and seedling blight. US Patent No. 4996049.

Hingley, S.T., Hastie, A.T., and Kueppers, F. 1986. Effect of ciliostatic factors from *Pseudomonas aeruginosa* on rabbit respiratory cilia. *Infection and Immunnity*, 51:254–258.

Jourdan, E., Henry, G., Duby, F., Dommes, J., Barthelemy, J.P., Thonart, P., and Ongena, M. 2009. Insights into the defense-related events occurring in plant cells following perception of surfactin-type lipopeptide from *Bacillus subtilis*. *Molecular Plant–Microbe Interactions*, 22:456–468.

Kim, B.S., Lee, J.Y., and Hwang, B.K. 2000. *In vivo* control and *in vitro* antifungal activity of rhamnolipid B, a glycolipid antibiotic, against *Phytophthora capsici* and *Colletotrichum orbiculare*. *Pest Management Science*, 56:1029–1035.

Kim, S.K., Kim, Y.C., Lee, S., Kim, J.C., Yun, M.Y., and Kim, I.S. 2011. Insecticidal activity of rhamnolipid isolated from *Pseudomonas* sp. EP-3 against green peach aphid (*Myzus persicae*). *Journal of Agriculture and Food Chemistry*, 59:934–938.

Kosaric, N. 2001. Biosurfactants for soil bioremediation. *Food Technology and Biotechnology*, 39:295–304.

Kruijt, M., Tran, H., and Raaijmakers, J.M. 2009. Functional, genetic and chemical characterization of biosurfactants produced by plant growth-promoting *Pseudomonas putida* 267. *Journal of Applied Microbiology*, 107:546–556.

Krzyzanowska, D.M., Potrykus, M., Golanowska, M., Polonis, K., Gwizdek-Wisniewska, A., Lojkowska, E., and Jafra, S. 2012. Rhizosphere bacteria as potential biocontrol agents against soft rot caused by various *Pectobacterium* and *Dickeya* spp strains. *Journal of Plant Pathology*, 94:353–365.

Ma, Y., Prasad, M.N.V., Rajkumar, M., and Freitas, J. 2011. Plant growth promoting rhizobacteria and endophytes accelerate phytoremediation of metalliferous soils. *Biotechnology Advances*, 29:248–258.

Mackey, D. and McFall, A.J. 2006. MAMPs and MIMPs: Proposed classifications for inducers of innate immunity. *Molecular Microbiology*, 61:1365–1371.

Marchant, R. and Banat, I.M. 2012a. Biosurfactants: A sustainable replacement for chemical surfactants? *Biotechnology Letters*, 34:1597–1605.

Marchant, R. and Banat, I.M. 2012b. Microbial biosurfactants: Challenges and opportunities for future exploitation. *Trends in Biotechnology*, 30:558–565.

Miller, R.M. 1995. Biosurfactant-facilitated remediation of metal-contaminated soils. *Environmental Health Perspectives*, 103:59–62.

Nielsen, C.J., Ferrin, D.M., and Stanghellini, M.E. 2006. Efficacy of biosurfactants in the management of *Phytophthora capsici* on pepper in recirculating hydroponic systems. *Canadian Journal of Plant Pathology*, 28:450–460.

Nielsen, T.H., Christophersen, C., Anthoni, U., and Sorensen, J. 1999. Viscosinamide, a new cyclic depsipeptide with surfactant and antifungal properties produced by *Pseudomonas fluorescens* DR54. *Journal of Applied Microbiology*, 86:80–90.

Nielsen, T.H., Sørensen, D., Tobiasen, C., Andersen, J.B., Christophersen, C., Givskov, M., and Sørensen, J. 2002. Antibiotic and biosurfactant properties of cyclic lipopeptides produced by fluorescent *Pseudomonas* spp from the sugar beet rhizosphere. *Applied and Environmental Microbiology*, 68:3416–3423.

Ongena, M. and Jacques, P. 2008. *Bacillus* lipopeptides: Versatile weapons for plant disease biocontrol. *Trends in Microbiology*, 16:115–125.

Ongena, M., Jourdan, E., Adam, A., Paquot, M., Brans, A., Joris, B., Arpigny, J.L., and Thonart, P. 2007. Surfactin and fengycin lipopeptides of *Bacillus subtilis* as elicitors of induced systemic resistance in plants. *Environmental Microbiology*, 9:1084–1090.

Passeri, A., Schmidt, M., Haffner, T., Wray, V., Lang, S., and Wagner, F. 1992. Marine biosurfactants. IV. Production, characterization and biosynthesis of an anionic glucose lipid from the marine bacterial strain MM1. *Applied Microbiology and Biotechnology*, 37:281–286.

Perfumo, A., Smyth, T.J.P., Marchant, R., and Banat, I.M. 2010. Production and roles of biosurfactants and bioemulsifiers in accessing hydrophobic substrates. In: *Handbook of Hydrocarbon and Lipid Microbiology*. K.N. Timmis, Editor, Berlin: Springer-Verlag, 1501–1512.

Perneel, M., Heyrman, J., Adiobo, A., De Maeyer, K., Raaijmakers, J.M., De Vos, P., and Höfte, M. 2007. Characterization of CMR5c and CMR12a, novel fluorescent *Pseudomonas* strains from the cocoyam rhizosphere with the biocontrol activity. *Journal of Applied Microbiology*, 103:1007–1020.

Perneel, M., D'Hondt, L., De Maeyer, K., Adiobo, A., Rabaey, K., and Hofte, M. 2008. Phenazines and biosurfactants interact in the biological control of soil-borne diseases caused by *Pythium* spp. *Environmental Microbiology*, 10:778–788.

Rahman, K.S.M., Banat, I.M., Thahira, J., Thayumanavan, Tha., and Lakshmanaperumalsamy, P. 2002. Bioremediation of gasoline contaminated soil by bacterial consortium with poultry litter, coir pith and rhamnolipid biosurfactant. *Bioresource Technology*, 81:25–32.

Rahman, K.S.M., Rahman, T.J., Banat, I.M., Kane, G., Lord, R., and Street, G. 2006. Bioremediation of petroleum sludge using bacterial consortium with biosurfactant. In: *Environmental Bioremediation Technologies*. S.N. Singh and R.D. Tripathi, Editors, New York: Springer-Verlag, 391–408, ISBN: 3540347909.

Rahman, K.S.M., Rahman, T.J., Kourkoutas, Y., Petsas, I., Marchant, R., and Banat, I.M. 2003. Enhanced bio-remediation of *n*-alkane in petroleum sludge using bacterial consortium amended with rhamnolipid and micronutrients. *Bioresource Technology*, 90:159–168.

Romero, D., de Vicente, A., Olmos, J.L., Da'vila, J.C., and Perez-Garcia, A. 2007. Effect of lipopeptides of antagonistic strains of *Bacillus subtilis* on the morphology and ultrastructure of the cucurbit fungal pathogen *Podosphaera fusca*. *Journal of Applied Microbiology*, 103:969–976.

Ron, E.Z. and Rosenberg, E. 1999. Natural roles of biosurfactants. *Environmental Microbiology*, 3:229–236.

Rosenberg, E. and Ron, E.Z. 1998. Surface active polymers from the genus *Acinetobacter*. In: *Biopolymers from Renewable Resources*. D.L. Kaplan, editor, Berlin: Springer, 281–291.

Rostas, M. and Blassmann, K. 2009. Insects had it first: Surfactants as a defence against predators. *Proceedings of the Royal Society B*, 276:633–638.

Sachdev, D.P. and Cameotra, S.S. 2013. Biosurfactants in agriculture. *Applied Microbiology and Biotechnology*, 97:1005–1016.

Sastoque-Cala, L., Cotes-Prado, A.M., Rodríguez-Vázquez, R., and Pedroza-Rodríguez, A.M. 2010. Effect of nutrients and conditions of fermentation on the production of biosurfactants using rhizobacteria isolated from fique. *Universitas Scientiarum*, 15:251–264.

Schofield, M.H., Thavasi, R., and Gross, R.A. 2013. Modified sophorolipids for the inhibition of plant pathogens. US Patent No. 20130085067 A1.

Schulz, D., Passeri, A., Schmidt, M., Lang, S., Wagner, F., Wray, V., and Gunkel, W. 1991 Marine biosurfactants, I. Screening for biosurfactants among crude oil degrading marine microorganisms from the North Sea. *Zeitschrift fuer Naturforschung C: Journal of Biosciences*, 46:197–203.

Singh, P.B., Sharma, S., Saini, H.S., and Chadha, B.S. 2009. Biosurfactant production by *Pseudomonas* sp. and its role in aqueous phase partitioning and biodegradation of chlorpyrifos. *Letters in Applied Microbiology*, 49:378–383.

Snook, M.E., Mitchell, T., Hinton, D.M., and Bacon, C.W. 2009. Isolation and characterization of leu7-surfactin from the endophytic bacterium *Bacillus mojavensis* RRC 101, a biocontrol agent for *Fusarium verticillioides*. *Journal of Agricultural and Food Chemistry*, 57:4287–4292.

Stanghellini, M.E. and Miller, R.M. 1997. Biosurfactants: Their identity and potential efficacy in the biological control of zoosporic plant pathogens. *Plant Disease*, 81:4–12.

Thavasi, R., Jayalakshmi, S., Balasubramaian, T., and Banat, I.M. 2007. Biosurfactant production by *Corynebacterium kutscheri* from waste motor lubricant oil and pea nut oil cake. *Letter in Applied Microbiology*, 45:686–691.

Thavasi, R., Jayalakshmi, S., Balasubramaian, T., and Banat, I.M. 2011a. Effect of biosurfactant and fertilizer on biodegradation of crude oil by marine isolates of *Bacillus megaterium, Corynebacterium kutscheri* and *Pseudomonas aeruginosa*. *Bioresource Technology*, 102:772–778.

Thavasi, R., Jayalakshmi, S., and Banat, I.M. 2011b. Application of biosurfactant produced from peanut oil cake by *Lactobacillus delbrueckii* in biodegradation of crude oil. *Bioresource Technology*, 102:3366–3372.

Thavasi, R., Subramanyam, N.V.R.M., Jayalakshmi, S., Balasubramanian, T., and Banat, I.M. 2009. Biosurfactant production by *Azotobacter chroococcum* isolated from the marine environment. *Marine Biotechnology*, 11:551–556.

Thimon, L., Peypoux, F., Wallach, J., and Michel, G. 1995. Effect of the lipopeptide antibiotic iturin A, on morphology and membrane ultrastructure of yeast cells. *FEMS Microbiology Letters*, 128:101–106.

Thrane, C., Nielsen, T.H., Nielsen, M.N., Sorensen, J., and Olsson, S. 2000. Viscosinamide-producing *Pseudomonas fluorescens* DR54 exerts a biocontrol effect on *Pythium ultimum* in sugar beet rhizosphere. *FEMS Microbiology Ecology*, 33:139–146.

Tran, H., Ficke, A., Asiimwe, T., Hofte, M., and Raaijmakers, J.M. 2007. Role of the cyclic lipopeptide massetolide A in biological control of *Phytophthora infestans* and in colonization of tomato plants by *Pseudomonas fluorescens*. *New Phytology*, 175:731–742.

Varnier, A.L., Sanchez, L., Vatsa, P., Boudesocque, L., Garcia-Brugger, A., Rabenoelina, F., Sorokin, A. et al. 2009. Bacterial rhamnolipids are novel MAMPs conferring resistance to *Botrytis cinerea* in grapevine. *Plant Cell and Environment*, 32:178–193.

Velho, R.V., Medina, L.F., Segalin, J., and Brandelli, A. 2011. Production of lipopeptides among *Bacillus* strains showing growth inhibition of phytopathogenic fungi. *Folia Microbiologica (Praha)*, 56:297–303.

Wattanaphon, H.T., Kerdsin, A., Thammacharoen, C., Sangvanich, P., and Vangnai, A.S. 2008. A biosurfactant from *Burkholderia cenocepacia* BSP3 and its enhancement of pesticide solubilization. *Journal of Applied Microbiology*, 105:416–423.

Wicke, C., Hüners, M., Wray, V., Nimtz, M., Bilitewski, U., and Lang, S. 2000. Production and structure elucidation of glycoglycerolipids from a marine sponge-associated *Microbacterium* species. *Journal of Natural Products*, 63:621–626.

Yoo, D.S., Lee, B.S., and Kim, E.K. 2005. Characteristics of microbial biosurfactant as an antifungal agent against plant pathogenic fungus. *Journal of Microbiology and Biotechnology*, 15:1164–1169.

16 Biosurfactants and Soil Bioremediation

Edwan Kardena, Qomarudin Helmy, and Naoyuki Funamizu

CONTENTS

16.1 Introduction 327
16.2 Type of Bioremediation 329
    16.2.1 Intrinsic In Situ Bioremediation 329
    16.2.2 Engineered In Situ Bioremediation 330
    16.2.3 Engineered Ex Situ Bioremediation 334
16.3 Designing a Bioremediation Project 337
    16.3.1 Site and Soil Characterization 337
        16.3.1.1 Contaminants Characteristics 337
        16.3.1.2 Physicochemical Environmental Characteristics 339
    16.3.2 Selecting Remediation Method and Technology 339
    16.3.3 Treatability Studies 340
        16.3.3.1 Phase I: Biodegradability and Toxicity Assessment 341
        16.3.3.2 Phase II: Preliminary Confirmation 342
        16.3.3.3 Phase III: Pilot-Scale/Field Trial 342
    16.3.4 Field Remediation Activities 342
    16.3.5 Site Closure 344
16.4 Biosurfactants-Enhanced Bioremediation Process 346
    16.4.1 Role of Biosurfactants in Soil Bioremediation 352
        16.4.1.1 Increasing the Surface Area of Poorly Water-Soluble Hydrophobic
                Compounds 352
        16.4.1.2 Increasing the Bioavailability of Hydrophobic Compounds 352
    16.4.2 Utilizing Biosurfactants in Soil Bioremediation 354
16.5 Conclusions 356
References 356

16.1 INTRODUCTION

Soil contamination occurs every moment of our life. It happens when a hazardous solid or liquid substance mixes with the naturally occurring soil. In most cases, the contaminants are attached to the soil either physically or chemically. Sometimes they are trapped in between the particles of the soil. The contaminants usually come in contact with the soil by one of two ways; they are either spilled into the soil or are buried directly. One of the most common substances spilled into soil is oil (Figure 16.1). It often gets into the soil by leaking from tanks, being spilled during the transportation process, during loading/unloading, and leaks from pipelines. The severity of the contamination depends on the type of oil involved. The heavier the oil, the slower it is able to spread. However, if light oil is involved, it is able to seep through the top soil quite quickly. It then continues to move quickly through the layers of soil. The faster the cleanup operator responds to the spill/leak, the better the chance of stopping the contamination from spreading to the surrounding areas. Biological

FIGURE 16.1   Untreated oily waste from abandoned sludge pit (left) and rising oily waste originally from oil sludge burial site (right). (Author's project documentation with permission from LAPI-ITB consultant.)

cleanup processes are currently in vogue as a promising cost-effective and performance-effective technology to address numerous environmental pollution problems. These pollutants range from industrial wastes (e.g., polychlorinated biphenyls, trichloroethylene, pentachlorophenol, and dioxin), polyaromatic hydrocarbons, refined petroleum products (e.g., jet fuel, gasoline, diesel fuel and the benzene, toluene, ethylbenzene, and xylene cluster), acid mine drainage, pesticides, munitions compounds (e.g., trinitrotoluene), and inorganic heavy metals to crude oil (Finnerty, 1994; Mulligan, 2005). The major organic chemical waste categories include organic aqueous waste (pesticides), organic liquids (solvents from dry cleaning), oils (lubricating oils, automotive oils, hydraulic oils, fuel oils), and organic sludges/solids (painting operations, tars from dyestuffs intermediates). Most soil contamination is the result of accidental spills and leaks, originating from cleaning of equipment, residues left in used containers, and outdated materials (Vogel, 1996).

Bioremediation is the application of microorganisms or microbial processes or products to remove or degrade contaminants from an area. A more rigorous definition is the intentional use of biological degradation procedures to remove or reduce the concentration of environmental pollutants from sites where they have been released. The concentrations of pollutants are reduced to levels considered acceptable to site owners and/or regulatory agencies. Bioremediation most often addresses treatment of contaminated solids, such as groundwater aquifers, soils, and sediments. Advantages of using bioremediation processes compared with other remediation technologies are as follows:

1. Biologically based remediation detoxifies hazardous substances instead of merely transferring contaminants from one environmental medium to another.
2. Bioremediation is generally less disruptive to the environment than excavation-based processes.
3. The cost of treating a hazardous waste site using bioremediation technologies can be considerably lower than that for conventional treatment methods: vacuuming, absorbing, burning, dispersing, or moving the material.

Bioremediation overcomes the main deficiency of conventional cleanup approaches that rely on water flushing: dissolution or desorption of the contaminants into water is too slow, since the water's contaminant-carrying capacity is very limited. With bioremediation, high densities of active microorganisms locate themselves close to the nonaqueous source of contamination.

Although degrader microorganisms are present in the environment (soils, sediments, and aquifers), their naturally occurring numbers may be too small to bring about the rapid reaction needed to have enhanced degradation processes. In that case, the strategy is to add to the contaminated environment the materials needed to allow growth of the microorganisms to actively degrade the contaminant. The most commonly added materials are normally selected from among an electron-acceptor substrate, an electron-donor substrate, inorganic nutrients, and material to increase the solubility of the contaminant (Rittmann and McCarty, 2001).

16.2 TYPE OF BIOREMEDIATION

The possible types of bioremediation activities fall into three main categories: intrinsic in situ, engineered in situ, and engineered ex situ. The distinction between in situ and ex situ indicates whether the contaminated solids remain in place during the bioremediation (soil, groundwater, or other environment without removal of the contaminated material) while ex situ bioremediation entails the removal of all or part of the contaminated material (excavated and transferred to an aboveground treatment system). Another way to categorize bioremediation is by the degree of human intervention. Engineered or sometimes call-accelerated bioremediation is one end of the scale, where there is a high degree of human intervention.

16.2.1 INTRINSIC IN SITU BIOREMEDIATION

Intrinsic in situ bioremediation is a method of applying in situ remediation with essentially no human intervention. Intrinsic bioremediation is passive, it relies on the naturally occurring rates of supply of substrates and nutrients, as well as the intrinsic population of degrader microorganisms to prevent further spread of the contaminants from its source. Intrinsic bioremediation is also called unmanipulated, unstimulated, or unenhanced biological remediation of an environment; that is, biological natural attenuation of contaminants in the environment. Although intrinsic bioremediation does not accelerate the rate of source decontamination, it can be an effective form of biological containment. Intrinsic bioremediation involves no engineered measures to increase the supply rate of oxygen, nutrients, or other stimulants.

Intrinsic bioremediation is sometimes confused with natural attenuation, which is defined as the reduction of contaminant concentrations due to naturally occurring processes such as biodegradation, sorption, advection, dilution, dispersion, volatilization, photo-oxidation, and other chemical reaction. Rittmann and McCarty (2001) suggest that intrinsic bioremediation is a more stringent subset of natural attenuation in which microbial transformation reactions (biological processes) are responsible for the decrease in contaminant concentration. Meanwhile, natural attenuation referred to the naturally occurring processes (biological, physical, and chemical) in the remediation processes of the contaminated area. The U.S. EPA issued a comprehensive directive for natural attenuation (U.S. EPA, 1997a). The term natural attenuation or monitored natural attenuation refers to the reliance on natural attenuation processes (within the context of a carefully controlled and monitored site cleanup approach) to achieve site-specific remedial objectives within a time frame that is reasonable compared to that offered by other more active methods. The natural attenuation processes that are at work in such a remediation approach include a variety of physical, chemical, or biological processes that, under favorable conditions, act without human intervention to reduce the mass, toxicity, mobility, volume, or concentration of contaminants in soil or groundwater. These in situ processes include biodegradation, dispersion, dilution, sorption, volatilization, and chemical or biological stabilization, transformation, or destruction of contaminants. Natural attenuation processes are typically occurring at all sites, but to varying degrees of effectiveness depending on the types and concentrations of contaminants present and the physical, chemical, and biological characteristics of the soil and groundwater (Borden, 1994; U.S. EPA, 1997b). Natural attenuation processes may reduce the potential risk posed by site contaminants in three ways:

1. The contaminant may be converted into a less toxic form through destructive processes such as biodegradation or abiotic transformations.
2. Potential exposure levels may be reduced by lowering concentration levels (through destructive processes, or by dilution or dispersion).

3. Contaminant mobility and bioavailability may be reduced by sorption to the soil or rock matrix.

Other terms associated with natural attenuation in the literature include intrinsic remediation, intrinsic bioremediation, passive bioremediation, natural recovery, and natural assimilation. While some of these terms are synonymous with natural attenuation, others refer strictly to biological processes, excluding chemical and physical processes.

16.2.2 ENGINEERED IN SITU BIOREMEDIATION

When the rate of natural bioremediation is limited by nutrient and electron-acceptor availability, enhanced in situ bioremediation may be applied. Engineered bioremediation refers to employing engineering tools to greatly increase the input rates for the stimulating materials where the operators take an active role in promoting or carrying out the bioremediation process. In situ bioremediation is the enhancement of in-place biodegradation of organic compounds within the subsurface or existing waste-holding lagoons/site. The goal of in situ bioremediation was to manage and manipulate the subsurface environment to optimize microbial degradation. Engineering requires an understanding of the microbial processes that affect the biodegradation of the target contaminant and the soil physical, chemical, and hydrological interaction. The major engineering challenge is the design of a delivery and recovery system that provides for a responsive control of the subsurface environment, and a monitoring system that provides data for process optimization. Biological processes require close monitoring of the microbial environment to maintain control of the overall degradation of the target contaminant (Cookson, 1995).

In situ bioremediation has several advantages and disadvantages in its operation methods as described in Table 16.1 (NRC, 2000; U.S. EPA, 2006).

Because engineered in situ bioremediation is used to accelerate biologically driven removal of contaminants trapped in the solid matrix/phase, its success depends on being able to achieve substantially increased inputs of stimulating materials. Characteristics of the ideal candidate sites

---

TABLE 16.1

Advantages and Disadvantages of In Situ Bioremediation

Advantages:
- Minimum site disturbance because few surface structures are required
- Minimum contaminant exposure (volatile) reduced potential for cross-media transfer of contaminant and also reduced risk of human exposure to contaminated media
- Less cost for material removal to remediation sites
- Allow treatment under and around buildings
- As an in situ technology, there is typically little secondary waste generated
- The areal zone of treatment can be larger than with other remedial technologies because the treatment moves with the plume and can reach areas that would otherwise be inaccessible

Disadvantages:
- Extensive and costly site characterization and monitoring
- Difficult to control reaction conditions, if biotransformation halts at an intermediate compound, the intermediate may be more toxic and/or mobile than the parent compound
- Difficult to accurately predict end points, depending on the particular site some contaminants may not be completely transformed to innocuous products
- Careful monitoring of process required, long-term monitoring and periodic reevaluation of the remedy effectiveness will generally be necessary
- Ineffective for metal contaminant mixed with organic compounds
- When inappropriately applied, injection wells may become clogged from profuse microbial growth resulting from the addition of nutrients, electron donor, and/or electron acceptor

for successful implementation of engineered in situ bioremediation have the following features (Cookson, 1995; Rittmann and McCarty, 2001; Hazen, 2010):

1. The hydraulic conductivity is relatively homogenous, isotropic, and greater than 10^{-5} cm/s (Table 16.2).
2. Residual concentrations are not excessive; NAPLs (nonaqueous phase liquids) concentrations greater than about 10 ppm reduce aquifer permeability and microorganism access to the biodegradable contaminants.
3. Contaminant originating from a single source.
4. The contaminated zone is relatively shallow in order to minimize costs of drilling and sampling and should be completely isolated.

Although most sites will not meet all of these criteria, bioremediation strategies may be developed for the nonideal site. The bioremediation strategy may be developed with site-specific geological and microbiological data, combined with knowledge concerning the chemical, physical, and biochemical fate of the contaminants present (Sims et al., 1992). In situ bioremediation in the saturated zone may fail due to the lack of adequate mass transport of the electron acceptor (usually oxygen). Site setting factors, such as average permeability, and the scale and degree of heterogeneity are the main factors governing the advective and diffusional transport rates in the subsurface. If transport rates are too low, in situ bioremediation of saturated zones is not a viable option. In general, bioremediation processes are more difficult to apply to clayey and other low-permeability soils (van Cauwenberghe and Roote, 1998).

Engineered in situ bioremediation strategies will be limited by the ability to deliver the stimulus to the environment (i.e., degrader microorganisms, inorganic nutrients, electron acceptor, and electron donor). Hydraulic conductivity or permeability of the contaminated soil must be sufficient enough to allow perfusion of the stimulus through the soil. The minimum hydraulic conductivity for an effective process of in situ bioremediation is generally considered greater than 10^{-5} cm/s (Rittmann and McCarty, 2001) or 10^{-4} cm/s (Thomas and Ward, 1989). Moreover, the stimulants required must be compatible with the environment. For example, hydrogen peroxide is an excellent source of electron acceptor, but it can cause precipitation of metals in soils, and dense microbial growth around the injection site that all soil pores get plugged. It is also toxic to bacteria at high concentrations, >100 ppm (Thomas and Ward, 1989). Ammonia also can be problematic, because it adsorbs rapidly to clays, causes pH changes in poorly buffered environments, and can cause clays to swell, decreasing permeability around the injection point. It is generally

TABLE 16.2

Typical Values of Saturated Hydraulic Conductivity for Soils

Soil Description	Hydraulic Conductivity, k (cm/s)	Bioremediation Effectiveness
Clean gravel	1–100	Effective
Sand gravel	10^{-2}–10	Effective
Clean coarse sand	10^{-2}–1	Effective
Fine sand	10^{-3}–10^{-1}	Moderate
Silty sand	10^{-3}–10^{-2}	Moderate
Clayey sand	10^{-4}–10^{-2}	Limited
Silt	10^{-8}–10^{-3}	Ineffective
Clay	10^{-10}–10^{-6}	Ineffective

Source: Adapted from Coduto, D.P. 1999. *Geotechnical Engineering Principles and Practices.* Prentice Hall, Englewood Cliffs, NJ.

accepted that soil bacteria need a C:N:P ratio of 30:5:1 for unrestricted growth (Paul and Clark, 1989). The actual injection ratio used is usually slightly higher (a ratio of 100:10:1) (Cookson, 1995), since these nutrients must be bioavailable, a condition that is much more difficult to measure and control in the terrestrial subsurface. It may also be necessary to remove light nonaqueous phase liquid (LNAPL) contaminants that are floating on the water table or smearing the capillary fringe zone, hence bioslurping (Keet, 1995). This strategy greatly increases the biostimulation response time by lowering the highest concentration of contaminant the organisms are forced to transform (Hazen, 2010).

An important stage of engineered in situ bioremediation is selecting the technique for isolating and controlling the contamination. Hydraulic control is necessary to move or halt groundwater flow, raise or drop the water table, and control movement of the contaminated plume. Hydrological intervention is the most frequent procedure for isolating the contaminated plume since it is less costly compared to physical containment structure and more flexible in pumping rates operation, and flow patterns can be changed as needed over the period of treatment. Ideally, the contaminated zone should be isolated and relatively shallow in order to minimize costs of drilling and sampling. In addition to isolation of the contaminated zone, the hydraulic system is the key to the in situ bioremediation process control. Fluid flow provides delivery of appropriate stimulants (degrader microbe, nutrients, substrates, and electron acceptors within the subsurface) (Cookson, 1995). The design goal is to provide a system of injection wells, recovery wells, and possibly barriers that allows control of mass transfer into and out of the contaminated area. The injected water may be supplemented with nutrients, electron acceptor, and electron donor or have additional degrader microbial seed added before being re-circulated or re-injected into the subsurface. Injection systems consist of gravity and force delivery methods (Table 16.3).

Engineered in situ bioremediation system can be classified according to the means by which stimulatory materials are added. Which approach is taken depends upon the relationship between the type of contamination and the type of microorganisms already present at the contamination site. For example, if the microorganisms already present are appropriate to break down the type of contamination, cleanup operators may only need to stimulate these microorganisms by the addition of fertilizers, nutrients, oxygen, phosphorus, and so on. The following are three frequently used methods for engineered in situ bioremediation.

TABLE 16.3
Subsurface Injection System

Gravity systems:
- Ponding
- Flooding
- Trench
- Infiltration gallery
- Infiltration bed
- Surface spraying
- Subsurface drain

Forced systems:
- Pump injection
- Air injection
- Air vacuum

Source: Adapted from Cookson, J.T. 1995. *Bioremediation Engineering: Design and Application.* McGraw-Hill, New York.

Bioventing: The bioventing biological system treats contaminated soil in situ by injecting atmospheric air into unsaturated soil. The air provides a continuous oxygen source, which enhances the growth of microorganisms naturally present in the soil. First, injection wells must be dug into the contaminated soil. How many wells, how close together they go, how deep they are dug all depends on the factors affecting the rate of degradation (type of contamination, type of soil, nutrient levels, concentration of contaminants, etc.). Once all of the injection wells are dug, an air blower is used to control the supply of air that is given to the microorganisms. These injection wells can also be used to add nitrogen and phosphorus, maximizing the rate of degradation. Bioventing is typically used to treat contaminated soil and removes and/or degrades a number of contaminants that are biodegradable under aerobic conditions. Bioventing treats contaminants and combinations of contaminants with varying degrees of success. Volatile, nonbiodegradable constituents can be treated, but off gas treatment costs may be incurred or increased (Figure 16.2). Bioventing is most applicable where the depth to water exceeds 3 m and the surficial soils do not require treatment or are being treated by another method. Shallower soils and sites with shallower water tables can be treated if the surface is capped.

When extraction wells are used for bioventing, the process is similar to air sparging and soil vapor extraction (SVE). However, while SVE removes constituents primarily through volatilization, bioventing systems promote biodegradation of constituents and minimize volatilization (generally by using lower air flow rates than for air sparging and SVE). In practice, some degree of volatilization and biodegradation occurs when either SVE or bioventing is used.

Biosparging/Air Sparging: Originally developed in Europe, air sparging has become a popular means of engineered bioremediation in North America for aerobic biodegradation. Air sparging involves injecting a gas (usually air/oxygen) under pressure into the saturated zone to transfer volatile compounds to the unsaturated zone for biodegradation. The term biosparging is used to highlight the bioremediation aspect of the treatment process or refers to a situation where biodegradation is the dominant remedial process, with volatilizing playing a secondary role. The air injected below the water table increases the oxygen concentration and enhances the rate of biological degradation of organic contaminants by naturally occurring microorganisms. The addition of bioremediation processes makes the application of air sparging more favorable for the remediation of less volatile contaminants like diesel fuel and waste oils (Miller, 1996). Nutrients and other amendments can be added from injection wells or an infiltration gallery (Figure 16.3). Similar to bioventing, some nutrients can be added in a gaseous state. In most cases, a gas-recovery system is needed to capture volatile components and prevent off-site contaminant transport in the gas phase (Rittmann and McCarty, 2001).

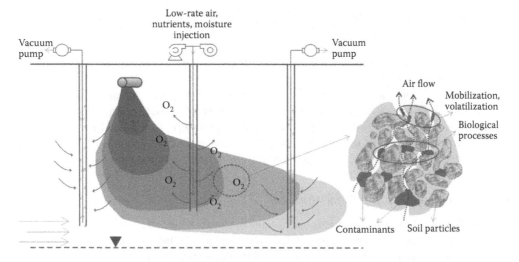

FIGURE 16.2 Typical of a bioventing system and degradation process scheme.

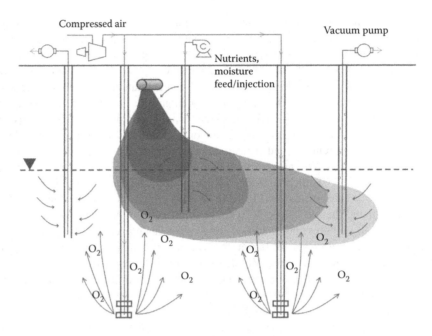

FIGURE 16.3 Typical of a biosparging system for aerobic treatment of contamination below the water table.

Liquid Circulating Systems: Liquid circulating systems are used for in situ bioremediation of contamination in the saturated zone. The delivery system, consisting of wells or trenches, is designed to circulate adequate amounts of nutrients and oxygen through the zone of contamination to maximize contaminant biodegradation. Groundwater is extracted, treated above ground if necessary, and then disposed or amended with nutrients and recirculated. Aboveground treatment can include air stripping, biological treatment, activated carbon sorption, or a combination. The recirculation system is designed to hydraulically isolate the target area and minimize contaminant migration out of the treatment zone. Oxygen is provided by sparging with air or pure oxygen or by adding hydrogen peroxide to injected water. If the groundwater is fairly close to the surface, the hydrogen peroxide can be administered through sprinkler systems. If the groundwater is fairly deep beneath the surface, injection wells are used (Figure 16.4).

There are some specific limitations when hydrogen peroxide is used as electron acceptor to enhance the bioremediation, including the following:

1. Concentrations of hydrogen peroxide greater than 100 ppm in groundwater are inhibiting to microorganisms.
2. Microbial enzymes and high iron content of subsurface materials can rapidly reduce concentrations of hydrogen peroxide and reduce the zone of influence. Amended hydrogen peroxide can be consumed very rapidly near the injection well, which may limit the biological growth to the region near the injection well, limiting adequate contamination/microorganisms contact and can cause biofouling of wells which can retard the input of nutrients.

16.2.3 ENGINEERED EX SITU BIOREMEDIATION

Engineered ex situ bioremediation is a biological process in which excavated soil is placed in a lined aboveground treatment area and aerated following processing to enhance the degradation of organic contaminants by the indigenous microbial population. Under aerobic conditions, specific microorganisms can utilize organic contaminants such as petroleum hydrocarbon mixtures,

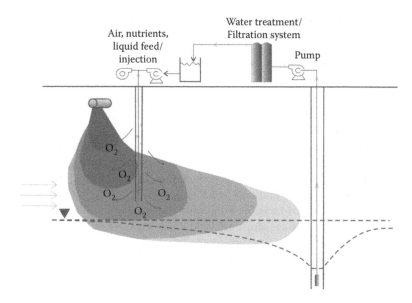

FIGURE 16.4 Typical of a liquid circulating system equipped with water treatment facility before recirculation.

polycyclic aromatic hydrocarbons (PAH), phenols, cresols, and some pesticides as a source of carbon and energy and degrade them ultimately to carbon dioxide and water. Contaminated soils can be excavated and treated in aboveground facility treatment systems. Aboveground bioremediation for contaminated solids includes land farming, composting, biopile, and slurry reactors.

All engineered bioremediation can be characterized as either biostimulation (addition of nutrients to stimulate naturally occurred degrader microorganisms), or bioaugmentation (addition of specific degrader microorganisms), or processes that use both. Ex situ engineered bioremediation techniques are destruction or transformation techniques directed toward stimulating the microorganisms to grow and use the contaminants as a food and energy source by creating a favorable environment for the microorganisms. Generally, this means providing some combination of oxygen, nutrients supply, and moisture, and controlling the temperature and pH. Sometimes, microorganisms adapted for degradation of the specific contaminants are applied to enhance the process. Although not all organic compounds are amenable to biodegradation, bioremediation techniques have been successfully used to remediate soils, and sludges contaminated by petroleum hydrocarbons, solvents, pesticides, wood preservatives, and other organic chemicals. Bioremediation is not yet commonly applicable for treatment of inorganic contaminants. There are two main types of engineered ex situ bioremediation.

Solid Phase: Solid-phase treatment consists of placing the excavated materials into an aboveground enclosure. Inside this enclosure, the contaminated soil is spread onto a treatment bed. This treatment bed usually has some kind of built-in aeration system. Using this system, operators are able to control the nutrients, moisture, oxygen, and pH. This allows them to maximize the efficiency of the bioremediation. The soil can also be tilled like farmland, helping to provide oxygen and enable additional aerobic biodegradation of the contamination. Solid-phase treatment is especially effective if the contaminants are fuel hydrocarbons.

There are three solid-phase bioremediation techniques. They are: land farming, biopile, and composting. *Land farming* is a full-scale bioremediation technology, which usually incorporates liners and other methods to control leaching of contaminants, which requires excavation and placement of contaminated soils, sediments, or sludges. Contaminated media is applied into lined beds and periodically turned over or tilled to aerate the soils. Nutrients (N, P, and K), moisture (usually by spraying or irrigation), aeration (by tilling the soils), pH adjustment (buffered near neutral pH by adding agricultural lime), and bulking agents can be added initially and throughout the treatment period to optimize condition for microbial growth. Contaminated media is usually treated in lifts that are up to 50 cm thick (Figure 16.5). When the desired level of treatment is achieved, the lift is removed and a new lift is constructed. It may be desirable to only remove the top of the remediated lift, and then construct the new lift by adding more contaminated media to the remaining material and mixing. This serves to inoculate the freshly added material with an actively degrading microbial culture, and can reduce treatment times.

Biopile is a bioremediation technology in which excavated soils are mixed with soil amendments and placed in aboveground enclosures that include an impermeable liner to minimize the risk of contaminants spreading and leachate collection system. Biopile is an aerated static pile composting process in which compost is formed into piles and aerated with blowers or vacuum pumps. Enclosed operations improve the control of moisture from rainfall, evaporation, volatilization, dust emission, and also promote solar heating. Moisture, heat, nutrients, oxygen, and pH can be controlled to enhance biodegradation. Soil piles are constructed up to 6 m high (Cookson, 1995); however, a height of 2–3 m is generally applied for optimum biodegradation process. The aeration system consists of a series of perforated pipes running underneath each pile and connected to blowers that draw air through the pile.

Composting is a controlled biological process by which organic contaminants (e.g., PAHs) are converted by microorganisms (under aerobic and anaerobic conditions) to innocuous, stabilized by-products. Typically, thermophilic conditions (54–65°C) must be maintained to properly compost soil contaminated with hazardous organic contaminants. The increased temperatures result from heat produced by microorganisms during the degradation of the organic material in the waste. In most cases, this is achieved by the use of indigenous microorganisms. Soils are excavated and mixed with bulking agents and organic amendments, such as wood chips, animal, and vegetative wastes, to enhance the porosity of the mixture that is to be decomposed. Maximum degradation efficiency is achieved through maintaining oxygenation (e.g., daily windrow turning), irrigation as necessary, and closely monitoring moisture content and temperature.

Slurry Phase: Slurry phase bioremediation treatment involves the controlled treatment of excavated soil in a bioreactor (lagoons, open vessels, and closed system). Aqueous slurry is created by

FIGURE 16.5 Typical of a permanent bioremediation treatment facility in oil and gas industry to treat crude oil and/or oil-sludge-contaminated soil. (Author's project documentation with permission from LAPI-ITB consultant, Indonesia.)

combining soil, sediment, or sludge with water, nutrients, and other additives. The contaminated soil is excavated and removed from the site as completely as possible. The slurry is mixed to keep solids suspended and microorganisms in contact with the soil contaminants. Cleanup crews use this bioreactor to mix the contaminants and the microorganisms. This mixing process keeps the microorganisms in constant contact with the contaminants. Since the cleanup operators have complete control of the conditions in the bioreactor, they can adjust things until they achieve the optimal conditions for the degradation of the contaminants. Since the degradation can be kept at or very close to optimal conditions, it does not take very long to break down the contaminants.

16.3 DESIGNING A BIOREMEDIATION PROJECT

This chapter discusses and focuses on the most popular engineered ex situ bioremediation technology. Land farming has been proven to be most successful in treating organic contaminants including petroleum hydrocarbon (crude oil and its refinery products), oily sludge, pesticides, coke waste, and wood-preserving waste (creosote and PCP). As a rule of thumb, the higher the molecular weight (and the more rings with a PAH), the slower the degradation rate. Also, the more chlorinated or nitrated the compound, the more difficult it is to degrade.

Most bioremediation projects are performed in steps/phases, but these phases often overlap and one impacts on the other. A bioremediation project can be grouped in the following steps (Cookson, 1995):

1. Performing site and soil characterization
2. Selecting remediation method and technology
3. Treatability studies
4. Performing field remediation activities
5. Obtaining site closure

16.3.1 SITE AND SOIL CHARACTERIZATION

A bioremediation process must be well managed for success. A well-managed process requires control, and this is achieved only when site evaluations and characterization have been properly performed. Site characterization must define the potential biological system, the site characteristics that impact on biological reactions, and provide information for process control. Site characterization must provide adequate data for designing a facility that provides control of the degradation process. This includes the delivery of necessary chemicals and nutrients at appropriate rates throughout the treatment area. The applicability of bioremediation is established after considering two aspects: contaminants characteristics and physicochemical environmental characteristics.

16.3.1.1 Contaminants Characteristics

Prior to designing an engineered bioremediation system, the feasibility of biodegradation should be carefully evaluated. This evaluation should include the ease or difficulty of degrading the target contaminants, the ability to achieve total mineralization, and the environmental conditions necessary to implement the process. Pollutant toxicity is often used as justification for a bioremediation process because this toxicity could inhibit the degradative activity of indigenous microorganisms. Although few sites with obvious toxicity have been reported, the sites that have been described are of clear potential for bioremediation if the added microorganisms can also resist the toxicity (Grasset and Vogel, 1995). There are various factors that should be incorporated into this evaluation process.

- Biodegradability of contaminants: Years of experience and research has established the degradation pathways of many specific contaminants. Contaminant characteristics and structure also will provide answers in terms of biodegradability. Table 16.4 describes compilation of biodegradability and transformability of some contaminants (ICSS, 2006).

- Mineralization potential of the compounds: A review of pertinent reaction pathways will provide insight as to whether the contaminant will be utilized as a primary substrate or whether cometabolic reactions are necessary.
- Specific endogenous microbial, substrate, and other conditions: Of prime importance is the availability of carbon and energy in the contaminated environment. Electron-acceptor availability and the redox condition should be carefully determined. In addition, the

TABLE 16.4
Biodegradability and Preferential Degradation Conditions of Some Contaminants

Contaminants	Biodegradability	Preferred Conditions
Mineral oil hydrocarbon		
Short chain	Easy	Aerobic
Long chain	Moderately easy	Aerobic
Cycloalkanes	Moderately easy	Aerobic
Monoaromatic hydrocarbons		
AHs	Easy	Aerobic
Phenols	Easy	Aerobic
Cresols	Easy	Aerobic
Catechols	Easy	Aerobic
PAHs		
2- to 3-ring PAHs	Moderately easy	Aerobic
4- to 6-ring PAHs	Difficult	Aerobic
Chlorinated aliphatic hydrocarbons		
Tetrachloroethylene, trichloroethane	Moderate	Anaerobic
Trichloroethylene, dichloroethane	Moderate	Aerobic/anaerobic
Dichloroethylene, vinyl chloride	Moderate	Aerobic
Chlorinated AHs		
Cl-phenols (superchlorinated)	Difficult	Anaerobic
Cl-phenols (low-chlorinated)	Moderate	Aerobic/anaerobic
Cl-benzenes (superchlorinated)	Difficult	Aerobic/anaerobic
Cl-benzenes (low-chlorinated)	Moderate	Aerobic/anaerobic
PCBs (superchlorinated)	Difficult	Anaerobic
PCBs (low-chlorinated)	Moderate	Aerobic/anaerobic
Nitroaromatic compounds		
Mono- and dinitroaromatics	Moderate	Aerobic/anaerobic
Trinitrotoluene	Moderate	Aerobic/anaerobic
Trinitrophenol	Difficult	Aerobic/anaerobic
Pesticides		
g-Hexachlorocyclohexane	Moderate	Aerobic/anaerobic
Atrazins	Moderate	Aerobic
Dioxins, 2, 3, 7, 8-PCDD/PCDF	Very difficult	Anaerobic
Inorganic compounds		
Free cyanide	Difficult	Aerobic
Complex cyanide	Difficult	
Ammonium	Easy	Aerobic/anaerobic
Nitrate	Easy	Anaerobic
Heavy metals[a]	Difficult	
Radioisotopes[a]	Difficult	

[a] Biotransformable, yet not degradable.

presence of microorganisms capable of degrading the contaminants, in sufficient numbers, should be evaluated. Total plate counts, specific microbial/degraders counts, and laboratory and in situ respiration tests can be utilized to perform this evaluation.

Contaminants biodegradability is associated with many factors including being related to compound structure and its physicochemical characteristics such as solubility and bioavailability (which itself is not intrinsic to the compound, but related to the interactions between the compound, the microorganisms, and the soil). Reports about the limited use of bioaugmentation relative to biostimulation often study compounds that are either known for their nonavailability (e.g., low concentrations of polycyclic aromatic hydrocarbons/PAHs) (Vogel, 1996) or relatively easy to degrade when other limiting conditions (e.g., nutrients) are provided such as petroleum hydrocarbons (McGugan et al., 1995; Neralla et al., 1995) or crude oil (Venosa et al., 1995).

16.3.1.2 Physicochemical Environmental Characteristics

Environmental characteristics play an important role in determining biological activity, whether of indigenous microorganisms, added microorganisms, or cultured indigenous microorganisms returned to the soil. These conditions fall into two general categories: those that reduce the microbial activity, such as temperature, water-holding capacity/humidity, available oxygen, redox potential, ionic strength, and those that restrict the mass transfer of the compound to the microorganism, such as soil type, particle size, permeability, and organic-matter content (Margesin and Schinner, 2001).

- Availability of nutrients: In general, the concentration levels of only N and P are determined. Nutrient content has an effect on microbial community composition and often limit degradation efficiency (Leckie et al., 2004; Welander, 2005).
- Bioavailability is the key issue in bacterial degradation. Thus, a bioavailable organic compound is accessible for microbial uptake. Organic compounds are normally in a soluble form for substrate uptake while sorption onto soil particles creates a barrier for degradation. Some specific bacterial species adapt to interact with humic substance-sorbed compounds, though this ability is considered rare (Vacca et al., 2005). Degradation may then take place even if the organic compound is bound to humic substances. Some soil type binds organic compounds tightly, thus making contaminants less bioavailable for microbial degradation (Theng et al., 2001). This means that soil may immobilize contaminants due to strong sorption especially if the contaminants are hydrophobic (Reichenberg et al., 2010; Yang et al., 2010). Because organic compounds have a strong affinity to organic matter naturally occurring in the soil, bioavailability of contaminants is heavily influenced by soil organic matter.
- Site's hydrogeologic characteristics: Hydraulic conductivity, thickness of the saturated zone, homogeneity, and depth to the water table are parameters that should be factored into the design of the system. Distribution and transport of added nutrients and electron acceptors also leachate control of the bioremediation cell, which will be influenced by the site hydrogeology data.
- Extent and distribution of contaminants: Successful bioremediation requires best available historical knowledge of the site, and this knowledge should always be backed by thorough pre-analyses of the soil. Contaminants are most likely spread unevenly in the soil, which makes remediation work even more challenging (Muckian et al., 2007).

16.3.2 SELECTING REMEDIATION METHOD AND TECHNOLOGY

Characterization and evaluation of the site are necessary to select an appropriate remedy and identify the source and nature of the contaminants. The process to determine whether bioremediation

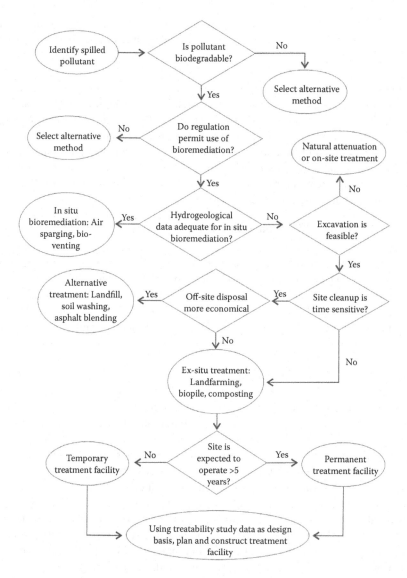

FIGURE 16.6 Bioremediation treatment selection decision tree.

may be feasible for particular contaminated sites is illustrated in Figure 16.6. Decisions to use bioremediation should be made after applicable regulatory policies, potential environmental impacts, operational feasibility, logistical coordination, and other pertinent issues have been evaluated.

16.3.3 TREATABILITY STUDIES

Before a bioremediation project is attempted at field scale, treatability studies of contaminated soil and its contaminants may need to be performed (Whyte et al., 2001; Kim et al., 2005; Lamichhane et al., 2012). The general goal of treatability studies is to ensure that the planned treatment has a high success rate. The number and type of treatability studies depend on the characteristics of the site and the contaminants. The level of effort that needs to be expected on treatability studies is proportional to the uncertainty about site and contaminant characteristics; low uncertainty means little effort may ensure success whereas high uncertainty may need extensive effort to ensure success of the treatment applied (Rittmann and McCarty, 2001). Treatability studies consist of laboratory

FIGURE 16.7   Treatability study of oil-contaminated soil: flask and laboratory microcosm (above) and pilot demonstration (below). (Author's project documentation.)

evaluation, pilot studies, and field demonstrations (Figure 16.7). Laboratory evaluations yield information for selecting the best treatment program. Pilot studies yield information to the expected performance, process control parameters, and design criteria. After the biodegradability of the contaminants is established, treatability studies can be used to evaluate the environmental conditions necessary for efficient degradation of the target compounds.

Treatability tests can be done by various methods, from the simplest way to a larger system such as a pilot plant. Treatability studies can be done through three phases. Phase I determines whether the contaminated soil and the contaminants can be treated biologically. Phase II is intended to prepare the design criteria for the application of bioremediation at the field scale, while phase III is the stage of pilot-scale bioremediation applications. For the in situ bioremediation technique, the characteristic of the field will determine the application of bioremediation processes.

16.3.3.1   Phase I: Biodegradability and Toxicity Assessment

In phase I, characterization activities include the following:

- Chemical analysis of contaminants and contamination levels that occur, such as concentrations of contaminants and the composition of the contaminants (fingerprint analysis).
- Determining the composition and natural microorganism populations (indigenous bugs) that exist in the contaminated soil.

- Testing to predict the toxicity of the material that may interfere with the process of biodegradation as TCLP analysis and LD_{50}.
- Determining the availability of nutrients in the contaminated soil, such as nitrate (NO_3) and phosphate (PO_4).
- Determining the physical characteristics of contaminated soil, such as permeability, soil texture analysis, desorption, acidity, and the hydraulic conductivity.

16.3.3.2 Phase II: Preliminary Confirmation

Phase II includes tests on the following:

- Abiotic desorption to determine the solubility of contaminants.
- Biodegradation of contaminated material to determine the ability of endogenous microorganisms to decompose contaminants. This test will determine whether biostimulation or bioaugmentation approach that will be applied best in the field.
- Reaction kinetics that will be applied in the field whether it is the soil column, slurry reactor, or "pan" as the simulated treatment method.

16.3.3.3 Phase III: Pilot-Scale/Field Trial

In phase III, field pilot trials are conducted to determine the success rate of bioremediation projects based on the results of biodegradability and toxicity tests that have been carried out in the laboratory as well as preliminary confirmation assessment before all process parameters was applied in a full-field scale. Evaluation of the biological processes is determined from a decrease in the concentration of contaminants in the treated soil and an increase in the growth of microorganisms in line with a decrease in the concentration of contaminants. This suggests the process of degradation of contaminants caused by the activity of microorganisms.

Several factors that should be taken into consideration in the phase III test are as follows:

- Test is normally performed on a small portion of the contaminated site.
- The pilot location should be representative of the overall site in terms of hydrogeology and contamination.
- Condition for the pilot test should simulate the proposed bioremediation treatment method.
- The pilot site must be large enough that realistic complications are encountered.
- Simulation modeling of the transport and microbiological aspect of the bioremediation is recommended as part of the design and evaluation of the pilot study.

16.3.4 Field Remediation Activities

After the site characterization and treatability study was conducted, a field-scale bioremediation treatment process (land farming) was designed with the following considerations:

1. Contaminants degradation can be done neither by using native microorganisms nor by specific augmented microorganisms.
2. Do not place the treatment facility/bioremediation cells near residential areas so as to avoid noise, odor, health problems, and aesthetic possibilities.
3. Bioremediation cells equipped with a collection system for water and drainage.
4. The addition of a "bulking agent" can improve soil texture and is more economical.
5. "Runoff" and leachate collected from the bioremediation cells can be used as an irrigation-containing fertilizer on bioremediation cells in the dry season.

Guidelines/criteria for site selection of contaminants processing facility with land farming and its planning layout of the location are shown in Table 16.5 and Figure 16.8.

TABLE 16.5

Guideline/Criteria for Site Selection of Bioremediation Facility

Parameters	Definition	Recommendation
Geography	Treatment site	1. Choose a location with good drainage and is outside the flood area. 2. Choose a location with a long distance (>500 m) from the nearest village/residential. 3. Sites with adequate facilities and infrastructure are preferred.
Accessibility	Access to and exit from the bioremediation treatment facility	1. Access road to the treatment facility equips with minimum compacted gravel, asphalt, or concrete. 2. Clay-coated roads are acceptable if the treatment facility is located in remote area. 3. Make sure the tonnage capacity of the bridge meets all the heavy equipment used. 4. Treatment facility close to the contaminated soil is preferred, make sure accessibility to and from treatment facility throughout the project.
Area requirement	Area required in the bioremediation treatment facility	Including access to main roads, soil-staging/storage areas, contaminated soil processing areas, buffer zone, water storage, nutrition, microbial seed tank.
Utility	Electricity and water availability	Ensure availability of electricity and water in the treatment facility location.
Logistic of contaminated soil	Transportation, handling, and contaminated soil storage	1. Trucks transporting contaminated soil requiring access route and maneuvering space within the treatment location. 2. Ensure the availability of handling equipment for the transfer, mixing, and scarify the polluted soil. 3. Treatment area should be covered by the coating/impermeable liner.
Security	Access control to the treatment facility	1. Treatment facility fenced with access door. 2. Posting warning sign and contact addresses. 3. Lock all equipment when not in use.

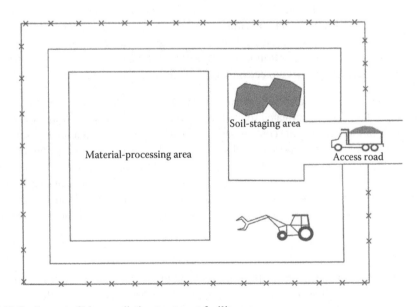

FIGURE 16.8 Layout of bioremediation treatment facility area.

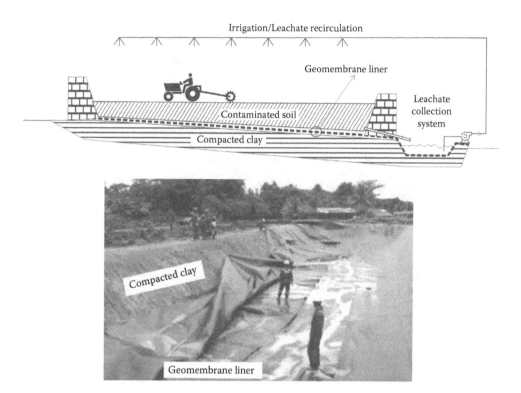

FIGURE 16.9 Design of land farming bioremediation cells (above) and illustration of geomembrane layer installation (below). (Author's project documentation with permission from LAPI-ITB consultant.)

Design of treatment area or bioremediation cells (Figure 16.9) can be described as follows:

- Treatment area is a stretch that is made specifically with slope 2–3% to allow the collection of leachate or runoff from treatment area.
- To protect the ground water, treatment area must be coated by compacted clay layer or impermeable geotextile liners and thus serves as a waterproof coating.
- In anticipation of heavy rainfall, the drainage system is provided around the treatment area.
- Provide a collection pit to accommodate the leachate and runoff water.
- To keep the contaminated soil moisture, water it periodically using a vacuum truck with sprayers or portable pump connected to a plastic hose as a water sprayer.

Bioremediation of contaminated soil in land farming is generally divided into a three-phase operation: (1) Haul in/fill in the contaminated soil in the treatment cell. The operation starts from the preparation of cell conditions, loading the contaminated soil in the stock pile, contaminated soil preparation before being transferred to the processing cells, as well as the spreading and distribution of contaminated soil into the processing cell. (2) The bioremediation and monitoring process. A routine operation in the treatment of contaminated soil such as mixing, fertilizing, watering, maintenance and monitoring (sampling), as well as quality control process. (3) Haul out/taking out of treated soil. Example guideline/procedure for bioremediation of petroleum oil-contaminated soil process is illustrated in Figure 16.10.

16.3.5 SITE CLOSURE

All hazardous waste management facilities must eventually cease their treatment, storage, or disposal activities. When such operations cease, the owner and operator must close the facility in a way that ensures it will not pose a future threat to human health and the environment. Site closure is a

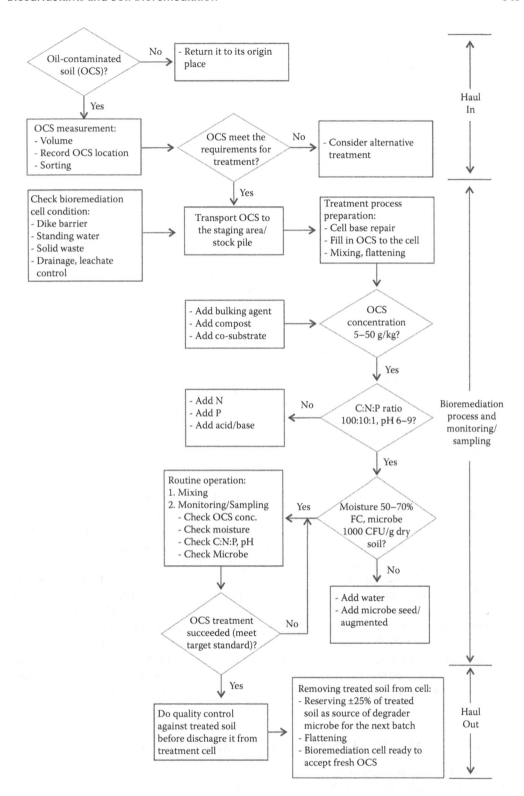

FIGURE 16.10 Guideline/procedure for bioremediation of petroleum oil-contaminated soil process.

TABLE 16.6
End-Point Criteria for Bioremediation of Petroleum Oil-Contaminated Soil in Indonesia

Parameters	Unit	Value
Hydrocarbons		
TPH	μg/g	10,000
Benzene	μg/g	1
Toluene	μg/g	10
Ethylbenzene	μg/g	10
Xylene	μg/g	10
Total PAHs	μg/g	10
TPLC		
Pb	ppm	5
As	ppm	5
Ba	ppm	150
Cd	ppm	1
Cr	ppm	5
Cu	ppm	10
Hg	ppm	0.2
Se	ppm	1
Zn	ppm	50
Toxicology test		
LD50	g/kg BW	>15

Source: Adapted from Indonesia EPA. 2003. Technical Procedure and Requirement for
           Petroleum Oil Contaminated Soil Treatment by the Biological Process. I-EPA
           Decree No. 128/2003. Ministry of Environmental Office, Jakarta, Indonesia.

milestone achieved when the remaining contamination in the soil meets a risk or cleanup threshold determined not to pose a threat to human health or the environment. Determining the end point of a corrective action at a contaminated soil/site may involve reaching a targeted concentration for certain contaminants or reducing the risk of contamination to a specific threshold. Depending on the jurisdiction overseeing the remedial activity, completion of corrective action at a contaminated site may be based on the remaining and foreseeable risk to human health and the environment (U.S. EPA, 2005). Generally, two types of closure are allowed: closure by removal or decontamination, referred to as clean closure, and closure with the waste in place. If all hazardous waste and contaminants, including contaminated soils and equipment, can be removed from the site or unit at closure, the site or unit can be clean closed and post-closure care is not required. Closure is the final step in the assessment, cleanup, and monitoring process. It provides a mechanism where the department/operator/site owner can allow assessment, cleanup, and monitoring to stop if certain conditions are met. An example of end-point criterion requirements for bioremediation of petroleum oil-contaminated soil in Indonesia are shown in Table 16.6.

16.4   BIOSURFACTANTS-ENHANCED BIOREMEDIATION PROCESS

Surfactants of microbial origin, referred to as biosurfactants, have potential use in several industries such as cosmetics (Brown, 1991), pharmaceutical (Singh and Cameotra, 2004), food and beverage (Nitschke and Costa, 2007), textile (Kesting et al., 1996), soap/detergent (Mukherjee, 2007), petrochemical (Banat et al., 2000), mining and manufacturing processes (Zouboulis et al., 2003) and also potential for applications in crude oil recovery and bioremediation of contaminated sites

(Finnerty, 1994; Pekdemir et al., 1999; Christofi and Ivshina, 2002). Biosurfactants are produced on microbial cell surfaces or excreted extracellularly and contain both hydrophilic and hydrophobic moieties (Makkar and Cameotra, 2002). Biosurfactants have several advantages over the chemical surfactant, such as higher biodegradability, lower toxicity, better environmental compatibility, high selectivity and specific activity at extreme temperature, pH, salinity, and the ability to be synthesized from renewable feedstock (Kosaric, 1992; Desai and Banat, 1997). Biosurfactants possess the characteristic property of reducing the surface and interfacial tensions using similar mechanisms as that of chemical surfactants. Microbial surfactants constitute a diverse group of surface-active molecules and are known to occur in a variety of chemical structures, such as glycolipids, lipopeptides, fatty acids, neutral lipids, phospholipids, polymeric, and particulate structures. Several biosurfactants have high surface-active compounds and promising substitutes for synthetic surfactant. Biosurfactants are particularly suited for environmental application due to their biodegradability (Huszcza and Burczyk, 2003). Compared to chemically manufactured surfactants, biosurfactants also offer several advantages of little or no environmental impact, the possibility of in situ production, and produced from renewable and cheaper substrates (Rashedi et al., 2006; Helmy et al., 2011).

The principal mediators of bioremediation are degrader microbe and their products as they transform or mineralize pollutants, thereby decreasing their masses and toxicities in contrast to most other components of the environment. The success of bioremediation is governed by three important factors: availability of microbes, accessibility of contaminants, and a conducive environment. The efficiency of bioremediation is dependent on the microbial ability to degrade these complex mixtures and their rate-limiting kinetics (Cameotra and Makkar, 2010). Bioremediation of contaminants in the environment is motivated by their ubiquitous distribution, their low bioavailability and high persistence in soil, and their potentially deleterious effect on human health. Due to high hydrophobicity and solid–water distribution ratios, these contaminants tend to interact with nonaqueous phases and soil organic matter and, as a consequence, become potentially unavailable for microbial degradation since bacteria are known to degrade chemicals only when they are soluble in water. Conversion of the chemicals during bioremediation by the microbial cells is governed by the rate of uptake and metabolism (the intrinsic activity of the cell) and the rate of transfer to the cell (mass transfer). These factors regulate the so-called bioavailability of a chemical. For example, the hydrocarbons are hydrophobic compounds with low water solubility, thus, microorganisms have developed several mechanisms to increase the bioavailability of these compounds in order to utilize them as potential carbon and energy source. One of the main reasons for the prolonged persistence of hydrophobic hydrocarbons in contaminated environments is their strong adsorption even on coarse-grained and organic free soils by microporosity, so that they are no longer available for hydrocarbon degrading microorganisms and remain even after bioremediation. Hence, for efficient and complete biodegradation, solubilization of these hydrocarbons with biosurfactants prior to bioaugmentation is advantageous. Moreover, use of biosurfactant-producing hydrocarbon degrading microorganisms for bioaugmentation to enhance hydrocarbon degradation offer the advantage of a continuous supply of a nontoxic and biodegradable surfactant at a low cost (Moran et al., 2000; Rahman et al., 2003).

Various research, and reports on biosurfactant-enhanced biodegradation of hydrocarbons in soils have been thoroughly reviewed (Kosaric, 2001; Mulligan et al., 2001; Thompson et al., 2005; Lebeau et al., 2008; Rahman and Gakpe, 2008; Mulligan, 2009; Cameotra and Makkar, 2010; Mukherjee and Das, 2010; Megharaj et al., 2011; Bustamante et al., 2012; Ławniczak et al., 2013), which justify the use of biosurfactants and biosurfactants-producing organisms for bioremediation of hydrocarbons. In the last decades, the research focus has been on bio-/surfactant-mediated bioremediation. The increasing interest is attributable to the fact that surfactants can enhance the solubilization of pollutants from contaminated soil and increase their solubility, which in turn improves their bioavailability. Many bio-/surfactants of different kinds have been so far investigated for their possible applications in facilitating the biodegradation of organic and inorganic contaminants such as petroleum oil and its products, PAHs, pesticides, and heavy metals. The various studies and findings by researchers regarding biosurfactant-enhanced biodegradation of pollutants in the last decade are summarized in Table 16.7.

TABLE 16.7

An Overview of Various Studies on Biosurfactant-Enhanced Bioremediation Process

Contaminants	Agent/Degrader Microbe	Biosurfactants	Results	Reference
			Petroleum Hydrocarbon	
Diesel oil and biodiesel	Microbial consortium	Rhamnolipids	The addition of rhamnolipids into samples containing either diesel or biodiesel fuel did not enhance the biodegradation efficiency of petroleum hydrocarbons. However, rhamnolipids were readily degraded by a soil-isolated consortium of hydrocarbon degraders in all samples, under both aerobic and nitrate-reducing conditions.	Chrzanowski et al. (2012)
Crude oil	*Pseudomonas* sp. BP10 and *Rhodococcus* sp. NJ2	Glycolipids	*Pseudomonas* sp. BP10 degraded 60.6% of TPH (total petroleum hydrocarbons) while *Rhodococcus* sp. NJ2 metabolized only 49.5% after 30 days of incubation in MSM (minimal salt media) with 2% of crude oil at their optimum conditions.	Kumari et al. (2012)
Crude oil	*Lactobacillus delbrueckii*	Glykolipid from *L. delbrueckii*	Addition of 0.1% biosurfactants significantly enhanced crude oil biodegradation by 75% compared to control (54,75%, without BS).	Thavasi et al. (2011)
Hexadecane	*Candida tropicalis*	Rhamnolipids	Removal at 93% after 4 days compared to 78% without biosurfactants.	Zeng et al. (2011)
Oil-based drill cuttings	Mixed culture	Rhamnolipids	Approximately 83% of organics were removed after washing under optimal conditions (liquid/solid ratio, 3:1; washing time, 20 min; stirring speed, 200 rpm; rhamnolipid concentration, 360 mg/L; temperature, 60°C), and the total petroleum hydrocarbon concentration of the cuttings dropped from 85,000–12,600 mg/kg. In the bioremediation stage, concentrations of saturated and aromatic hydrocarbons decreased to 2140 and 1290 mg/kg, respectively, after 120 days.	Yan et al. (2011)
Hydrocarbons and crude oil	Indigenous soil microflora	Sophorolipid	30% of 2-methylnaphthalene was effectively washed and solubilized with 10 g/L of sophorolipid with similar or higher efficiency than that of commercial surfactants. Addition of sophorolipid in soil increased biodegradation of model compounds: 2-methylnaphthalene (95% degradation in 2 days), hexadecane (97%, 6 days), and pristane (85%, 6 days). Also, effective biodegradation method of crude oil in soil was observed by the addition of sophorolipid, resulting in 80% biodegradation of saturates and 72% aromatics in 8 weeks.	Kang et al. (2010)
Oil spill	Indigenous marine microflora	Biosurfactant from *Gordonia* sp. strain JE-1058	The addition of JE1058BS (biosurfactant from *Gordonia* sp. strain JE-1058) to seawater stimulated the degradation of weathered crude oil (ANS 521) via the activity of the indigenous marine bacteria. Over the 28-day period, the alkane and aromatic hydrocarbon fractions in the JE1058BS-treated flasks were reduced by 93% and 39%, respectively, while those in the nutrient-treated flasks were reduced by 67% and 9%, respectively.	Saeki et al. (2009)

Diesel oil and biodiesel	Microbial consortium	Rhamnolipids	Removal at 77% after 7 days compared to 58% without biosurfactants for blends.	Owsianiak et al. (2009)
Oil sludge	*Pseudomonas aeruginosa* and *Rhodococcus* sp.	Crude biosurfactant (not specified)	Removal at 98% after 8 weeks was observed upon the addition of biosurfactants and nutrients compared to the inoculation with the mixed culture without any additives (52%).	Cameotra and Singh (2008)
Petroleum hydrocarbon	Indigenous soil microflora	Rhamnolipids	Removal at 85% after 20 days for TPH with addition of 1 mg biosurfactant/g soil.	Benincasa (2007)
Diesel oil	Indigenous soil microflora	Surfactin	Addition of 40 mg/L of surfactin significantly enhanced biomass growth (2500 mg VSS/L) as well as increased diesel removal (94%), compared to batch experiments with no surfactin addition (1000 mg VSS/L and 40% removal). Addition of surfactin more than 40 mg/L, however, decreased both biomass growth and diesel biodegradation efficiency, with a worse diesel biodegradation percentage (0%) at 400 mg/L of SF addition.	Whang et al. (2007)
Diesel oil	Indigenous soil microflora	Rhamnolipids	Addition of rhamnolipid to diesel/water systems from 0 to 80 mg/L substantially increased biomass growth and diesel biodegradation percentage from 1000 to 2500 mg VSS/L and 40–100%, respectively.	Whang et al. (2007)
Crude oil	Indigenous marine microflora	Rhamnolipids	Removal up to 25% for alkanes after 5 days with biosurfactant alone and 59% when used with nutrients.	McKew et al. (2007)
Crude oil	Microbial consortium	Rhamnolipids	The addition of 500 mg/L rhamnolipids accelerates the biodegradation of total petroleum hydrocarbons from 32% to 61% at 10 days of incubation.	Abalos et al. (2004)
Kerosene	Not specified	Mannosylerythritol lipids (MELs)	Degradation rate of kerosene by addition of MEL and BS-UC reached 87 and 90% at 15 h.	Hua et al. (2004)

Mono-AH

BTEX	*Alcaligenes piechaudii* and *Ralstonia picketti*	3-Hydroxy fatty acids	*Alcaligenes piechaudii* removed 96% and 97% of toluene and m+p-xylenes, respectively, after 2 days of the incubation but it took 30 days to degrade 59% benzene while *Ralstonia picketti* removed toluene and m+p-xylenes slower than *A. piechaudii* but degraded almost 100% of benzene, after 5 days of the incubation.	Plaza et al. (2007)

PAHs

Phenanthrene	*Glomus etunicatum*	Lipopeptides (*Bacillus subtilis* BS1)	Mycorrhizal colonization or BS1 inoculation improved the tolerance to stress of phenanthrene and increased the plant biomass. Biosurfactant secreted by BS1 strain considerably enhanced the solubility of phenanthrene, favoring its enrichment in rhizosphere soil and plant roots. The co-inoculation of BS1 and *G. etunicatum* significantly decreased the residual concentrations of phenanthrene in soil, and resulted in higher soil enzyme activities of catalase and polyphenol oxidase.	Xiao et al. (2012)

continued

TABLE 16.7 (continued)
An Overview of Various Studies on Biosurfactant-Enhanced Bioremediation Process

Contaminants	Agent/Degrader Microbe	Biosurfactants	Results	Reference
PAHs				
Phenanthrene	Mixed culture	Biosurfactants from *Acinetobacter calcoaceticus*	About 78.7% of phenanthrene was degraded in 30 days at 25°C; and addition of biosurfactant did not affect the biodegradation. However, addition of the biosurfactant or inoculation of *A. calcoaceticus* BU03 at 55°C significantly enhanced the biodegradation by increasing the K desorption.	Zhao et al. (2011)
Phenanthrene Phenanthrene	*Sphingomonas* sp. *Brevibacillus parabrevis* PDM-3	Rhamnolipids Glycolipids	Removal at 99% after 10 days compared to 84% without biosurfactant. 93% of phenanthrene was degraded in 6 days.	Pei et al. (2010) Reddy et al. (2010)
PAHs	Activated sludge microflora	Rhamnolipids	At a rhamnolipids/RD of 15 mg/L aerobic treatment for 25 days SRT was enough to remove over 90% of the total PAHs, 88% of the COD originating from the inert organics (CODinert) and 93% of the COD originating from the inert soluble microbial products (CODimp). At this SRT and RD concentration, about 96–98% of the RD was biodegraded by the AASR system, 1.2–1.4% was accumulated in the system, 1.1–1.3% was released in the effluent, and 1.2–1.4% remained in the waste sludge.	Sponza and Gok, (2011)
PAHs	Alfalfa + arbuscular mycorrhizal fungi + microbial consortium of PAH degraders	Rhamnolipids	Removal at 61% after 90 days compared to 17% with only phytoremediation.	Zhang et al. (2010)
Benzo [a] Pyrene, Phenantrene	*Bacillus subtilis* B-UM.	Biosurfactants from *Acinetobacter calcoaceticus*	Within 42 days of composting period, the degradation of PHE and B[a]P in the absence of the biosurfactants was 71.2% and 16.4%, respectively. Therefore, inoculation of *A. calcoaceticus* BU03 or biosurfactants from BU03 together with inoculation of *B. subtilis* B-UM increased the degradation of B[a]P to 83.8% and 65.1%, respectively, while PHE was almost completely removed with these two treatments.	Wong et al. (2010)
Pyrene	*Pseudomonas fluorescens*	Rhamnolipids	Removal at 98% after 10 days compared to 91% without emulsan.	Husain (2008)
Anthracene	*Sphingomonas* sp. and *Pseudomonas* sp.	Rhamnolipids	Removal at 52% after 18 days compared to 32% without biosurfactant for *Pseudomonas*.	Cui et al. (2008)

Contaminant	Microorganism	Biosurfactant	Description	Reference
Anthracene	Pseudomonas sp.	Rhamnolipids	Removal at 72% after 48 days	Santos et al. (2008)
Fluoranthene	Pseudomonas alcaligenes PA-10	Rhamnolipids	Inoculation of fluoranthene-contaminated soil microcosms with P. alcaligenes PA-10 resulted in the removal of significant amounts (45 ± 5%) of the PAH after 28 days compared to an uninoculated control. Addition of 0.5 g/L biosurfactant increased the initial rate of fluoranthene degradation to 96.61 ± 0.81%.	Hickey et al. (2007)
Chlorinated Ahs				
PCBs	Mixed culture	Rhamnolipids	A 42-d combined chemical–biological treatment supplemented by biosurfactant addition resulted in an average 47–50% of polychlorinated biphenyls removal—independent of the hydrogen peroxide carrier used.	Viisimaa et al. (2013)
4-Chlorophenol	Activated sludge microflora	Biosurfactant (not specified)	Complete removal of 4-CP in the reactor which 2× critical micelle concentration (CMC) eventuated on the end of the 1st day compared to 14 days in the control reactor.	Uysal and Turkman (2007)
PCBs (2, 3, and 4 Cl)	Mixed culture	Rhamnolipids	Complete removal of 2 Cl, 78% of 3 Cl, and 12% of 4 Cl in reactor with 200 mg/L BS addition compared to100% (2 Cl), 43% (3 Cl), 11% (4 Cl) in control reactor.	Fiebig et al. (1997)
Nitroaromatic Compounds				
Carbazole	Pseudomonas sp. strain GBS.5	Rhamnolipids	The specific activity of carbazole degradation was found to be 11.36 mmol/min/g dry cells. Study also confirmed that in addition to carbazole, bacterium has the ability to degrade other polycyclic aromatic hydrocarbons such as fluoranthene, fluorene, naphthalene, phenanthrene, and pyrene.	Singh et al. (2013)
Pesticides				
Beta-cypermethrin	Pseudomonas aeruginosa CH7	Rhamnolipids	About 90% of the beta-cypermethrin could be degraded by Pseudomonas aeruginosa CH7 within 12 days.	Zhang et al. (2011)
Heavy Metals				
Cadmium, lead	Pseudomonas sp. LKS06	Rhamnolipids	The maximum biosorption capacity of Pseudomonas sp. LKS06 biomass for Cd(II) and Pb(II) was found to be 27.5 and 77.8 mg/g, respectively, at the optimum pH of 6.0.	Huang and Liu (2013)
Cadmium, lead	–	Rhamnolipids	Washing of contaminated soil with tap water revealed that 2.7% of Cd and 9.8% of Pb in contaminated soil was in freely available or weakly bound forms, whereas washing with rhamnolipid removed 92% of Cd and 88% of Pb after 36 h of leaching.	Juwarkar et al. (2007)
Copper, zinc, and nickel	–	Rhamnolipids	The removal of heavy metals from sediments was up to 37% of Cu, 13% of Zn, and 27% of Ni when rhamnolipid without additives was applied. Adding 1% NaOH to 0.5% rhamnolipid improved the removal of copper by up to 4 times compared with 0.5% rhamnolipid alone.	Dahrazma and Mulligan (2007)

16.4.1 ROLE OF BIOSURFACTANTS IN SOIL BIOREMEDIATION

The main issue which directly influences the efficiency of biological treatment is the bioavailability of the pollutant. Possible sorption of molecules into the soil matrix, formation of nonaqueous
phases, interactions with organic matter, biotransformation, and contaminant aging—these naturally occurring processes often result in limited bioavailability, thus decreasing the efficiency of
bioremediation. There are at least two ways in which biosurfactants are involved in bioremediation:
increasing the surface area of low to insoluble hydrophobic substrates and increasing the bioavailability of hydrophobic compounds. Biodegradation of hydrocarbons in soil can also be efficiently
enhanced by the addition or in situ production of biosurfactants. It was generally observed that the
degradation time, and particularly the adaptation time, for microbes was shortened (Kosaric, 2001;
Ławniczak et al., 2013).

16.4.1.1 Increasing the Surface Area of Poorly Water-Soluble Hydrophobic Compounds

The concentration of cells in an open system, such as an oil-polluted body of water, never reaches
a high enough value to effectively emulsify oil. Furthermore, any emulsified oil would disperse in
the water and not be more available to the emulsifier-producing strain than to competing microorganisms. One way to reconcile the existing data with these theoretical considerations is to
suggest that the emulsifying agents do play a natural role in oil degradation, but not in producing
macroscopic emulsions in the bulk liquid. If emulsion occurs at, or very close to, the cell surface
and no mixing occurs at the microscopic level, then each cluster of cells creates its own micro
environment and no overall cell-density dependence would be expected (Ron and Rosenberg,
2002). As amphiphiles, biosurfactants exhibit the tendency to deposit at the oil/water interface.
Biosurfactants may facilitate the transport of hydrophobic contaminants into the aqueous phase
through specific interaction resulting in solubilization and micellization. Increased mobilization
allows for subsequent removal of such pollutants either by soil flushing or potentially makes them
more susceptible to biodegradation. Apart from interactions with the pollutants, biosurfactants
may also directly influence the efficiency of the degrader microorganisms. Biosurfactants exhibit
strong biological activity, especially at the cellular membrane level. These modifications may
result in enhanced hydrophobicity, which is considered to be relevant in terms of biodegradation
efficiency, or change the permeability of cellular membranes, which would potentially be beneficial during biodegradation (Ławniczak et al., 2013).

16.4.1.2 Increasing the Bioavailability of Hydrophobic Compounds

Bioavailability of contaminants in soil to the metabolizing organisms is influenced by factors such
as desorption, diffusion, and dissolution. Biosurfactants are produced to decrease the tension at the
hydrophobic-water interface aiming to pseudosolubilize the hydrophobic compound, thus increasing mobility, bioavailability, and consequent biodegradation. One of the major reasons for the prolonged persistence of high molecular weight hydrophobic compounds is their low water solubility,
which increases their sorption to surfaces and limits their availability to biodegrading microorganisms. When organic molecules are bound irreversibly to surfaces, biodegradation is inhibited.
Biosurfactants can enhance growth on bound substrates by desorbing them from surfaces or by
increasing their apparent water solubility (Ron and Rosenberg, 2002). When dissolved in water in
very low concentrations, biosurfactants are present as monomers. In such conditions, the hydrophobic tail, unable to form hydrogen bonding, disrupt the water structure in its vicinity, thus causing
an increase in the free energy of the system. At higher concentrations, when this effect is more
pronounced, the free energy can be reduced by the aggregation of the biosurfactant molecules into
micelles, where the hydrophobic tails are located in the inner part of the cluster and the hydrophilic
heads are exposed to the bulk water phase. At concentrations above the critical micelle concentrations (CMC), additional quantities of biosurfactants in solution will promote the formation of more
micelles. The formation of micelles leads to a significant increase in the apparent solubility of

hydrophobic organic compounds, even above their water solubility limit, as these compounds can partition into the central core of a micelle. The effect of such a process is the enhancement of mobilization of organic compounds and of their dispersion in solution. This effect is also achieved by the lowering of interfacial tension between immiscible phases. In fact, this contributes to the creation of additional surfaces, thus improving the contact between different phases (Franzetti et al., 2010). The mechanism of biosurfactants-enhanced biodegradation/bioremediation of hydrophobic compounds is shown in Figure 16.11.

The bioavailability of hydrophobic organic compounds can be enhanced by biosurfactants through the following mechanisms: emulsification of nonaqueous phase liquid contaminants, enhancement of the apparent solubility of the pollutants, facilitated transport of the pollutants from the solid phase, and help microorganisms adsorb to soil particles occupied by the contaminant, thus decreasing the diffusion path length between the sites of adsorption and the site of biouptake by the microorganisms. These mechanisms may cause enhanced mass transport and their relative contributions strongly depend on the physical state of the pollutants. The use of biosurfactants can improve the bioremediation processes by mobilization, solubilization, or emulsification. The mobilization and solubilization mechanisms are promoted low-molar mass biosurfactants, at below and above the CMC, respectively. Whereas, the emulsification processes is promoted by high molar mass biosurfactant (Bustamante et al., 2012).

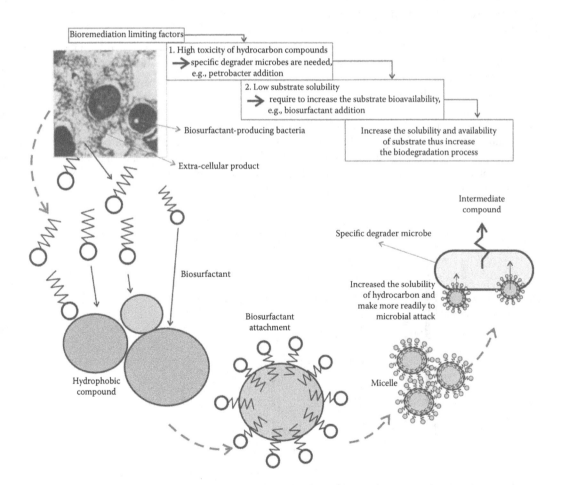

FIGURE 16.11 Mechanisms of biosurfactants-enhanced biodegradation of hydrophobic compounds.

16.4.2 UTILIZING BIOSURFACTANTS IN SOIL BIOREMEDIATION

The use of biosurfactants as an additive in bioremediation applications to soil and groundwater contaminated by insoluble organic pollutants were well studied at a laboratory and pilot-scale based. However, few reports were found in the literature regarding large-scale field application of this approach, for example, Kosaric (2001) performed field application of bioremediation at several contaminated sites in Canada and the Middle East with biosurfactant addition to the culture medium. These sites represented soil and sand contaminated by heavy hydrocarbons, primarily of industrial origin. Bioremediation was accelerated when glycolipid biosurfactants were added (0.5 kg/ton of soil) to the nutrient which was applied to the soil. Martienssen et al. (2003) also demonstrated the efficiency of surfactant (BioVersal FW) for the in situ remediation of a highly contaminated site at Halle/Saale (Germany). They could achieve 50 g hydrocarbons per kg soil elimination during the field-scale investigation over a period of 15 months. Recent report from Kardena et al. (2013), in their on-going bioremediation project showed that addition of both specific degrader microbe and biosurfactant-producing bacteria succeed in removing 46 g total petroleum hydrocarbon per kg soil from 4883 cubic meter of oil-sludge-contaminated soil during 16 months of treatment (Figure 16.12).

Figure 16.13 shows laboratory data when oil-sludge-contaminated soil with high oil content (320 g TPH/kg soil) was treated with soil washing method first to reduce oil content before continuing with biodegradation process. Microcosm-1 (-◊-) was washed with tap water, microcosm-2 (-□-) with biosurfactant from *Azotobacter* sp. at tenfold CMC value, and microcosm-3 (-○-) with Tween80 also at tenfold CMC value. Significant TPH reduction was found in microcosm-2 with 37% reduction efficiency followed with microcosm-3 and −1 with 25% and 8% reduction, respectively. Recovered oil was removed from the microcosm reactors and then continued with biodegradation process for 30 days. At the end of the treatment, TPH reduction in the microcosm reactor was measured with highest removal efficiency found at microcosm-2 (85%) followed with microcosm-3 and −1 with 79% and 46% reduction, respectively.

Regarding the actual application of biosurfactants in bioremediation processes, the biological agents may either be added externally, that is, by spraying the microbe and its products into the contaminated soil (Figure 16.14) or produced on-site, which seems especially promising in the case of in situ treatment. In the latter case, the production of biosurfactants may be obtained by bioaugmentation with appropriate microorganisms, since endogenous microorganisms rarely exhibit satisfying efficiency.

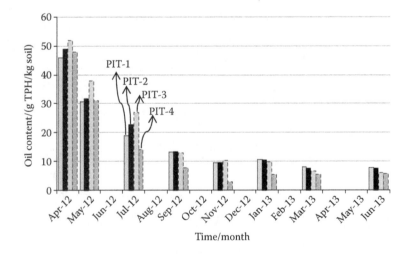

FIGURE 16.12 Performance of field application of oil-sludge-contaminated soil bioremediation process with average initial oil content of 48.7 g TPH/kg soil from total 4883 cubic meter of contaminated soil treated.

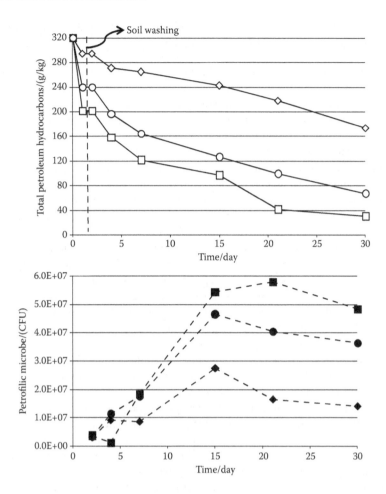

FIGURE 16.13 Biodegradation of an oil-sludge-contaminated soil (open symbol) and growth profile of petrofilic bacteria/PB (close symbol) in a laboratory microcosms with addition of PB alone (-◊/♦-), PB + biosurfactant from *Azotobacter* sp. (-□/■-), PB + Tween 80 (-○/●-).

FIGURE 16.14 Large-scale on-site production of degrader and biosurfactant-producing bacteria to improve the bioremediation process performance. (Author's project documentation with permission from LAPI-ITB consultant.)

Successive steps which should be taken into consideration during the design of biosurfactant-mediated bioremediation processes include the following:

1. Selection of a biosurfactant and biosurfactant-producing bacteria. Biosurfactants are a wide group of structurally diverse surface-active compounds produced by a variety of microorganisms which are mainly classified by their chemical structure and their microbial origin. Selections of appropriate biosurfactant producers preferentially are from native microflora (autochthonous soil microbes, rhizobacteria, etc.). Alternatively, use of known producing isolates which may be genetically modified to secrete biosurfactants or application of microbial consortia with high bioaugmentation potential (high similarity between consortium members and autochthonous microorganisms).

2. Assessment of potential biosurfactant-induced toxicity. Surface-active compounds themselves can represent a contamination when introduced in the environment. Toxicity of bio-/surfactants could be both toward the whole ecosystems or the degrading microorganisms, thus inhibiting pollutant biodegradation. Disruption of cellular membranes by interaction with lipid components and reactions of surfactant molecules with proteins that are essential to the functioning of the cell are reported as one of the main toxicity mechanisms of bio-/surfactants. For these reasons, the toxicity of such compounds should always be assessed, especially when an in situ application is planned. This assessment could take into consideration due to antimicrobial properties of some biosurfactants produced to others microbial diversity in the soil environment.

3. Evaluation of efficiency for biosurfactant-amended remediation. Evaluation of biocompatibility between biosurfactant producers and the biofactor relevant for the treatment process are important to avoid antagonistic interaction.

4. Selection of an introduction method. Direct spraying of the whole cultivated broth with free-living cells is more advantageous since purification process is costly.

5. Establishment of an optimal biosurfactant production method. This step includes assessment of potential carbon source for biosurfactant production, preferably from low-cost raw material (waste material); optimization of the production process; and availability of nutrients and other environmental factors that influence the production process.

16.5   CONCLUSIONS

The role of biosurfactants as an enhancer agent in bioremediation applications of soil contaminated by low soluble organic pollutants has so far become a promising technique to be applied. This application can potentially increase the biodegradation rate and reduce contaminant minimum concentration. This is due to their ability to enhance the pseudosolubilization and emulsification of the immiscible fractions of the contaminants, thus enhancing their bioavailability to degrading microorganisms. In either case, they can stimulate the growth of oil-degrading bacteria thereby providing co-substrate and improving their ability to utilize hydrophobic compounds. Determination of the potential success of biosurfactant-mediated soil bioremediation requires an understanding of the bioavailability of the pollutant, the survival and activity of the added microorganisms (both pollutant degrader and biosurfactant producer) or its genetic material, interaction between biosurfactant or its producer with degrader microbe, and the general environmental conditions that control soil bioremediation rates.

REFERENCES

Abalos, A., Vinas, M., Sabate, J., Manresa, M.A., and Solanas, M. 2004. Enhanced biodegradation of Casablanca crude oil by a microbial consortium in presence of a rhamnolipid produced by *Pseudomonas aeruginosa* AT10. *Biodegradation* 15, 249–260.

Banat, I.M., Makkar, R.S., and Cameotra, S.S. 2000. Potential commercial applications of microbial surfactant, *Appl. Microbiol. Biotechnol.* 53(5), 495–508.

Benincasa, M. 2007. Rhamnolipid produced from agroindustrial wastes enhances hydrocarbon biodegradation in contaminated soil. *Curr. Microbiol.* 54(6), 445–449.

Borden, R.C. 1994. Natural bioremediation of hydrocarbon-contaminated ground water. In *Handbook of Bioremediation*. Norris, R.D., Hinchee, R.E., Brown, R., McCarty, P.L., Semprini, L., Wilson, J.T., Kampbell, D.H., Reinhard, M., Bouwer, E.J., Borden, R.C., Vogel, T.M., Thomas, J.M., and Ward C.H. (eds). Lewis Publishers, Boca Raton, FL, pp. 177–199.

Brown, M.J. 1991. Biosurfactants for cosmetic applications. *Int. J. Cosmetic Sci.* 13(2), 61–64.

Bustamante, M., Durán, N., and Diez, M.C. 2012. Biosurfactants are useful tools for the bioremediation of contaminated soil: A review. *J. Soil Sci. Plant Nutr.* 12(4), 667–687.

Cameotra, S.S. and Makkar, R.S. 2010. Biosurfactant-enhanced bioremediation of hydrophobic pollutants. *Pure Appl. Chem.* 82(1), 97–116.

Cameotra, S.S. and Singh, P. 2008. Bioremediation of oil sludge using crude biosurfactants. *Int. Biodeter. Biodegr.* 62, 274–280.

Chrzanowski, L., Dziadas, M., Ławniczak, L., Cyplik, P., Białas, W., Szulc, A., Lisiecki, P., and Jelen, H. 2012. Biodegradation of rhamnolipids in liquid cultures: Effect of biosurfactant dissipation on diesel fuel/B20 blend biodegradation efficiency and bacterial community composition. *Bioresour. Technol.* 111, 328–335.

Christofi, N. and Ivshina, I.B. 2002. Microbial surfactants and their use in field studies of soil remediation. *J. Appl. Microbiol.* 93(6), 915–929.

Coduto, D.P. 1999. *Geotechnical Engineering Principles and Practices*. Prentice-Hall, Englewood Cliffs, NJ.

Cookson, J.T. 1995. *Bioremediation Engineering: Design and Application*. McGraw-Hill, New York.

Cui, C-Z., Zeng, C., Wan, X., Chen, D., Zhang, J-Y., and Shen, P. 2008. Effect of Rhamnolipids in degradation of anthracene by two newly isolated strains, *Sphingomonas* sp. 12A and *Pseudomonas* sp. 12B. *World J. Microb. Biot.* 18, 63–66.

Dahrazma, B. and Mulligan, C.N. 2007. Investigation of the removal of heavy metals from sediments using rhamnolipid in a continuous flow configuration. *Chemosphere*, doi:10.1016/j.chemosphere.2007.05.037.

Desai, J.D. and Banat, I.M. 1997. Microbial production of surfactants and their commercial potential. *Microbiol. Mol. Biol. Rev.* 61(1), 47–64.

Fiebig, R., Schulze, D., Chung, J-C., and Lee, S.T. 1997. Biodegradation of polychlorinated biphenyls (PCBs) in the presence of a bioemulsifier produced on sunflower oil. *Biodegradation* 8, 67–75.

Finnerty, W.R. 1994. Biosurfactants in environmental biotechnology. *Curr. Opin. Biotechnol.* 5(3), 291–295.

Franzetti, A., Gandolfi, I., Bestetti, G., and Banat, I.M. 2010. (Bio)surfactant and bioremediation, successes and failures. In *Trend in Bioremediation and Phytoremediation*. Plaza, G. (ed.). Research Signpost Kerala, India.

Grasset, B.F. and Vogel, T.M. 1995. Bioaugmentation: Biotreatment of contaminated soil by adding adapted bacteria. In *Bioaugmentation for Site Remediation*. Hinchee, R.E., Fredrickson, J., Alleman, B.C. (eds). Battelle Press, Columbus, OH.

Hazen, T.C. 2010. In situ: Groundwater bioremediation. In *Handbook of Hydrocarbon and Lipid Microbiology*. Timmis, K.N. (ed.). Springer-Verlag, Berlin.

Helmy, Q., Kardena, E., Funamizu, N., and Wisjnuprapto. 2011. Strategies toward commercial scale of biosurfactant production as potential substitute for it's chemically counterparts. *Int. J. Biotechnol.*, 12(1/2), 66–86.

Hickey, A.M., Gordon, L., Dobson, A.D.W., Kelly, C.T., and Doyle, E.M. 2007. Effect of surfactants on fluoranthene degradation by *Pseudomonas alcaligenes* PA-10. *Appl. Microbiol. Biotechnol.* 74, 851–856.

Hua, Z., Chen, Y., Du, G., and Chen, J. 2004. Effects of biosurfactants produced by *Candida antarctica* on the biodegradation of petroleum compounds. *World J. Microbiol. Biotechnol.* 20(1), 25–29.

Huang, W. and Liu, Z-M. 2013. Biosorption of Cd(II)/Pb(II) from aqueous solution by biosurfactant-producing bacteria: Isotherm kinetic characteristic and mechanism studies. *Colloids Surf. B: Biointerfaces* 105, 113–119.

Husain, S. 2008. Effect of surfactants on pyrene degradation by *Pseudomonas fluorescens* 29 L. *World J. Microb. Biot.* 24, 2411–2419.

Huszcza, E. and Burczyk, B. 2003. Biosurfactant production by *Bacillus coagulans*. *J. Surf. Detergents*, 6(1), 61–64.

ICSS (International Centre for Soil and Contaminated Sites). 2006. *Manual for Biological Remediation Technique*. ICSS at the Federal Environmental Agency, Dessau, Germany.

Indonesia EPA. 2003. *Technical Procedure and Requirement for Petroleum Oil Contaminated Soil Treatment by the Biological Process. I-EPA Decree No. 128/2003*. Ministry of Environmental Office, Jakarta, Indonesia.

Juwarkar, A.A., Nair, A., Dubey, K.V., Singh, S.K., and Devotta, S. 2007. Biosurfactant technology for remediation of cadmium and lead contaminated soils. *Chemosphere* 68, 1996–2002.

Kang, S-W., Kim, Y-B., Shin, J-D., and Kim, E-K. 2010. Enhanced Biodegradation of hydrocarbons in soil by microbial biosurfactant, sophorolipid. *Appl. Biochem. Biotechnol.* 160(3), 780–790.

Kardena, E., Widayatno, R.L., and Helmy, Q. 2013. Bioremediation of petroleum oil contaminated soil in Bula Block, West Seram, Indonesia. Project Progress Report to Indonesian EPA (unpublished).

Keet, B.A. 1995. Bioslurping state of the art. In *Applied Bioremediation of Petroleum Hydrocarbons*. Hinchee, R.E., Kittel, J.A., Reisinger, H.J. (eds). Battelle Press, Columbus, OH, pp. 329–334.

Kesting, W., Tummuscheit, M., Schacht, H., and Schollmeyer, E. 1996. Ecological washing of textiles with microbial surfactants. *Progr. Colloid Polymer Sci.* 101, 125–130.

Kim, S-J., Choi, D-H., Sim, D-S., and Oh, Y-S. 2005. Evaluation of bioremediation effectiveness on crude oil-contaminated sand. *Chemosphere* 59, 845–852.

Kosaric, N. 1992. Biosurfactant in industry. *J. Am. Oil Chem. Soc.* 64(11), 1731–1737.

Kosaric, N. 2001. Biosurfactants and their application for soil bioremediation. *Food Technol. Biotechnol.* 39(4), 295–304.

Kumari, B., Singh, S.N., and Singh, D.P. 2012. Characterization of two biosurfactant producing strains in crude oil degradation. *Process Biochem.* 47, 2463–2471.

Lamichhane, K.M., Babcock, R.W.Jr., Turnbull, S.J., and Schenck, S. 2012. Molasses enhanced phyto and bioremediation treatability study of explosives contaminated Hawaiian soils. *J. Hazard Mater.* 243, 334–339.

Ławniczak, Ł., Marecik, R., and Chrzanowski, Ł. 2013. Contributions of biosurfactants to natural or induced bioremediation. *Appl. Microbiol. Biotechnol.* 97(6), 2327–2339.

Lebeau, T., Braud, A., and Jézéquel, K. 2008. Performance of bioaugmentation-assisted phytoextraction applied to metal contaminated soils: A review. *Environ. Pollut.* 153, 497–522.

Leckie, S.E., Prescott, C.E., Grayston, S.J., Neufeld, J.D., and Mohn, W.W. 2004. Characterization of humus microbial communities in adjacent forest types that differ in nitrogen availability. *Microb. Ecol.* 48, 29–40.

Makkar, R.S. and Cameotra, S.S. 2002. Effect of various nutritional supplements on biosurfactant production by strain of *Bacillus subtilis* at 45°C. *J. Surf. Detergents* 5(1), 11–17.

Margesin, R. and Schinner, F. 2001. Biodegradation and bioremediation of hydrocarbons in extreme environments. *Appl. Microbiol. Biotechnol.* 56, 650–663.

Martienssen, M., Reichel, O., and Schirmer, M. 2003. Use of surfactants to improve the biological degradability of petroleum hydrocarbons. *Chem. Ing. Tech.* 75(11), 1749–1755.

McGugan, B.R., Lees, Z.M., and Senior, E. 1995. Bioremediation of an oil contaminated soil by fungal intervention. In *Bioaugmentation for Site Remediation*. Hinchee, R.E., Fredrickson, J., Alleman, B.C. (eds). Battelle Press, Columbus, OH, pp. 149–156.

McKew, B.A., Coulon, F., Yakimov, M.M., Denaro, R., Genovese, M., Smith, C.J., Osborn, A.M., Timmis, K.N., and McGenity, T.J. 2007. Efficacy of intervention strategies for bioremediation of crude oil in marine systems and effects on indigenous hydrocarbonoclastic bacteria. *Environ. Microbiol.* 9, 1562–1571.

Megharaj, M., Ramakrishnan, B., Venkateswarlu, K., Sethunathan, N., and Naidu, R. 2011. Bioremediation approaches for organic pollutants: A critical perspective. *Environ. Int.* 37, 1362–1375.

Miller, R., 1996. Air sparging. Technology Overview Report. Ground-Water Remediation Technologies Analysis Center (GWRTAC). Pittsburgh, PA, 10pp.

Moran, A.C., Olivera, N., Commendatore, M., Esteves, J.L., and Sineriz, F. 2000. Enhancement of hydrocarbon waste biodegradation by addition of a biosurfactant from *Bacillus subtilis* O9. *Biodegradation* 11, 65–71.

Muckian, L., Grant, R., Doyle, E., and Clipson N. 2007. Bacterial community structure in soils contaminated by polycyclic aromatic hydrocarbons. *Chemosphere* 68, 1535–1541.

Mukherjee, A.K. 2007. Potential application of cyclic lipopeptide biosurfactants produced by *Bacillus subtilis* strains in laundry detergent formulations. *Lett. Appl. Microbiol.* 45(3), 330–335.

Mukherjee, A.K. and Das, K. 2010. Microbial surfactants and their potential applications: An overview. *Adv. Exp. Med. Biol.* 672, 54–64.

Mulligan, C.N. 2005. Environmental applications for biosurfactants. *Environ. Pollut.* 133, 183–198.

Mulligan, C.N. 2009. Recent advances in the environmental applications of biosurfactants. *Curr. Opin. Colloid Interface Sci.* 14, 372–378.

Mulligan, C.N., Yong, R.N., and Gibbs, B.F. 2001. Surfactant-enhanced remediation of contaminated soil: A review. *Eng. Geol.* 60, 371–380.

National Research Council. 2000. *Natural Attenuation for Groundwater Remediation*. National Academy Press, Washington, DC.

Neralla, S., Wright, A.L., and Weaver, R.W. 1995. Microbial inoculants and fertilization for bioremediation of oil in wetlands. In *Bioaugmentation for Site Remediation*. Hinchee, R.E., Fredrickson, J., Alleman, B.C. (eds). Battelle Press, Columbus, OH, pp. 31–38.

Nitschke, M. and Costa, S.G.V.A.O. 2007. Biosurfactants in food industry. *Trends Food Sci. Technol.* 18(5), 252–259.

Owsianiak, M., Chrzanowski, L., Szulc, A., Staniewski, J., Olszanowski, A., Olejnik-Schmidt, A.K., and Heipieper, H.J. 2009. Biodegradation of diesel/biodiesel blends by a consortium of hydrocarbon degraders: Effect of the type of blend and the addition of biosurfactants. *Bioresour. Technol.* 100, 1497–1500.

Paul, E.A. and Clark, F.G. 1989. *Soil Microbiology and Biochemistry*. Academic Press, San Diego, CA.

Pei, X-H., Zhan, X-H., Wang, S-M., Lin, Y-S., and Zhou, L-X. 2010. Effects of a Biosurfactant and a synthetic surfactant on phenanthrene degradation by a *Sphingomonas* strain. *Pedosphere* 20, 771–779.

Pekdemir, T., Ishigami, Y., and Uchiyama, H. 1999. Characterization of aescin as a biosurfactant for environmental remediation. *J. Surf. Detergents* 2(3), 337–341.

Plaza, G.A., Wypych, J., Berry, C., and Brigmon, R.L. 2007. Utilization of monocyclic aromatic hydrocarbons individually and in mixture by bacteria isolated from petroleum contaminated soil. *World J. Microbiol. Biotechnol.* 23, 533–542.

Rahman, K.S.M. and Gakpe, E. 2008. Production, characterisation and applications of biosurfactants—Review. *Biotechnology* 7, 360–370.

Rahman, K.S.M., Rahman, T.J., Kourkoutas, Y., Petsas, I., and Banat, I.M. 2003. Enhanced bioremediation of n-alkane in petroleum sludge using bacterial consortium amended with rhamnolipid and micronutrients. *Bioresour. Technol.* 90(2), 159–168.

Rashedi, H., Jamshidi, E., Assadi, M.M., and Bonakdarpour, B. 2006. Biosurfactant production with glucose as a carbon source. *Chem. Biochem. Eng. Q.* 20(1), 99–106.

Reddy, M.S., Naresh, B., Leela, T., Prashanthi, M., Madhusudhan, N.Ch., Dhanasri, G., and Devi, P. 2010. Biodegradation of phenanthrene with biosurfactant production by a new strain of *Brevibacillus* sp. *Bioresour. Technol.* 101, 7980–7983.

Reichenberg, F., Karlson, U.G., Gustafsson, O., Long, S.M., Pritchard, P.H., and Mayer P. 2010. Low accessibility and chemical activity of PAHs restrict bioremediation and risk of exposure in a manufactured gas plant soil. *Environ. Pollut.* 158, 1214–1220.

Rittmann, B.E and McCarty, P.L. 2001. *Environmental Biotechnology: Principles and Applications*. McGraw-Hill, New York.

Ron, E.Z. and Rosenberg, E. 2002. Biosurfactants and oil bioremediation. *Curr. Opin. Biotechnol.* 13, 249–252.

Saeki, H., Sasaki, M., Komatsu, K., Miura, A., and Matsuda, H. 2009. Oil spill remediation by using the remediation agent JE1058BS that contains a biosurfactant produced by *Gordonia* sp. strain JE-1058. *Bioresour. Technol.* 100, 572–577.

Santos, E.C., Jacques, J.S., Bento, F.M., Peralba, M.C.R., Selbach, P.A., Sa, E.L.S., and Camargo, F.A.O. 2008. Anthracene biodegradation and surface activity by an iron-stimulated Pseudomonas sp. *Bioresour. Technol.* 99, 2644–2649.

Sims, J.L., Suflita, J.M., and Russell, H.H. 1992. Ground water issue: In situ bioremediation of contaminated ground water, EPA/540/S-92/003, U.S. Environmental Protection Agency Office of Solid Waste and Emergency Response, February 1992, 11pp.

Singh, G.B., Gupta, S., and Gupta, N. 2013. Carbazole degradation and biosurfactant production by newly isolated *Pseudomonas* sp. strain GBS.5. *Int. Biodeter. Biodegr.* 84, 35–43.

Singh, P. and Cameotra, S.S. 2004. Potential application of microbial surfactants in biomedical sciences. *Trends Biotechnol.* 22(3), 143–146.

Sponza, D.T. and Gok, O. 2011. Effects of sludge retention time (SRT) and biosurfactant on the removal of polyaromatic compounds and toxicity. *J. Hazard. Mater.* 197, 404–416.

Theng, B.K.G., Aislabie, J., and Fraser, R. 2001. Bioavailability of phenanthrene intercalated into an alkylammonium-montmorillonite clay, *Soil Biol. Biochem.* 33, 845–848.

Thavasi, R., Jayalakshmi, S., and Banat, I.M. 2011. Application of biosurfactant produced from peanut oil cake by *Lactobacillus delbrueckii* in biodegradation of crude oil. *Bioresour. Technol.* 102, 3366–3372.

Thomas, J.M. and Ward, C.H. 1989. In situ biorestoration of organic contaminants in the subsurface. *Environ. Sci. Technol.* 23, 760–766.

Thompson, I.P., Van Der Gast, C.J., Ciric, L., and Singer, A.C. 2005. Bioaugmentation for bioremediation: The challenge of strain selection. *Environ. Microbiol.* 7, 909–915.

U.S. EPA. 1997a. Use of Monitored Natural Attenuation at Superfund, RCRA Corrective Action, and Underground Storage Tank Sites OSWER Directive 9200.4-17 U.S. Environmental Protection Agency. Office of Solid Waste and Emergency Response, Washington, D.C.

U.S. EPA. 1997b. *Proceedings of the Symposium on Natural Attenuation of Chlorinated Organics in Ground Water EPA/540/R-97/504*. U.S. Environmental Protection Agency, Office of Research and Development, Washington, DC.

U.S. EPA. 2005. Introduction to closure/post-closure. EPA530-K-05-009. U.S. Environmental Protection Agency, Office of Solid Waste and Emergency Response, Washington, DC.

U.S. EPA. 2006. In situ and ex situ biodegradation technologies for remediation of contaminated sites EPA/625/R-06/015. U.S. Environmental Protection Agency, Office of Research and Development, National Risk Management Research Laboratory, Cincinnati, OH.

Uysal, A. and Turkman, A. 2007. Biodegradation of 4-CP in an activated sludge reactor: Effects of biosurfactant and the sludge age. *J. Hazard. Mater.* 148, 151–157.

Vacca, D.J., Bleam, W.F., and Hickey, W.J. 2005. Isolation of soil bacteria adapted to degrade humic acid-sorbed phenanthrene. *Appl. Environ. Microbiol.* 71, 3797–3805.

van Cauwenberghe, L. and Roote, D.S. 1998. In situ bioremediation. technology overview report. Ground-Water Remediation Technologies Analysis Center (GWRTAC), Pittsburgh, PA, 24pp.

Venosa, A.D., Suidan, M.T., Haines, J.R., Wrenn, B.A., Strohmeier, K.L., Eberhart, B.L., Kadkhodayan, M., Holder, E., King, D., and Anderson, B. 1995. Field bioremediation study: Spilled crude oil on Fowler Beach, Delaware. In *Bioaugmentation for Site Remediation*. Hinchee, R.E., Fredrickson, J., and Alleman, B.C. (eds). Battelle Press, Columbus, pp. 49–56.

Viisimaa, M., Karpenko, O., Novikov, V., Trapido, M., and Goi, A. 2013. Influence of biosurfactant on combined chemical–biological treatment of PCB-contaminated soil. *Chem. Eng. J.* 220, 352–359.

Vogel, T.M. 1996. Bioaugmentation as a soil bioremediation approach. *Curr. Opin. Biotechnol.* 7, 311–316.

Welander, U. 2005. Microbial degradation of organic pollutants in soil in a cold climate. *Soil Sediment Contam.* 14, 281–291.

Whang, L-M., Liu, P-W., Maa, C-C., and Cheng, S.S. 2007. Application of biosurfactants, rhamnolipid, and surfactin, for enhanced biodegradation of diesel-contaminated water and soil. *J. Hazard. Mater.* 151, 155–163.

Whyte, L.G., Goalen, B., Hawari, J., Labbé, D., Greer, C.W., and Nahir, M. 2001. Bioremediation treatability assessment of hydrocarbon-contaminated soils from Eureka. Nunavut. *Cold Regions Sci Technol.* 32, 121–132.

Wong, J.W.C., Zhao, Z., and Zheng, G. 2010. Biosurfactants from *Acinetobacter calcoaceticus* BU03 enhance the bioavailability and biodegradation of polycyclic aromatic hydrocarbons. *Proceedings of the Annual International Conference on Soils, Sediments, Water and Energy*, University of Massachusetts at Amherst, MA, October 19–22, 2009: Vol.15, Article 5, pp. 36–51.

Xiao, X., Chen, H., Si, C., and Wu, L. 2012. Influence of biosurfactant-producing strain *Bacillus subtilis* BS1 on the mycoremediation of soils contaminated with phenanthrene. *Int. Biodeter. Biodegr.* 75, 36–42.

Yan, P., Lu, M., Guan, Y., Zhang, W., and Zhang, Z. 2011. Remediation of oil-based drill cuttings through a biosurfactant-based washing followed by a biodegradation treatment. *Bioresour. Technol.* 102, 10252–10259.

Yang, Y., Zhang, N., Xue, M., and Tao, S. 2010. Impact of soil organic matter on the distribution of polycyclic aromatic hydrocarbons (PAHs) in soils. *Environ. Pollut.* 158, 2170–2174.

Zeng, G., Liu, Z., Zhong, H., Li, J., Yuan, X., Fu, H., Ding, Y., Wang, J., and Zhou, M. 2011. Effect of monorhamnolipid on the degradation of n-hexadecane by *Candida tropicalis* and the association with cell surface properties. *Appl. Microbiol. Biotechnol.* 90, 1155–1161.

Zhang, C., Wang, S., and Yan, Y. 2011. Isomerization and biodegradation of beta-cypermethrin by *Pseudomonas aeruginosa* CH7 with biosurfactant production. *Bioresour. Technol.* 102, 7139–7146.

Zhang, J., Yin, R., Lin, X., Liu, W., Chen, R., and Li, X. 2010. Interactive effect of biosurfactant and microorganism to enhance phytoremediation for removal of aged polycyclic aromatic hydrocarbons from contaminated soils. *J. Health Sci.* 56, 257–266.

Zhao, Z., Selvam, A., and Chung Wong, J.W. 2011. Synergistic effect of thermophilic temperature and biosurfactant produced by *Acinetobacter calcoaceticus* BU03 on the biodegradation of phenanthrene in bioslurry system. *J. Hazard. Mater.* 190, 345–350.

Zouboulis, A.I., Matis, K.A., Lazaridis, N.K., and Golyshin, P.N. 2003. The use of biosurfactants in flotation: Application for the removal of metal ions. *Miner. Eng.* 16(11), 1231–1236.

17 Biosurfactant Use in Heavy Metal Removal from Industrial Effluents and Contaminated Sites

Andrea Franzetti, Isabella Gandolfi, Letizia Fracchia,
Jonathan Van Hamme, Panagiotis Gkorezis, Roger Marchant,
and Ibrahim M. Banat

CONTENTS

17.1 Introduction 361
17.2 Mechanisms of Biosurfactant–Metal Interactions.. 361
17.3 Removal of Metals from Industrial Effluents 363
17.4 Removal of Metals from Contaminated Sites... 364
17.5 Conclusions and Future Perspectives.. 366
References...366

17.1 INTRODUCTION

Remediation of metal-contaminated environments is particularly challenging given that, unlike organic molecules, metals cannot be biodegraded or mineralized. As such, remediation approaches must focus either on changing the redox state of a metal contaminant to a less toxic form, or on physically removing the metal from the environment. Biological processes can play a central role in the remediation of metal-contaminated water, soil, and sludge as microbes are well known to interact with, and change the properties of, a wide range of toxic and nontoxic metals. For example, metals may be used as electron donors or electron acceptors for energy production within a cell, may be used to shuttle electrons between organisms in syntrophic relationships, or possibly used as cofactors for intracellular and extracellular enzymatic reactions (Croal et al., 2004; Haferburg and Kothe, 2007). Microorganisms have, therefore, evolved mechanisms to oxidize, reduce, transport, bind, and sequester metals to either avoid toxic effects or to assist with basic cellular processes. These physiological responses to metals can be harnessed for bioremediation (Gadd, 2010; Singh et al., 2007; Van Hamme et al., 2006), and the focus of this chapter is on the use of biosurfactants for mobilizing and removing metals from contaminated environments. Specifically, the mechanisms underlying biosurfactant–metal interactions will be described together with applied examples of the use of biosurfactants for treatment of metal-contaminated industrial effluents and contaminated sites.

17.2 MECHANISMS OF BIOSURFACTANT–METAL INTERACTIONS

Compared with their synthetic counterparts, biosurfactants hold more promise for being effective in remediation schemes targeting metal removal or mobilization (Sriram et al., 2011). This is, in part, due to the fact that biosurfactants have been reported to have higher selectivity for a greater number of metal ions and organic compounds, greater tolerance to variations in pH, salt concentrations and temperature, and can generally be produced from widely available, low-cost, and renewable resources (Aşçi et al., 2007; Makkar et al., 2011). In order to save on purification costs, whole microbial cells or microbial exopolymers have been used to concentrate or precipitate metals from liquid waste streams (Banat et al., 2010). Analogous remediation of metal-contaminated soils is more complex because microbial cells and large exopolymers cannot move easily through the soil. In such cases, purified or partially purified biosurfactants avoid that constraint due to their small size, which generally appears to be smaller than 1500 Da (Miller, 1995), in addition to their surface activity that may allow them further penetration of soil particles. On the whole, biosurfactants are less toxic and more biodegradable than chemical surfactants that are being commonly used for *in situ* remediation strategies designed specifically for metal-contaminated soil (Juwarkar et al., 2008).

According to Miller (1995), biosurfactants enhance metal desorption from soils in two ways. First, biosurfactants are able to form complexes with the free, non-ionic forms of metals in solution. This complexation reduces the solution phase activity of metals and speeds desorption following Le Chatelier's principle. Second, biosurfactants can make direct contact with absorbed metal at the solid–solution interface under conditions of reduced interfacial tension, which allows biosurfactants to accumulate at the solid–solution interface.

The mechanisms driving biosurfactant–metal binding include ion exchange, precipitation-dissolution, counter-ion association, and electrostatic interaction (Rufino et al., 2012). However, it appears that the ability of biosurfactants to form complexes with metals is the main reason for their utility in remediation of heavy metal-contaminated soil. Specifically, anionic biosurfactants form ionic bonds with metals, generating nonionic complexes with stronger stabilizing forces than those between the metals bonds and soil. Once formed, metal–biosurfactant complexes desorb from the soil matrix and move into the soil solution due to the neutral charge of the complex with a subsequent incorporation of the metal into micelles. In more detail, this mechanism presumes either electrostatic attraction between negatively charged surfaces and metals that form an outer-sphere surface complex, or chemical bonding in which hydroxide groups serve as ligands to form an inner-sphere surface complex with metals. Both cases are facilitated when oxide groups are easily protonated or deprotonated and surrounded by water molecules.

Unlike anionic biosurfactants, cationic biosurfactants can replace charged metal ions on the surface of soil particles by competition for some but not all of the negatively charged surfaces (i.e., ion exchange). Remarkably, the mono-rhamnolipid biosurfactant produced by *Pseudomonas aeruginosa* has been demonstrated to have a strong affinity for metals such as Cd^{+2}, Zn^{+2}, and Pb^{+2} through its carboxyl group (Herman et al., 1995; Juwarkar et al., 2007). The strength of these charge interactions means that metal ions can be removed from soil surfaces even in the absence of biosurfactant micelles. A study focusing on the kinetics of Cd^{2+} desorption from Na-feldspar after rhamnolipid application at a concentration of 77 mM has suggested that the rate-controlling step was correlated with the surface reaction mechanism (Aşçi et al., 2012). Given the uncharged and hydrophobic nature of the micellar core, presumably metal removal was due to charge interactions with the charged polar head groups on the surface of the biosurfactant micelles. Indeed, Das and colleagues reported that lipopeptides produced by *Bacillus circulans* were able to remove lead and copper from water both below and above the CMC concentration, showing that the metals, being positively charged, bind to the outer hydrophilic surface of the biosurfactants, rather than being incorporated within surfactant micelles (Das et al., 2009). It has been suggested that the high content of uronic acids provide exopolymers produced by *Pseudoalteromonas* sp. strain TG12 with an ability to complex with metal species such as Na^+, followed by Mg^{2+}, K^+, Mn^{2+}, $Fe^{2+/3+}$, and Al^{3+} (Gutierrez et al., 2008). Interestingly, the authors showed that the presence of 0.6 M NaCl drastically reduced the polymer's ability to desorb these metals from sediment by competition from Na^+ for binding sites on the polymer. Similarly, high salinity led to reduction in Cu^{2+} and Pb^{2+} binding by extracellular polymeric substance (Bhaskar and Bhosle, 2006).

Although micelle formation may not be critical for metal removal, it is important to consider when dealing with bioremediation of environments co-contaminated with organic pollutants. When biosurfactants are released into the environment, they can partition into different abiotic and biotic phases such soil particles, water, air, immiscible liquid, and organic matter, altering the physicochemical conditions at the interfaces (Marchant and Banat 2012a,b). Some of these alterations directly affect interactions between bacteria and contaminants such as hydrocarbons and metals (i.e., micelle formation and emulsification of contaminants, interaction with sorbed contaminants, sorption to soil particles and the alteration of cell-envelope composition and hydrophobicity) (Paria, 2008; Volkering et al., 1997). Therefore, it is necessary to take into account all the processes affecting the fate of biosurfactants before deploying them into a contaminated environment (Franzetti et al., 2006, 2012). Furthermore, environmental conditions can influence the mobility and sorption properties of biosurfactants toward metals. Dahrazma et al. (2008) highlighted the importance of pH on the morphology of rhamnolipid aggregates in heavy metal solutions. Specifically, larger aggregates (>200 nm) formed at more basic conditions than acidic conditions (55–60 nm), thus affecting the mobility of the micelles in porous media like soil. The rhamnolipids were observed to form star-like microstructures at pH 5 and 6 while at pH 7 or 8 regular vesicular dispersions dominated. Raza et al. (2010) hypothesized that the addition of Sr^{2+}/Pb^{2+} might regulate micelle formation in rhamnolipid solutions due to electrostatic bonding between Sr^{2+} and negatively charged rhamnolipid carboxylate groups while it can influence the chelating activity via a bidentate ligand in the case of Pb^{2+}.

17.3   REMOVAL OF METALS FROM INDUSTRIAL EFFLUENTS

Industrial wastewater may contain significant concentrations of toxic metals, which require removing or recovering prior to discharge. The use of biosurfactants to enhance metal removal from industrial effluents has been proposed by some researchers. A biosurfactant produced by *Flavobacterium* sp. grown on used vegetable oil was simply added to water in stirred batch reactors. This compound was able to remove over 75% of lead from 100 mg/L lead-contaminated water at $10 \times$ CMC, and was more effective than the synthetic surfactants Triton X-100 and SDS (Kim and Vipulanandan, 2006). The plant-derived surfactant saponin was also used for the removal of metals from water and industrial sludge (Gao et al., 2012; Pekdemir et al., 2000). The effectiveness of saponin from different origins was compared with other biological amphiphilic compounds in the removal of Cd and Pb from contaminated water, but the percentage removal achieved was lower than that obtained with tannic acid (Pekdemir et al., 2000).

In comparison with sophorolipid, nonionic saponin, used in batch experiments, was more effective in removing Pb, Ni, and Cr from a contaminated sludge taken from an industrial water treatment plant (Gao et al., 2012). Huang and Liu (2013) suggested the biosurfactant-producing *Pseudomonas* sp. strain LKS06, rather than its purified product, may be used to remove Cd and Pb from industrial wastewater. In this case, biomass would act as a biosorbent, on which both physical and chemical sorption can take place simultaneously.

Biosurfactants have also been exploited as ion collectors in wastewater treatment using a foam flotation process. This two-stage technique is based on the application of a surface-active material or compound to adsorb the metals from the water and a subsequent separation by flotation of the resulting foam. Such a method has been applied to different metals and by using different biosurfactants. For example, Zouboulis et al. (2003) investigated the removal of Zn and Cr ions from aqueous solutions. They concluded that the application of the biosurfactants Surfactin-105 and Lichenysin-A as flotation collectors for the separation of the metal-loaded sorbents, resulted in better float abilities of metal-laden sorbents compared with chemically produced surface active compounds such as SDS or dodecylamine. Similar results were obtained by Chen et al. (2011), who observed a higher Hg removal from artificially contaminated water with surfactin than with SDS and Tween-80, when all were used at a concentration of $10 \times$ CMC.

The potential of tea-derived saponin to remove Cd, Pb, and Cu through foam flotation was also investigated. The maximum total removal was 81.8% when all operating conditions were properly optimized (Yuan et al., 2008). Recently, more efforts have been devoted to process optimization and modeling. Rangarajan and Sen (2013) studied the removal of Ca, Mg, and Fe(II) in bubble column experiments using the lipopeptide produced by a marine *Bacillus megaterium* strain. They compared the effect of a simulated biosurfactant solution and of real culture broths, with or without cells, and found that the presence of proteins in the broth without cells increased the overall stability of lipopeptide-enriched foam. However, the presence of cells adversely affected the foam stability. Bodagh et al. (2013) investigated the best conditions for the removal of Cd, Zn, and Cu from wastewater through the use of rhamnolipid produced by *P. aeruginosa* MA01, although the percentages of removal were not particularly high compared with other papers (not more than 57% for cadmium and even lower for other metals). Interestingly, they also found that Cd removal was affected by the rhamnolipid congener used (di- or mono-rhamnolipid).

Another promising method to remove metals present at low concentrations from large volumes of polluted water is micellar-enhanced ultrafiltration (Baek et al., 2003; Ferella et al., 2007). The application of biosurfactants to this method had already been proposed by Hong et al. (1998), who used spiculisporic acid from *Penicillium spiculisporum* and derived compounds to remove Cu, Zn, Cd, and Ni from aqueous solutions, both metal mixtures and single component solutions. Recently, El Zeftawy and Mulligan (2011) used a rhamnolipid biosurfactant in micellar-enhanced ultrafiltration application targeting metals from contaminated water and concluded that the rhamnolipid-based ultrafiltration technique is an efficient technique for the removal of Cd, Pb, Cu, Zn, and Ni ions from contaminated industrial wastewater.

New approaches and attempts for metal stabilization using biosurfactants have also been recently addressed by Gnanamani et al. (2010) who showed bioremediation of Cr(VI) by a biosurfactant-producing marine isolate *Bacillus* sp. MTCC 5514. Removal of Cr(VI) was obtained in two steps: reduction of Cr(VI) to the trivalent form Cr(III) followed by entrapment in the micelles of the biosurfactant. Biosurfactants have also been reported for the ability to mobilize arsenic from mine tailings (Wang and Mulligan, 2009). Particularly, the mobilization of As(V) under alkaline conditions using rhamnolipid was found to be positively correlated with the mobilization of other metals (i.e., Fe, Cu, Pb, and Zn).

Rhamnolipids were also used to extract copper from a mining residue containing 8950 mg Cu per kg ore. A 2% rhamnolipid solution was required to extract about 28% of copper from the ore, while adding 1% NaOH to the biosurfactant solution dramatically improved the copper extraction up to 42% (Dahrazma and Mulligan, 2007). Menezes et al. (2011) treated aqueous effluent produced by acid mine drainage by dissolved air flotation using biosurfactants produced by *Candida lipolytica* and *Candida sphaerica*. High percentages of removal of Fe(III) and Mn(II), above 94%, were obtained and the values were found to be similar to those obtained with the use of the synthetic surfactant sodium oleate.

Recently, the use of Quillaja bark saponin was proposed to develop a new method of Cr removal from tannery sludge to achieve a cost-effective and environmentally acceptable remediation solution (Kiliç et al., 2011). However, the maximum Cr removal obtained was only 24% at pH 2, while the chemical oxidative process carried out with H_2O_2 and sulfuric acid at the same pH recovered approximately 70% of the chromium. The low efficiency of saponin was ascribed to the presence of high organic content in the tannery sludge, which affects chromium mobility primarily by its high sorptive capacity and therefore represents a major constraint to the biosurfactant-assisted removal of the metal.

17.4   REMOVAL OF METALS FROM CONTAMINATED SITES

These processes have been exploited to achieve the removal of heavy metals from soil for the past 20 years. Since the first demonstration by Miller (1995) showing the ability of biosurfactants to facilitate removal of heavy metal from soil, numerous investigations have been published assessing their potential. In recent years, however, some authors have reported concerns regarding the possible toxic effects to autochthonous soil microorganisms, which should be taken into consideration as well as the overall effect of releasing surface active compounds into the environments (Bondarenko et al., 2010; Franzetti et al., 2006).

Mulligan et al. (2001) evaluated the performance of surfactin from *Bacillus subtilis*, rhamnolipids from *P. aeruginosa*, and sophorolipid from *Torulopsis (Starmerella) bombicola* and concluded that removal of metals from sediments by use of a solution containing these biosurfactants is feasible. Juwarkar et al. (2008) have shown that a di-rhamnolipid produced by *P. aeruginosa* BS2 selectively removed chromium, lead, cadmium, copper, and nickel from a multi-element contaminated soil in the order of Cd = Cr > Pb = Cu > Ni. Wen et al. (2009) studied the behavior of rhamnolipids in soils contaminated by Cd and Zn, and suggested that rhamnolipids enhanced metal phytoextraction without the possible increase of metal mobility in the long term.

The extent of metal removal by biosurfactants is related to the soil type, pH, cation exchange capacity (CEC), and the nature and concentrations of contaminants and co-contaminants (Singh and Cameotra, 2004). Aşçi et al. (2010) found that metal ions could be efficiently recovered from quartz using rhamnolipid treatment and reported that 91.6% of sorbed Cd (0.31 mmol Cd(II)/kg) and 87.2% of sorbed Zn (0.672 mmol/kg) were recovered.

Commonly, soils are co-contaminated with metals and organic pollutants that may be treated with biosurfactants for simultaneous removal of both contaminants. For example, surfactin and fengycin from *B. subtilis* A21 were effective at removing high concentrations of petroleum hydrocarbons (64.5% with an initial concentration of 1,886 mg/kg) and metals (cadmium, cobalt, lead, nickel, copper, and zinc) resulting in reduced soil phytotoxicity (Singh and Cameotra, 2013b). A washing agent composed of (bio)surfactant and an inorganic ligand was found to be useful for removal of Cd and phenanthrene from soil (Lima et al., 2011). In this study, the removal of Cd^{2+} increased with increasing iodine concentration, particularly in solutions containing biosurfactants produced by *B. subtilis* LBBMA155 (lipopeptide) and *Flavobacterium* sp. LBBMA168 (a mixture of flavolipids) in combination with Triton X-100. Biosurfactant produced by the yeast *C. lipolytica* was also used for the removal of heavy metals and petroleum derivatives using a soil barrier. Biosurfactant significantly reduced soil permeability, demonstrating its applicability as an additive in reactive barriers allowing the removal of around 96% Zn and Cu and the reduction of Pb and Cd concentrations in groundwater (Rufino et al., 2011, 2012).

Much research has been directed toward developing biosurfactant formulations for improving metal removal from soil and water. In this respect, both the addition of other chemicals to biosurfactant solutions as well as the physical form of the formulations (i.e., liquid or foam) have been considered. Liu and colleagues investigated the ability of rhamnolipids, ethylenediaminetetraacetic acid (EDTA), and citric acid to remove metals. The authors concluded that rhamnolipids removed less metals compared with EDTA and citric acid, but increased the effect of the other two kinds of chelating agent in Cu leaching (Liu et al., 2013). Similarly, the rhamnolipid biosurfactant blend (JBR-425) was found to be effective in removing Zn, Cu, Pb, and Cd from soil, both alone and when amended with citric acid and EDTA (Slizovskiy et al., 2011). With respect to the physical form of biosurfactant formulations, Wang and Mulligan (2004) reported that the application of rhamnolipid foam was more effective than rhamnolipid solution for removal of Cd and Ni from a sandy soil. Chen et al. (2011) evaluated the separation of mercury ions from artificially contaminated water using a foam fractionation process with a surfactin, compared with the chemical surfactants (SDS and Tween-80). They concluded that using the biodegradable, nontoxic, and cost-effective anionic biosurfactant surfactin resulted in a higher mercury recovery compared with the synthetic counterpart, probably due to the presence of two carboxylate groups in the surfactin molecule.

Given the interest in using plants for phytoremediation of metal-contaminated soils, there is potential to exploit microorganisms that produce both biosurfactants and plant-growth promoting compounds to enhance metal phytoextraction (Rajkumar et al., 2012). In this context, researchers have reported both the isolation of plant-associated biosurfactant-producing bacteria and metal-resistant microorganisms with potential applications in phytoremediation (Becerra-Castro et al., 2012; Singh and Cameotra, 2013a). A *Pseudomonas* sp. strain produced both mucoid biofilm and biosurfactant and could tolerate high levels of chromate (Bramhachari et al., 2012). The recently discovered lipopeptide produced by a multi-metal-resistant *Escherichia fergusonii* strain showed stability at extremes of temperature, pH, and osmotic concentrations (Sriram et al., 2011). Despite the promise, there have been negative impacts reported for the effects of biosurfactants on phytoextraction (Jensen et al., 2011; Wen et al. 2010). Jensen and colleagues found that rhamnolipid treatment was unsuitable because of insufficient mobilization of Cu and Zu during phytoextraction. Similarly, Wen et al. (2010) suggested that neither low nor high concentrations of rhamnolipid are likely to consistently assist Cd phytoextraction using maize or sunflower.

Beyond microbial biosurfactants, saponin, a plant (Quillaja)-derived biosurfactant has been evaluated for its ability to remove metals from soil. Chen et al. (2008) observed that 2000 mg L^{-1} of saponin could remove 83% of the Cu and 85% of the Ni from kaolin containing 0.45 mg copper/g kaolin and 0.14 mg nickel/g kaolin. Gao et al. (2012) concluded that (a) plant-derived nonionic saponin is more efficient than sophorolipid for the removal of metals from polluted sludge; and (b) saponin interacts very well with metals bound to carbonates and Fe-Mn oxides in soils. Indeed, Gusiatin and Klimiuk (2012) found that saponin effectively decreased the total metal concentration in soils contaminated with Cu, Cd, and Zn during soil-washing experiments. More recently, saponin was found to effectively remove high levels of copper, lead, and zinc from soil using foam fractionation (Maity et al., 2013).

17.5   CONCLUSIONS AND FUTURE PERSPECTIVES

Biosurfactants are a very diverse group of biomolecules ranging from the low molecular weight glycolipids to the high molecular weight compounds such as extracellular polymeric substance or lipopolysaccharides. Not surprisingly, the physicochemical properties of these molecules also vary greatly, which accounts for the wide range of results that have been achieved using different combinations of biosurfactants and metal ions. What is clear is that high removal of metals can be achieved from both contaminated liquid industrial effluents and from contaminated solid materials and soils. The use of foam fractionation clearly provides an exciting prospect for dealing with the often large volume of effluents produced from industrial processes without the need for recourse to complex technology. One of the major advantages regularly bandied about for the use of biosurfactants is their green credentials, that is, lack of toxicity, biodegradability, and relative stability under a wide range of physicochemical environments. However, we must bear in mind that biosurfactants have been shown to have significant biocidal and biostatic effects on certain groups of microorganisms and therefore the unrestricted addition of biosurfactants to natural environments may have unforeseen consequences. We must also be aware that biosurfactants may not be without deleterious effects on plants, particularly on their roots. Bearing in mind the above caveats biosurfactants are clearly effective in remediation of metals from effluents and environments and may in the future have a significant role on a large scale.

REFERENCES

Aşçi, Y., Açikel, U., and Açikel, Y.S. 2012. Equilibrium, hysteresis and kinetics of cadmium desorption from sodium-feldspar using rhamnolipid biosurfactant. *Environmental Technology* 33: 1857–1868.

Aşçi, Y., Nurbaş, M., and Açikel, Y.S. 2007. Sorption of Cd(II) onto kaolin as a soil component and desorption of Cd(II) from kaolin using rhamnolipid biosurfactant. *Journal of Hazardous Materials* 139: 50–56.

Aşçi, Y., Nurbaş, M., and Açikel, Y.S. 2010. Investigation of sorption/desorption equilibria of heavy metal ions on/from quartz using rhamnolipid biosurfactant. *Journal of Environmental Management* 91: 724–731.

Baek, B.K., Cho, H.J., and Yang, J.W. 2003. Removal characteristics of anionic metals by micellar-enhanced ultrafiltration. *Journal of Hazardous Material* B99: 303–311.

Banat, I.M., Franzetti, A., Gandolfi, I., Bestetti, G., Martinotti, M.G., Fracchia, L., Smyth, T.J., and R. Marchant. 2010. Microbial biosurfactants production, applications and future potential. *Applied Microbiology and Biotechnology* 87: 427–444.

Becerra-Castro, C., Monterroso, C., Prieto-Fernández, A., Rodríguez-Lamas, L., Loureiro-Viñas, M., Acea, M.J., and Kidd, P.S. 2012. Pseudometallophytes colonising Pb/Zn mine tailings: A description of the plant-microorganism-rhizosphere soil system and isolation of metal-tolerant bacteria. *Journal of Hazardous Materials* 217–218: 350–359.

Bhaskar, P.V. and Bhosle, N.B. 2006. Bacterial extracellular polymeric substance (EPS): A carrier of heavy metals in the marine food-chain. *Environment International* 32: 191–198.

Bodagh, A., Khoshdast, H., Sharafi, H., Zahiri, H.S., and Noghabi, K.A. 2013. Removal of cadmium(II) from aqueous solution by ion flotation using rhamnolipid biosurfactant as an ion collector. *Industrial & Engineering Chemistry Research* 52: 3910–3917. doi: 10.1021/ie400085t.

Bondarenko, O., Rahman, P.K.S.M., Rahman, T.J., Kahru, A., and Ivask, A. 2010. Effects of rhamnolipids from *pseudomonas aeruginosa* DS10-129 on luminescent bacteria: Toxicity and modulation of cadmium bioavailability. *Microbial Ecology* 59: 588–600.

Bramhachari, P.V., Ravichand, J., Deepika, K.V., Yalamanda, P., and Chaitanya, K.V. 2012. Differential responses of marine sediment bacteria Pseudomonas stutzeri strain VKMO14 to chromate exposures. *Research Journal of Microbiology* 7: 114–122.

Chen, H.-R., Chen, C.-C. A., Reddy, S., Chen, C.-Y.,Li, W. R.,Tseng, M.-J., Liu, H.-T., Pan, W., Maity, J.P., and Atla, S.B. 2011. Removal of mercury by foam fractionation using surfactin, a biosurfactant. *International Journal of Molecular Sciences* 12: 8245–8258. doi: 10.3390/ijms12118245.

Chen, W.-J., Hsiao, L.-C., and Chen, K.K.-Y. Metal desorption from copper(II)/nickel(II)-spiked kaolin as a soil component using plant-derived saponin biosurfactant (2008) *Process Biochemistry* 43: 488–498.

Croal, L.R., Gralnick J.A., Malasarn, D., and Newman, D.K. 2004. The genetics of geochemistry. *Annual Review of Genetics* 38: 175–120.

Dahrazma, B. and Mulligan, C.N. 2007. Investigation of the removal of heavy metals from sediments using rhamnolipid in a continuous flow configuration. *Chemosphere* 69: 705–711.

Dahrazma, B., Maulligan, C.N., and Nieh, M.P. 2008. Effects of additives on the structure of rhamnolipid (biosurfactant): A small-angle neutron scattering (SANS) study. *Journal of Colloid and Interface Science* 319: 590–593.

Das, P., Mukherjee, S., and Sen, R. 2009. Biosurfactant of marine origin exhibiting heavy metal remediation properties. *Bioresource Technology* 100: 4887–4890.

El Zeftawy, M.A.M. and Mulligan, C.N. 2011. Use of rhamnolipid to remove heavy metals from wastewater by micellar-enhanced ultrafiltration (MEUF). *Separation and Purification Technology* 77: 120–127.

Ferella, F., Prisciandaro, M., Michelis, I.D., and Veglio, F. 2007. Removal of heavy metals by surfactant-enhanced ultrafiltration from wastewater. *Desalination* 207: 125.

Franzetti, A., Di Gennaro, P., Bevilacqua, A., Papacchini, M., and Bestetti, G. 2006. Environmental features of two commercial surfactants widely used in soil remediation. *Chemosphere* 62: 1474–1480.

Franzetti, A., Gandolfi, I., Raimondi, C., Bestetti, G., Banat, I.M., Smyth, T.J., Papacchini, M., Cavallo, M., and Fracchia, L. 2012. Environmental fate, toxicity, characteristics and potential applications of novel bioemulsifiers produced by *Variovorax paradoxus* 7bCT5. *Bioresource Technology* 108: 245–251.

Gadd, G.M. 2010. Metals, minerals and microbes: Geomicrobiology and bioremediation. *Microbiology* 156: 609–643.

Gao, L., Kano, N., Sato, Y.,Li, C., Zhang, S., and Imaizumi, H. 2012. Behavior and distribution of heavy metals including rare earth elements, thorium, and uranium in sludge from industry water treatment plant and recovery method of metals by biosurfactants application. *Bioinorganic Chemistry and Applications*, Article ID 173819, doi: 10.1155/2012/173819.

Gnanamani, A., Kavitha, V., Radhakrishnan, N., Suseela Rajakumar, G., Sekaran, G., and Mandal, A.B. 2010. Microbial products (biosurfactant and extracellular chromate reductase) of marine microorganism are the potential agents reduce the oxidative stress induced by toxic heavy metals. *Colloids and Surfaces B: Biointerfaces* 79: 334–339.

Gusiatin, Z.M. and Klimiuk, E. 2012. Metal (Cu, Cd and Zn) removal and stabilization during multiple soil washing by saponin. *Chemosphere* 86: 383–391.

Gutierrez, T., Shimmield, T., Haidon, C., Black, K., and Green, D.H. 2008. Emulsifying and metal ion binding activity of a glycoprotein exopolymer produced by *Pseudoalteromonas* sp. strain TG12. *Applied and Environmental Microbiology* 74: 4867–4876.

Haferburg, G. and Kothe, E. 2007. Microbes and metals: Interactions in the environment. *Journal of Basic Microbiology* 47: 453–467.

Herman, D.C., Artiola, J.F., and Miller, R.M. 1995. Removal of cadmium, lead, and zinc from soil by a rhamnolipid biosurfactant. *Environmental Science and Technology* 29: 2280–2285.

Hong, J.J., Yang, S.M., Lee, C.H., Choi, Y.K., and Kajiuchi, T. 1998. Ultrafiltration of divalent metal cations from aqueous solution using polycarboxylic acid type biosurfactant. *Journal of Colloid and Interface Science* 202: 63–73.

Huang, W. and Liu, Z.M. 2013. Biosorption of Cd(II)/Pb(II) from aqueous solution by biosurfactant-producing bacteria: Isotherm kinetic characteristic and mechanism studies. *Colloids and Surfaces B: Biointerfaces* 105: 113–119.

Jensen, J.K., Holm, P.E., Nejrup, J., and Borggaard, O.K. 2011. A laboratory assessment of potentials and limitations of using EDTA, rhamnolipids, and compost-derived humic substances (HS) in enhanced phytoextraction of copper and zinc polluted calcareous soils. *Soil and Sediment Contamination* 20: 777–789.

Juwarkar, A.A., Dubey, K.V., Nair, A., and Singh, S.K. 2008. Bioremediation of multi-metal contaminated soil using biosurfactant—A novel approach. *Indian Journal of Microbiology* 48: 142–146.

Juwarkar, A.A., Nair, A., Dubey, K.V., Singh, S.K., and Devotta, S. 2007. Biosurfactant technology for remediation of cadmium and lead contaminated soils. *Chemosphere* 68: 1996–2002.

Kiliç, E., Font, J., Puig, R., Çolak, S., and Çelik, D. 2011. Chromium recovery from tannery sludge with saponin and oxidative remediation. *Journal of Hazardous Materials* 185: 456–462.

Kim, J. and Vipulanandan, C. 2006. Removal of lead from contaminated water and clay soil using a biosurfactant. *Journal of Environmental Engineering-Asce* 132: 777–786.

Lima, T.M.S., Procópio, L.C., Brandão, F.D., Carvalho, A.M.X., Tótola, M.R., and Borges, A.C. 2011. Simultaneous phenanthrene and cadmium removal from contaminated soil by a ligand/biosurfactant solution. *Biodegradation* 22: 1007–1015.

Liu, X., Wang, J.-T., Zhang, M., Wang, L., and Yang, Y.-T. 2013. Remediation of Cu-Pb-contaminated loess soil by leaching with chelating agent and biosurfactant. *Huanjing Kexue/Environmental Science* 34: 1590–1597.

Maity, J.P., Huang, Y.M., Hsu, C.-M., Wu, C.-I., Chen, C.-C., Li, C.-Y., Jean, J.-S., Chang, Y.-F., and Chen. C.-Y. 2013. Removal of Cu, Pb and Zn by foam fractionation and a soil washing process from contaminated industrial soils using soapberry-derived saponin: A comparative effectiveness assessment. *Chemosphere* 92: 1286–1293.

Makkar, R.S., Cameotra, S.S., and Banat, I.M. 2011. Advances in utilization of renewable substrates for biosurfactant production. *Applied Microbiology and Biotechnology Express* 1: 5. doi: 10.1186/2191-0855-1-5

Marchant, R. and Banat, I.M. 2012a. Biosurfactants: A sustainable replacement for chemical surfactants? *Biotechnology Letters* 34: 1597–1605.

Marchant, R. and Banat, I.M. 2012b. Microbial biosurfactants: Challenges and opportunities for future exploitation. *Trends in Biotechnology* 30: 558–565.

Menezes, C.T.B., Barros, E.C., Rufino, R.D., Luna, J.M., and Sarubbo, L.A. 2011. Replacing synthetic with microbial surfactants as collectors in the treatment of aqueous effluent produced by acid mine drainage, using the dissolved air flotation technique. *Applied Biochemistry and Biotechnology* 163: 540–546.

Miller, R.M. 1995. Biosurfactant-facilitated remediation of metal-contaminated soils. *Environmental Health Perspectives* 103: 59–62.

Mulligan, C.N., Yong, R.N., and Gibbs, B.F. 2001. Heavy metal removal from sediments by biosurfactants. *Journal of Hazardous Materials* 85: 111–125.

Paria, S. 2008. Surfactant-enhanced remediation of organic contaminated soil and water. *Advances in Colloid and Interface Science* 138: 24–58.

Pekdemir, T., Tokunaga, S., Ishigami, Y., and Hong, K.J. 2000. Removal of cadmium or lead from polluted water by biological amphiphiles. *Journal of Surfactants and Detergents* 3: 43–46.

Rajkumar, M., Sandhya, S., Prasad, M.N.V., and Freitas, H. 2012. Perspectives of plant-associated microbes in heavy metal phytoremediation. *Biotechnology Advances* 30: 1562–1574.

Rangarajan, V. and Sen, R. 2013. An inexpensive strategy for facilitated recovery of metals and fermentation products by foam fractionation process. *Colloids and Surfaces B-Biointerfaces* 104: 99–106.

Raza, Z.A., Khalid, Z.M., Khan, M.S., Banat, I.M., Rehman, A., Naeem, A., and Saddique, M.T. 2010. Surface properties and sub-surface aggregate assimilation of rhamnolipid surfactants in different aqueous system. *Biotechnology Letters* 32: 811–816.

Rufino, R.D., Luna, J.M., Campos-Takaki, G.M., Ferreira, S.R.M., and Sarubbo, L.A. 2012. Application of the biosurfactant produced by *Candida lipolytica* in the remediation of heavy metals. *Chemical Engineering Transactions* 27: 61–66.

Rufino, R.D., Rodrigues, G.I.B., Campos-Takaki, G.M., Sarubbo, L.A., and Ferreira, S.R.M. 2011. Application of a yeast biosurfactant in the removal of heavy metals and hydrophobic contaminant in a soil used as slurry barrier. *Applied and Environmental Soil Science.* Article ID 939648, doi:10.1155/2011/939648.

Singh, A., Van Hamme, J.D., and Ward, O.P. 2007. Surfactants in microbiology and biotechnology: Part 2. Application aspects. *Biotechnology Advances* 25: 99–121.

Singh, A.K. and Cameotra, S.S. 2013a. Rhamnolipids production by multi-metal-resistant and plant-growth-promoting rhizobacteria. *Applied Biochemistry and Biotechnology* 170: 1038–1056.

Singh, A.K. and Cameotra, S.S. 2013b. Efficiency of lipopeptide biosurfactants in removal of petroleum hydrocarbons and heavy metals from contaminated soil. *Environmental Science and Pollution Research* 20: 7367–7376.

Singh, P. and Cameotra, S.S. 2004. Potential applications of microbial surfactants in biomedical sciences. *Trends in Biotechnology* 22: 142–146.

Slizovskiy, I.B., Kelsey, J.W., and Hatzinger, P.B. 2011. Surfactant-facilitated remediation of metal-contaminated soils: Efficacy and toxicological consequences to earthworms. *Environmental Toxicology and Chemistry* 30: 112–123.

Sriram, M.I., Gayathiri, S., Gnanaselvi, U., Jenifer, P.S., Mohan Raj, S., and Gurunathan, S. 2011. Novel lipopeptide biosurfactant produced by hydrocarbon degrading and heavy metal tolerant bacterium *Escherichia fergusonii* KLU01 as a potential tool for bioremediation. *Bioresource Technology* 102: 9291–9295.

Van Hamme, J.D., Singh, A., and Ward, O.P. 2006. Physiological aspects. Part 1 in a series of papers devoted to surfactants in microbiology and biotechnology. *Biotechnology Advances* 24: 604–620.

Volkering, F., Breure, A.M., and Rulkens, W.H. 1997. Microbiological aspects of surfactant use for biological soil remediation. *Biodegradation* 8: 401–417.

Wang, S. and Mulligan, C.N. 2004. Rhamnolipid foam enhanced remediation of cadmium and nickel contaminated soil. *Water, Air, and Soil Pollution* 157: 315–330.

Wang, S. and Mulligan, C.N. 2009. Arsenic mobilization from mine tailings in the presence of a biosurfactant. *Applied Geochemistry* 24: 928–935.

Wen, J., McLaughlin, M.J., Stacey, S.P., and Kirby, J.K. 2010. Is rhamnolipid biosurfactant useful in cadmium phytoextraction? *Journal of Soils and Sediments* 10: 1289–1299.

Wen, J., Stacey, S.P., McLaughlin, M.J., and Kirby, J.K. 2009. Biodegradation of rhamnolipid, EDTA and citric acid in cadmium and zinc contaminated soils. *Soil Biology and Biochemistry* 41: 2214–2221.

Yuan, X.Z., Meng, Y.T., Zeng, G.M., Fang, Y.Y., and Shi, J.G. 2008. Evaluation of tea-derived biosurfactant on removing heavy metal ions from dilute wastewater by ion flotation. *Colloids and Surfaces a-Physicochemical and Engineering Aspects* 317: 256–261.

Zouboulis, A.I., Matis, K.A., Lazaridis, N.K., and Golyshin, P.N. 2003. The use of biosurfactants in flotation: Application for the removal of metal ions. *Minerals Engineering* 16: 1231–1236.

Index

A

Absorbance units (AU), 103
Acinetobacter, 78, 174
    radioresistens, 302
ACP, *see* Acyl-carrier protein (ACP)
Active transport, 23
Acyl-carrier protein (ACP), 93
    β-hydroxyacyl-, 94
Adsorptive chromatography, 146; *see also*
        Chromatography
Agricultural BS, 257–259, 278, 313, 314; *see also*
        Applications of BS
    agrochemical formulation, 320–321
    antimicrobial mechanism, 317
    biopesticides, 317
    in disease control, 315–317
    in immunity development, 318
    in pest control, 318–319
    in plant–microbe interactions, 320
    plant–pathogen elimination, 279
    soil quality improvement, 279
    in soil remediation, 319–320
Agrochemical market, 314
Air sparging, 333
Alasane, 11; *see also* High-molecular
        weight BS
Alcanivorax borkumensis, 77
Alkyl polyglucosides (APGs), 3
Amphiphilics, 3; *see also* Microbial surfactants
Anidulafungin, 248; *see also* Industrial
        applications of BS
ANN, *see* Artificial neural network (ANN)
ANN-GA-based models, 62; *see also* Biosurfactant
        production
Antibiofilm biosurfactants, 298
Antifoaming agents, 119–120
APGs, *see* Alkyl polyglucosides (APGs)
Apolipoprotein, 51; *see also* Biosurfactant (BS)
Applications of BS, 271, 288; *see also* Agricultural BS;
        Food industry BS; Industrial applications;
        Medicinal BS; Production enhancement
        of BS
    in biodegradation processes, 271–274
    biological applications, 270
    in cosmetic industry, 275
    heavy metal desorption, 274
    metals remediation, 274–275
    microbial enhanced oil recovery, 275–276
    oil recovery and processing, 275
    V2-7 bioemulsifier, 273
Aqueous-two-phase-systems (ATPS), 143
Artificial neural network (ANN), 62
ATPS, *see* Aqueous-two-phase-systems
        (ATPS)
AU, *see* Absorbance units (AU)

B

Bacillus genus, 170–171
    mojavensis, 316
    subtilis, 78
Bacteria, 176–177
Bacterial production of BS, 73, 79
    analytical method standardization, 74
    glycolipids, 77
    high molecular mass bioemulsifiers, 78–79
    lipopeptide biosurfactants, 78
    rhamnolipids, 75–77
    sophorolipids, 73
    yeasts, 73
Batch cultivation, 117–118; *see also* Bioreactors for
        BS production
β-(1,3)-ᴅ-glucan, 248
Biochemical oxygen demand (BOD), 304
Biodegradation of oil-contaminated soil, 355
Bioemulsifiers, 245; *see also* Biosurfactants
    in cosmetic industry, 251
Biofilm
    content, 296
    development, 296
    microorganisms, 297
    -producing bacteria, 300
Biofilm control, 296
    antibiofilm biosurfactants, 298
    cellular hydrophobicity, 296
    on food-processing surfaces, 297–298
    rhamnolipids, 298, 299
    surfaces sanitization, 297
    surface types, 298
    surfactin, 298, 299
Bioprocess development, 105
Bioreactors for BS production, 117, 125–126
    batch cultivation, 117–118
    bubble-free membrane bioreactor, 121–123
    continuous cultivation, 118
    dispersing, 118
    fed-batch cultivation, 118
    fermentation strategies, 117
    foam avoidance, 121
    foam disruption, 119
    foam fractionation, 123–125
    foaming, 119
    obstacles, 118
    rotating disc bioreactor, 121
    rushton turbines, 118–119
    solid state cultivation, 123
Bioremediation, 49, 271, 328, 329; *see also*
        Applications of BS
    advantages of, 328
    air sparging, 333
    biopile, 336
    biosparging, 333

Bioremediation (*Continued*)
  bioventing, 333
  composting, 336
  of contaminants, 347
  engineered ex situ, 334
  engineered in situ, 330–334
  intrinsic in situ, 329–330
  land farming, 335–336
  liquid circulating systems, 334
  mediators of, 347
  procedure, 345
  slurry phase bioremediation, 336–337
  solid-phase treatment, 335
Bioremediation project designing, 337
  bioremediation procedure, 345
  contaminant degradation conditions, 338
  contaminants characteristics, 337–339
  decision tree, 340
  environmental characteristics, 339
  field activities, 342
  method selection, 339
  through phytoremediation, 319
  site and soil characterization, 337
  site closure, 344
  treatment, 344
  treatment facility, 343, 336
Biosurfactant (BS), 3, 49, 117, 130, 153, 245, 270, 295,
      313, 346, 366; *see also* Biofilm control;
      Biosurfactant production; Microbial
      surfactants
  advantages, 313, 347
  antiadhesive property, 295
  antimicrobial properties of, 300–301
  applications of, 65, 66–67, 186
  aspects of, 271
  -based production companies, 168–169
  beneficial properties, 4
  cationic, 4
  cHAL, 255
  challenges and opportunities, 65, 67
  chemical structures of, 38, 52
  classes of, 129
  classification, 50–54, 246, 271
  cost analysis, 130–131
  downstream processing, 130, 131
  as emulsifiers, 301–303
  fermentation media conditions, 41
  flavolipids, 52
  glycolipid, 51
  heavy metal desorption, 274
  industrial applications of, 273
  limitation in, 13
  lipopeptide, 299
  lipopeptides and lipoproteins, 51
  manufacturing companies, 153–154, 168
  market volume, 153
  micelles, 51
  microbial, 272, 316
  particulate, 51–52
  phospholipids and fatty acids, 51
  polymeric, 51
  properties, 84, 165–166, 246
  publications on, 130
  raw materials, 102

  in soil bioremediation, 352, 354
  upstream processing, 131
  uses of, 44
  in various applications, 272
Biosurfactant from food wastes, 303
  in dairy industry, 304–305
  in edible oil industry, 305
  end-point criteria, 346
  in starch industry, 306
  in sugar and fruit-processing industry, 304
Biosurfactant production, 54; *see also*
      Biosurfactant (BS)
  alternate substrates, 181, 280
  ANN-GA-based models, 62
  chain of events in, 54
  environmental factors, 57
  extraction and purification, 63–65
  factors affecting, 54, 63
  GMO and mutants for, 59
  high-quality, 50
  low-cost carbon sources for, 55
  lyophilizer, 63
  media optimization technique, 62
  nutritional factors, 55–57
  recombinant DNA technology and GMOs, 57–63
  recovery methods, 63
  substrates for, 133, 184–186
Biosurfactants-enhanced bioremediation
      process, 346
  bioavailability increase, 352
  biodegradation of hydrophobic compounds, 353
  hydrophobic compound bioavailability, 353
  in soil bioremediation, 352, 354
  studies on, 347, 348–355
  surface area increase, 352
Biosurfactants vs. chemical surface-active agents,
      37, 45–46
  alkyl polyglycoside production, 43
  biodegradability, 45
  carbohydrates to alkyl polyglycosides, 41–42
  chemical structures, 38, 39
  comparison approaches, 40
  ecotoxicity, 45
  environmental impact, 45
  FAMEE, 42–43
  from fatty alcohol to lauryl ether sulfate, 42
  LES synthesis, 43
  nonylfenol ethoxylate, 45
  production, 40
  sophorolipids, 44
  substance selection, 37
  synthesis, 41
  usage, 43–45
Biuret test, 113
BOD, *see* Biochemical oxygen demand (BOD)
Brevibacillus brevis, 279
BS, *see* Biosurfactant (BS)
Bubble-free membrane bioreactor, 121–123; *see also*
      Bioreactors for BS production
  membranes, 121
  parameter of, 123
  PP module, 122
Burkholderia, 76
  thailandensis, 75

C

CAGR, *see* Compound annual growth rate (CAGR)
cAMP, *see* Cyclic adenosine monophosphate (cAMP)
Capital expenditure, 155; *see also* Cost analysis of BS production
Carbohydrate-based substrates, 181
Carotenoid glycosiden, 9
Caspofungin, 248; *see also* Industrial applications of BS
Cassava wastewater, 285
Cationic biosurfactants, 4
Cationic metal species, 274
CCAJ, *see* Clarified cashew apple juice (CCAJ)
Cellobioselipids, 7, 8; *see also* Glycolipid
Cell separation, 140–141; *see also* Purification of BS
Cellular hydrophobicity, 296
Cellulose, 88
Centrifugal partition chromatography, 146; *see also* Chromatography
CESIO, *see* European Committee for Surfactants and Their Organic Intermediates (CESIO)
CF, *see* Clearance factor (CF)
cHAL biosurfactant, 255
Chemically derived surfactants, 38, 245
    chemical structures of, 39
    cocamidopropyl betaine, 40
    in cosmetic industry, 249
    cumene sulfonate, 39–40
    esterquat, 39
    feedstocks and reaction conditions for, 41
Chemical oxygen demand (COD), 284
Chemical surfactants, 50, 153; *see also* Biosurfactant (BS)
    industrial applications of, 273
    worldwide consumption of, 73
Chromatography, 145; *see also* Purification of BS
    adsorption material, 145
    adsorptive, 146
    centrifugal partition chromatography, 146
    disadvantage, 146
    for rhamnolipids, 145
Clarified cashew apple juice (CCAJ), 282
Clearance factor (CF), 137; *see also* Downstream processing
CLP, *see* Cyclic lipopeptide (CLP)
CMC, *see* Critical micelle concentration (CMC)
CNS, *see* Coagulase-negative *Staphylococci* (CNS)
Coagulase-negative *Staphylococci* (CNS), 248
Cocamidopropyl betaine, 40; *see also* Chemically derived surfactants
COD, *see* Chemical oxygen demand (COD)
Colopsinol A, 8; *see also* Polyketideglycosids
Compound annual growth rate (CAGR), 314
Concentration factor (CTF), 137; *see also* Downstream processing
Consumables 157; *see also* Cost analysis of BS production
Contaminant degradation condition, 338
Contaminants characteristics, 337–339
Continuous cultivation, 118; *see also* Bioreactors for BS production
Cord factor, 6
Corn steep liquor, 89
Cost analysis of BS production, 153; *see also* Production costing of BS
    commercial production strategies, 159

    consumables, 157
    cost estimation, 155
    lipopeptide biosurfactants, 154
    operating costs distribution, 158
    R&D vs. commercial, 158–159
    volumetric productivity, 155
    yields and cost parameters, 154
Critical micelle concentration (CMC), 20, 295, 352
Crude oil-based substrates, 181
Crystallization, 144–145; *see also* Purification of BS
CTAB/methylene-blue agar, 113
CTF, *see* Concentration factor (CTF)
Cumene sulfonate, 39–40; *see also* Chemically derived surfactants
Cyclic adenosine monophosphate (cAMP), 277
Cyclic lipopeptide (CLP), 318

D

Dairy wastewater, 283–284
Daptomycin, 247–248; *see also* Industrial applications of BS
Design of experiment (DOE), 107
Dextrin 100, 254
D-Glucose-6P, 93
Di-rhamnolipids, 84
Dispersing agents, *see* Biosurfactants
Distillery wastes (DWs), 280
DOE, *see* Design of experiment (DOE)
Downstream processing, 130, 131, 147; *see also* Purification of BS
    clearance factor, 137
    concentration factor, 137
    cost of substrates, 134
    feeding of waste streams, 135
    influence process steps, 132
    metabolic engineering, 136
    microbial production influence, 131
    optimized cultivation operation, 136
    patents on, 190–191
    purification factor, 137
    space–time yield, 135
    unit operations, 137–138
DWs, *see* Distillery wastes (DWs)

E

EB-162, 171, 185; *see also* Patents; Surfactin
Echinocandins, 248; *see also* Industrial applications of BS
ED, *see* Entner–Doudoroff (ED)
Elaiophylin, 8, 9; *see also* Polyketideglycosids
Emulsan, 11, 52; *see also* Biosurfactant (BS); High-molecular weight BS
Emulsifier, 301, *see* Biosurfactant (BS)
    lecithin, 301
    nanoemulsions, 303
    producing microorganisms, 302
Emulsion, 276
Endoplasmic reticulum (ER), 23
Enhanced oil recovery (EOR), 256; *see also* Industrial applications of BS
Entner–Doudoroff (ED), 93
EOR, *see* Enhanced oil recovery (EOR)
EPS, *see* Exopolysaccharides (EPS)

ER, *see* Endoplasmic reticulum (ER)
Esterquat, 39; *see also* Chemically derived surfactants
European Committee for Surfactants and Their Organic
        Intermediates (CESIO), 42
Exopolysaccharides (EPS), 78–79, 296
Extraction, 142–143; *see also* Purification of BS

F

Face-centered central composite design (FCCCD), 109
FAMEE, *see* Fatty alcohol methyl ester ethoxylate (FAMEE)
Fatty acids, 5, 51; *see also* Biosurfactant (BS); Microbial
        surfactants
Fatty alcohol methyl ester ethoxylate (FAMEE), 38, 42
    ethoxylation reaction for formation of 43
    production of, 44
FCCCD, *see* Face-centered central composite design
        (FCCCD)
Fed-batch cultivation, 118; *see also* Bioreactors for BS
        production
Field remediation activities, 342
Filamentous fungi, 181
Flavolipids, 52; *see also* Biosurfactant (BS)
Foamate, 123
Foam disruption, 119; *see also* Bioreactors for BS
        production
    antifoaming agents, 119–120
    mechanical, 120
    minimum velocity, 120
Foam fractionation, 123; *see also* Bioreactors for BS
        production
    columns, 124
    drawback of, 125
    enrichment factor, 125
    recovery, 125
Foam separation, 139–140; *see also* Purification of BS
Food emulsifier, 276
Food industry BS, 276, 295; *see also* Applications of BS;
        Biofilm control; Biosurfactant from food wastes
    emulsifier, 276
    quorum-sensing factors, 299
    rhamnolipid surfactants, 298
    stabilizer, 276
Food stabilizer, 276
FPW, *see* Fruit-processing waste (FPW)
fru-1,6-2P, *see* Fructose-1,6-bisphosphate (fru-1,6–2P)
fru-6P, *see* Fructose-6-phosphate (fru-6P)
Fructose-1,6-bisphosphate (fru-1,6–2P), 93
Fructose-6-phosphate (fru-6P), 93
Fruit-processing waste (FPW), 284

G

GA, *see* Genetic algorithm (GA)
gap, *see* Glyceraldehyde-3-phosphate (gap)
Generally regarded as safe (GRAS), 62
Genetic algorithm (GA), 62
Genetically modified organisms (GMOs), 57, 302
Gibbs–Marangori-Elasticity, 119; *see also* Bioreactors for
        BS production
GISA, *see* Glycopeptide-intermediate-susceptible
        S. aureus (GISA)
Glucoselipids, 7; *see also* Glycolipid
Glyceraldehyde-3-phosphate (gap), 93

Glycerol, 86–88
Glycolipid, 5; *see also* Microbial surfactants
    cellobioselipids, 7, 8
    cord factor, 6
    glucoselipids, 7
    MELs, 7
    mycol acids, 6
    from nonpathogenic bacteria, 77
    rhamnolipids, 6
    sophorolipid, 6
    trehalose-dicorynomycolate, 7
Glycolipid biosurfactants, 51; *see also* Biosurfactant (BS);
        Sophorolipid (SL)
    in cosmetic industry, 250
Glycopeptide-intermediate-susceptible *S. aureus*
        (GISA), 248
GMOs, *see* Genetically modified organisms (GMOs)
Gram-negative species, 9
Gram-positive species, 9, 21
GRAS, *see* Generally regarded as safe (GRAS)

H

HAA, *see* 3-(3-Hydroxyalkanoyloxy) alkanoic acid (HAA)
Heavy metal removal by BS, 361
    biosurfactant–metal interactions, 361–363
    from contaminated sites, 364–366
    metal removal, 363–364
Hemicellulose, 88–89
Herpes simplex virus (HSV), 277
Hexokinase, 93
High molecular mass bioemulsifiers, 78
High-molecular weight BS, 11; *see also* Microbial
        surfactants
    alasane, 11
    emulsan, 11
    lipoheteropolysaccharides, 11
    RAG-1 emulsan, 12
HSV, *see* Herpes simplex virus (HSV)
Hydrophobic substrates, 136
Hydrotropes, 39
3-(3-Hydroxyalkanoyloxy) alkanoic acid (HAA), 93

I

Immobilization, 140; *see also* Purification of BS
Induced systemic resistance (ISR), 318
Industrial applications of BS, 245, 259–260, 273;
        see also Agricultural BS; Applications
        of BS; Biosurfactant (BS); Lipopeptides;
        Rhamnolipid; Surfactin
    as antioxidant agents, 253
    biofilm preventive strategies, 249
    as biological control agents, 257–258
    biomedical and pharmaceutical applications, 247–249
    cHAL biosurfactants, 255
    in cosmetics industry, 249–251
    in detergent formulations, 251–252
    in enhanced oil recovery, 256–257
    in food industry, 252–254
    glycolipid biosurfactants, 250
    Lactobacillus biosurfactants, 249
    manufacturing industries, 259
    MELs, 249

prospect for, 246–247
rhamnolipids compositions, 248
sophorolipids, 250
sugar-based biosurfactants, 249
surfactant vendors, 247
in textile industry, 254–255
trans-membrane ion influxes, 247
Industrial biotechnology, *see* White biotechnology
Ionophorics, 8; *see also* Polyketideglycosids
ISR, *see* Induced systemic resistance (ISR)
Iturin A, 276–277; *see also* Medicinal BS

K

KEGG, *see* Kyoto Encyclopedia of Genes and Genomes
        (KEGG)
Klebsiella pneumoniae, 107
Kyoto Encyclopedia of Genes and Genomes (KEGG), 92

L

Lactobacillus biosurfactants, 249
Lactonic diacetylated sophorolipid, 20; *see also*
        Biosurfactant
Land farming, 337
Lauryl ether sulfate (LES), 38
    two-stage synthesis of, 43
Lauryl polyglycoside, 38
    formation reaction of, 42
Lecithin, 301
LES, *see* Lauryl ether sulfate (LES)
Leu7-Surfactin, 258
Light nonaqueous phase liquid (LNAPL), 332
Lignin, 89
Lignocellulose, 88, 285
    biomass, 88–89
    classification, 89
    drawback of, 92
Lipofection, 277; *see also* Medicinal BS
Lipoheteropolysaccharides, 11
Lipopeptides, 5, 9, 51, 247, 298; *see also* Biosurfactant
        (BS); Industrial applications of BS; Microbial
        surfactants; Surfactin
    antifungal activity, 258
    antimicrobial action, 248
    BS, 78, 154
    cosmetics-containing, 251
    cyclic lipopeptides, 258–259
    daptomycin, 247–248
    echinocandins, 248
    as emulsifiers, 251
    -like surfactin, 154
    micro organisms synthesizing, 246
    in oil recovery, 256
    polymyxines, 10, 247
    serratamolide, 10
    serrawettin, 10
    surfactants, 300
    surfactin, 9–10
    synthetically modified, 248
    viscosin, 10
Lipopolysaccharide (LPS), 30
Lipoproteins, 51; *see also* Biosurfactant (BS)
Liquid circulating systems, 334

Listeria monocytogenes, 297
LNAPL, *see* Light nonaqueous phase liquid (LNAPL)
Low-molecular weight BS, 5; *see also* Microbial
        surfactants
LPS, *see* Lipopolysaccharide (LPS)
L-Rhamnose, 276
Lyophilizer, 63; *see also* Biosurfactant production

M

MAMPs, *see* Microbe-associated molecular patterns
        (MAMPs)
Mannosylerythritol lipids (MELs), 7, 50, 139, 246, 300;
        see also Biosurfactant (BS); Glycolipid;
        Industrial applications of BS
    as cosmetic ingredients, 250
    lipid-A, 174
MAP, *see* Mitogen-activated protein (MAP)
Mass spectrometry (MS), 74
Medicinal BS, 276; *see also* Applications of BS
    antiadhesive agents, 277–278
    antimicrobial activity, 277, 278
    genetic manipulation, 277
    immune modulatory action, 277
    iturin A, 276–277
    lipofection, 277
    motility and biofilm formation, 278
MEOR, *see* Microbial-enhanced oil recovery (MEOR)
Methyl tertiary-butyl ether (MTBE), 64, 142
Micelles, 51; *see also* Biosurfactant (BS)
Microbe-associated molecular patterns (MAMPs), 318
Microbial carotenoids, 9; *see also* Microbial surfactants
Microbial-enhanced oil recovery (MEOR), 256–257,
        275–276; *see also* Applications of BS
Microbial metabolites, 3
Microbial surfactants, 3, 12–13, 84, 101, 270, 301, 347; *see*
        also Biosurfactants
    advantages, 153
    application fields, 183, 192
    benefits, 84
    fatty acids and phospholipids, 5
    glycolipids, 5–8
    high-molecular weight BS, 11
    isoprenoide and carotenoid glycolipids, 9
    lipopeptides, 9
    low-molecular weight BS, 5
    microbial BSs, 295
    polyketideglycosids, 8
    proteins, 12
    spiculisporic acid, 11
Microemulsion systems, 302
Microorganisms in food industries, 297
Mitogen-activated protein (MAP), 318
Molasses, 86, 284, 304
Molecular weight cut-offs (MWCO), 141
MS, *see* Mass spectrometry (MS)
MTBE, *see* Methyl tertiary-butyl ether (MTBE)
MWCO, *see* Molecular weight cut-offs (MWCO)
Mycol acids, 6

N

Nanoemulsions, 303
National Center for Biotechnology Information (NCBI), 92

NCBI, *see* National Center for Biotechnology Information (NCBI)
Nonpolar solvents, 142
Nonylfenol ethoxylate, 45

O

OFAT method, *see* One-factor-at-a-time method (OFAT method)
Oil-based substrates, 181
Oil-in-water (O/W), 301
One-factor-at-a-time method (OFAT method), 106
Oral secretion (OS), 319
OS, *see* Oral secretion (OS)
Osmotolerant, 22
O/W, *see* Oil-in-water (O/W)

P

PAH, *see* Polycyclic aromatic hydrocarbons (PAH)
PAHs, *see* Poly aromatic hydrocarbons (PAHs)
Palm oil mill effluent (POME), 102
Palm sludge for BS production, 101, 113
    biuret test, 113
    co-substrate determination, 107–108
    identification of BS, 112–113
    Klebsiella pneumoniae, 107
    media requirement optimization, 108–112
    medium component selection, 103–104
    medium constituent validation, 112
    microbial isolation and screening, 103
    microbial isolation to media screening, 102
    multivariate optimization, 106–107, 108–112
    nonsignificant medium components, 105
    effect of nutritional constituent, 104
    optimization study, 105, 109
    quadratic model validation, 112
    response surface plots, 110–111
    significant medium components, 104–105
    SPO as substrate, 102–103
    univariate optimization, 106, 108
Particulate biosurfactants, 51–52; *see also* Biosurfactant (BS)
Patents, 165, 220–221
    Acinetobacter genus, 174, 175
    activity on immune system, 194–195
    in agriculture applications, 202–204
    alternative substrates, 181
    on antiadhesion activity, 193
    on antimicrobial activity, 192
    on antitumor activity, 193–194
    applicants, 167
    Bacillus genus, 170–171
    bacteria, 176–177
    in bioremediation applications, 192–197, 204–209
    BS-based product producing companies, 168–169
    BS producer organisms, 169
    BS producer screening, 177, 180–181
    on BS production by GM strains, 182–183
    on chemical modification, 189
    in chemical production applications, 215–220
    commercial activity, 168–169
    in cosmetic applications, 198–201
    country of origin, 167
    criteria for patent search, 166
    in detergent applications, 213–215
    distribution with respect to applicant types, 168
    on downstream processes, 190–191
    filamentous fungi, 181
    on foam removal apparatus, 189–190
    genetic modifications, 181
    on immobilization, 189
    of medical biosurfactants, 195–197
    in oil recovery applications, 210–212
    patenting activity, 166–167
    on processes and reactors, 187–189
    producer microorganisms, 169–177
    production processes, 183
    Pseudomonas genus, 172–173
    purification processes, 183
    related to application, 192
    on simultaneous production/separation, 190
    techniques, 177
    yeast, 178–180
PEG, *see* Polyethylene-glycol (PEG)
PF, *see* Purification factor (PF)
PHA, *see* Polyhydroxyalkanoate (PHA)
Phase separation, 139; *see also* Purification of BS
Phosphatidylethanolamine, 5
Phospholipase A2 (PLA2), 277
Phospholipids, 5, 51, 52; *see also* Biosurfactant (BS); Microbial surfactants
Phytoremediation, 49
PLA2, *see* Phospholipase A2 (PLA2)
Plant sophorose glycosides, 31; *see also* Biosurfactant
PLL, *see* Poly-L-Lysine (PLL)
Polar solvents, 142
Poly aromatic hydrocarbons (PAHs), 273
Polycyclic aromatic hydrocarbons (PAH), 335
Polyethylene-glycol (PEG), 143
Polyhydroxyalkanoate (PHA), 93
Polyketideglycosids, 8; *see also* Microbial surfactants
    colopsinol A, 8
    elaiophylin, 8, 9
    ionophorics, 8
Poly-L-Lysine (PLL), 277
Polymeric biosurfactants, 51; *see also* Biosurfactant (BS)
Polymyxines, 10; *see also* Lipopeptides
    polymyxin A, 247
Polypropylene (PP), 122
Polyvinyl chloride (PVC), 20, 254
POME, *see* Palm oil mill effluent (POME)
PP, *see* Polypropylene (PP)
Precipitation, 143–144; *see also* Purification of BS
Production costing of BS, 155; *see also* Cost analysis of BS production
    capital cost, 155–156
    consumables and utilities, 157
    cost and market price, 156
    labor, 157
    operating cost, 156
    raw material cost, 157
    sophorolipid cost, 157–158
Production enhancement of BS, 279, 288; *see also* Applications of BS
    alternative substrates, 280
    bioprocess development, 286
    Candida lipolytica, 281

cassava wastewater, 285
cheap substrates, 280
dairy wastewater, 283–284
downstream processing, 287
lactic whey and distillery wastes, 283–285
molasses, 284
mutant and recombinant strains, 287
process optimization, 286
production economy, 280
production from lignocellulosic waste, 285–286
soap stock, 283
starchy substrates, 285
vegetable oils and oil wastes, 280–283
Productive oil wells, 256
Proteins, 12; *see also* Microbial surfactants
Pseudomonas, 172–173, 174, 316; *see also* Biosurfactant
    production
  biocontrol activities, 259
  fluorescens, 297, 316
  putida, 316
Pseudomonas aeruginosa, 61, 90, 316; *see also*
    Rhamnolipids (RL)
  biofilm, 298
  catabolism of, 93
  regulatory network, 94
Purification factor (PF), 137; *see also* Downstream
    processing
Purification of BS, 129, 146–148; *see also* Downstream
    processing; Sophorolipid (SL)
  cell separation, 140–141
  chromatography, 145–146
  crystallization, 144–145
  extraction, 142–143
  foam separation, 139–140
  immobilization, 140
  isolation by phase separation, 139
  microbial influence, 131–137
  precipitation, 143–144
  ultrafiltration, 141–142
PVC, *see* Polyvinyl chloride (PVC)
Pyolipic acid, 60

Q

Quorum-sensing factors, 299
  regulatory network, 94

R

RAG-1 emulsan, 12, 174; *see also* High-molecular
    weight BS
Reactive oxygen species (ROS), 318
Remediation method selection, 339
Renewable feedstocks, 85, 95; *see also* Rhamnolipids (RL)
  carbon sources, 86
  corn steep liquor, 89
  economic assessment, 89–90
  glycerol, 86–88
  lignocellulosic biomass, 88–89
  maximum product concentrations, 87
  molasses, 86
  nitrogen source, 89
  oils and free fatty acids, 88
  soap stock, 88

spent wash, 86
  substrate-to-product conversion yields, 87
  sugars, and industrial waste streams, 86
  traditional and alternative substrates, 85–86
  waste streams, 85
  whey, 86
Renewable resources, 133
Response surface methodology (RSM), 106
Response surface plots, 110
Rhamnolipids (RL), 6, 60, 38, 298, 299, 300; *see also*
    Biosurfactant (BS); Glycolipid; Industrial
    applications of BS; Renewable feedstocks
  antiadhesive activity of, 253–254
  antifungal properties, 358
  bacterial strains producing, 75–77
  Burkholderia species producing, 75, 76
  chromatographic operations, 145
  constituent, 85
  in cosmetic industry, 275
  di-rhamnolipids, 84
  in healthcare products, 250
  liposomes, 249
  Pseudomonas aeruginosa, 75, 77
  precipitation of, 143
  production, 61, 83, 84, 95
  in wound healing, 248
Rhamnosyltransferase 1 complex (RhlAB), 61
RhlAB, *see* Rhamnosyltransferase 1 complex (RhlAB)
Rhodococcus species, 77
ROS, *see* Reactive oxygen species (ROS)
Rotating disc bioreactor, 121; *see also* Bioreactors for BS
    production
RSM, *see* Response surface methodology (RSM)
Rushton turbines, 118–119; *see also* Bioreactors for BS
    production

S

SCS ratio, *see* Substrate cost-to-sales ratio (SCS ratio)
SDS, *see* Sodium dodecylsulfate (SDS)
Self-assembly (SA), 21, 249
Serratamolide, 10; *see also* Lipopeptides
Serrawettin, 10; *see also* Lipopeptides
SF, *see* Surfactants(SF)
SIE, *see* Sugar industry effluent (SIE)
Site closure, 344
Soap stock, 88, 283, 305
Sodium dodecylsulfate (SDS), 250
Soil bioremediation by biosurfactants, 319–320; *see also*
    Bioremediation; Biosurfactants
  saturated hydraulic conductivity for soils, 331
  soil contamination, 327
Soil contamination, 327
Soil vapor extraction (SVE), 333
Solid state cultivation, 123; *see also* Bioreactors for BS
    production
Solubilizers, 39
Sophogreen, 29; *see also* Biosurfactant
Sophorolipid (SL), 6, 19, 38, 130; *see also* Purification
    of BS; Glycolipid; Glycolipid biosurfactants;
    Industrial applications of BS
  acidic, 21, 133
  advantages, 131
  bioremediation, 30

Sophorolipid (SL) (*Continued*)
  biosynthesis, 23–24
  in cleaning, 44, 29, 252
  in cosmetic industry, 250, 275
  cost of substrates, 28, 134
  crystalline sophorolipids, 144
  crystallization of lactonic congeners of, 144
  dermatological applications, 29
  dimeric and trimeric, 22
  as emulsifiers, 253
  enzyme induction, 31
  as feed and food, 30
  fermentation parameters, 25
  genes responsible for biosynthesis, 136
  lactonic, 133
  lactonic diacetylated, 20
  medical applications, 30–31
  metabolic engineering for improved production, 136
  nanotechnology, 32
  nonacetylated, 20
  patents concerning, 250
  in plant protection, 29
  plant sophorose glycosides, 31
  producing organisms, 21–23
  production methods with respect to yields, 26
  production process, 25
  production with *Candida bombicola*, 134
  product phase of, 139
  product recovery, 27–28
  regulation of biosynthesis, 24
  sophorolipid-rich phase, 131
  as source of chemical compounds, 31–32
  strain modification, 28–29
  structure and properties, 19–21
  substrates for, 27, 133
  supramolecular assembly, 21
  toxicity, 20–21
Sophorose, 19; *see also* Sophorolipid (SL)
Spent wash, 86
Spiculisporic acid, 11; *see also* Microbial surfactants
Starch, 285
Strain engineering for rhamnolipid production, 90, 95; *see also* Rhamnolipids (RL)
  alternative strains, 90–91
  by-products synthesis, 94
  carbon yield, 93–94
  influencing regulatory mechanisms, 94–95
  metabolic spectrum and pathways, 93
  potential targets, 91–92
  precursors, 93, 95
  rhamnolipid biosynthesis, 94
  rhamnolipid precursor molecule synthesis, 94
  substrate spectrum, 92–93
Substrate
  alternative, 280
  for BS production, 133, 184–186
  carbohydrate-based, 181
  cheap, 280
  cost of, 134
  crude oil-based, 181
  hydrophobic, 136
  for rhamnolipid production, 92–93
  for sophorolipid, 27, 133
  SPO as, 102–103

starchy, 285
  -to-product conversion yields, 87
  traditional and alternative, 85–86
Substrate cost-to-sales ratio (SCS ratio), 134; *see also* Purification of BS
  dependency on, 135
Sugar-based biosurfactants, 249; *see also* Industrial applications of BS
Sugar industry effluent (SIE), 284
Superstructures, 21
Surfaces sanitization, 297
Surface types, 298
Surfactants (SF), 38, 129
  application, 83–84, 129
  vendors, 247
  worldwide total annual production of, 83
Surfactin, 9–10, 52, 58; *see also* Biosurfactant (BS); Lipopeptides; Industrial applications of BS
  antiadhesive activity of, 253–254
  application in medicine, 277
  biology, 159
  in cosmetic industry, 251
  Leu7-Surfactin, 258
  multiple biologic activity of, 171
  precipitation of, 143
  producing microbes, 317
SVE, *see* Soil vapor extraction (SVE)

T

Textile finishing industry, 254; *see also* Industrial applications of BS
Trans-membrane ion influxes, 247
Treatability studies, 340
  biodegradability and toxicity assessment, 341
  decision tree, 340
  guideline for site selection, 340,
  pilot-scale/field trial, 342
  preliminary confirmation, 342
Trehalose-dicorynomycolate, 7; *see also* Glycolipid
Trehalose lipids, 77
Triton X-100, 50; *see also* Biosurfactant (BS)

U

Ultrafiltration, 141–142; *see also* Purification of BS
Upstream processing, 131; *see also* Purification of BS
  influence of downstream process recovery, 132
  production cost reduction, 133
Ustilagin acids, *see* Cellobioselipids
Ustilago maydis sporidia, 257

V

V2–7 bioemulsifier, 273; *see also* Applications of BS
Viscosin, 10; *see also* Lipopeptides

W

Waste streams, 86, 85
Whey, 86, 284
  in biosurfactant production, 280, 306
  lactic, 283
  recycling, 304–305

for sophorolipid production, 133
  as source of carbon, 92
Whey wastes (WWs), 280
  for rhamnolipids production, 85
White biotechnology, 129
WWs, *see* Whey wastes (WWs)

Y

Yeast, 178–180
  biosurfactant producers, 154, 169, 174, 178–180

biosurfactants against, 253, 278, 300
emulsifiers from, 174
extract, 89
fermentation parameters, 25
growth inhibition, 301
MELs from, 7, 13, 65
patents on, 182
phospholipids from, 51
sophorolipids from, 19, 20, 23
species producing biosurfactants, 62
sugar conversion, 92